Reconsidering the Impact of Climate Change on Global Water Supply, Use, and Management

Prakash Rao
Symbiosis International University, India

Yogesh Patil
Symbiosis International University, India

A volume in the Advances in Environmental Engineering and Green Technologies (AEEGT) Book Series

IGI Global
DISSEMINATOR OF KNOWLEDGE

www.igi-global.com

Published in the United States of America by
IGI Global
Information Science Reference (an imprint of IGI Global)
701 E. Chocolate Avenue
Hershey PA, USA 17033
Tel: 717-533-8845
Fax: 717-533-8661
E-mail: cust@igi-global.com
Web site: http://www.igi-global.com

Library of Congress Cataloging-in-Publication Data

Names: Rao, Prakash, editor. | Patil, Yogesh, 1970- editor.
Title: Reconsidering the impact of climate change on global water supply, use, and management / Prakash Rao and Yogesh Patil, editors.
Description: Hershey : Information Science Reference, [2017] | Series: Advances in environmental engineering and green technologies | Includes bibliographical references and index.
Identifiers: LCCN 2016032942| ISBN 9781522510468 (hardcover) | ISBN 9781522510475 (ebook)
Subjects: LCSH: Water-supply--Management. | Water-supply--Environmental aspects. | Climatic changes. | Global environmental change.
Classification: LCC HD1691 .R425 2016 | DDC 333.91--dc23 LC record available at https://lccn.loc.gov/2016032942

This book is published in the IGI Global book series Advances in Environmental Engineering and Green Technologies (AEEGT) (ISSN: 2326-9162; eISSN: 2326-9170)

British Cataloguing in Publication Data
A Cataloguing in Publication record for this book is available from the British Library.

All work contributed to this book is new, previously-unpublished material. The views expressed in this book are those of the authors, but not necessarily of the publisher.

For electronic access to this publication, please contact: eresources@igi-global.com.

Advances in Environmental Engineering and Green Technologies (AEEGT) Book Series

ISSN:2326-9162
EISSN:2326-9170

Mission

Growing awareness and an increased focus on environmental issues such as climate change, energy use, and loss of non-renewable resources have brought about a greater need for research that provides potential solutions to these problems. Research in environmental science and engineering continues to play a vital role in uncovering new opportunities for a "green" future.

The **Advances in Environmental Engineering and Green Technologies (AEEGT)** book series is a mouthpiece for research in all aspects of environmental science, earth science, and green initiatives. This series supports the ongoing research in this field through publishing books that discuss topics within environmental engineering or that deal with the interdisciplinary field of green technologies.

Coverage

- Renewable Energy
- Policies Involving Green Technologies and Environmental Engineering
- Alternative Power Sources
- Sustainable Communities
- Green Technology
- Waste Management
- Industrial Waste Management and Minimization
- Electric Vehicles
- Air Quality
- Cleantech

IGI Global is currently accepting manuscripts for publication within this series. To submit a proposal for a volume in this series, please contact our Acquisition Editors at Acquisitions@igi-global.com or visit: http://www.igi-global.com/publish/.

The Advances in Environmental Engineering and Green Technologies (AEEGT) Book Series (ISSN 2326-9162) is published by IGI Global, 701 E. Chocolate Avenue, Hershey, PA 17033-1240, USA, www.igi-global.com. This series is composed of titles available for purchase individually; each title is edited to be contextually exclusive from any other title within the series. For pricing and ordering information please visit http://www.igi-global.com/book-series/advances-environmental-engineering-green-technologies/73679. Postmaster: Send all address changes to above address.

Titles in this Series

For a list of additional titles in this series, please visit: www.igi-global.com

Oceanographic and Marine Cross-Domain Data Management for Sustainable Development
Paolo Diviacco (Istituto Nazionale di Oceanografia e di Geofisica Sperimentale (OGS), Italy) Adam Leadbetter (Marine Institute, Ireland) and Helen Glaves (British Geological Survey, UK)
Information Science Reference • copyright 2017 • 425pp • H/C (ISBN: 9781522507000) • US $205.00 (our price)

Examining the Development, Regulation, and Consumption of Functional Foods
Sailas Benjamin (University of Calicut, India)
Medical Information Science Reference • copyright 2017 • 344pp • H/C (ISBN: 9781522506072) • US $185.00 (our price)

Applying Nanotechnology for Environmental Sustainability
Sung Hee Joo (University of Miami, USA)
Information Science Reference • copyright 2017 • 558pp • H/C (ISBN: 9781522505853) • US $225.00 (our price)

Exploring the Nutrition and Health Benefits of Functional Foods
Hossain Uddin Shekhar (University of Dhaka, Bangladesh) Zakir Hossain Howlader (University of Dhaka, Bangladesh) and Yearul Kabir (University of Dhaka, Bangladesh)
Medical Information Science Reference • copyright 2017 • 523pp • H/C (ISBN: 9781522505914) • US $200.00 (our price)

Smart Cities as a Solution for Reducing Urban Waste and Pollution
Goh Bee Hua (National University of Singapore, Singapore)
Information Science Reference • copyright 2016 • 362pp • H/C (ISBN: 9781522503026) • US $190.00 (our price)

Food Science, Production, and Engineering in Contemporary Economies
Andrei Jean-Vasile (Petroleum-Gas University of Ploiesti, Romania)
Information Science Reference • copyright 2016 • 473pp • H/C (ISBN: 9781522503415) • US $240.00 (our price)

Smart Grid as a Solution for Renewable and Efficient Energy
Ayaz Ahmad (COMSATS Institute of Information Technology, Pakistan) and Naveed Ul Hassan (LUMS School of Science & Engineering, Pakistan)
Information Science Reference • copyright 2016 • 415pp • H/C (ISBN: 9781522500728) • US $220.00 (our price)

IGI GLOBAL
DISSEMINATOR OF KNOWLEDGE
www.igi-global.com

701 E. Chocolate Ave., Hershey, PA 17033
Order online at www.igi-global.com or call 717-533-8845 x100
To place a standing order for titles released in this series, contact: cust@igi-global.com
Mon-Fri 8:00 am - 5:00 pm (est) or fax 24 hours a day 717-533-8661

Editorial Advisory Board

Table of Contents

Detailed Table of Contents

Climate change impacts are being felt in many parts of the world and have become an issue of major concern. Tropical countries particularly those in the Asian region are at greater risk and vulnerable to the impacts of climate change as indicated by the report of IPCC. With regard to India there are several impacts forecast which could have adverse consequences on the natural resources and ecosystems of the country making them vulnerable and reducing their capacity to cope with a changing climatic regime. This introductory chapter of the book provides an insight to the recent trends, issues and challenges in water resource development in context to the global climate change.

The physical characteristics and geographical isolation of mountain ecosystems in Himachal Pradesh has contributed towards a practice of cooperation and development of traditional knowledge among local communities. Over the centuries these traditional knowledge has been used to manage resource scarcities and adapt to vulnerabilities by the local population for over many generations. This paper describes one of such traditional 'Farmers Managed Irrigation System' (FMIS) practice, Kuhl, it's an open channel irrigation system to capture and transfer fresh snowmelt water from the glaciers to the agricultural fields in the valley. The characteristic of the local water management institution (structure, participation, process and governance) has been studied to understand their performance and success in managing water rights over the centuries. Lastly, the paper discusses the relevance of traditional water management against the backdrop of climate change, present status and issues related to the continuance of the Kuhl system and recommendations thereof.

One of the most important and visible indicators of global climate change is the recession of glaciers in many parts of the world. The findings of the IPCC Assessment Report (2007) suggest that there has been a significant decline in mountain glaciers and snow cover, which has contributed to the increased sea levels and water flow changes in river basins. In the latter half of 20th century, an increase in the rate of retreat has been observed in Himalayan Glaciers since advent of industrialization. The paper attempts to present a critical review and understanding of the recent changes in retreat of selected Himalayan glaciers from potential climate change. The study explores scientific evidence and analysis of observed data from a large and a small glacier in the Indian Himalaya. This will help in understanding the vulnerability of retreat of the glaciers based on their size. The larger glaciers are unlikely to disappear in the near future, due to its large mass balance and large response time due to climatic changes. Future adaptation needs, research and action for downstream communities, ecosystems and impacts on power sector are discussed.

Coastal Maharashtra is in transition. Growing coastal tourism and allied developmental activities along southern Maharashtra coast needs integration of various stakeholders to address the various issues and concerns. Integrated Coastal Zone Management (ICZM), which can cater to the needs of people while preserving the environment is thus need of the hour. The effects on natural resources, including water, will change the socioeconomic as well as the cultural fabric of coastal communities. This chapter explores a holistic approach to the developmental issues and the impact of climate change on the coastal region with specific cases of villages of Devbag and Tarkarli, coastal Maharashtra, India. It explores both physical and socioeconomic landscapes with special attention given to water resources in the context of changing dynamics of coastal communities and coastal tourism. The chapter discusses the issues and concerns of villages of Devbag and Tarkarli and proposes solutions for a sustainable development.

Himalayan Glaciers are the largest freshwater resource on earth and the rivers originating from them are an important source of water. They significantly modify stream flow both in quantity and timing as annual basin run-off is enhanced or decreased in years of negative or positive glacier mass balance respectively. Although glacial advances and retreats are a part of its natural cyclic phenomenon, the rate of de-glaciations has accelerated in recent times due to climatic changes and global warming caused by anthropogenic activities. Some of the important glaciers of Himalayas are receding at an alarming

Pallikaranai is one of the last remaining natural wetlands of Chennai. This marsh collects floodwater and increases groundwater levels in the region. The present study characterizes the water sources available around Pallikaranai Marsh. Groundwater was found to be the main source of water in the study area, extracted through domestic wells as well as commercially through a large number of agricultural wells. Direct surface water extraction from wetlands by private tankers was also observed in some areas. Acute water shortage and inefficient water supply by the government has led to thriving of tanker market in this area. Shrinking of the marsh and surrounding water bodies owing to construction, dumping of waste and encroachment accompanied by over-extraction of groundwater is driving this area towards extreme water crisis especially in event of climate change. Conservation of wetlands and evolving norms for sustainable water extraction of groundwater especially by commercial entities is recommended.

With time there has been an unpredictable climate change affecting the requirement of water for agriculture. Survival of agriculture has become a matter of concern with an efficient management of water. The aim of the chapter is to design a conceptual framework in sustaining water for agriculture in the era of climate change. The authors in the present chapter have used secondary data from previous research work and critically analysed the cases on water management for agriculture. The chapter discusses on practices like drip irrigation, new adaptations, mitigation technologies, vapour pressure, agro-ecological zone model and other water management strategies and the agricultural practices which though increases the yield but is leaving an adverse impact on climate. The chapter designs a conceptual framework to sustain water for agricultural use in the era of climate change and discusses it. This must be dealt through in order to attain sustainability.

This chapter assesses energy, water, and food resource systems based on their inter- and intra-sectoral imperatives of large scale development investments at the institutional level (including private and public activities) and how to achieve security of resource supplies. It identifies key interrelated processes, practices, and factors that underpin integrated resource management (IRM) and their attendant benefits. Applying the E4 framework concerned with energy, economy, environment, and equity to identify the main threats to these systems, the chapter evaluates their institutional, political, economic, cultural and behavioral components, and characterizes the forces that drive each of them at different governance scales. The chapter is guided by political economy, economic, and sociological theories that suggest that institutional structures affect economic factors and processes (i.e. production, distribution, and consumption processes). A case study of energy, water, and food (EWF) conflicting sectoral imperatives in Delaware is discussed in detail to better understand how these policy and institutional processes occur, which forms they take, and in which ways they define the quality and quantity of EWF resource systems in the State. In order to verify these parameters, the analysis considered the advantages of a sectorally balanced, E4 framework, in particular to evaluate the valency and magnitude of cross-sectoral connections, balance competing needs, and identify policy options that address various trade-offs.

Climate is the primary determinant of agricultural productivity and evidence shows possibility of shifts in earth's climate. Concern over the potential effects of long-term climatic change on agriculture has been raised over the past decade. Change in the climatic conditions on the globe created threat to the availability water for agriculture production. The present chapter is an attempt to distil what is known about the likely effects of climate change on water availability to agriculture for food security and nutrition in coming decades. Apart from few exceptions, the likely impacts of climate change on agriculture water resources in the future are not understood in any great depth. There are many uncertainties as to how changes in various environmental parameters will interact with the availability of water and further agriculture production. The future consequences of water resources on agriculture are discussed and summarized. Possible mitigation and adaptations to changing water availability for agriculture are also discusses.

The Sustainability value parity framework is proposed to deepen the understanding of the importance of 'inter-relatedness' of water-energy-waste with the goal of balancing water usage, aligning energy intensity and optimizing waste utilization. Generic waste that is burgeoning is a deterrent to the practice of sustainability that aligns water, energy, infrastructure, health, food, and lifestyle. This chapter delineates the gap between globalization at the macro-level and global citizenry at the grassroots-base and posits a value bridge assessed by appropriate thresholds of water - energy - waste. The emergent need to strengthen climate resilience and to usher into the sustainable pathway of climate-proofed development needs tuning of processes, lifestyle, hazardous substances and consumption. Climate change manifests as an over-arching risk that is strewn with unpredictability, multiple dimensions, uncertainties, spikes, imbalances leading to inequity.

Rainfall time series data from three sites (Kinshasa, Luki, and Mabali) in the western Democratic Republic of Congo were analyzed using regression analysis; rainfall intensities decreased in all three sites. The Congo Basin waters will follow the equation $y = -20894x + 5483.16$; $R2 = 0.7945$. The model suggests

18%-loss of the Congo Basin water volume and 7%-decrease for fish biomasses by 2025. Financial incomes generated by fishing will decrease by 11% by 2040 compared with 1998 levels. About 51% of women (N= 408,173) from the Lake Tumba Landscape fish; their revenues decreased by 11% between 2005 and 2010. If this trend continues, women's revenues will decrease by 59% by 2040. Decreased waters will severely impact women (e.g. increasing walking distances to clean waters). Increasing populations and decreasing waters will lead to immigrations to this region because water resources will remain available and highly likely ignite social conflicts over aquatic resources.

Chapter 13

Ali Reja Osmani, Karimganj Law College, India

Moving ahead from the freshwater reservoir versus climate change debate, the Indo-Bangla controversy over the Tipaimukh Project exists over the right of riparian states. India needs more energy to propel its economic growth, whereas Bangladesh is worried about downstream impact. The concerns of Bangladesh are based on the experience of severe water shortage and other impacts of Farakka Barrage and Teesta Barrage and also Himalayan Component of the Interlinking of River Project. Over the years some progress was made at bilateral level. But the major problem remains unaddressed i.e. without reconciling the issues of indigenous people a big dam cannot be constructed. This paper highlighted the existing scenario of Bangladesh and the indigenous people of Manipur in India in one hand and ecological, socio-economic concern in other hand i.e. obligation not to cause significant harm. There is no straight way answers available to be choose between a 'Yes' or 'No'; neither depends on the issues of 'might' over 'right' or 'development' over 'destruction', but on the circumstances to come.

Chapter 14

Sneha Kumari, Symbiosis International University, India
Yogesh Patil, Symbiosis International University, India
Prakash Rao, Symbiosis International University, India

Western Maharashtra consists of drought prone villages, which around seventy years ago were facing drastic challenges regarding water requirements. People living in such areas started to migrate from these villages in need of food and water. The chapter focuses on few case studies of drought prone areas of Western Maharashtra which made a move towards sustainability and has created an excellent model of watershed projects. Several practices like ban on cutting trees, terrace farming, contour bunding, drip irrigation, sprinkle irrigation, introduction of cooking stoves, rotational grazing, efficient water distribution and change in cultivation of agricultural crops according to the climate were followed which led the village to become a model for Aadarsh Gaon (An Ideal Village). The author has used secondary data from previous research work and critically analyses the case studies on water management in drought prone areas of Western Maharashtra which managed to become a model for sustainable use of water.

The topic of bottled water is particularly timely since water crisis and waste plastics are shifting from one of problem-framing to what is much more concerned to climate agenda. As plastics are everywhere to make life possible for a faster and convenient pace of society, bottled water has become one of the most disposable products to be consumed and disposed of, on a regular basis, both at and away from home. These eventually translate to consumer wastes causing a profound impact to the environment. The more bottles mean more greenhouse gases from production, transportation, waste bottles management, including fresh water supply. These adverse impacts are compounding a framework to justify as to how producer and consumer apply, individually and within the whole product life cycle, to sustainable consumption and production practice (SCP).

Climate change and wastewater control are one of the foremost demanding situations for Indian cities. Urbanization and unparalleled growth of cities across India continue to create immense pressure on land and water resources. This uncontrolled growth continues to produce growing volumes of wastewater. Climate change, impacts inclusive of, intense storm events in summer time or extended moist periods in wintry weather are quite visible in India. In urban and peri-urban areas, wastewater use for agriculture is an emerging precedence. Due to susceptible enforcement of regulatory, most of the wastewater generated is permitted off untreated/ partially treated. While many previous studies have checked out the global modifications and associated impacts of climatic variations on water resources, few have targeted at the evaluation of the particular effects and adaptation priorities for water systems in towns. Proper reuse of wastewater for irrigation will significantly lessen the shortage, offer a sustainable water source, improving farming productiveness, lessen pollution, generate livelihood potential for low earnings city households along with contributing to their each day food needs. There are tradeoffs which need to understand which includes problems to individual's health, and surroundings. Through suitable treatment methods, water users' cooperatives, policy shift and the introduction of market based approaches, treated wastewater use in agriculture can be enhanced and all associated risks can also be curtailed. This chapter focuses on use of treated urban wastewater and its management for agriculture in selected Indian cities.

Climate change is having a significant impact on water resources globally; more so in developing countries due to lower resilience and weakened coping mechanisms. Water resource and supply management systems need to be adaptive to the uncertainty and unpredictability posed by this challenge. The current water crisis is mainly that of governance within the water sector. On this premise, this chapter seeks to explore the adaptiveness of water management institutions and systems to climate change impacts, with a focus on Kenya's case. Kenya continues to face numerous challenges relating to water supply, which are further exacerbated by climate change. Prior to the development of key legislative and institutional frameworks, through adoption of the Water Act 2002, the water sector grappled with gross mismanagement, manifested via inefficiency in water service provision and resource degradation. Deploying the conceptual model of adaptive (co)-management institutional prescription of polycentric governance, experimentation, public participation, and management of bioregional scale, this chapter stands to contribute novel insights into designing sustainable, adaptive water governance regimes.

This chapter is an attempt to study the impact of climate change on water and agricultural production in India and abroad. While analyzing best practices in climate change adaptation and water management, the chapter examines regional issues and challenges. Policy interventions, success stories and new initiatives to tackle drought, boost rained agriculture as well as increase the irrigation potential have been studied together with the need for necessary course corrections. Leveraging technology for crop forecasting, inter-State river water disputes and measures needed to resolve them in the light of international experience are other areas of focus. In fine, the chapter calls for a comprehensive water policy that will not only recognize water as a national resource but also help bridge all differences for making world a worthy place to live in. The research methodology adopted in this chapter is primarily historical-analytical. Research papers, journal articles, official reports and newspaper clippings have all been consulted for analysis and interpretation.

Common duckweeds – Lemna – are free-floating aquatic macrophytes belonging to Lemnaceae family. They occur universally, in tropical and temperate zones and are enabled to grow in stagnant, slow-flowing, nutrient-enriched waters. They have an ability to grow over a wide range of temperatures (6 – 33 0C) and pH (optimum 5.5 – 7.5). Lemna species also manifest a characteristic macrophyte community in association with plants functioning as a dominant primary producer. Asexual mode of reproduction is exclusive and therefore, all resources are directed towards vegetative growth. Structure wise they show leaflets (frond) and root-like diffuse form. This simple morphological and physiological form offers special scientific and engineering properties – reflected in easy handling and manipulation under laboratory conditions. As a result they are regarded as a model plant system for a number of chemical and biogeochemical studies. Climate change affects plant growth and physiology. For example, increasing atmospheric concentrations of carbon dioxide (CO2) and ozone (O3) or increase in ultraviolet (UV) radiation (due to the thinning of the ozone layer) can have pronounced effects on the growth and development of plants. In many instances the decrease in growth rate can create disturbances in photosynthetic processes. In the recent past there has been a surge in the interest in looking for alternative remediation technologies to meet formidable demands of polluting materials and situations. One such approach is phytoremediation – the use of plants and associated microbes for environmental cleanup. Cardinal points in favour of such an approach are – cost-effectiveness, natural energy-driven and minimal capital and running costs. Potency of Lemna spp. as a phytoremediation agent as well as its limitations haven dealt with in details in literature. Moreover, Lemna spp. are accepted and highly standardized as test organisms in aquatic ecotoxicology. The authors have shown earlier that Lemna spp. can play a vital role in combating pollution burden in a glaring local situation of river pollution. However, the recent trend in research indicates that Lemna spp (duckweeds at large) can play a critical and sensitive role in being an indicator of overt and subtle climate change. The following is an attempt to take a review of such recent trends.

Preface

In recent times, energy, environment and sustainability related issues have gained considerable importance as some of the most critical issues threatening the survival of life on earth revolve around managing natural resources and energy security. Ever since the advent of the modern industrial era about two centuries ago, the world has witnessed several phases of industrial growth and societal transformation. Some of these are exemplified in the development of an industry centric growth along with transformational changes e.g. information revolution, globalization, urbanization, market based economy. However, much of this has come at the cost of severe consequences and impacts on our natural resources at the global level. The industry based economic model adopted across global and regional geographies has meant that planet earth has had to bear the burden of a rising population and an unsustainable consumption pattern. According to the Living Planet Report 2014 (WWF, 2014), the consequences of resource and consumption driven economy will lead to increasing stress on natural resources, ecosystems and conservation of biodiversity across land, water, air, soil, etc. While population pressures, rapid urbanization, environmental pollution, transboundary conflicts, poaching and trade could be detrimental at regional levels, from a global perspective view the issue of climate change perhaps would take centre stage as a globally sensitive environmental problem. This needs to be understood and addressed through rational use of our environmental resources, systems and processes. The reliance on a fossil fuel based economy to bring about rapid growth in our economy and to ensure security continues to be a conundrum within the larger climate change and development debate.

The recent UNFCCC Conference held at Paris, France in November 2015 has rekindled hopes of a unified platform amongst developed and developing countries to accomplish a fair and equitable global climate deal. Building a climate smart economy and knowledge process though appropriate mitigation and adaptation measures without compromising on economic development could be the answer for a sustainable growth path. The signing of the Paris agreement also signals a new era in the world community coming together in implementing a framework based on a sustainable resource oriented growth addressing key sustainable development measures aligned with the Sustainable Development Goals (SDG) of the UN.

The impact of climate change on ecosystems and natural resources particularly water has clearly been highlighted as a major concern at various platforms and amongst academicians, researchers, policy makes, economists, business and industry, civil society and local communities across the world (Immerzeel, van Beek, & Bierkens, 2010; Joshi & Joshi, 2011; Loucks, Barber-Meyer, Hossain, Barlow, & Chowdhury, 2010). Loss of critical ecosystem habitats due to rising GHG emissions and resultant changes in weather patterns has not only caused severe stress to a vulnerable ecosystem (freshwater and marine) but also threatens the survival of several million communities living in densely populated regions of the world (Jianchu et al, 2009; McGranahan, Balk, & Anderson, 2007).

The Intergovernmental Panel on Climate Change (IPCC) in its recent report (IPCC, 2014) has projected that impact of climate change on water resources is likely to have significant implications for resource conservation, businesses dependent on water, agriculture, and livelihoods, river ecosystems and habitats, species diversity, etc. Global climate variability is expected to further exacerbate the vulnerability of various regions in terms of reduced freshwater availability, food shortages, health related risks, reduced adaptation of the poor and vulnerable. Regionally, the repercussions could be critical for emerging economies like Brazil, India, China, etc. as well as industrialized nations.

In the face of climate vulnerability, adopting appropriate mitigation strategies for limiting climate change can be easier said than done and requires careful intervention coupled with the right blend of alternate technology as well as providing proper economic safeguards for the citizens (Wise et al., 2014). Regional cooperation amongst various countries will be necessary to provide a balanced economic growth model aimed at minimizing the long term impacts on climate change on water with transboundary linkages.

This edited book, *Reconsidering the Impact of Climate Change on Global Water Supply, Use, and Management*, explores some of the key challenges of climate change and its impacts on water resources in a globalized economy and in regional geographies through a diverse mix of chapters. In order to provide readers with an understanding of the importance of climate change and its impacts on water resources, supply and management, the book first discusses the impact of climate indicator based mountain and coastal ecosystems (Chapters 2-5).

The next set of chapters (Chapters 6-8) deal with the issue of linking water quality and climate change in developing economies like India and Pakistan as well as a focus on environmental flow in agricultural ecosystems and irrigation. The dimensions of energy-water nexus is an important aspect to be considered while dealing with climate impacts on water resources in the light of mitigation and adaptation based responses. Three chapters (Chapters 9-11) focus on the inter-linkages between use of water and energy for business and industry and vice versa and provide a view on the challenges and opportunities.

The book also seeks to delve into the interesting association of climate and watershed management from a regional perspective particularly with reference to local communities and developmental concerns (Chapters 12-14). Chapters in this section discuss the potential impact of hydropower facilities in NE India as well as studies on watersheds and communities in the Congo region of Africa.

While Industry plays a major role as a causal agent of global climate change, we have included two chapters (Chapters 15-16) which attempt to find a business case to the management of industrial water as an adaptation response as also water sustainability and consumption in the bottled water industry. The final part of this book brings in chapters (Chapters 17-19) on water governance paradigm and throws light on a policy based approach to managing water resources in a changing climate scenario. The last chapter on Duckweeds offer a novel model plant system comparable to *Escherichia* and *Saccharomyces* in the unicellular world to investigate and indicate not only climate change induced phenomena but also unique biochemical mechanisms and processes.

The book will be of use to readers interested in climate change, water management, development sector, ecology, biodiversity and resource conservation and other inter disciplinary areas. We hope this edited book will provide a knowledge base to graduate and post graduate students, researchers, academicians, civil society groups, industry and show a direction for integrating dimensions of environmental, social, and economic issues that drive water management globally in the face of adverse impacts of climate change (Russi et al., 2013).

This book perhaps took longer than required for completion, but nevertheless provided a new set of knowledge to the growing discourse on climate change and ecosystems impacts will add value for science and society. We acknowledge the support and patience of our contributing authors in the course of preparation of this volume. At the same time, we also take this opportunity to sincerely thank all our reviewers for sparing their valuable time in thoroughly reviewing the chapters.

Prakash Rao
Symbiosis International University, India

Yogesh Patil
Symbiosis International University, India

REFERENCES

Immerzeel, W. W., van Beek, L. P. H., & Bierkens, M. F. P. (2010). Climate change will affect the Asian Towers. *Science*, *328*(5984), 1382–1385. doi:10.1126/science.1183188 PMID:20538947

IPCC. (2014). [*Synthesis report.* Contribution of Working Groups I, II and III to the Fifth Assessment Report of the Intergovernmental Panel on Climate Change. Geneva: IPCC.]. *Climatic Change*, ▪▪▪, 2014.

Joshi, A., & Joshi, P. K. (2011). A rapid inventory of indicators of climate change in the middle Himalaya. *Current Science*, *100*(6), 831–833.

Loucks, C., Barber-Meyer, S., Hossain, M. A. A., Barlow, A., & Chowdhury, R. M. (2010). Sea level rise and 450 tigers: Predicted impacts to Bangladeshs Sundarbans mangroves. *Climatic Change*, *98*(1-2), 291–298. doi:10.1007/s10584-009-9761-5

McGranahan, G., Balk, D., & Anderson, B. (2007). The rising tide: Assessing the risks of climate change and human settlements in low elevation coastal zones. *Environment and Urbanization*, *19*(1), 17–37. doi:10.1177/0956247807076960

Russi, D., ten Brink, P., Farmer, A., Badura, T., Coates, D., & Förster, J. … Davidson, N. (2013). The economics of ecosystems and biodiversity for water and wetlands. London: IEEP.

Wise, R. M., Fazey, I., Stafford Smith, M., Park, S. E., Eakin, E. C., Van Garderen, A., & Campbell, B. (2014). Reconceptualising adaptation to climate change as part of pathways of change and response. *Global Environmental Change*, *28*, 325–336. doi:10.1016/j.gloenvcha.2013.12.002

WWF. (2014). *Living planet report*. WWF International.

Chapter 1
Recent Trends, Issues, and Challenges in Water Resource Development and Global Climate Change

Prakash Rao
Symbiosis International University, India

Yogesh Patil
Symbiosis International University, India

ABSTRACT

Climate change impacts are being felt in many parts of the world and have become an issue of major concern. Tropical countries particularly those in the Asian region are at greater risk and vulnerable to the impacts of climate change as indicated by the report of IPCC. With regard to India there are several impacts forecast which could have adverse consequences on the natural resources and ecosystems of the country making them vulnerable and reducing their capacity to cope with a changing climatic regime.

This introductory chapter of the book provides an insight to the recent trends, issues and challenges in water resource development in context to the global climate change.

INTRODUCTION

The impacts of Climate Change are currently being felt across many parts of the world never before. It is pertinent to note that tropical nations particularly those in the Asian region are at risk and vulnerable to the impacts of climate change both in terms of loss of life, productivity, economic growth and changes in the natural ecosystem. A recent report by the Intergovernmental Panel on Climate Change (IPCC) has indicated that some of the worst impacts of climate change are likely to be faced by tropical nations. With regard to India there are several impacts forecast which could have adverse consequences on the

DOI: 10.4018/978-1-5225-1046-8.ch001

natural resources and ecosystems of the country making them vulnerable and reducing their capacity to cope with a changing climatic regime.

In recent times, the rapid expansionist approach of humans coupled with the strong developmental economy in many urban centers has had a deleterious effect on the natural resource base of the Indo Gangetic Plain affecting biodiversity and livelihoods. The change in climate variability has further exacerbated the changes in the region increasing the vulnerability of the local people. The rapid retreat of mountain glacier systems which are considered the lifeline of river basins and ecosystems needs to be checked in order to provide adequate sustenance for a large majority of people living in the fertile tracts of the region. According to the Fourth Assessment Report of the Intergovernmental Panel on Climate Change average global temperatures may rise by 1.1-6.4°C by the next century. Such changes in annual surface temperatures can have serious consequences on the stability of the glacial systems.

It is a well-known fact that Climate Change and freshwater issues are closely linked through ecological systems. In 2005, WWF brought out a report highlighting the impacts of Climate Change on glaciers and subsequent implications for river basins in the Indian subcontinent. Regionally, the repercussions can be quite devastating affecting the economy of countries like subcontinent nations like India, China, Nepal, Pakistan, and Bangladesh. For a region where more than a billion inhabitants depend on agricultural and water as a primary source of livelihood the implications are profound.

The Millions at Risk report by Martin Parry, co-chair of the IPCC's working group two outlines this clearly and these issues are further highlighted in the "Millennium Ecosystems Report" launched in March 2005.

The South Asian region is considered to be one of the area's most prone to degradation of natural resources due to intense human activities. The predominance of agriculture as major source of livelihood activity across much of South Asia is a key factor in assessing economic growth across the region. Given regional climate scenarios, it is perceived that future food and water security can be seriously undermined in the entire region.

GLACIERS AND CLIMATE CHANGE: PERSPECTIVES FROM THE HIMALAYAS

One of the most important and visible indicators of global climate change is the recession of glaciers in many parts of the world. About three-quarters of the Earth's fresh water is held in ice sheets and mountain glaciers. Glaciers serve as a natural regulator of regional water supplies. The Himalaya mountain range is covered by 33,000 km^2 area of glaciers which provides 8.6 x 10^6 m^3 of freshwater to rivers draining from the mountains annually (Dyurgerov and Meier, 1997).

The Himalayan range of mountains stretching across an arc from Afghanistan in the West to Myanmar in the East is no exception. The youngest chain of mountain ranges is also home to some of the world's loftiest mountain peaks which have for centuries been the cynosure of the adventurer and mountain lover. The Himalayan ecosystems are perhaps one of the most fragile ecosystems in the world with a wide and diverse range of habitats and floral and faunal values. According to projections by glaciologists and climatologists who have been studying the glacial systems of the Himalayas, the freshwater shortages projected for the future is likely to have long term implications for regional food security. Apart from this there are also potential implications for hydropower generation given the large capital investments being made in the sector.

The impact of climate change on freshwater resources needs to be seen in the context of net consumption of water over the coming decades. A warmer climate scenario with delayed or uncertain onset of monsoons will not only have an effect on rain fed crops but also on water storage putting addition stress on the region's irrigation. Any change in climate variability is likely to lead to shifts in distribution of ecosystems including species movements. This can lead to changes in diverse natural systems including water based systems and consequently will also have implications on the livelihoods of vast majority of people living in densely populated river basins of the world. Some of these threats could be manifested as:

- Glacier lake outburst floods (GLOFs) with glacial lakes in the Himalayan region leading to flooding of towns and villages in their path.
- Increased incidences of health related issues and spread of vector borne diseases (Parry et al., 2001)
- Declining agricultural yields for major food crops due to increasing climatic variability with longer dry season, more intense rainfall (Millennium Ecosystem Assessment, 2005)
- Insufficient flows for irrigation systems
- Shortage of drinking water supplies as result of changes in hydrologic regimes and related water flow diversions
- Roll back in economic investments made by countries towards attainment of their national economic developments and priorities.
- Investments in hydropower generation and civil infrastructure (bridges, roads, power lines, etc.) will be increasingly impacted by flooding, landslides and/or increased sedimentation.

As glacier melt water contributes significantly to the large rivers systems, freshwater scarcity in the summer months will affect the livelihoods of river dependent communities (drinking water supplies, irrigation supplies, subsistence agriculture and artisanal fisheries) and biodiversity (e.g. the Ganga river dolphin *Platanista gangetica*, Turtles.).

Recent reviews have documented many instances of shifts in species distributions upwards in altitude, progressively earlier migrations and breeding due to change in climate (Hughes, 2000; McCarty 2001; Parmesan and Yohe, 2003; Root et al., 2003; Walther et al., 2002). The relatively modest climatic changes over the past century have already had significant impacts on the distribution, abundance, and physiology of a wide range of species globally.

IMPACT OF CLIMATE CHANGE ON FISHERY SECTOR

Fishing is a major source of food to mankind and provides employment and economic returns to millions of people around the globe. Around a billion people rely on fish as the major source of protein particularly in the poorest and populous countries on the planet. In recent times fishes are increasingly being threatened by global warming and it has been noticed that climate change has economic consequences that are benign and disadvantageous. The mean sea level has increased at the rate of 0.0705 mm, 1.086 mm and 3.772 mm per year in the Arabian Sea, Bay of Bengal and Andaman Sea during 1992 – 2005 (IPCC Climate Change, 2007, Synthesis Report). The knowledge on the impact of climate change on coastal ecosystems and more specifically on fisheries is very limited and the inherent unpredictability of climate change and the impact on fisheries are very complex. Most of the fishes have a narrow range of optimum temperature related to their basic metabolism and availability of food organisms. As a result of climate change the areas they inhabit may expand, shrink or be relocated. Fishes respond to these changes in a variety of ways like changes in distribution pattern, growth and reproduction. The inter-annual changes in the ocean patterns may affect the reproduction and recruitment of the species.

The changes in climate will definitely influence the distribution and abundance of fish stocks and have a profound influence on the fishing community and their fishing practices.

The fishes having a high fecundity show rapid demographic responses to the changing climate patterns and has been evident in the distribution of Cod in the North Sea, Californian Sardine and Atlantic Herring. In the Indian seas a similar trend has been noticed in the distribution pattern of Indian oil sardines and Indian Mackerel.

In India marine fisheries sector ensures food security to more than 3 million people inhabiting the coastal belt of India (fishing community, fish processors, traders and ancillary workers) and their dependents live in areas vulnerable to environmental change, or depend for a major part of their livelihood on resources whose distribution and productivity are known to be influenced by climate variation.

It is hard to imagine life without the Ganga in full flow as it has been for centuries. The threat of global climate change is perhaps a certainty and it is up to us to act sensibly and responsibly if we are to ensure a sustainable future for our citizens.

Figure 1. Impact of climate change on rivers with special reference to river-linking
(Nianthi & Husain, 2004)

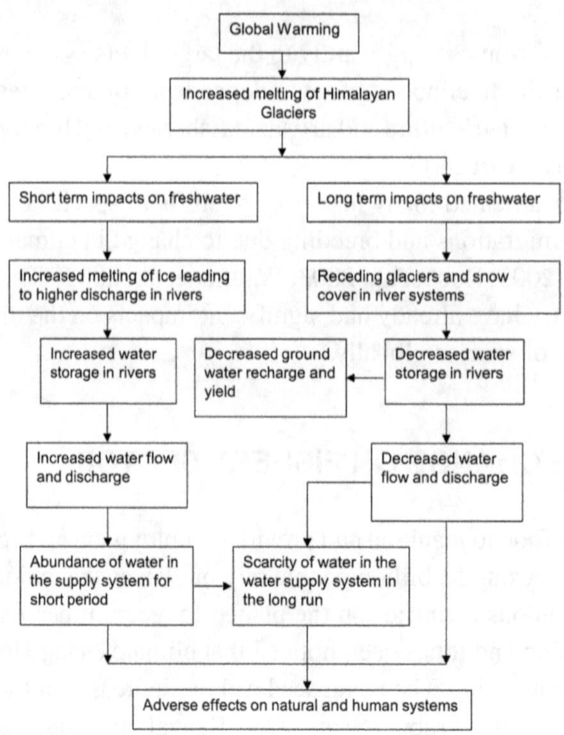

CLIMATE CHANGE IMPACTS ON WATER RESOURCES IN INDIA

The most significant sectors that will be negatively impacted in India are agriculture and water. In addition, forests and biodiversity will also be negatively affected, apart from the impact on human security due to extreme weather events including floods, droughts, cyclones, hail storms etc. In many cases, these sectors are interconnected. Reduced availability of water, for example, would affect agriculture as well as forests and biodiversity.

India contains at least 12 major rivers with a catchment area of about 253 mha. Many of these rivers are glacier fed. The importance of glaciers to availability of water in India can be gauged from the fact that the average water yield per unit area of the rivers of the Himalayan river systems (Indus, Ganga and Brahmaputra) is nearly double those of the peninsular rivers (Mall et al., 2006). The IPCC (2007) has suggested that some regions are expected to show a reduction in precipitation, especially during summer months and may lead to frequent severity and widespread droughts. It also adds that gross per capita water availability in India will decline from 1820 m3/yr in 2001 to as low as 1140 m3/yr in 2050.

In India the trends in average surface temperatures has shown an increase of about 0.57°C in the past century (Rupa Kumar et al., 1994). While the regional precipitation trends may not show any significant changes, the higher frequency of intense extreme events such as droughts or rainfall in a warmer atmosphere can lead to the possibility of increased flash floods in parts of India, Nepal and Bangladesh (Lal, 2000).

A study by WWF (2005) indicates that meltwater constitutes as much as 40% of the summer flows in the Ganga providing water to several million people and supporting their livelihoods. Over the past two to three decades various studies have shown a continuous and steady pattern in the retreat of glaciers including key glaciers in Uttarakhand like Gangotri, Milam & Pindari (WWF, 2005).

Studies on Himalayan glaciers by several scientists using techniques like mass balance indicate reduced water flows in many glaciers in the region (Berthier et al., 2006; Dyurgerov and Meier, 2005; Kumar and Dobhal, 1997,). Major river systems like the Ganga, Brahmaputra and the Indus are expected to be more vulnerable to climate change because of significant contribution from snow and glaciers. A hydrological study of the major river basins in the country using simulation models without considering current land use changes suggest that there could be a general reduction in the quantity of the available runoff under present GHG scenarios (Gosain et al., 2006).

Coastal ecosystems like the mangroves would also be under threat. For example in the Sundarbans, satellite imageries have shown that sea level is rising at an average rate of 3.14 mm a year. Projections indicate that a one metre rise is expected to inundate about 1000 Sq. Km. of the delta. In the past two decades four islands (Bedford, Lohachara, Kabasgadi and Suparibhanga) have been submerged with 6000 families rendered homeless. Scientists from University of Kolkata and Jadavpur University have predicted that one of the largest islands (Sagar Island) will lose at least 15% of its habitat area by 2020.

Although the embankments in the coastal zone act as a line of defense against wave action and coastal surges, the drainage system in the area is greatly hindered by such construction. In some areas the freshwater discharge is totally cut off, making the adjacent area hyper saline in nature. The change in the salinity gradient may disrupt the floral and the faunal community structure in the adjacent area. The entire Sunderbans ecosystem is defined by its hydrological regime and making alterations in this through construction of embankments, digging of salt pans, dredging of river beds, etc., without understanding ecological impacts can have long term negative impacts on both biodiversity and livelihoods. Indigenous mangrove species like *Heritiera fomes* have declined considerably over the past few years and have become endangered due to alteration of salinity levels. Other species like Horse shoe crabs which were commonly recorded in the Sundarbans have drastically declined due to increased salinity levels.

Scientific studies have also shown that several freshwater species of planktons and other invertebrate fauna have declined or been replaced by species which are better adapted to a marine ecosystem. The changes in the land use pattern are also expected to have implications for the survival of several rare and endangered species in the region including the Royal Bengal Tiger.

IMPACTS OF CYCLONES, EXTREME RAINFALL ON URBAN INFRASTRUCTURE

In the recent past there have there have been several instances of extreme rainfall events with events of devastating consequences on urban infrastructure in the Indian region. On 27 July 2005, parts of Mumbai received about 944 mm rainfall on a single day, flooding the metro. In August 2006, the desert area of Barmer in Rajasthan recorded 577 mm of rainfall in just three days—more than doubles the average rainfall it usually receives for an entire year, changing the landscape in many villages. More recently, the city of Chennai on the east coast of India faced one of the worst flooding in the past 100 years with extreme rainfall recorded of over 1000 mm within three days with two peaks occurring in November and December 2015. With urban infrastructure being severely affected, the loss on industrial infrastructure and economic opportunities is expected to lay stress on regional development. Around the same time large parts of western India are facing a severe drought situation with large tracts of the western state of Maharashtra facing critical water shortages. Large urban conglomerations like Mumbai and Pune have also resorted to rationing of water supplies and resorting to optimising their water needs. It would seem that in the future, such extremes would tend to occur with increasing frequency. A study carried out by Goswami et al. (2006) from the Indian Institute of Tropical Meteorology, published in Science (Goswami et al., 2006) noted that extreme rainfall events – more than 100 mm or less than 5 mm – are becoming more frequent and moderate events less frequent in central India. It would seem therefore, that India could be a place where the spectre of drought and flood would keep playing out with greater and greater frequency.

REGIONAL COOPERATION

At a policy level, treaties like the South Asian Association for Regional Cooperation (SAARC), ROPME, trade blocs, and regional institutions like ICIMOD need to develop frameworks for water and food usage by integrating global environmental factors like Climate Change in their developmental planning process. The developmental path that is currently being promoted in various countries particularly developing nations has implications for nations in the region threatening sectors like agriculture, tourism and industry. Regional cooperation amongst south Asian nations is therefore the need of the hour in order provide a balanced growth aimed at minimizing the long term impacts on water and other sectors which have trans boundary linkages. Through the forum of SAARC, governments of the region should consider evolving mechanisms for better coordination at scientific and policy levels through regional scientific seminars and intergovernmental meetings primarily focusing on these twin issues.

Studies by global environmental organizations (WWF, 2005) has identified temperature rise as the cause of increasing glacier melt. The rate and magnitude of temperature rise, along with, changes in water availability from precipitation and runoff are the two critical factors determining the ecological and social consequences of climate change (Poff et al., 2002). These drivers will affect the seven most important rivers of Asia that originate in the Himalaya.

The recent COP 21 negotiations have brought forth the need for limiting global climate temperatures below a 2° C scenario. The failure to do so may have repercussions on a change climate with effects on agricultural productivity. Adopting new management practices in a changed climate scenario is being seen as an adaptation response. However, given complex agro climatic conditions, such practices are difficult to take up primarily because of low technological and financial adaptability of local communities. Decrease in river flows will affect power generation and irrigation potential to sustain economic growth in South Asian countries.

Perhaps some autonomous responses are already taking place at the levels of individuals and communities to climate change. There are anecdotal reports of farmers shifting from a water intensive crop to ones which require less water. Several civil society organisations have been working with local communities to develop water harvesting technologies as well as promoting local decentralised mechanisms for watershed development .e g. In Rajasthan for some years now NGOs like Tarun Bharat Sangh have been involved in channelizing water to farmers through local adaptive measures despite a history of drought in the Alwar district of Rajasthan. Such local community initiatives have been commendable and have occurred even without the support of local or national governments. However, state level policy measures like mandatory water harvesting laws in some cities (e.g. Chennai) have been seen as a timely adaptation response in the face of climate variability .

Adaptation options must be explored in the light of the changing climate scenario. It is therefore important that we consider both socio economic and political considerations while addressing the issue of adaptation.

REFERENCES

Berthier, E., Arnaud, Y., Kumar, R., Ahmad, S., Wagnon, P., & Chevallier, P. (2007). Remote sensing estimates of glacier mass balances in the Himachal Pradesh (Western Himalaya, India). *Remote Sensing of Environment*, *108*(3), 327–338. doi:10.1016/j.rse.2006.11.017

Bhatnagar, A. K., & Koul, M. (2004). Impact of Climate Change on medicinal plants- an assessment. In S. K. Dash & P. Rao (Eds.), *Assessment of Climate Change in India and Mitigation Policies* (p. 264). WWF India Publication.

Dyurgerov, M. B., & Meier, M. F. (2005). *Glaciers and the changing earth system: a 2004 snapshot* (Vol. 58). Boulder, CO: Institute of Arctic and Alpine Research, University of Colorado.

Gosain, A. K., Rao, S., & Basuray, D. (2006). Climate change impact assessment on hydrology of Indian river basins. *Current Science*, *90*(3), 346–353.

Goswami, B. N., Venugopal, V., Sengupta, D., Madhusoodanan, M. S., & Xavier, P. K. (2006). Increasing trend of extreme rain events over India in a warming environment. *Science, 314*(5804), 1442–1445. doi:10.1126/science.1132027 PMID:17138899

Hughes, L. (2000). Biological consequences of global warming: Is the signal already apparent? *Trends in Ecology & Evolution, 15*(2), 56–61. doi:10.1016/S0169-5347(99)01764-4 PMID:10652556

Kumar, R., Kumar, K. K., & Pant, G. B. (1994). Diurnal symmetry of surface temperature trends over India. *Geophysical Research Letters, 21*(8), 677–680. doi:10.1029/94GL00007

Kumar, S., & Dobhal, D. P. (1997). Climatic effects and bedrock control on rapid fluctuations of Chhota Shigri glacier, northwest Himalaya, India. *Journal of Glaciology, 43*(145), 467–472.

Lal, M. (2000). Climatic change-implications for India's water resources. *Journal of Social and Economic Development, 3*, 57–87.

Mall, R. K., Gupta, A., Singh, R., Singh, R. S., & Rathore, L. S. (2006). Water resources and climate change: An Indian perspective. *Current Science, 90*(12), 1610–1626.

McCarty, J. P. (2001). Ecological consequences of recent climate change. *Conservation Biology, 15*(2), 320–331. doi:10.1046/j.1523-1739.2001.015002320.x

Millennium Ecosystem Assessment. (2005). Ecosystems and human well-being: biodiversity synthesis. World Resources Institute.

Nianthi, R. K. W. J., & Husain, Z. (2004). Impact of Climate Change on Rivers with Special Reference to River-Linking Project. Academic Press.

Pachauri, R. K., & Reisinger, A. (Eds.). (2007). *Climate change 2007 synthesis report: Summary for policymakers*. IPCC Secretariat.

Parmesan, C., & Yohe, G. (2003). A globally coherent fingerprint of climate change impacts across natural systems. *Nature, 421*(6918), 37–42. doi:10.1038/nature01286 PMID:12511946

Parry, M., Arnell, N., McMichael, T., Nicholls, R., Martens, P., Kovats, S., & Fischer, G. et al. (2001). Millions at risk: Defining critical climate threats and targets. *Global Environmental Change, 11*(3), 181–183. doi:10.1016/S0959-3780(01)00011-5

Poff, N. L., Brinson, M. M., & Day, J. W. (2002). *Aquatic ecosystems and global climate change*. Arlington, VA: Pew Center on Global Climate Change.

Root, T. L., Price, J. T., Hall, K. R., Schneider, S. H., Rosenzweig, C., & Pounds, J. A. (2003). Fingerprints of global warming on wild animals and plants. *Nature, 421*(6918), 57–60. doi:10.1038/nature01333 PMID:12511952

W. W. F. (2005). *An overview of glaciers, glacier retreat and its subsequent impacts in Nepal, India and China*. Kathmandu: WWF Nepal Program.

Walther, G. R., Post, E., Convey, P., Menzel, A., Parmesan, C., Beebee, T. J., & Bairlein, F. et al. (2002). Ecological responses to recent climate change. *Nature, 416*(6879), 389–395. doi:10.1038/416389a PMID:11919621

Chapter 2
Traditional Water Management System for Climate Change Adaptation in Mountain Ecosystems

Bhaskar Shrinivasulu Padigala
Centre for Environmental Planning and Technology University, India

ABSTRACT

The physical characteristics and geographical isolation of mountain ecosystems in Himachal Pradesh has contributed towards a practice of cooperation and development of traditional knowledge among local communities. Over the centuries these traditional knowledge has been used to manage resource scarcities and adapt to vulnerabilities by the local population for over many generations. This paper describes one of such traditional 'Farmers Managed Irrigation System' (FMIS) practice, Kuhl, it's an open channel irrigation system to capture and transfer fresh snowmelt water from the glaciers to the agricultural fields in the valley. The characteristic of the local water management institution (structure, participation, process and governance) has been studied to understand their performance and success in managing water rights over the centuries. Lastly, the paper discusses the relevance of traditional water management against the backdrop of climate change, present status and issues related to the continuance of the Kuhl system and recommendations thereof.

INTRODUCTION

Lot of factors like social, political, financial, and ecological that governs the lifestyle, culture, livelihoods and ultimately the life of mountain communities across the world. These communities are already vulnerable as a result of resource paucity, poor connectivity and fragile geographic conditions, to complicate the situation further global climate change has become an additional issue that will or in fact is affecting mountain communities directly and is also aggravating the impacts of other aspects of change (Aizebeokhai, 2009; Hansen et al., 2000; Toprak et al., 2013).

DOI: 10.4018/978-1-5225-1046-8.ch002

Variations in local climate patterns and resultant phenomenon's have impending severe consequences for the fragile mountain ecological systems and services and people dependent on them. Yet, very little information is available about the impacts of climate change on the resources (water resources and glaciers), livelihoods, and community's awareness and adaptive capacity to these climate variations (Ensor & Berger 2009; Eriksson et al., 2009).

Glaciers and mountain ecosystem and communities in the Indian Himalayas have been in the global climate change discussion arena for quite some time (Shrestha et al., 2012; Singh et al., 2006). The impacts of climate change are not only important in the context of ecosystem services, but also in connection with the overall water regime and associated hazards (Richardson & Reynolds, 2000).

One of the most perceptible consequences of climate change in the Himalayan region is the glacial retreat and resultant impacts on water regime of the area. Studies suggest that this constant deglaciation could have a deep impacts on the water regime of various river basins originating in the Himalayan region (Barnett et al., 2005; Singh & Bengtsson, 2004; Xu et al., 2009). Increasing glacial discharges are likely to rise for some time, but the water flow is then expected to reduce with decreasing glacier size. The effects of this phenomenon are most likely to be felt by communities occupying the arid parts of the region and who are heavily dependent on snow melt water for their livelihood.

At the same time, owning to their resource constraints these communities have also developed wide range of adaptive mechanisms ingrained in their cultural, livelihood and societal practices (Chettri et al., 2010). So far discussion on vulnerability of mountain communities and their adaptive capacity to climate change impacts traditional mechanisms and strategies have not been given due importance as a tool to adapt to climate change.

However, in a more positive development, the issue of acknowledgment and integration of traditional knowledge has been also documented in the Paris Agreement signed during recently concluded COP21 in Paris. Under the Article 7, the agreement encourages countries to develop climate change adaptation plan taking into consideration appropriate traditional knowledge, knowledge of indigenous peoples and local knowledge systems action as an integral components with a view to integrating adaptation into relevant socio-economic and environmental policies and actions. The agreement has also requested Global Climate Fund (GCF) to take adaptation (particular the principles of traditional knowledge amongst many other components) into account in its programmatic priorities (UNFCCC, 2015). On the water front, the COP21 also saw a creation of the international Paris Pact on Water and Climate Change Adaptation under the Lima to Paris Action Agenda (LPAA). This coalition includes nations, together with almost 290 water basin organizations, business and civil society parties. The international Paris Pact on Water and Climate Change Adaptation includes promises to implement adaptation plans, strengthening water monitoring and measurement systems in river basins and promoting financial sustainability and new investment in water systems management. Under the ambit of this pact, India announced a financial commitment to build climate resilience through improved groundwater management.

Given the importance of traditional knowledge in climate change adaptation and limited documentation of the same in India, this study has tried to investigate the traditional water management practices followed by local communities in Miyar valley of Himachal Pradesh, India.

Understanding traditional water resource management will considerably make climate change adaptation strategies at the community level much more efficient and effective, furthermore it's also useful for developmental agencies, NGO's and local authorities' to understand the environments changes better and device credible adaption plans so as to safeguard locals against economic, environmental and societal impacts of climate change. Incorporating traditional knowledge and strategies into mainstream scientific studies; not only can augment climate resilience but can also support sustainable development of the community.

CLIMATE ADAPTATION POTENTIAL AND TRADITIONAL KNOWLEDGE

Increasing awareness and deliberations on climate variability and its resultant impact on natural as well as human systems have resulted in rising interest amongst grassroots organizations, policymakers, researchers etc across the globe to study, evaluate and document community's vulnerability and adaptation potential to climate change (Burton et al., 2005; Duerden, 2004; Kelly & Adger, 2000; Rao & Patil, 2015; Smith et al., 2000).

According to Brooks (2003) adaptation potential can be defined as "reductions in social vulnerability as arising from the realization of adaptive capacity as adaptation. The term adaptation means adjustments in a system's behaviour and characteristics that enhance its ability to cope with external stresses", which means that, a system or a community over the time develops certain characteristics like institutions (Padigala, 2014) and resource management techniques. (Tompkins & Adger, 2004) that helps it in reducing vulnerability to external risks upto some degree. Likewise, several other researchers in course of their theoretical and field studies have also attempted to define and understand the complex theme and aspects of adaptation potential (Grothmann & Patt, 2005; Folke et al., 2002; Nelson et al., 2007; Yohe & Tol, 2002). But, it is essential to understand that adaptation potential of an individual or a community would depend on multiple factors like financial capital, human capital and social capital and would vary across scale and geographies.

Yet, determining a community's adaptation potential is fraught with difficulties and uncertainties, as it involves forecasting the adaptive capacity for a possible risk in the future. Furthermore, the adaptation potential to climate variability's is also influenced by other crucial factors like social and political structure, economic, technological and institutional capacities and traditional knowledge, as a consequence making evaluation of adaptation potential of an community all the way more complex (Few et al., 2007; Van et al., 2008).

Various scientific literature have indicated that climate induced variability's and hazards like droughts, melting of ice caps, sea level rise etc. have happened in the past with considerable impacts on the natural ecology as well as human societies (Crowley, 2000; Pandey et al., 2003). Equally, societies have depended on the knowledge collected over the course of their histories for their survival and livelihood (Menzies 2006; Norgaard, 1984). Even though, a few characteristic of traditional knowledge and practices based on them seem to be old and outdated, yet one must understand the fact that like local situations traditional knowledge accumulation is an active and continuously developing process that helps in communities to deal with new and changing conditions (Klein & Maciver, 1999).

The concept of 'indigenous, traditional or local knowledge' explored by many researchers in their research ascribe to traditional experiences, practices and skills that are accumulated and passed on by a community over many generations by various means like verbal stories, songs and community customs, etc (Correa, 2001; McGregor, 2004). According to Berkes (2012) traditional knowledge can be defined as "a cumulative body of knowledge, practice, and belief, evolving by adaptive processes and handed down through generations by cultural transmission, about the relationship of living beings (including humans) with one another and with their environment".

Traditional knowledge and practices have been developed by communities across the globe by studying and observing their surrounding environment and are being used to manage and adapt to various resource scarcities and risks like traditional weather forecasting or ethnoclimatology (Galacgac & Balisacan, 2009; Weatherhead et al., 2010), traditional knowledge on flora and fauna uses or ethnobotany (Balée, 1994) and ethnozoology (Pieroni et al., 2011), for agricultural management or ethnoagronomy (Barlett, 2013), management of local biodiversity and forest resources (Nazarea, 1999; Pandey, 1998), traditional knowledge of medicinal practices or ethnomedicine (Anyinam, 1995) and several others.

All over the globe, across scale and geographies, water is considered integral component for natural ecosystems and human society's development and resilience. Water resources in an area, whether spring water or rainwater has been an important factors in determining the local communities' livelihoods', lifestyle and culture. Therefore, in many places, especially in geographically challenging areas communities have developed various intricate technological interventions and institutional framework for sustainable utilization of water resources as well as to mitigate and adapt to water related hazards. For instance, the traditional water irrigation system in Oman referred to as Falaj irrigation systems are still being used in this arid country to meet majority of the agricultural communities irrigation water demand (Tourenq, 2011). Similarly, in Hindukush-Karakoram-Himalaya valleys decentralized water irrigation and management through irrigation canal networks and institutional mechanism have been effectively used by locals to manage water resources and at the same time reduce water related hazards (Chalise, 2000; Khan et al., 2010; Kreutzmann, 2011). Another, traditional fresh water management strategy i.e. rainwater harvesting has been or is still being successfully used in various continents/regions like Africa (Vohland & Barry, 2009) and South America (Bates et al., 2008) by farmers to manage water resources and at the same time increase resilience to environmental risks like droughts.

Looking at these examples and many other similar traditional water management systems and their institutional framework it is observed that these strategies not only serve the main objective i.e. water management but at the same time have multiple environmental, economic and social co-benefits that contribute towards climate resilience. Traditional water management usually requires local resources and community engagement resulting in local environmental sustainability and increased social capital in the community. This traditional practice reduces overexploitation of the water resources since in this arrangement water is treated as a common property resource that is governed by villagers themselves. Thus, it becomes important to study and evaluate these traditional water management practices to understand their role in community's adaptation potential and how it can be integrated into mainstream climate change adaptation discussions and interventions.

Traditional Water Management in India

Since time immemorial there has been a close relationship between humans, nature and its elements. Water is a crucial ingredient for livelihood, sustenance and cultural activities, and a key element in the growth and nourishment of all the communities, throughout the world. But, with increasing population, food demand and unsustainable water consuming activities, water resource management has become a critical issue.

India has numerous indigenous water management systems, developed by local communities to harness water resources for irrigation and drinking purposes (Singh, 1991) (refer table 1). Since these traditional water management systems are primarily constructed and managed by farmers, they may be referred to as farmer managed irrigation system (FMIS) (Wai, 1998).

Table 1. Different FIMS found in India

S.No	FMIS Name	Description	Practised in
1	**Trans-Himalayan Region**		
	Zing	Tanks for collecting water from melted ice	Ladakh
2	**Western Himalayas**		
	Kul	Water channels in mountain areas	Jammu, Himachal Pradesh
	Naula	Small ponds	Uttaranchal
	Kuhl	Headwall across a ravine to divert water from a natural stream for irrigation	Himachal Pradesh
	Khatri	Chambers carved in hard rock for storing water	Himachal Pradesh
3	**Eastern Himalayas**		
	Apatani	Terraced plots connected by inlet and outlet channels	Arunachal Pradesh
4	**North-Eastern Hill Ranges**		
	Zabo	Impounding runoff	Nagaland
	Cheo-oziihi	Channels from rivers	Nagaland
	Bamboo drip irrigation	Water from streams in the hills is brought to the plains via bamboo pipes for drip irrigation	Meghalaya
5	**Brahmaputra Valley**		
	Dongs	Ponds	Assam
	Dungs/ jampois	Small irrigation canals linking rice fields and a stream	West Bengal
6	**Indo-Gangetic Plain**		
	Ahar-pynes	Embanked catchment basin and channels	S. Bihar
	Bengal's inundation channels	Inundation canals	W. Bengal
	Dighis	Small square or circular reservoir fed by canals from rivers	Delhi
	Baolis	Stepwells	Delhi
7	**Thar Desert**		
	Kunds/ kundis	Underground storage	W.Rajasthan
	Kuis/ beris	Deep pits near tanks	W.Rajasthan
	Baoris/ bers	Community wells	Rajasthan
	Jhalaras	Tank	Rajasthan, Gujarat
	Nadi	Village ponds	Jodhpur, Rajasthan
	Tankas	Underground tank	Bikaner, Rajasthan
	Khadins	Embankment across lower hill slopes	Jaisalmer, W. Rajasthan
	Vav / Vavdi / Baoli / Bavadi	Stepwells	Gujarat, Rajasthan
	Virdas	Shallow wells	Rann of Kutch, Gujarat
	Paar	Area where water has percolated, accessed by kuis	-
8	**Central Highlands**		
	Talab / Bandhis	Reservoirs	Bundelkhand, Madhya Pradesh
	Saza Kuva	Open well	Mewar, E. Rajasthan

continued on following page

Table 1. Continued

S.No	FMIS Name	Description	Practised in
	Johads	Earthen check dams	Alwar district, Rajasthan
	Naada / bandh	Stone check dam	Mewar, Thar desert
	Pat	Diversion bund across stream	Jhabua district, Madhya Pradesh
	Rapat	Percolation tank	Rajasthan
	Chandela tank	Tank	Rajasthan
	Bundela tank	Tank	Rajasthan
9	**Eastern Highlands**		
	Katas / Mundas / Bandhas	Earthen embankments across drainage lines	Orissa & Madhya Pradesh
10	**Deccan Plateau**		
	Cheruvu	Reservoirs to store runoff	Chitoor, Cuddapah districts Andhra Pradesh
	Kohli tanks	Tanks	Maharashtra
	Bhandaras	Check dams	Maharashtra
	Phad	Check dams and canals	North western Maharashtra
	Kere	Series of tanks	Central Karnataka
	Ramtek Model	Intricate network of groundwater and surface water bodies, connected through surface and underground canals	Ramtek, Maharashtra
11	**Western Ghats**		
	Surangam	Horizontal well	Kasargode, Kerala
12	**Western Coastal Plains**		
	Virdas	Shallow wells	Rann of Kutch, Gujarat
13	**Eastern Ghats**		
	Korambu	Temporary wall of brushwood, grass and mud laid across channels to raise the level of water	Kerala
14	**Eastern Coastal Plains**		
	Yeri	Tank	Tamilnadu
	Ooranis	Pond	Tamilnadu
15	**The Islands**		
	Jackwells	Bamboo pipes are used to lead water into shallow pits	Great Nicobar Island

(Adapted from CGWB, 2011)

The technology and construction of the FMIS differs, depending on their purpose (drinking or irrigation), geography, and availability of resources (materials and human capital), social and cultural aspects (Shivakoti & Shrestha, 2008). Agricultural irrigation systems are generally spread over large open areas, and consist of intricate network channels and waterways for collecting and distributing water. Whereas, drinking water system are generally smaller and covered structures, to ensure manual collection of water for household needs and to reduce chances of water contamination.

Despite their prevalence and successful utilization by communities in different regions of the country, the role of FMIS has not been suitably acknowledged (Narain, 2006). This neglect is reflected in government policies. For example, in the National Irrigation Statistics, FMIS's have been merged with numerous similar irrigation practices under the head 'other sources of irrigation' without any effort to study and classify these traditional techniques (Sengupta, 1993). Similarly Irrigation schemes serving agricultural area less than 2000 hectare are clubbed as "minor irrigation schemes" (MoWR, 2001) in the minor irrigation census conducted by the Indian Ministry of Water Resources. According to (Chand, 1994) the rationale for this disregard seems to be the scale of irrigation system. While these systems are successful at a regional watershed level, but their input at national level remains proportionally small as compared to the other sources of irrigation like bore wells. Decreasing participation by the local population (Baker, 1996) is leading to an overall decrease in popularity, implementation and use of these traditional water management systems.

In recent decades, communities and grassroots organizations working on watershed projects are realizing the drawbacks of modern irrigation techniques. Also hurdles in government led irrigation systems like finance, resources and geographical limitation, etc. (Koul et al., 2012; Lal & Verma, 2008; Pattanaaik et al., 2012), have called attention to FIMS. This trend is visible both in research in India (Borthakur, 2009; Coward, 1990; Pant & Verma, 1983; Sharma & Kanwar, 2009) and also in different parts of the world (Bagadian, 1987; Uphoff, 1985).

OBJECTIVES AND SCOPE

Miyar valley, (32°42'36" N - 33°15'24"N latitude to 76°40'12"E – 77°1'15" E longitude) in Lahaul and Spiti district, lies in a rain shadow area of the middle Himalayas (See fig 1). North West disturbance causes precipitation in form of snow. Average rainfall is about 694 mm per annum (Saini, 2008). The basin lies in between the north of the Pir Panjal range and south of Zanskar range. Altitude varies from 2600 meters a.s.l to 6448 meter a.s.l, covering 963km^2 areas with maximum width and length of 25 and 60 kilometers respectively. Miyar is the largest & longest of a 6th order basin fed by 26 micro watersheds forming dendritic pattern of drainage (Saini, 2008).

The inhospitable geomorphology and climatic conditions of Miyar watershed sustains a small population in the district of Lahaul and Spiti, in 16 villages. The study area includes 5 villages, Tingrit, Ghumpa, Urgos, Sukto and Khanjar (Figure 2, 3 and 4). The traditional water management, socio-economic aspects, land use dynamics and community institutions of the area were studied.

The physical characteristics and geographical isolation of Miyar valley in Lahaul and Spiti, has contributed towards a practice of cooperation and development of traditional knowledge among local communities. Over the centuries this has been used in overcoming and managing resource scarcities by the local population from generation to generation.

Like most of the settlements in Himachal Pradesh region, villages in Miyar Valley are primarily situated on the foothills or in valleys. To overcome difficulties unique to their situation, for agricultural and pastoral activities and with shortage of human capital, communities in the valley have developed fresh water management practices involving channelling and transfer systems to provide fresh water for drinking and agriculture. Even today farmers in the valley depend on these traditional practices.

The main livelihood source in the valley is agriculture, being a rain shadow area the only source of irrigation water is snow melt water along with some amount of glacial water. Also, the area receives

Figure 1. Showing location of study area, Miyar valley, India

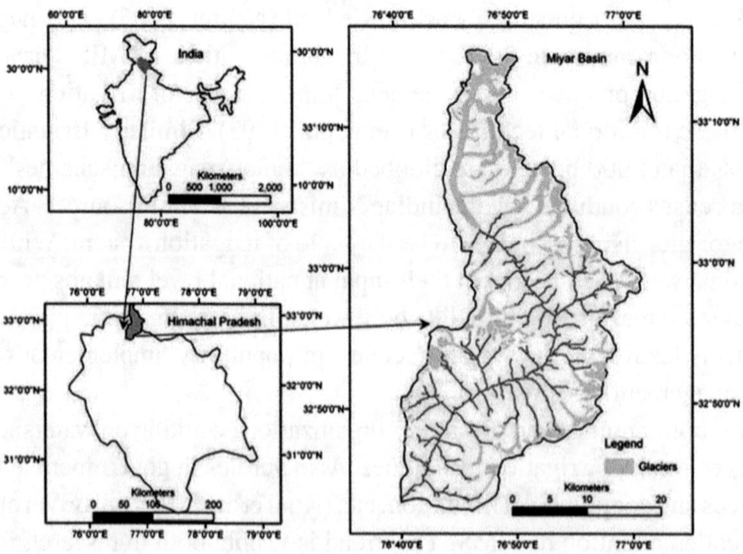

Figure 2. Urgos watershed, Miyar Valley

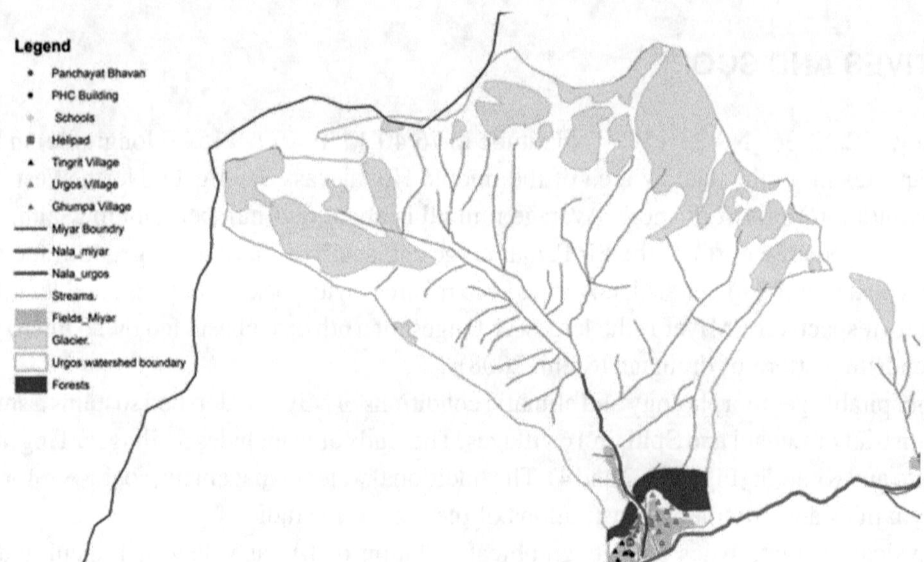

Figure 3. Sukto watershed, Miyar Valley

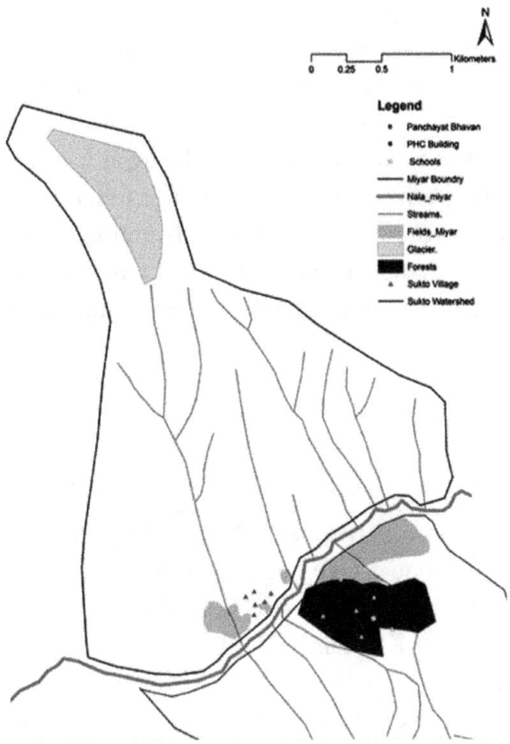

Figure 4. Khanjar watershed, Miyar Valley

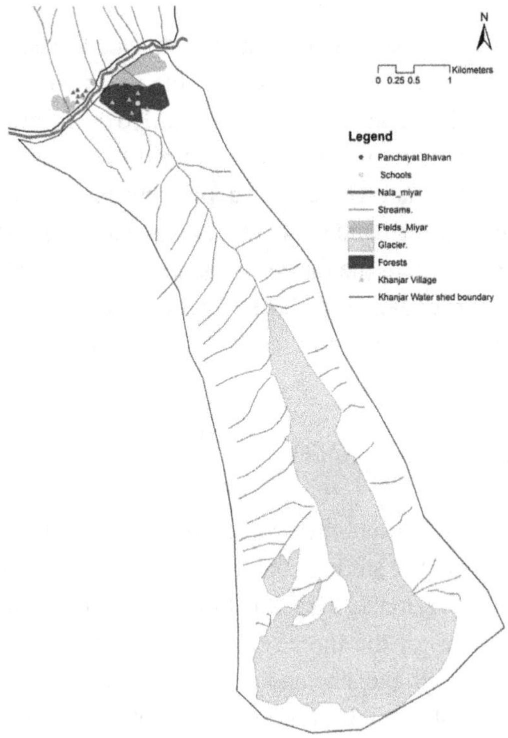

heavy snowfall during winters from October to march causing complete halt to agricultural activities and hindrance to the normal life, this means that people get only remaining six months to do agriculture and gather food and fodder for themselves and for their livestock's.

From last few years the area is witnessing change in the economic and land use pattern due to the improved road or transport services, irrigation systems and introduction of cash crops like peas, seed potato etc which has lead to increased economic stability in community and also change in the land use of the area like conversion of small patches of forest, grazing lands into agricultural fields by villagers (Padigala et al., 2011).

The sites were selected considering the above mention two circumstances (Climate dependency and changing economic pattern), and how these situations and community will behave in the light of current global environmental change. This area also makes a classic example to study traditional adaptation methods, since the mountain communities have developed their own methods to tackle with the harsh environmental conditions. And it would be interesting to see can these traditional methods are still viable against current environmental changes?

Despite the fact that there is an insufficient data to evaluate current socio-economic and environmental conditions, there is no doubt that the recent development and commercialization in the valley has improved the economic well-being of the local population. Thus, it becomes significant to understand as to whether the community's gradual shift from traditional practices towards more commercialization is adding to the climate vulnerability. Since traditional practices have been developed by the community over the centuries keeping in mind the local conditions and their past experiences.

The present article tries to explore the nature of local society-environment interactions, particularly the role of social institutions and social capital in adaptive capacity to possible climate change impacts in Miyar watershed situated in the north-eastern part of Lahaul and Spiti district of Himachal Pradesh, India. The first part of the article deals with climate change perception among local communities and the following part deals with detailed description of the characteristic *Kuhl* system (design, construction, etc.), governance mechanism for the management of the system (characteristics of the local institution of water management (structure, participation, process and governance have been studied to understand their performance and success in managing water rights over the centuries) and distribution of water rights among the farmers. The last section of the paper discusses the relevance of traditional water management against the backdrop of climate change, present status and issues related to the continuance of the *Kuhl* system and recommendations thereof.

METHODOLOGY

To identify, document and analyze the traditional water management practices in Miyar valley, Himachal Pradesh, the study adopted a research framework similar to the one used by Pattanaaik et al. (2012) and Lal & Verma (2008). Most of the people in these villages are farmers. The selection of villages was on the basis of the three selected sub-watersheds Urgos, Sukto and Khanjar. Other reasons were, ease of access, village size and availability of an interpreter.

A total of 80 households out of 120 households from the five villages were randomly selected to participate in the interview and focus group discussion exercise, with a special focus on village elders. The occupations of participants in this study included farmers, agro pastoralists and service providers.

For data collection, focus group discussions (FGD's), key informant interviews and Participatory Rural Appraisal (PRA) methods were employed. Local respondents were provided with the questionnaires, in which a check list based on the themes like traditional water management practices, local institution and community perception about FMIS were incorporated. A transact walk was undertaken with village farmers and elders to observe the design, network and applications of the Kuhl system. Statistical Package for Social Science (SPSS) and Excel computer software's were used to analyze the collected data.

RESULTS AND DISCUSSION

Environmental Change Perception

Agriculture as their main source of livelihood and limited time window to practice it, life in Miyar region depends a lot on local weather and availability of natural resources like fresh water. Large number of scientific literature based on climatic variability and climatic modelling on Himalayan region suggesting increasing risk of climate induced variability's like increasing temperatures and fluctuation in availability of irrigation water (Ageta and Kadota 1992; Bhutiyani et al. 2007; IPCC, 2007; Singh & Bengtsson, 2005). It becomes crucial to understand the community's awareness and perception to these predicted changes before any meaningful exercise can be carried out to understand their adaptive capacity and role of traditional knowledge in it.

On enquiring about observations on temperature changes, more than 65% of the interviewees agreed that there has been an observable temperature variation in the valley and that temperature has increased in the valley in the last two to three decades. In addition to temperature increase interviewees also perceived a small raise in the period of the summer season.

Linking to temperature variation, interviewees were enquired about modifications in the agriculture practices in terms of cropping time, crop variety, etc. and it was found that more than 90% of the farmers across the villages (in couple of villages it was up to 100% like Sukto and Tingrit) have not modified the cropping numbers. Similarly, crop varieties have not witnessed any change as a result of temperature changes with 100% of the interviewees respond that they have not adopted any new crop varieties. However, crop varieties being grow in the valley is changing rapidly from traditional crops like Kuth (*Saussrea lappa*), Barely and Buckwheat (*Fagopurum esculentum* loc. kathu) to more cash crops like peas etc. According to (Sen et al., 2011), this major diversification in the cropping varieties from traditional and resilient crops to economically rewarding, yet climatically risky cash crops cannot be directly related to climate change only. And that the key reason for commercialization of agricultural practices can be attributed to improved road or transport and communications services. Since, with greater access to markets famers both small as well as large farmers have incentive to go for risky yet profitable cash crop varieties.

With rain being more or less unknown and unpredictable; annual snow fall becomes the only source for the irrigation water availability in the valley. And with scientific assessments suggesting changes in snowfall pattern in Himalayas (Shekhar et al., 2010), it becomes significant to identify, recognize and analyse community's observations with respect to snowfall inconsistency in the valley.

More the 50% of the Interviewees from across the study area (expect for the respondents from village Khanjar) agreed that snowfall intensity has increased in the area over the year with small decrease in the time of the season of snowfall. Despite very few instances of water related risks in the area such as cloud burst induced floods or GLOF's (Glacial lake outburst floods), The only recent documented case of glacial lake outburst flood was in year 2000, where only one small village (Changut) was partially affected by the flood. However, interviewees felt that increasing unpredictability of the local climate may increase the chances of such events happening in the near future.

It was observed that, in spite of apparent temperature rise and extended cropping season perceived by the villagers, none of them made any significant alterations in their farming activities. This phenomenon could be ascribed to following two reasons, foremost is that the change observed by the villagers in temperature is small and not that considerable to which villagers would like to adjust their agricultural practices. The second reason was that villagers have started experiencing increased and irregular erratic annual snowfall which increases the risk of crop failure if the villagers attempt to produce a second harvest given that the first crop harvesting period lasts up to October month and second would require another 2-3 months' time before it could be harvested and thus the harvesting time for the second crop would overlap with the beginning of the winter spell with snowfall that could create a huge risk to the standing second harvest if that year the snowfall takes place little early.

Most of the literature studies looking into impacts of climate change on water regime in Himalayas have suggested increasing drinking water scarcity in the region (Jianchu 2009; Karki 2009; Pathak 2009; Rao 2008). However, It was surprising to observe that greater part of the interviewees from the local community (>80% across all villages) were not aware of any alteration in the drinking water availability. The main supply for drinking water for the village comes from the spring water, sporadic cases of drying of springs have been noticed, with less snowfall in the previous year, to which villagers have no difficulty in adjusting by finding a new source for water supply.

Nevertheless, these scientifically expected alterations in water regime due to possible future climate change variability's will without doubt modify and may even to some extent disturb the local market and natural resource accessibility. With changes in resources availability and accessibility significant transformation in the functioning of local economy, institutions and governance is likely to happen and how well natives adapt to these changes would depend a lot on the traditional knowledge, social capital and communal actions in the region (Padigala, 2014).

Traditional Water Management in Miyar Valley

The Community Water Management Systems or FMIS's, are prevalent in the western Himalayan eco region (Coward, 1990) and so in the Miyar region. The majority of the villages lie on the plateaux on both sides of the river and get water from the streams which trickle down from the cliffs.The terrain limits the access of people to water. Rain is almost unknown and erratic; melting snow in late March provides freshwater for irrigation marking the start of the agricultural period in the valley. All five villages cultivate Rabi during summer months, using 'Kuhls' a network of water channels for irrigation. Water rights are distributed based on 'Bari' or turn.

Kuhl System

Kuhl is a channel which harnesses water from snowmelt or glacier fed, streams which pour in to rivers reaching downstream. Water coming from the mountains is diverted in to a built channel which connects to the fields. The mouth of the Kuhl is at the highest reachable contour and is generally built using rocks. (See Figure 5) Water from the stream is diverted towards the mouth of the Kuhl by slightly diverting a small stream of water with the help of boulders and rocks. This water is diverted to fields through small common water channel. (See Figure 6) The channel is drained in to the main stream or river of the valley.

Generally, the length of a *Kuhl* varies between 1 to 15 kms. It has a trapezoidal cross section, 0.1 to 0.2 m^2 in area, and usually has water discharge rates varying between 15 to 100 liters per second. The snow melt freshwater water discharge varies with the season. A single *Kuhl* can irrigate an area of 80 to 400 ha through distributaries (40 cm X 50 cm width small channels for farms) or by flooding. The system is normally found at altitudes varying from 350 to 3000 m in the outer and middle mountain ranges, where there is a considerable elevation drop; which is used by farmers to construct the *Kuhl* and its allied machineries like *"Gharat"* (traditional water mills used by the locals in the western Himalayas, especially in Himachal Pradesh. This system is connected with the Kuhl system, wherein the water flowing in the channel is diverted and utilized to rotate wooden gear. This gear is connected to the flour mill and this whole machinery is used to grind dried peas and wheat locally). A total of 3 Kuhls were observed in the study area, harnessing the snow melt water from the 3 glaciers Urgos, Khanjar and Menthosa in their respective watersheds.

Kuhl has been a mainstay for agricultural practices over the ages for the farmers and local communities and continues to be managed by the local farmers. Since the 1970's Himachal Pradesh, Irrigation and Public Health (IPH) has taken active role in rejuvenating dysfunctional Kuhls, lining of Kuhls with cement and providing funds to the farmers etc. The Census of Minor Irrigation, documents 17000 Kuhls

Figure 5. Kuhl Head, Village Tingrit, Miyar Valley

Figure 6. Kuhl section, Village Tingrit, Miyar Valley

in the State (MoWR, 1988), irrigating more than 80,000 hectares of agricultural land in the year 1988-89 (Chand, 1994). According to the last minor irrigation census, in Lahaul and Spiti district nearly Sixty per cent of these surface flow irrigation schemes were found to be in farmer managed domain, with government-owned schemes comprising 34 per cent of the total number of the works (MoWR, 2001) (Table 2). It was also observed that majority of the water distribution system in Lahaul and Spiti was through open channels, whereas only 2 drip irrigation system were recorded in the census (Table 3), clearly indicating the widespread utilization of the traditional water irrigation system.

Bari System

Bari (turn) refers to the cyclic opportunity that each farmer gets for irrigating his fields. The Bari system is a local water right distribution system developed by the local water management institution known

Table 2. Interviewees perception on climate variability in Miyar Valley

Param-eters / Village	Change in Temperature		Change Cropping Number		Change Cropping Type		Change Cropping Time		Change Snowfall		Change Drinking Water Availability	
	Yes (%)	No (%)	Yes (%)	No (%)	Yes (%)	No (%)	Yes (%)	No (%)	Yes (%)	No (%)	Yes (%)	No (%)
Thingrit	80.0	20.0	0	100.0	0	100	15.0	85.0	55.0	45.0	10.0	90.0
Ghumpa	70.0	30.0	10.0	90.0	0	100	40.0	60.0	60.0	40.0	10.0	90.0
Urgos	84.8	15.2	6.1	93.9	0	100	18.2	81.8	72.7	27.3	15.2	84.8
Sukto	66.7	33.3	0	100.0	0	100	16.7	83.3	50.0	50.0	16.7	83.3
Khanjar	72.7	27.3	9.1	90.9	0	100	9.1	90.9	45.5	54.5	18.2	81.8

Table 3. Surface flow schemes according to water distribution system

District	Govt.	Coop. Societies	Panchayat	Group of Farmers	Individual Farmers	Others	Total
Bilaspur	357	0	3	342	138	49	889
Chamba	267	1	3	332	10	4	617
Hamirpur	72	0	10	91	16	9	204
Kangra	738	5	82	1008	27	13	1873
Kinnaur	132	0	0	94	9	0	235
Kullu	44	1	33	38	1	2	119
Lahaul & Spiti	195	0	0	340	39	0	574
MAndi	2474	18	12	1331	195	7	4037
Shimla	798	0	15	498	37	3	1351
Siraur	618	0	1	1254	63	7	1943
Solan	1155	0	164	518	52	87	1976
Una	35	0	0	63	4	9	111
Total	6885	25	329	5909	591	190	13929

(MoWR, 2000-01)

Table 4. Distribution of surface flow schemes according to ownership

District	Sprinkler	Drip Irrigation	Open Channel	Underground Channel	Others	Total
Bilaspur	3	2	858	19	7	889
Chamba	5	1	608	2	1	617
Hamirpur	1	2	190	1	10	204
Kangra	15	34	1726	92	6	1873
Kinnaur	0	0	234	0	1	235
Kullu	9	2	103	3	2	119
Lahaul & Spiti	0	2	572	0	0	574
Mandi	19	33	3927	26	32	4037
Shimla	7	5	1264	8	67	1351
Siraur	27	3	1855	45	13	1943
Solan	13	2	1906	6	49	1976
Una	0	0	109	2	0	111
Total	99	86	13352	204	188	13929

(MoWR, 2000-01)

as Kuhl committee. Under Bari system, the entire community is divided on the basis of number of farm families (those who have participated in the channel construction and repairs), and each family gets one full day to irrigate their fields turn wise (the distribution of the water rights is done in impartial lottery system, see Figure 7 & 8). But on mutual understanding, two families can share their water right for half a day and get their next turn after half cycle period instead of the full water right cycle period. The time allocation and amount of water are decided by the Kuhl committee based on the size of the land holdings. Though rare, any dispute related to water distribution is settled internally between the farmers and is facilitated by the village elders and Kuhl committee ensuring harmony and participation of the whole community.

Kuhl Committee

In earlier times Kuhls were constructed and managed by the village water-master "Kohli" who was responsible for annual Kuhl inspection and repairs by mobilizing the community members and resources. A revenue system was introduced by British's and princely estates to collect revenue from landowners using irrigation water from the Kuhls (Coward, 1987). This led to the formation of Kuhl committees to supervise and regulate water distribution. This institution is now prevalent in the entire western Himalayan region.

In all of the 5 villages surveyed, the Kuhl committees had been in existence since early pre independence times. Total of 3 committees have been formed in the area namely Kuhl committee Tingrit, Kuhl committee Urgos (Urgos and Ghumpa villages) and Kuhl committee Khanjar (Khanjar and Sukto villages).

Each household of the village has one representation in the Kuhl committee. The committee has an elected executive body of Pradhan (Chief Executive Officer), Up Pradhan (Deputy), Secretary and Treasurer from among the members. The term for an executive body is 1 year and after the completion of the term new body is selected unanimously, among the members.

The Kuhl committee has a fixed monthly membership fees ranging between INR 10 to INR 20 per member, which is utilized for Kuhl management and repairs. Kuhl committee also takes fees between INR 1000 to INR 1500 as a membership fees from any new member joining the committee. The main functions of the Kuhl committee are to address issues like Kuhl work and maintenance, distribution of water rights, resolving water disputes at village level only, and to procure government grants for maintenance and repair of Kuhls etc (see Figure 9).

Figure 7. Showing Bari system being followed, Village Tingrit, Miyar Valley

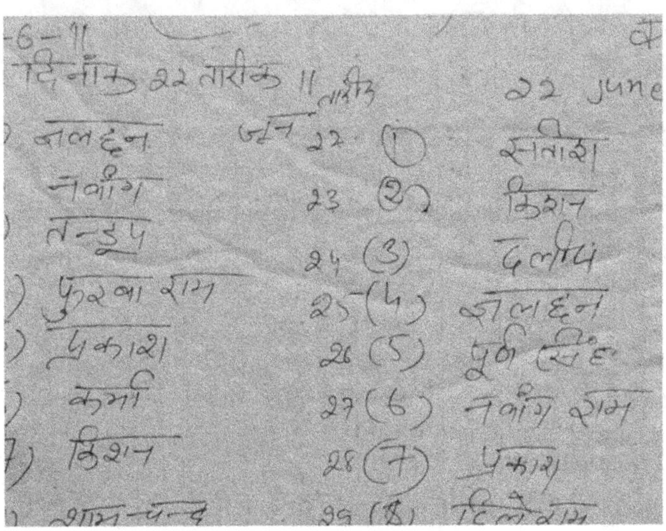

Figure 8. Showing Bari system being followed, Village Tingrit, Miyar Valley

Figure 9. Kuhl committee, Village Tingrit, Miyar Valley

Success of Traditional Kuhl and Bari system

As briefed in earlier sections, the Kuhl and Bari system has been and still is a major contributor for freshwater irrigation in the Miyar valley. The important factors that have led to the successful and prevalent use of the system in the region are listed below:

1. **Community Participation:** One of the main reasons for the success of this system is the community participation in constructing and maintaining Kuhls, irrespective of the caste, landholding size and economic status. All farmers, availing the water from the system annually come together under the guidance of Kuhl committee to collectively work and repair the channels. This in turn increases the ccommunities social capital (Padigala, 2014).

2. **Equitable Distribution of Water Rights:** Following equitable and transparent system of distribution of water rights among the farmers has over the years generated a harmonious and smoothly functioning distribution system.
3. **Cropping Pattern:** Uniformity in the crop type and activities in the valley helps in synchronization, which in turn leads to efficient use of alternating water rights among farmers.
4. **Resource Scarcity:** Resource scarcity and temporal availability of the resources like human capital and snowmelt freshwater during summers, has led to evolution of numerous successful community initiated resource management activities based on collective actions such as the Kuhl system and Jowari or Agricultural labour pooling (usually includes 4 to 5 families per group) activity undertaken during short span of summer season to counter human capital constraints (Padigala, 2014).
5. **Low Maintenance and Energy Requirement:** Since the system works on gravitational force and is chiefly constructed using local materials there is very less economic investment needed to construct and maintain a Kuhl system (see Figure 10).

CHALLENGES AND SUGGESTIONS

One of the prime challenges faced by traditional Kuhl system is the swiftly changing socio-economic scenario in the valley. Among other settlements in the valley, Miyar has undergone a higher degree of transformation over the years. The subsistence production of barley, kidney beans, seed potato etc. is being replaced by cash crops such as peas. Nearly 60% of the land is under pea cultivation, and one third of the area for potato seeds (Sen et al., 2011).

To meet the demand of water and fertilizer intensive cash crops like peas with larger monetary benefits, the farmer are using advance systems of irrigation like drip irrigation and sprinkler irrigation. This has been made possible by the technological and infrastructure (roads and telecommunication) improvement in the region in recent years.

Figure 10. A schematic overview of a typical Kuhl system, Miyar Valley

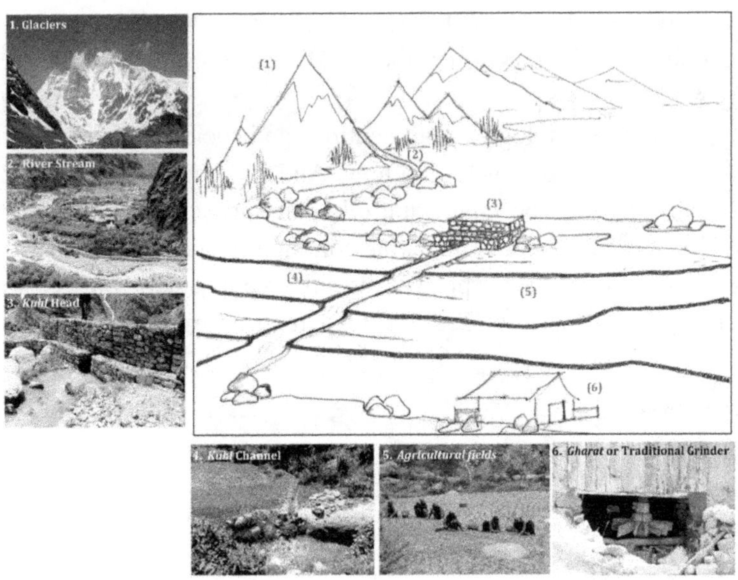

Over time there has been a decline in the local participation in kuhl system, which could be attributed largely to availability of new non-farm employment (Baker, 1996), migration and market influence, etc. Also land holding sizes have reduced because of land division due to migration and socio-cultural changes. Rising opportunities for employment under the Mahatma Gandhi National Rural Employment Guarantee Scheme, (MGNREGS) encourage the farmers to opt out of the community managed irrigation schemes.

Although, government interventions during 1970's helped traditional water systems to prevail in the region, but recent governmental activities and policies are proving harmful. For example, construction of Kuhls with cement is costly and more prone to damage due to cold weather as compared to traditional materials e.g. stone, soil mortar etc. Also use of laborers and sometimes locals in constructing kuhls under various government schemes like MGNREGS raises ownership issues. The work is normally sub-standard because local skills are disregarded and the government overseer normally does not belong to the region; and there is a lack of local participation and feeling of collective action in the work. Maintenance of such Kuhls is a problem due to lack of ownership and funds, and its poor quality.

As a remedial measure to protect the traditional water systems, external infrastructural interventions in the irrigation system must consider the local geographical conditions, socio-economic scenario, community involvement (decision making and project execution) and cost efficiency in order to implement successful interventions. More studies should be undertaken to understand these community managed systems and impacts of socio-economic changes. To improve efficiency in the water delivery mechanism and conserve the traditional knowledge, local Kuhl committee's should be proactively involved in decision making, project implementation and in capacity building.

CONCLUSION

The physical characteristics and geographical isolation of Miyar valley have contributed towards a tradition of cooperation and self-reliance in solving problems. The traditional *kuhl* system exhibits one such evolution of traditional knowledge, local institution, and sustainable utilization of resources, collective action and excellent social bonding to adapt community against resource scarcity and, to attain overall community development. Existing local institutions and adaptations strategies currently are efficiently managing natural resources and dealing with resource scarcities. And these strategies can play a vital role in providing a buffer zone for the community to adapt against any future climatic and social or economic variability's.

But, changing population, technological interventions and socio-economic dynamics in the valley, traditional water system practices are slowly declining. And with increasing climate induced risks there is an urgent need to study these impacts, and carefully plan interventions (awareness, funds and govt. policies etc.) considering local conditions and taking community into confidence, since these traditional practices like *Kuhls*, not only serve the purpose of resource management, but also brings together community to act collectively for greater social and ecological purposes.

ACKNOWLEDGMENT

Author is grateful to Prof. Milap Chand Sharma and Prof. Sucharita Sen for providing important inputs and guidance all through the project. Author would also like to express thanks to Mr. Sunil Kraleti for his support during field surveys and documentation. Finally, would like to appreciate the helpful participation and warm hospitality by the locals of Miyar valley during this in this project.

REFERENCES

Ageta, Y., & Kadota, T. (1992). Predictions of changes of glacier mass balance in the Nepal Himalaya and Tibetan Plateau: A case study of air temperature increase for three glaciers. *Annals of Glaciology, 16*, 89–94.

Aizebeokhai, A. P. (2009). Global warming and climate change: Realities, uncertainties and measures. *International Journal of Physical Sciences*, *4*(13), 868–879.

Anyinam, C. (1995). Ecology and ethnomedicine: Exploring links between current environmental crisis and indigenous medical practices. *Social Science & Medicine*, *40*(3), 321–329. doi:10.1016/0277-9536(94)E0098-D PMID:7899944

Bagadian, B. U. (1987). *Farmers' Participation in Irrigation Management: The Philippine Experience.* Paper for Workshop on People's Participation in Irrigation Management. Administrative Staff College of India (ASCI), Hyderabad, India.

Baker, M. J. (1996). *Changing Contexts, Steady Flows: Patterns of Institutional Change within the Communal Irrigation Systems (Kuhls) of Kangra Valley* (Vol. 15). Himachal Pradesh, India: Himalayan Research Bulletin.

Balée, W. (1994). *Footprints of the forest: Ka'apor ethnobotany. The historical ecology of plant utilization by an Amazonian people*. Columbia University Press.

Barlett, P. F. (Ed.). (2013). *Agricultural decision making: Anthropological contributions to rural development*. Academic Press.

Barnett, T. P., Adam, J. C., & Lettenmaier, D. P. (2005). Potential impacts of a warming climate on water availability in snow-dominated regions. *Nature*, *438*(7066), 303–309. doi:10.1038/nature04141 PMID:16292301

Bates, B. C., Kundzewicz, Z. W., Wu, S., & Palutikof, J. P. (Eds.). (2008). *Climate Change and Water, Technical Paper VI of the Intergovernmental Panel on Climate Change*. Geneva: IPCC Secretariat.

Berkes, F. (2012). *Sacred Ecology* (3rd ed.). New York: Routledge.

Bhutiyani, M. R., Kale, V. S., & Pawar, N. J. (2007). Long-term trends in maximum, minimum and mean annual air temperatures across the Northwestern Himalaya during the twentieth century. *Climatic Change*, *85*(1-2), 159–177. doi:10.1007/s10584-006-9196-1

Borthakur, S. (2009). Traditional rain water harvesting techniques and its applicability. *Indian Journal of Traditional Knowledge*, *8*(4), 525-530.

Brooks, N. (2003). *Vulnerability, risk and adaptation: A conceptual framework*. Tyndall Centre for Climate Change Research Working Paper 38.

Burton, I., Malone, E., & Huq, S. (2005). Adaptation policy frameworks for climate change: developing strategies, policies, and measures. Cambridge, UK: Cambridge University Press.

Central Ground Water Board. (2011). *Select Case Studies Rain Water Harvesting and Artificial Recharge*. New Delhi: Ministry of Water Resources.

Chalise, S. R. (2000). *Water resources management in the HKH: an overview. Water of Life: Perspectives of Water Harvesting in the Hindu Kush-Himalayas*. Kathmandu, Nepal: ICIMOD.

Chand, R. (1994). Role of Water Rights in Farmer-Managed Hill Irrigation Systems. *Economic and Political Weekly*, *29*(13), 26–30.

Chettri, N., Sharma, E., Shakya, B., Thapa, R., Bajracharya, B., Uddin, K., . . . Choudhury, D. (2010). *Biodiversity in the Eastern Himalayas: Status, Trends and Vulnerability to Environmental Change; Environmental Change Impact and Vulnerability in the Eastern Himalayas*. Technical report 2. Kathmandu: ICIMOD.

Correa, C. M. (2001). Traditional knowledge and intellectual property. Geneva: The Quaker United Nations Office (QUNO).

Coward, E. W. Jr. (1990). Property Rights and Network Order: The Case of Irrigation Works in Western Himalayas. *Human Organization*, *49*(1), 78–88. doi:10.17730/humo.49.1.g53435062n67k2g2

Coward, E. W., & Gilbert, L. (1987). *Studies of Farmer-Managed Irrigation Systems'*, Background Paper for the Workshop on People's Participation in Irrigation Management, Administrative Staff College of India.

Crowley, T. J. (2000). Causes of climate change over the past 1000 years. *Science*, *289*(5477), 270–277. doi:10.1126/science.289.5477.270 PMID:10894770

Duerden, F. (2004). Translating climate change impacts at the community level. *Arctic*, *57*(2), 204–212. doi:10.14430/arctic496

Ensor, J., & Berger, R. (2009). *Understanding climate change adaptation: lessons from community-based approaches*. Practical Action Publication. doi:10.3362/9781780440415

Eriksson, M., Jianchu, X., Shrestha, A. B., Vaidya, R. A., Nepal, S., & Sandström, K. (2009). *The changing Himalayas: impact of climate change on water resources and livelihoods in the greater Himalayas. International centre for integrated mountain development*. ICIMOD.

Few, R., Brown, K., & Tompkins, E. L. (2007). Public participation and climate change adaptation: Avoiding the illusion of inclusion. *Climate Policy*, *7*(1), 46–59. doi:10.1080/14693062.2007.9685637

Folke, C., Carpenter, S., Elmqvist, T., Gunderson, L., Holling, C. S., & Walker, B. (2002). Resilience and sustainable development: building adaptive capacity in a world of transformations. *AMBIO: A Journal of the Human Environment, 31*(5), 437-440.

Galacgac, E. S., & Balisacan, C. M. (2009). Traditional weather forecasting for sustainable agroforestry practices in Ilocos Norte Province, Philippines. *Forest Ecology and Management, 257*(10), 2044–2053. doi:10.1016/j.foreco.2009.01.002

Grothmann, T., & Patt, A. (2005). Adaptive capacity and human cognition: The process of individual adaptation to climate change. *Global Environmental Change, 15*(3), 199–213. doi:10.1016/j.gloenvcha.2005.01.002

Hansen, J., Makiko, S., Ruedy, R., Lacis, A., & Oinas, V. (2000). Global warming in the twenty-first century: An alternative scenario. *Proceedings of the National Academy of Sciences of the United States of America, 97*(18), 9875–9880. doi:10.1073/pnas.170278997 PMID:10944197

IPCC. (2007). *Impacts, Adaptation, and Vulnerability*. Geneva, Switzerland: Working Group II Contribution to the Intergovernmental Panel on Climate Change Fourth Assessment Report. Summary for Policy Makers. IPCC.

Jianchu, X., Shrestha, A., & Eriksson, M. (2009). *Climate change and its impacts on glaciers and water resource management in the Himalayan Region. Assessment of Snow, Glaciers and Water Resources in Asia*. Koblenz, Germany: International Hydrological Programme of UNESCO and Hydrology and Water Resources Programme of WMO.

Karki, M., Mool, P., & Shrestha, A. (2009). Climate Change and its Increasing Impacts in Nepal. *The Initiation, 3*, 30–37.

Kelly, P. M., & Adger, W. N. (2000). Theory and practice in assessing vulnerability to climate change andFacilitating adaptation. *Climatic Change, 47*(4), 325–352. doi:10.1023/A:1005627828199

Khan, A. N., Shaw, R., Pulhin, J., & Pereira, J. (2010). Climate change adaptation and disaster risk reduction in Pakistan. Climate change adaptation and disaster risk reduction. *Asian Perspective*, 5.

Klein, R. J., & Maciver, D. C. (1999). Adaptation to climate variability and change: Methodological issues. *Mitigation andAdaptation Strategies for Global Change, 4*(3-4), 189–198. doi:10.1023/A:1009690729283

Koul D, N., Singh, S., Neelam, G., & Shukla, G. (2012). Traditional water management systems - An overview of Ahar-pyne system in South Bihar plains of India and need for its revival. *Indian Journal of Traditional Knowledge, 11*(2), 266-272.

Kreutzmann, H. (2011). Scarcity within opulence: Water management in the Karakoram Mountains revisited. *Journal of Mountain Science, 8*(4), 525–534. doi:10.1007/s11629-011-2213-5

Lal, C., & Verma, L. R. (2008). Indigenous Technological Knowledge on soil and water management from Himachal Himalaya. *Indian Journal of Traditional Knowledge, 7*(3), 485-493.

McGregor, D. (2004). Coming full circle: Indigenous knowledge, environment, and our future. *American Indian Quarterly, 28*(3), 385–410. doi:10.1353/aiq.2004.0101

Menzies, C. R. (Ed.). (2006). *Traditional ecological knowledge and natural resource management*. U of Nebraska Press.

Ministry of Water Resources (MoWR) State Level Report. (1988). *1st National census of minor irrigation schemes, Government of India*. Author.

Ministry of Water Resources (MoWR) State Level Report (2001). *3rd National census of minor irrigation schemes*. GoI.

Narain, S. V. (2006). *Community-led Alternatives to Water Management: India Case Study, Human development report*. UNDP.

Nazarea, V. D. (1999). *Ethnoecology: Situated Knowledge/Located Lives*. Tucson, AZ: University of Arizona Press.

Nelson, D. R., Adger, W. N., & Brown, K. (2007). Adaptation to environmental change: Contributions of a resilience framework. *Annual Review of Environment and Resources*, *32*(1), 395–419. doi:10.1146/annurev.energy.32.051807.090348

Norgaard, R. B. (1984). Traditional agricultural knowledge: Past performance, future prospects, and institutional implications. *American Journal of Agricultural Economics*, *66*(5), 874–878. doi:10.2307/1241018

Padigala, B. (2014). Social Capital and Local Institutions: A Perspective to Assess Communities Adaptation Potential to Climate Change. In Handbook of Climate Change Adaptation. Springer. doi:10.1007/978-3-642-40455-9_114-1

Padigala, B., Sen, S., Sharma, M. C., & Kraleti, S. (2011). Social Capital and Local Institutions- A Perspective to Assess Communities Adaptation Potential to Climate Change. *Second National Research Conference on Climate Change*. IIT Delhi.

Pandey, D. N. (1998). *Ethnoforestry: Local Knowledge for Sustainable Forestry and Livelihood Security*. Udaipur, India: Himanshu Publications.

Pandey, D. N., Gupta, A. K., & Anderson, D. M. (2003). Rainwater harvesting as an adaptation to climate change. *Current Science*, *85*(1), 46–59.

Pant, N., & Verma, R. K. (1983). *Farmers Organisation and Irrigation Management*. New Delhi: Ashish Publishing House.

Pathak, M. (2010). Climate Change: Uncertainty for Hydropower Development in Nepal. *Hydro Nepal*, (6), 31-34.

Pattanaaik, S.K., Sen, D., Kumar, N., Moyong, O., & Debnath, P. (2012). Traditional system of water management in watersheds of Arunachal Pradesh. *Indian Journal of Traditional Knowledge*, *11*(4), 719-723.

Pieroni, A., Giusti, M. E., & Quave, C. L. (2011). Cross-cultural ethnobiology in the Western Balkans: Medical ethnobotany and ethnozoology among Albanians and Serbs in the Pešter Plateau, Sandžak, South-Western Serbia. *Human Ecology*, *39*(3), 333–349. doi:10.1007/s10745-011-9401-3

Rao, P., Areendran, G., & Sareen, R. (2008) Potential impacts of climate change in the Uttarakhand Himalayas. Mountain Forum Bulletin, 8(1).

Rao, P., & Patil, Y. (2015). Climate Resilience in Natural Ecosystems in India: Technology Adoption and the Use of Local Knowledge Processes and Systems. In W. L. Filho (Ed.), *Handbook on Climate Change Adaptation* (pp. 2063–2077). Springer-Verlag Berlin Heidelberg. doi:10.1007/978-3-642-38670-1_52

Richardson, S. D., & Reynolds, J. M. (2000). An overview of glacial hazards in the Himalayas. *Quaternary International, 65*, 31–47. doi:10.1016/S1040-6182(99)00035-X

Saini, R. (2008). *Glacier dynamics water resource assessment and landscape evolution in Miyar basin, Lahaul Himalayas, Himachal Pradesh.* (Unpublished M Phil Dissertation), JNU, New Delhi, India.

Sen, S., Sharma, M., Padigala, B., Kraleti, S., & Deswal, S. (2011). *Response and adaptability to climatic variability in agro-ecosystems: a case of Miyar basin, Lahaul Himalaya, India.* Unpublished paper.

Sengupta, N. (1993). *User Friendly Irrigation Designs*. New Delhi: Sage Publications.

Sethi, A. S., & Kanwar, B. S. (1989). Irrigation Potential of Kuhls and its Impact on Crops Productivity — A Case Study of Himachal Pradesh. In V. S. Mahajan (Ed.), *Agriculture and Rural Economy in India* (pp. 170–179). New Delhi: Deep and Deep Publications.

Sharma, N., & Kanwar, P. (2009). Indigenous water conservation systems—A rich tradition of rural Himachal Pradesh. *Indian Journal of Traditional Knowledge, 8*(4), 510-513.

Shekhar, M. S., Chand, H., Kumar, S., Srinivasan, K., & Ganju, A. (2010). Climate-change studies in the western Himalaya. *Annals of Glaciology, 51*(54), 105–112. doi:10.3189/172756410791386508

Shivakoti, G. P., & Shrestha, S. G. (2008). FMIS of Nepal: Balancing Water Use and Environment Conservation for Sustainable Livelihood. *Himalayan Journal of Sociology and Anthropology, 1*, 64–74. doi:10.3126/hjsa.v1i0.1554

Shrestha, U. B., Gautam, S., & Bawa, K. S. (2012). Widespread climate change in the Himalayas and associated changes in local ecosystems. *PLoS ONE, 7*(5), e36741. doi:10.1371/journal.pone.0036741 PMID:22615804

Singh, K. K. (1991). Irrigation Management by Farmers; The Indian Experience. In Farmer Participation and Irrigation Organisation. Westview Press.

Singh, P., Arora, M., & Goel, N. K. (2006). Effect of climate change on runoff of a glacierized Himalayan basin. *Hydrological Processes, 20*(9), 1979–1992. doi:10.1002/hyp.5991

Singh, P., & Bengtsson, L. (2004). Hydrological sensitivity of a large Himalayan basin to climate change. *Hydrological Processes, 18*(13), 2363–2385. doi:10.1002/hyp.1468

Singh, P., & Bengtsson, L. (2005). Impact of warmer climate on melt and evaporation for the rainfed, snowfed and glacierfed basins in the Himalayan region. *Journal of Hydrology (Amsterdam), 300*(1), 140–154. doi:10.1016/j.jhydrol.2004.06.005

Smith, B., Burton, I., Klein, R. J., & Wandel, J. (2000). An anatomy of adaptation to climate change and variability. *Climatic Change, 45*(1), 223–251. doi:10.1023/A:1005661622966

The United Nations Framework Convention on Climate Change (UNFCCC). (2015). *Adoption of the Paris Agreement*. Available at: https://unfccc.int/resource/docs/2015/cop21/eng/l09r01.pdf

Tompkins, E. L., & Adger, W. (2004). Does adaptive management of natural resources enhance resilience to climate change? *Ecology and Society*, *9*(2), 10.

Toprak, Z. F., Hamidi, N., Toprak, Ş., & Şen, Z. (2013). Climatic identity assessment of the climate change. *International Journal of Global Warming*, *5*(1), 30–45. doi:10.1504/IJGW.2013.051480

Tourenq, C., Brook, M., Knuteson, S., Shuriqi, M. K., Sawaf, M., & Perry, L. (2011). Hydrogeology of Wadi Wurayah, United Arab Emirates, and its importance for biodiversity and local communities. *Hydrological Sciences Journal*, *56*(8), 1407–1422. doi:10.1080/02626667.2011.631139

Uphoff, N. (1985). People's Participation in Water Management: Gal Ova, Sri Lanka. In L. C. Garcia-Zamor (Ed.), Public Participation in Development Planning. Westview Press.

Van Aalst, M. K., Cannon, T., & Burton, I. (2008). Community level adaptation to climate change: The potential role of participatory community risk assessment. *Global Environmental Change*, *18*(1), 165–179. doi:10.1016/j.gloenvcha.2007.06.002

Vohland, K., & Barry, B. (2009). A review of in situ rainwater harvesting (RWH) practices modifying landscape functions in African drylands. *Agriculture, Ecosystems & Environment*, *131*(3-4), 119–127. doi:10.1016/j.agee.2009.01.010

Wai, F. L. (1998). *Governing Irrigation Systems in Nepal: Institutions, Infrastructure, and Collective Action*. Oakland, CA: ICS Press.

Weatherhead, E., Gearheard, S., & Barry, R. G. (2010). Changes in weather persistence: Insight from Inuit knowledge. *Global Environmental Change*, *20*(3), 523–528. doi:10.1016/j.gloenvcha.2010.02.002

Xu, J., Grumbine, R. E., Shrestha, A., Eriksson, M., Yang, X., Wang, Y. U. N., & Wilkes, A. (2009). The melting Himalayas: Cascading effects of climate change on water, biodiversity, and livelihoods. *Conservation Biology*, *23*(3), 520–530. doi:10.1111/j.1523-1739.2009.01237.x PMID:22748090

Yohe, G., & Tol, R. S. (2002). Indicators for social and economic coping capacity—moving toward a working definition of adaptive capacity. *Global Environmental Change*, *12*(1), 25–40. doi:10.1016/S0959-3780(01)00026-7

Chapter 3
Understanding Glacial Retreat in the Indian Himalaya:
Historical Trends and Field Studies from a Large Glacier

Rajesh Kumar
Sharda University, India

Prakash Rao
Symbiosis International University, India

G. Areendran
WWF, India

ABSTRACT

One of the most important and visible indicators of global climate change is the recession of glaciers in many parts of the world. The findings of the IPCC Assessment Report (2007) suggest that there has been a significant decline in mountain glaciers and snow cover, which has contributed to the increased sea levels and water flow changes in river basins. In the latter half of 20th century, an increase in the rate of retreat has been observed in Himalayan Glaciers since advent of industrialization. The paper attempts to present a critical review and understanding of the recent changes in retreat of selected Himalayan glaciers from potential climate change. The study explores scientific evidence and analysis of observed data from a large and a small glacier in the Indian Himalaya. This will help in understanding the vulnerability of retreat of the glaciers based on their size. The larger glaciers are unlikely to disappear in the near future, due to its large mass balance and large response time due to climatic changes. Future adaptation needs, research and action for downstream communities, ecosystems and impacts on power sector are discussed.

DOI: 10.4018/978-1-5225-1046-8.ch003

INTRODUCTION

The Cryosphere plays a major role in the Earth's climate system by impacting on the surface energy budget, the water cycle, sea level etc. and has control on the physical, biological and social environment over a major portion on the Earth. The cryosphere is most sensitive to temperature change hence capable of providing the most visible signatures of climate change. There are many indicators of climate change like changes in surface temperature, precipitation, extreme weather events, glaciers, ocean and land ice and sea level. Icesheets and glaciers changes are closely related with the global sea level changes impacting the global ocean circulation and marine ecosystems. The warming trend has been reported in the vertical profiles of temperature measured through the entire thickness of mountain glaciers, or through ice sheets climate over recent decades (Lüthi and Funk, 2001; Hoelzle *et al.*, 2011). The loss of glaciers and snow cover near populated areas have direct impacts on the livelihood of the local communities in terms of water resources for drinking, agriculture, hydropower etc.

Over the last few decades, changes in climate and local weather conditions have impacted the world's glaciers both in terms of structure and characteristics, reflected in the form of advancement or retreat of glacial snouts (IPCC, 2007; UNEP, 2007). The changes in the length, width, area, volume and mass balance of the glaciers are among the most directly visible signals of global warming and these changes are the primary reasons why glacial observations have been used for climate system monitoring for many years (Haeberli, 1990; Wood, 1990), especially in areas where time series data on climate (mainly temperature and precipitation) is difficult to get and where climate change signals are not yet clear (Yadav et al., 2004; Roy and Balling, 2005). Thirty reference glaciers that have been studied in detail since 1975 show an average annual mass loss of 0.58 metres (m) water equivalent in the past decade, which is four times the rate for the period 1976-85 (UNEP, 2007).The IPCC, 2013 report (The physical science basis) clearly reflects the warming trend on the planet and state that "The last three decades has been successively warmer at the Earth's surface than any preceding decade since 1850 while in the Northern Hemisphere, 1983–2012 was likely the warmest 30-year period of the last 1400 years (medium confidence)". Unnatural rates of glacial melting, related to the changing climate, can have serious implications on the hydrology of the associated river systems and consequently on the livelihoods of millions of people who are dependent on these rivers and their ecosystems.(Stern,2007). Overexploitation of water resources and changes in hydrological cycle could be vital to future developmental strategies of the fertile belt of Northern India where agricultural productivity, irrigation etc.is to a large extent dependent on availability of freshwater resources (Mall et al, 2006).

The right approach, thus, in addressing these impacts is to have a better scientific understanding through long-term observations and analysis of the interactions of the different components of the mountain ecosystems with their climate. Then the lager data can be helpful in predicting the future scenarios and utilizing the information to formulate effective adaptation and management strategies. There is also a pressing need for disaster management and rehabilitation policies in areas affected by the glacial retreat, particularly focusing on agriculture, meadows and other livelihood support structures of the local people.

A growing body of scientific knowledge solidly indicates that carbon concentrations are changing the global climate and that there would still be significant changes to the global climate even if emissions would drastically reduce and fall to zero. The world is anticipating a 2°C warming above the pre-industrial level, and this will result in some unavoidable changes, the effects of which need to be understood, especially in fragile and endangered ecosystems. Generating a deeper understanding of the climate impacts on critical indicators such as glaciers are needed as there is inadequate documented scientific evidence within India on the impact of climate change on the Himalayan ecosystems. This would help in developing regional climate projection models which are more accurate in providing information on future impacts. The recent observation of IPCC (2013) reveals that the globally average temperature show a warming of 0.85 (0.65 to 1.06)°C, over the period 1880 to 2012 as calculated on linear trend basis.

Glacial Retreat in the Himalayas

In the Himalayan region, glaciers and snow cover have been thinning since the end of 19th century in line with the global trends. With significant snout fluctuations, most of the glaciers in the Himalayan mountain ranges have been retreating and their rate of retreat is much faster than that of glaciers in other parts of the world (Cruz et al., 2007). These changes correspond to the rising surface temperature trends in the Himalayas which have been reported to be higher than the global average warming (UNESCO, 2007; Jianchu et al., 2007; Barnett et al., 2005). A study of the temperature trends in the North West Himalayan region (Bhutiyani et al., 2007) shows that a significant warming of 1.6°C has occurred over the last century with warming in winter taking place at a faster rate, with the highest warming rates recorded in the period 1991 -2002. Apart from warming, other factors like high human population density near these glaciers, deforestation and land use changes have also been responsible for the decline and shrinkage of glaciers (Cruz et al., 2007). At current rates of retreat, the smaller valley type glaciers are more likely to decline at a faster rate in the future, with uncertain impacts on the downstream areas.

The dynamics of a glacier are influenced by climatic factors including temperature, precipitation, amount of snowfall and wind. However, establishing firm linkages between climatic variability due to anthropogenic factors and glacier retreat requires long-term studies and monitoring. Mass balance and glacier snout monitoring have been important methods for the assessment of the volumetric and geometric changes in the glaciers brought about by advancement or recession (Dobhal et al., 2007).

In India, although high mountain glaciers occur across all the Himalayan range states of Jammu and Kashmir, Himachal Pradesh, Uttarakhand, Sikkim and Arunachal Pradesh, although only a few have been studied for long term monitoring or documented in terms of glacial retreat. Recent studies from 466 glaciers of the Indian Himalayas indicate that there has been a 21% reduction in the glaciered area from 2,077 square kilometers (sq km) in 1962 to 1,628 sq km in 2004 (Kulkarni et al., 2007). Smaller glaciers of less than one sq km have reduced in area by as much as 38% compared to a 12% retreat of the larger glaciers.

In the Indian Himalayas, studies on snout recession/mass balance measurements have been carried out on a few glaciers and some studies have been able to generate long-term monitoring data (Dobhal et al., 2007; Kumar et al., 2007; Wagnon et al, 2007). Studies by various institutions on the changing characteristics of glaciers in the Himalayas are limited (see box: Studies on some important glaciers in the Himalayas), generating an important debate on the role of climate change in the extent of glacier loss (Raina, 2009).

These glaciers have left behind imprints of their unnatural retreat, which is in tandem with the discrete but well defined phases of global warming. The melting process has two key adverse impacts on the environment. One is the loss in the reserve of freshwater and the other is a significant rise in sea levels (Kaser et al., 2006).

While the global view (IPCC, 2007) suggests that a large percentage of glaciers are declining worldwide, there are conflicting views also on the rate of retreat in the era of industrialization. Some studies suggest that a few glaciers in the Himalayas may be showing reduced retreat patterns or actually increasing. e.g., the reduction in the retreat rate of Gangotri Glacier for 2004-05 (Kumar et al., 2007).

STUDY AREA

The Gangotri glacier, is one of the largest mass of ice and snow in the western Himalayan region of India and is located in District Uttarkashi in the state of Uttarakhand in India (Figure 1-2). The location derives considerable religious significance as an important sacred site as a source of the river Ganga which flows through several North Indian states before flowing into the Bay of Bengal. As such the region and the shrine at Gangotri, attracts thousands of pilgrims every year. The glacier along with other tributary glaciers (Figure 3) is a vital source of freshwater water supply, for much of the northern plains which has a large human population living downstream. The water discharge from the glacier flows initially as the river Bhagirathi before meeting the Alaknanda river at the confluence at Devprayag to form the river Ganga. Snow and glacial melt water contribute nearly 29% to the annual flows of the Ganga (up to Devprayag) and hence potential impacts on these glaciers due to various factors are likely to affect this large river system (Singh et al., 2009).

The north-west facing glacier is a typical valley type glacier originating in the Chaukhamba group of peaks in the Himalaya. Other smaller glaciers act as tributaries and join the main stem of the glacier to form the larger Gangotri group of glaciers. Including the tributaries, this covers an area of 210.60 sq. km(ETM+1999). The area and length of the main trunk of the glacier is 56.59 sq. km and 29.13 km respectively. The average width of the glacier is 1.85 km. The glacier, lies between 79°4' 46.13" E-79°16' 9.45" E and 30°43'47.00" N-30°55' 51.05" N (ETM+1999) with elevation ranging from 4,015–6,145 meters above sea level (SRTM data analysis).

Figure 1. Location Map of Gangotri glacier

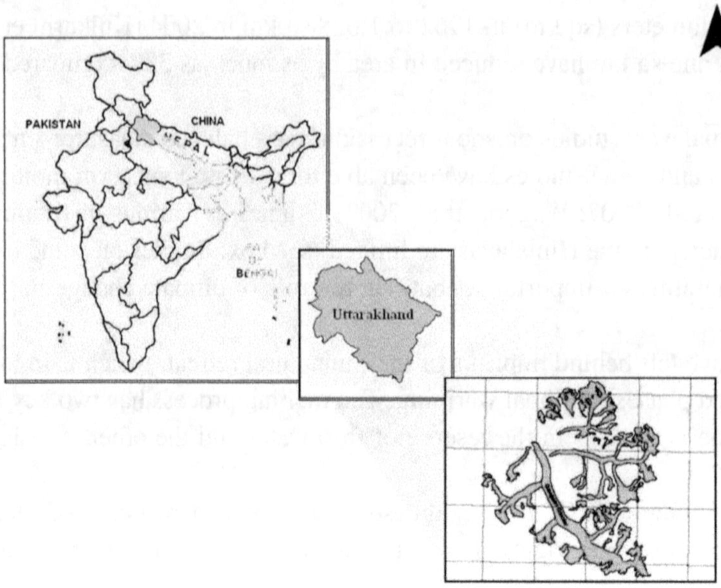

Figure 2. Location Map of Gangotri glacier and Snout of Kafni glacier

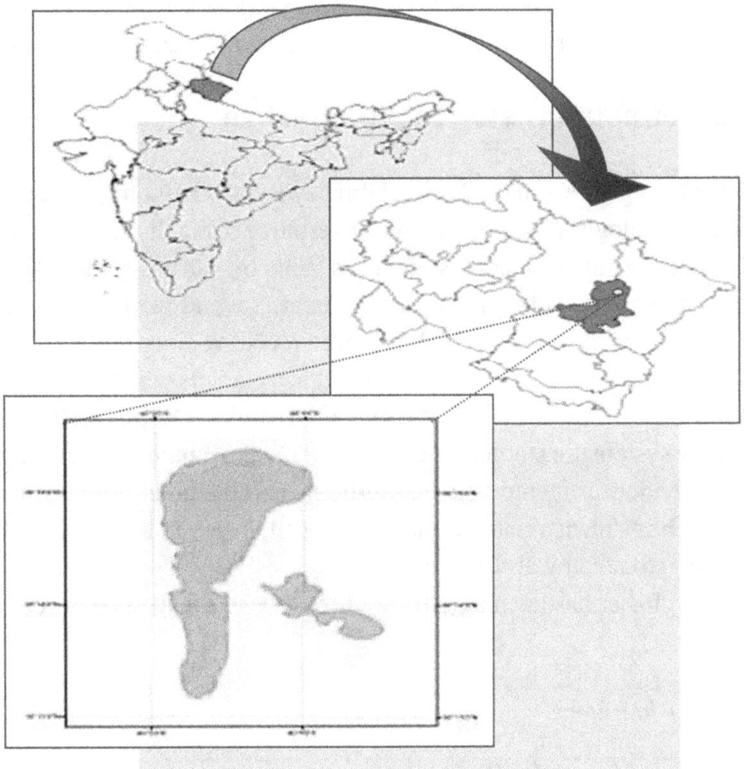

Figure 3. Location Map of Gangotri glacier

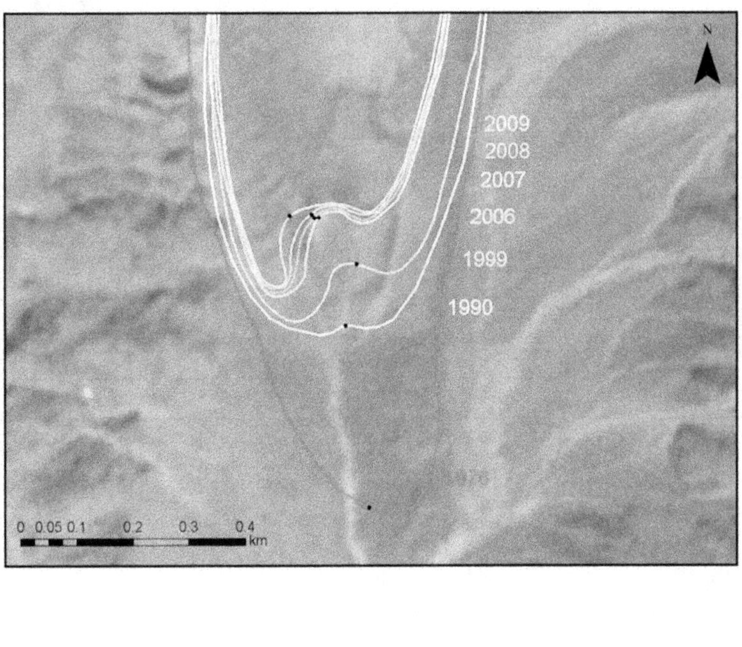

The snout of the glacier occurs at an altitude of about 3,949 m above sea level, and from where the Bhagirathi originates (GPS observation, 2008). The snout position (2008) is at 73°4' 47.26" E and30°55' 36.45" N. Table 1 provides an overview of the characteristics of the glacier system.

Gangotri Glacier Retreat: A Historical Perspective

As one of the largest known glaciers in the Indian Himalayas, the Gangotri glacier often evokes considerable interest among glaciologists, geologists, water resource specialists, and mountaineers. It is said that the region has been frequented for at least hundred years by various groups in trying to understand the glacial system and process including reasons for retreat. Several workers have carried field studies on the snout and other sites on the glacier which has been receding since the last 'Little Ice Age', which ended in the 19th century. Interestingly, many of the tributary glaciers which source the river Ganga have also reduced with a few even being separated from the main trunk of the glacier. According to Srivastava (2001), this is evident by systematic studies since 1935 on the movement of the glacier snout, and by the presence of recessional evidences features such as terminal and lateral moraines. In the past century, the retreat rate of the glacier has shown a rising trend. It has been observed by glaciologists that the snout of the Gangotri glacier has retreated by about two km in the last 100 years. Both the snout retreat and area vacated by the Gangotri glacier has been documented by several scientists (see Table 2).

Table 1. Gangotri glacier inventory

Glacier Coordinates	30°43"22" - 30°55'49" N 79°4'41" - 79°16'34" E
GSI identification number	5O131 06029
Highest elevation (m asl)	7138
Lowest elevation (m asl)	4000
Surface area (km^2)	143.58
Ablation (km^2)	92.48
Accumulation (km^2)	51.11
Mean width (km)	1.50
Maximum length (km)	30.20
Ablation zone length (km)	26
Accumulation zone length(km)	4.20
Mean depth(km)	0.20
Ice volume (km^3)	28.716
Tributaries: Chaturangi Glacier Raktvarn Glacier Kirti Glacier	length 22.45 km; area 67.70 km^2 length 15.90 km; area 55.30 km^2 length 11.05 km; area 33.14 km^2

Source: (GSI), 2001; Naithani et al., 2001; Singh et al., 2006)

Table 2. Snout retreat and area vacated by Gangotri glacier (1935-96)

Period	Annual Snout retreat (m)	Average annual surface area vacated (sqm /year)	Reference
1935- 1956	4.35 (small cave) 10.16 (large cave)	2500	Jangpani (1958)
1956-1971	27.33	10032	Vohra (1971)
1971-1974	27.34	594	Puri and Singh (1974)
1974-1975	35	13975	Puri (1984)
1975-1976	38		Puri (1984)
1976-1977	30	14875	GSI, 2001
1977-1990	28.08	15096	Puri (1991)
1990-1996	28.33	22370	Sangewar (1997)
Source: Srivastava et al., 2001(GSI, 2001)			

Local Climate Conditions Around the Gangotri Glacier

The combination of high altitude temperature and precipitation along with other climate factors like wind chill, humidity etc. often plays a key role in development of the state of glaciers and their associated hydrological features with impacts on water flow in the downstream region. Meteorological analysis of long-term data becomes a fundamental aspect in determining the structural changes taking place in glaciers. Various scientific institutions have been studying the Gangotri glacier for the past few years and have collected extensive data on the meteorological parameters (rainfall and temperature) through standard meteorological observatories installed near the glacier snout.

Rainfall

Precipitation is an important element in the evolution of a glacier worldwide and with reference to Gangotri past data suggests that the much of the region around Gangotri landscape receives less than 15 mm of daily rainfall during the summer season (Singh et al., 2005). A few exceptions in precipitation days in the form of extreme weather events like snow storms also occur. When studying precipitation trends on glaciers, it is noticed that August and September usually receive higher rainfall as compared to the other months. This is a continuous trend and even in recent years (2008-90 one can see higher precipitation patterns in those months from data obtained from weather stations installed at Bhojwasa. It has also been found that early morning and late evening are the most probable times for the occurrence of rainfall (Singh *et al.*, 2005).

Temperature

Temperature in the ablation season increases for a few months after which it starts decreasing. July has been found to be the warmest month on the basis of the mean-maximum and minimum temperatures. Daily variations in temperature show that the maximum temperature is observed around 1400 hours, while the minimum is observed in the early morning hours (Singh *et al.*, 2005). Changes in minimum temperature have been seen to be more significant than those in the maximum temperature. It has also been found that the 'maximum diurnal temperature range' occurs in May and October while August shows the least variation in temperature range. This is probably due to the presence of a cloud cover during the rainy season.

Hydrology

The major sources of runoff at Gangotri are melting snow and ice. As a general assumption, since this area receives low rainfall, there is unlikely to be a heavy runoff from the region. Stream flow at the Gangotri glacier shows a wide variation depending on various factors. Early on and towards the end of the ablation season, day and night time flow volume show little variation ; however as the peak melting time approaches during the day, there is a comparative reduction in the nighttime flow. Despite this,it is also also observed that there is a continuous but low water flow pattern even when night temperatures are lower than normal. For large glacier this is a reflection of strong meltwater storage characteristics. Monthly variation of flow shows that the discharge starts rising from May and maintains a high flow during June to August with maximum average discharge in August. The discharge starts reducing from September onwards. Singh *et al.*, (2006) have also reported similar results in the discharge flows from the glacier.

METHODOLOGY

Field studies at Gangotri glacier for monitoring local climate and hydrological conditions involved using primary field data through the use of i) Differential Global Positioning System (DGPS) for assessment of snout positions at different time periods and ii) use of Remote Sensing satellite imagery along with iii) existing secondary data from earlier studies. DGPS measurements have been extensively used for locating the snout position in glaciological studies including at Gangotri during various field visits starting from 2006. While the task of long term data collections and observations is often tedious, due to inconsistency and access to field sites due to inclement weather conditions, it still remains an important factor for accuracy in any glacial research. In this context, it is important to note that due to variability in local climate, the snout position is quite dynamic and inter-seasonal changes are seen in the exact position of the snout. Primary data was collected on snout position during the months of September and October since 2006. This was supplemented by meteorological and local climate data through an automatic weather station (AWS) installed at Bhojwasa near the snout of the glacier. Apart from the glacial and meteorological monitoring, data was collected about the discharge patterns of the glacial meltwater. Remote sensing data was collected using satellite imageries of 1976, 1990, 1999 and 2006 by delineating the glacier boundary using ERDAS 9.0 and auto calculating the area using GIS methodology.

RESULTS AND DISCUSSION

A comparative analysis of the glacier's snout position was carried out using data from secondary sources and interpretations from various satellite imageries over the past three decades. Satellite imageries available since 1976 formed the baseline for the analysis of the fluctuations in snout position together with

DGPS observations since 2006. A study of data from available sources suggests that the main trunk of the Gangotri glacier has been in a continuous state of recession and fragmentation during the past century. The length of the glacier has been computed for different years based on available data from various scientific agencies and workers. These trends show that the length of the glacier has reduced by about 0.59 km in 33 years, from 1976-2009, with an average retreat rate of 17.59 m/year (see Figure 4).

Based on the comparison of satellite imageries of the Gangotri glacier for the years 1976, 1990, 1999 and 2006, our analysis shows that the glacier is not only receding in length but also in terms of glaciated area from all the sides. The possible reasons behind this retreat may be linked with two main factors: (a) reduction in snowfall and (b) an increase in the temperature of the region. Cumulative analysis shows that between 1976 and 2006, the glacier area reduced by 15.5 sq.km, with an average loss of 0.51 sq.km per year. This reduction in glacier area is 22.9% over 1976 (see Figure 5). The glacier area reduced by 7.2 sq. km between 1976 and 1990, in fourteen years, with a 10.6% reduction in the glacial area. In nine years, between 1990 and 1999, the glacial area reduced by 4.3 sq.km and resulting in 7% reduction in area when compared to 1990. However, the rate has increased between 1999 and 2006, with glacial area

Figure 4. Length of Gangotri glacier in different years

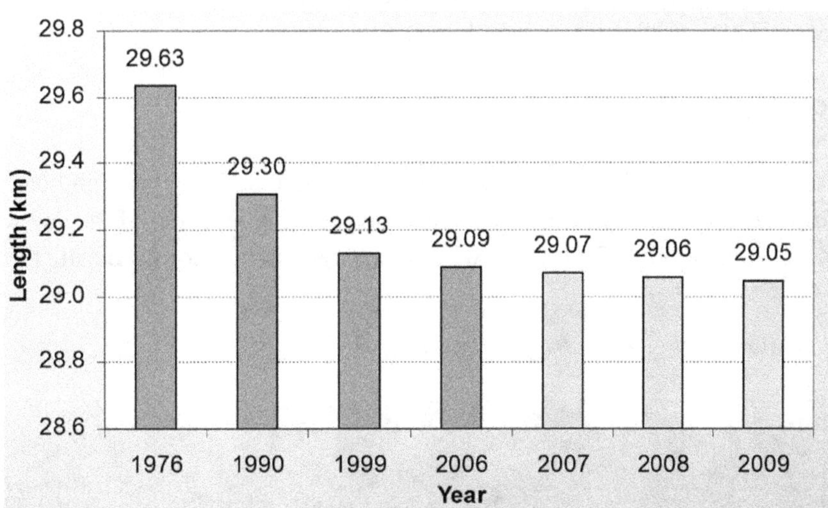

Figure 5. The cumulative and intermittent reduction in the area of Gangotri Glacier since 1976

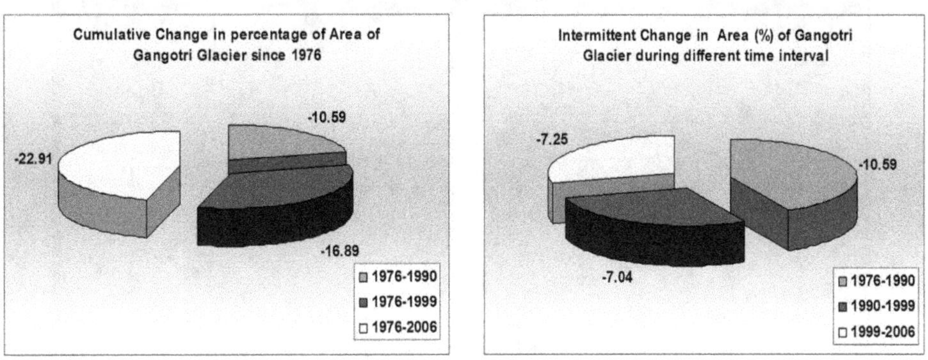

reducing by 4.1 sq.km and a 7.25% reduction in area over 1999. This also indicates that in some years there is a reverse trend in the rate of retreat of the glacier as compared to preceding years. It is possible therefore that local climate conditions and fluctuations may be responsible for the reduced rate of retreat at different time intervals.

The calculation of the snout retreat using a DGPS based system involves assessment of point to point snout data that gives the straight line distance of the two points that may include the shift of the water outlet point (left/right) in the curvilinear shape of the larger snout. Traditional methods using a simple GPS methodology often assess the shift of the snout position in the glacier flow direction through measuring the perpendicular distance on the curvilinear shape of the snout. This measurement normally gives a lesser value if the snout point has moved either left or right rather than in the glacier flow direction as compared to the previous position.This can be critical since fluctuations are often noticed in the snout positions for larger glaciers which have a significantly large area. Based on this measurement, we observed that the snout retreat during 2006-07 was 17.85 m when compared to snout position of 2006. In the subsequent years, the retreat rate has reduced 13.06 m in 2007-08 and 10.57 m in 2008-09. It can be inferred that retreat of the glacier between 2007 and 2009, is lower than the long term average. More importantly, while there is an overall average retreat noticed, fragmentation of the Gangotri glacier is likely to be of larger concern in the context of reduced water flows of the glacial system in the long term. The retreat of the Gangotri Glacier snout is shown by overlaying the glacier boundary on the Enhanced Thematic Mapper (ETM+) 1999 imagery (see Figure 6).

Hydro-Meteorological Trends

Assessment of summer season discharge data at the snout in the past few years indicates that there are variations on a diurnal and monthly basis. Flows recorded at different times of the day were found to be highest during the evening hours (1700 hours) while discharge rates are lowest during the morning hours

Figure 6. Snout position of the Gangotri glacier during 1976 to 2009

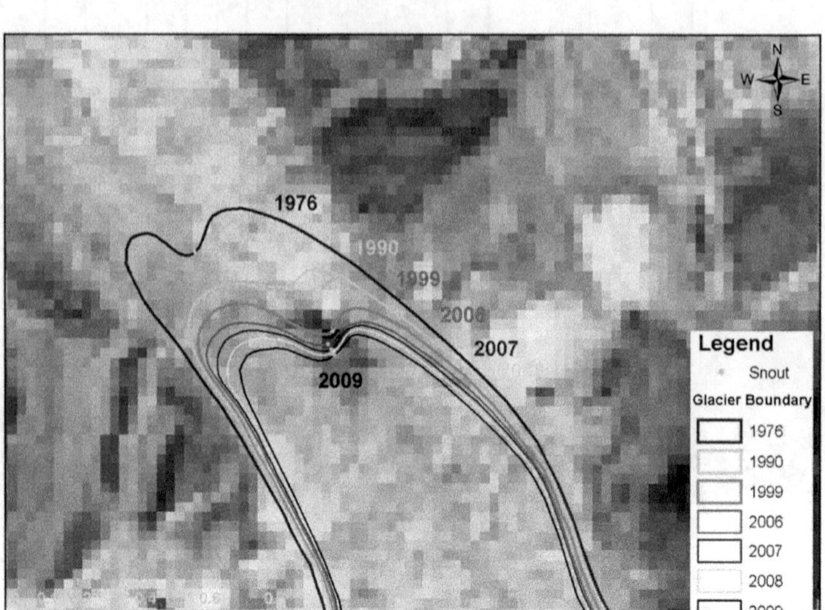

(see Figure 7). These results also corroborate earlier findings by Singh *et al.*,(2006) where high diurnal fluctuations in discharge rates during daytime and low rates at Bhojwasa were observed.

With the commencement of the summer season, increased water discharge reaches a peak during July and August due to the combined impact of more melting of ice, which is driven by higher average temperatures and rainfall in the catchments. The discharge starts reducing in September and reaches a very low level throughout the accumulation season. During winters, the flow is reduced and keeps almost at a constant value. The combined impact of low winter temperature (subzero) and higher albedo (90-95% for fresh snow) reduces the winter discharge rate to quite a low level.

The daily average temperature and average discharge for the ablation and accumulation season at Bhojwasa (Figure 8), is indicative of a positive correlation of 0.73 between the average temperature and average discharge during the ablation season. However, these parameters are negatively correlation (0.46) in the accumulation season, which shows a lower influence of temperature on melting because most of the solar radiation is reflected back to the atmosphere due to the snow covered valley that has a high albedo effect. The preliminary trends observed during the study in 2008 seemed to suggest that the volume of discharge and flow patterns is also being influenced by the temperature variations in the valley. This is in consonance with results obtained by other studies (Singh *et al.*, 2006) over the past decade. However, it may be mentioned here that large glaciers like Gangotri need to be closely monitored for water discharge during both ablation and accumulation season and only long term trends can suggest any significant variations in retreat patterns. For smaller glaciers, which do not have a clearly demarcated ablation and accumulation zone, retreat rates may need to be monitored critically. While these results are specific to the snout at Gangotri, comparisons with similar studies indicate a consistent pattern in the rate of discharge and temperature increase for the region.

With regard the impact of mid-day temperature on evening discharge in the glaciated basin, a positive correlation of 0.63 between the 1200 hours temperature and the 1700 hours discharge at Bhojwasa (see

Figure 7. Diurnal discharge pattern at Bhojwasa, Gangotri 2008

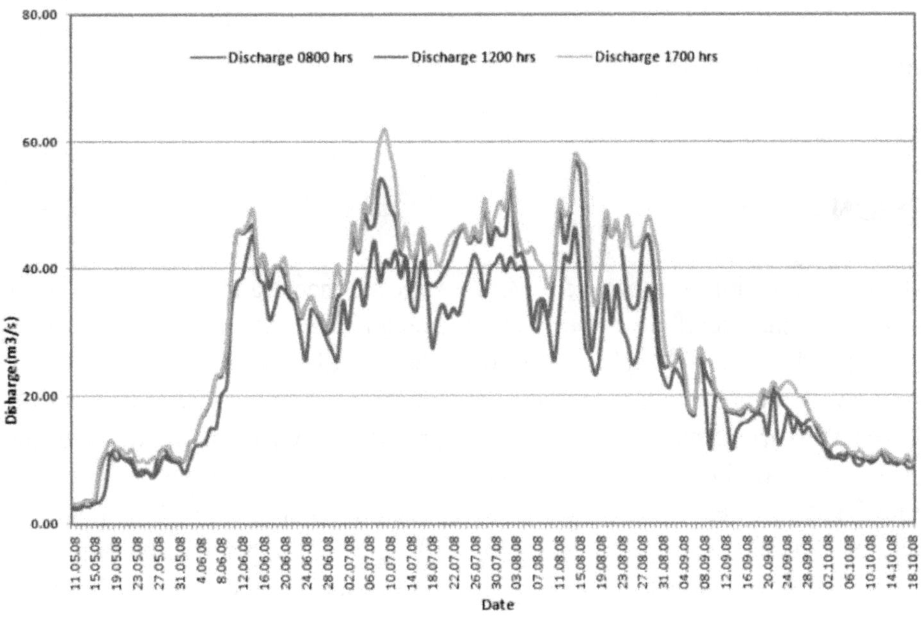

Figure 8. Average temperature and discharge during winter 2008-09 at Bhojwasa, Gangotri

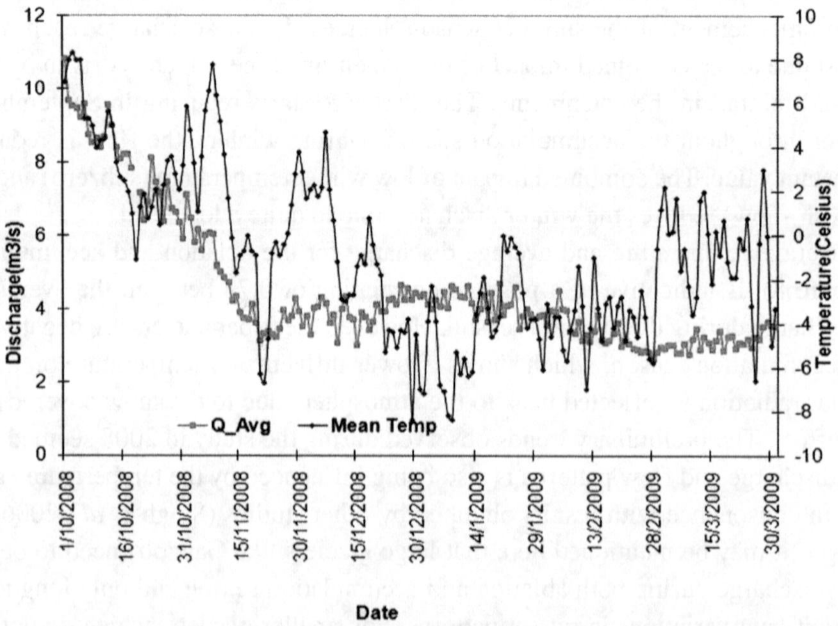

Figure 9) is noticed. However, during October the discharge rate declines as the average temperature starts reducing (see Figure 9).

Trends in the Gangotri region (see Figure 9) reflect that temperatures start reducing from September. The maximum temperature reaches up to 18° C in June and July and remains mostly around 14°C during May to August. The early morning temperature (0800 hours) remains at around 9° C till August and falls to zero in mid-September. As winter advances, the intraday temperatures in December also show a variation with morning temperature at sub-zero levels and moving down to - 15°C. During peak winter months, the morning and evening temperature is subzero while noon temperatures may be above 0° C. During March, the morning temperature remains below zero, while the noon and evening temperature varies around 5°C .

CONCLUSION

The entire Himalayan region is known to have the largest concentration of glaciers in the world excluding the polar region. According to the WWF, the region is called the "Water Towers of Asia" as it provides around nearly 86,000,000 cubic meters of water annually. These Himalayan glaciers feed seven of Asia's great rivers, namely the Ganga, Indus, Brahmaputra, Salween, Mekong, Yangtze and Huang Ho(WWF,2005).The current study has attempted to provide an understanding of the characteristics of large glacier like Gangotri, in the context of melting patterns and local hydrometeorological conditions. We also believe that the science of climate change is complex particularly while monitoring glaciated basins but also in studying large vs smaller glaciers. In long term, temperature variations may obviously be a key aspect to changes taking place in a glacier, but impacts of other climate/non climate stressors

Figure 9. Influence of mid day temperature (1200 hrs) on evening discharge (1700 hrs) at Bhojwasa, Gangotri

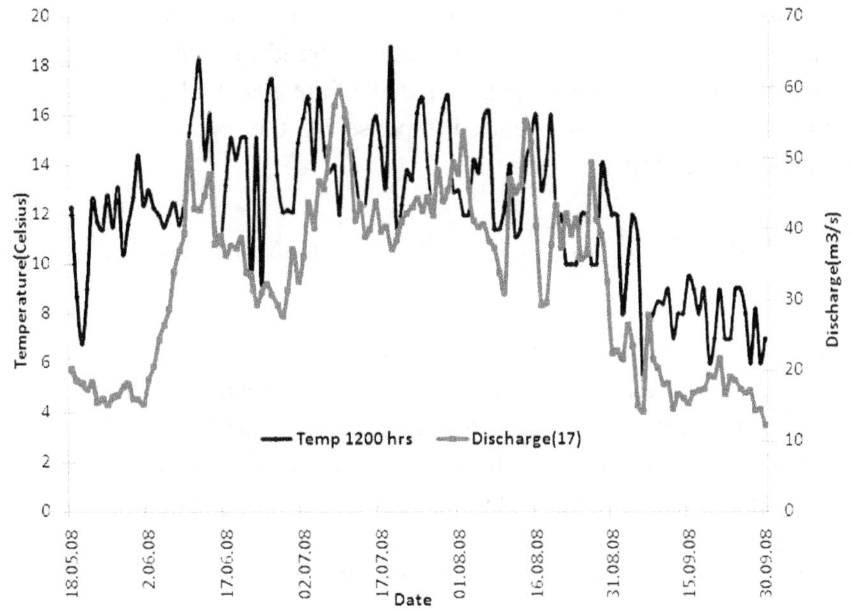

like local variability in monsoon, developmental concerns, industrial growth, tourism etc. may also play an equally critical role.

From a basin perspective, a combination of climate variability and non-climate stressors will have serious consequences for the freshwater ecosystems of the Ganga basin with long term impacts for ecosystem, livelihoods, agricultural and industry in the catchment of the basin. As the Gangetic plains are considered a major food basket to the economy of the northern Indian plains, changes in environmental flows along the basin ecosystem may also affect region's agricultural productivity and industrial activity, but also have consequent impacts on future economic growth in terms of Gross Domestic Product (GDP). In a carbon constrained world, where greenhouse gas emissions are rising, beyond the level considered unsustainable, the economic costs of taking mitigation actions on limiting climate impacts now, would have far lesser implications than if actions were to be taken at a later stage. (Stern, 2007).

Larger vs Smaller Glaciers

Field studies indicate that larger glaciers like Gangotri which occur a vast area are unlikely to show major declines in the near future, due to its relatively large mass balance. Although this study did not take into account changes occurring in a smaller valley type glaciers in the region, it may be worthwhile to mention that smaller glaciers with low mass balance and ice volume are likely to retreat at a faster rate. This could lead to increased losses in glaciated portion as well as their tributary glaciers,(see Figure 3) a trend which has been observed across the Himalayas for many other smaller glaciers (Dobhal *et al*, 2004). Regional climatic variations could threaten the fragile nature of these glaciers which are likely to disappear at a much faster rate or be considerably reduced in length as compared to the larger ice bodies. These glaciers are perhaps more vulnerable to local climate variations and hence long term and continuous assessment is required to monitor the hydro-meteorological parameters existing in their vicinity to develop predictive models for future water resource scenarios.

Larger glaciers like Gangotri show a continuous recessionary trend in recent years through this and other studies (Singh *et al.*, 2006). However, at present this trend may have a limited influence on the water flows downstream primarily due to the large extent of the glaciered area. The movement of larger glaciers is dynamic and snout retreat is a delayed response, but the fragmentation of the larger glaciers needs to be continuously monitored.

The Climate Change Angle

There is limited consensus and considerable degree of variations in studies on glaciers as reported amongst the scientific community and policy makers on the true extent of Himalayan glaciers retreating due to impacts of global climate change,(IPCC,2007,Raina,2009). The results of this study is a pointer to the fact that regional climatic variations along with changes in social and economic dimensions may also play a critical role in the behavior of these glaciers. Local communities living closer to Gangotri have indicated changes in snowfall levels in the winter months resulting in less soil moisture, which in turn is changing cropping patterns and availability of water.

Though glacial retreat is a natural phenomenon with the earth moving towards a warming phase, the accelerated rate of retreat could result from a combination of other impact factors also. Based on our study and past scientific data, it is clear that climate change has a visible impact on the status and condition of some of the glaciers as is evident through the increase in the net loss of ice mass year after year. Hence, it becomes difficult to predict the glacier retreat or advance scenarios with confidence. It is worthwhile to note that the phenomenon of global warming and climate change will impact not only temperature but also the other variables like intensity and quantity of precipitation, cloud cover, wind and radiation. The influence of other climate factors and stressors like precipitation, flooding and extreme weather events in the Himalayan region is well documented (Bhutiyani et al 2007, Raina, 2009, Mall, *et al*, 2006). However despite this the characteristics and nature of the Himalayan ecosystem and monsoonal effects remains less understood.

Since region specific climate models are limited, a long term monitoring of meteorological parameters is crucial before arriving at definitive conclusion regarding the impact of climate change.

Non-Climatic Stressors

Non-climatic stressors, such as rapid economic growth, including tourism and infrastructure development for industries like construction, large hydropower etc., are increasing pressure on the fragile Himalayan ecosystem. Unplanned development has already resulted in severe pressures on both local ecology and communities. The impacts of climate change and changes in climate variability would further create additional stress. This is exemplified by instances of extreme weather event in the region in the form of landslides, cloudburst, flooding etc. which has affected not only local communities, biodiversity but also industry. The decline in glacial area and variations in annual runoff patterns in the future gains importance in the context of hydro power planning and development of the Himalayan states. We propose that future studies should also focus on the impact of regional climate variables together with non- climatic stressors like local pollutants, changing urbanization trends, tourism etc. Implementing the Mission on Himalayan Ecosystem as part of the National Action Plan on Climate Change (NAPCC) of the Government of India would therefore be a key priority.

Way Forward

Continuing pursuits in scientific excellence has resulted in critical evidence of changes in the glaciers of the Himalayas by many workers, but probably not very accurately as there is inadequate record of organized, systematic and long term observed data monitoring. However, integrating the growing body of anecdotal evidence, observations of the local communities, policy directions along with the recent initiatives of the UNFCCC may provide the impetus to focusing on glaciological studies in the Himalayas.

The dynamics of the monsoon are influenced by Himalayan systems which act as a reservoir to sustaining agriculture, providing freshwater and groundwater recharge, and are home to a unique ecosystem with many endemic species (WWF, 2009). With the Himalayan region transcending geographical barriers glacier based studies may requires not just local action but also trans-boundary cooperative arrangements.

Future efforts in building the resilience of the local community and the ecosystems should take into account a concerted and integrated approach through local, regional and national policy. There is an urgent need by communities, scientists, industry groups and policymakers to take a closer look at the linkages between local impacts, scientific research, policy interventions and the larger understanding of using resource conservation technologies and practices for promoting societal benefits.

ACKNOWLEDGMENT

This study was conducted as part of a joint initiative between World Wide Fund for Nature (WWF) – India and the Birla Institute of Technology, Extension Centre (BITEC) and supported financially by WWF Netherlands as part of the Regional Programme on Himalayan glaciers. We are grateful to Mr. Ravi Singh, Secretary General & CEO, WWF- India and Dr. Sejal Worah, Programme Director, WWF- India for their kind encouragement and support for this study. Director, BITEC, Jaipur is acknowledged for support and facilities provided to carry out the field research. We gratefully acknowledge the contribution of several field assistants who provided the much needed logistical support at high altitude at Gangotri in difficult conditions for data collection and analysis.

We thank Dr Pratap Singh and Ms Esther Blom, reviewers for providing objective inputs and suggestions to this study. We are thankful to the support and cooperation provided by the staff of Gangotri National park and the Forest Department, Government of Uttarakhand during the course of this study.

REFERENCES

Barnett, T. P., Adam, J. C., & Lettenmaier, D. P. (2005). Potential impacts of a warming climate on water availability in snow-dominated regions. *Nature*, *438*(7066), 303–309. doi:10.1038/nature04141 PMID:16292301

Bhutiyani, M. R., Vishwas, S. K., & Pawar, N. J. (2007). Long-term trends in maximum, minimum and mean annual air temperatures across the Northwestern Himalaya during the twentieth century. *Climatic Change*, *85*(1-2), 159–177. doi:10.1007/s10584-006-9196-1

Cruz, R.V., Harasawa, H., Lal, M., Wu, S., Anokhin, Y., Punsalmaa, B., … NinhHuu, N.(2007). *Asia Climate Change 2007*: *Impacts, Adaptation and Vulnerability*. Contribution of Working Group II to the Fourth Assessment Report of the Intergovernmental Panel on Climate Change. Cambridge University Press.

Dobhal, D. P., Gergan, J. T., & Thayyen, R. J. (2004). Recession and morphogeometrical changes of Dokriani glacier (1962–1995) Garhwal Himalaya, India. *Current Science*, *86*(5), 692–696.

Dobhal, D. P., Gergan, J. T., & Thayyen, R. J. (2007). Mass balance and snout recession measurements (1991-2000) of Dokriani glacier, Garhwalhimalaya, India. Climatic and anthropogenic impacts on the variability of water resources. Technical Document in Hydrology No. 80. UNESCO.

Haeberli, W. (1990). Glacier and permafrost signals of 20th century warming. *Annals of Glaciology*, *14*, 99–101.

Hoelzle, M., Darms, G., Lüthi, M. P., & Suter, S. (2011). Evidence of accelerated englacialwarming in the Monte Rosa area, Switzerland/Italy. *Cryosphere*, *5*(1), 231–243. doi:10.5194/tc-5-231-2011

IPCC. (2007). *Climate Change 2007: The Physical Science Basis. Contribution of Working Group I to the Fourth Assessment Report of the Intergovernmental Panel on Climate Change(IPCC)*. New York: Cambridge University Press.

IPCC. (2013). *Climate Change 2013: The Physical Science Basis. Contribution of Working Group I to the Fifth Assessment Report of the IntergovernmentalPanel on Climate Change*. Cambridge, UK: Cambridge University Press.

Jianchu, X., Shrestha, A., Vaidya, R., Eriksson, M., & Hewitt, K. (2007). *The Melting Himalayas. Regional Challengesand Local Impacts of Climate Change on Mountain Ecosystems and Livelihoods. ICIMOD Technical Paper*. Kathmandu: ICIMOD.

Kaser, G., Cogley, J. G., Dyurgerov, M. B., Meier, M. F., & Ohmura, A. (2006). Mass balance of glaciers and ice caps: Consensus estimates for 19612004. *Geophysical Research Letters*, *33*(19), 5. doi:10.1029/2006GL027511

Kulkarni, A. V., Bahuguna, I. M., Rathore, B. P., Singh, S. K., Randhawa, S. S., Sood, R. K., & Dhar, S. (2007). Glacial retreat in Himalaya using Indian Remote Sensing satellite data. *Current Science*, *92*(1), 69–74.

Kumar, R., Hasnain, S. I., Wagnon, P., Arnaud, Y., Chevallier, P., Linda, A., & Sharma, P. (2007). Climate change signals detected through mass balance measurements on benchmark glacier, Himachal Pradesh, India. Climatic and anthropogenic impacts on the variability of water resources. Technical Document in Hydrology, 80. UNESCO.

Lüthi, M., & Funk, M. (2001). Modelling heat flow in a cold, high-altitude glacier:Interpretation of measurements from ColleGnifetti. *Swiss Alps. J. Glaciol.*, *47*(157), 314–324. doi:10.3189/172756501781832223

Mall, R. K., Gupta, A., Singh, R., Singh, R. S., & Rathore, L. S. (2006). Water resources and climate change: An Indian perspective. *Current Science*, *90*(12), 1610–1626.

Naithani, A. K., Nainwal, H. C., & Prasad, C. P. (2001). Geomorphological evidences of retreat of Gangotri glacier and its characteristics. *Current Science*, *80*, 87–94.

Raina, V. K. (2009). *Himalayan Glaciers – A State-of-Art Review of Glacial Studies, Glacial Retreat and Climate Change. Ministry of Environment and Forests, discussion paper*. Government of India.

Roy, S.S., & Balling, R.C., Jr. (2005). Analysis of trends in maximum and minimum temperature, diurnal temperature range, and cloud cover over India. *Geophysical Research. Letters, 32*, 4.

Singh, P., Haritashya, U. K., Kumar, N., & Singh, Y. (2006). Hydrological characteristics of the Gangotri Glacier, central Himalayas, India. *Journal of Hydrology (Amsterdam)*, *327*(1-2), 55–67. doi:10.1016/j.jhydrol.2005.11.060

Singh, P., Haritashya, U. K., Ramasastri, K. S., & Kumar, N. (2005). Prevailing weather conditions during summer seasons around Gangotri Glacier. *Current Science*, *88*(5), 753–760.

Singh, P., Polglase, L., & Wilson, D. (2009). Role of Snow and Glacier melt runoff modeling in Hydropower projects in the Himalayan Region.*International Conference on Water, Environment, Energy and Society*.

Srivastava, D. (2001). *Glaciology of Indian Himalaya: A Bilingual Contribution in 150 Year of geological survey of India. Special Publication No. 63*. GSI.

Stern, N. (2007). *Stern Review on the Economics of Climate Change*. Cambridge, UK: Cambridge University Press. doi:10.1017/CBO9780511817434

UNEP. (2007). *Global outlook for ice & snow*. UNEP/GRID-Arendal.

UNESCO. (2007). *Case Studies on Climate Change and World Heritage*. UNESCO World Heritage Centre.

Wagnon, P., Linda, A., Arnaud, Y., Kumar, R., Sharma, P., Vincent, C., & Chevallier, P. et al. (2007). Four years of mass balance on ChhotaShigri Glacier, Himachal Pradesh, India, a new benchmark glacier in the western Himalaya. *Journal of Glaciology*, *53*(183), 603–611. doi:10.3189/002214307784409306

Wood, F. B. (1990). Monitoring Global Climate Change: The Case of Greenhouse Warming. *Bulletin of the American Meteorological Society*, *71*(1), 42–52. doi:10.1175/1520-0477(1990)071<0042:MGCCTC>2.0.CO;2

WWF. (2005). *An Overview of Glaciers, glacier retreat and its subsequent impacts in the Nepal*. WWF.

WWF. (2009). *The Eastern Himalayas – Where World Collide*. WWF.

Yadav, R.R., Park, W.K., Singh, J., & Dubey, B. (2004). Do the western Himalayas defy global warming? *Geophysical Research Letters*, *31*(17).

Chapter 4
Issues, Concerns, and Local Stakes:
Future of Water Resources in Coastal Villages of Devbag and Tarkarli, Coastal Maharashtra, India

Navendu Chaudhary
Symbiosis International University, India

Yogesh Pisolkar
Symbiosis International University, India

ABSTRACT

Coastal Maharashtra is in transition. Growing coastal tourism and allied developmental activities along southern Maharashtra coast needs integration of various stakeholders to address the various issues and concerns. Integrated Coastal Zone Management (ICZM), which can cater to the needs of people while preserving the environment is thus need of the hour. The effects on natural resources, including water, will change the socioeconomic as well as the cultural fabric of coastal communities. This chapter explores a holistic approach to the developmental issues and the impact of climate change on the coastal region with specific cases of villages of Devbag and Tarkarli, coastal Maharashtra, India. It explores both physical and socioeconomic landscapes with special attention given to water resources in the context of changing dynamics of coastal communities and coastal tourism. The chapter discusses the issues and concerns of villages of Devbag and Tarkarli and proposes solutions for a sustainable development.

INTRODUCTION

The coast is the place where sea and land meet. It is a unique feature with properties that connect both lithosphere and hydrosphere. This interface is represented on geographic maps as a coastline. This interface is a dynamic system with varying size, boundaries and shape that are constantly under state of change due to natural as well as anthropogenic forces. At beaches, marshes, mangroves, wetlands, coral

DOI: 10.4018/978-1-5225-1046-8.ch004

reefs, these changes are more pronounced, at rocky shores are more resistant to change. These forces influence the coastal area so profoundly that in dealing with a number of practical and scientific problems one needs to consider the adjoining features such as a whole catchment basin of the river or creek along with the coastline. The importance of preserving the coastal zone is critical in maintaining the balance of coastal ecosystem as well as preserving the livelihood of communities residing near coastal zone. Most of these communities traditionally rely on fishing for their livelihood. Beaches have recently attracted the attention of the tourism industry and these days compete with traditional fishing in terms of provision of livelihood among coastal communities. This budding tourism is proving to both boon and bane for these communities. On one hand it provides necessary livelihood options on the other it is putting tremendous constraints on resources mainly drinking water.

Integrated coastal zone management is thus need of the hour, which can cater to the needs of people while preserving the environment. The coast is not only under threat by anthropogenic activities mainly through tourism, but natural forces are also at play, which poses a threat to its delicate balance. In light of the threat of climate change coasts are especially vulnerable to sea level change. Coastal areas all over the world are being subjected to high human pressures, which hold implications for local people from the hinterlands to coastal areas; this migration is continuing into the 21st Century with coastal zones developing into major attractions for living and working. Almost 40% of the world's population live within 100 Km of a coastline, a land type that accounts for only 22% of the earth's landmass (WRI 2000). This trend is the product not only of population growth, but also urbanization. World Resources 2000-2001 sounds the alarm bell with regard to the state of the world's ecosystems as do various reports of the Group of Experts on the Scientific Aspects of Marine Pollution over the last 20 years (GESAMP 1990).

The Intergovernmental Oceanographic Commission (IOC) of UNESCO organized an event centered on the outcomes of Ocean and Climate forum and Oceans ay at COP21. The event was a collective deliberations of UNFCCC negotiators and decision makers. The event deliberations include major ocean and climate issues along with special emphasis on most vulnerable communities and ecosystems. The participants laid out clear recommendations including; adaptation of stringent reductions in CO_2 emissions to avoid disastrous consequences on coastal and island communities, ecosystems. It recommended that temperature increase to be restricted to less than 2°C, It is proposed to incorporate oceans into the text of the Paris agreement. It suggests the implementation of ecosystem-based adaptation strategies through integrated coastal and ocean management institutions at national, regional, and local levels. The stakeholders need to be encouraged to develop and support measures to address the issues associated with the displacement of coastal and island populations as a result of climate change. Global finance mechanisms need to be formulated for catering adaptation and mitigation efforts in coastal communities. Encourage capacity building in coastal communities. It also advocated expansion of outreach programs and education or awareness efforts (IOC, 2015). In this context, significant researches are happening at various levels in terms of education for sustainable development in business schools (Rao et al., 2013; Rao et al., 2015), climate resilience in natural ecosystems (Rao & Patil, 2015) and smart methods/models for waste management in the era of climate change (Patil & Rao, 2015a; Patil & Rao, 2015b).

The coast in developing countries has today become a site for intense competition, both between and within nations. States vie with each other to offer the best in terms of location, infrastructure, markets and even aesthetics. Change is evident on a number of Asian coasts, from the west coast of India to that of Indonesia and Thailand to the east coast of China. Rapid changes are occurring on the African coast, especially from north Mombasa to Malindi, and on the coast of Latin America from Rio de Janeiro to Sao Paulo in Brazil and Chile's coast between Valparaiso and Concepcion, which are becoming large urban areas (Hinrichsen, 1996). Chief conflict in the use of coastal resources is between anthropogenic activities and the deterioration of natural resources. Recent economic developments which are highly fractured and uneven has led to progress that has expanded the scale of anthropogenic influence leading to numerous undesirable consequences for coastal ecosystems. These developments have led to increased coastal population density; intensification of the use of natural resources; increase solid waste and waste water, unregulated constructions, and disturbance of natural ecosystem through increase human and technological interventions.

In India in particular coastal and marine areas have been regions of conflict for decades, with the introduction of commercial trawlers and industrial-scale aquaculture being strongly opposed by local traditional fisher communities since the late 1970s. But the rapid rise in large-scale commercial exploitation is only one source of conflict. The coasts are increasingly being viewed as suitable sites for industries, tourism complexes and trade hubs (as ports). Once again, this period of globalization has dealt a heavy blow to fledgling attempts at conserving the ecological and livelihood integrity of this part of India, which sustains a third of its population (Shrivastava & Kothari, 2012).Considering the importance of sensitive ecosystems in coastal areas and increased frequencies of natural as well as human induced coastal hazards, to safeguard and regulate activities on the coast, costal Regulation Zone (CRZ) notification was promulgated in 1991. The original CRZ notification became a nuisance for industrial and commercial interest.

The proposal for a new Coastal Management Zone (CMZ) notification allows states to determine what should and should not be allowed in various zones along the coast and legitimizes all structures that have come up in violation of the CRZ notification. Civil Society Organizations (CSOs) and fisher communities (through networks like the National Filmmakers' Forum) have severely criticized the proposal for being a sell-out to commercial and industrial interest. Responding to this, the MoEF has allowed the CMZ draft notification to lapse and as of the time of writing, promised widespread consultations before coming up with a new notification (Shrivastava & Kothari 2012).The increase in tourist activity has only muddled he water.

Coastal Management now promotes the view that human actions should not destroy non-renewable resources, but should be sustainable. Impacts, such as resource depletion, environmental degradation and pollution, can often be seen to be the result of poor planning rather than being an independent problem (Kay & Alder, 1999). Increasing emphasis has been placed on integrated Coastal Zone Management (ICZM), the comprehensive assessment, setting of objectives, planning and management of coastal systems and resources, taking into account traditional, cultural and historical perspective and conflicting interests and uses (IPCC, 1994) Management decisions often need to be proactive, anticipating rather than responding to an issue (Salomans et. al., 1999) and greater emphasis is placed on capacity building, community involvement and coastal environment that future generations will inherit (Cicin-Sain, 1993; Hildebrand, 1997).In case of rapid geo-environment changes due to natural geomorphic processes or due to human induced acceleration of these processes can now be considered in terms of geomorphic hazards which increases vulnerability of coasts (the degree to which it can adopt to or unable to cope with changes). The adaptive capacity of the coastal (natural and socio-economically) has to be analyzed thoroughly so as to prepare strategies for sustainable development of the coast.

Over the years, especially in the last two to three decades regions and places in the West coast of Maharashtra (popularly known as Konkan) have changed very drastically. With about 10 on going and more proposed thermal projects from north to south, 2 proposed atomic power generation stations proposed at Madban, Jaitapur, mineral exploration from about 18 villages in Sindhudurg district and haphazard growth of tourist places due to Konkan railway and coastal highway Konkan definitely need systematic coastal zone management. A site classification for the development and management of coastal areas of Maharashtra is suggested by Karlekar (1996), which is based on his field study and the scheme suggested by Clark (1977). A convenient division of the coastal strip in the north, middle and south Konkan is suggested by Dikshit (1988). The policies and programs related to CEM in Konkan must be essentially sited specific. Each site on the Konkan coast is distinct as regards its ecological significance, its resistance to change and the Geoenvironment (Karlekar 1996).

PHYSICAL ENVIRONMENT

Over the years, sites and places in the Konkan are changing very drastically. A broad classification of sites can be done in terms of CONSERVATION SITES (from north and south of Mumbai), DEVELOPMENT SITES (from Raigad and north Ratnagiri districts) and PRESERVATION SITES (south Ratnagiri and Sindhudurg districts), (Figure 1) An environmental management of coastal sites in Konkan can be effective and meaningful if the CEM sites are identified and classified from these major areas characterized by specific environmental conditions (Karlekar, 1996). Tourism along Maharashtra Coast 720 km. The coastline of Maharashtra offers numerous spectacular sites for tourism. Maharashtra Tourism

Figure 1. Tarkarli – Devbag, Konkan

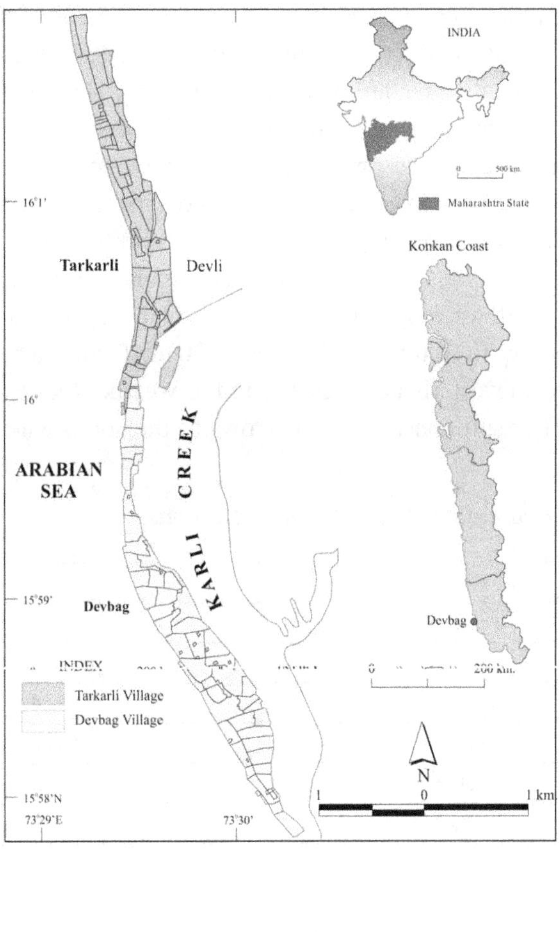

Development Corporation (MTDC)'s accommodation facilities are available at various destinations from North to South Konkan. (Table 1).

Apart from these government facilities, semi-government (MTDC rented) and private sectors have their resorts, motels from Dahanu to Terekhol which offer a wide range of tourist spots especially 'Forts, Temples and Beach Tourism' are major attractions.

With rapid development of Konkan in last two decades due to infrastructure development, industrialization and growing tourism, series of issues have emerged, which need proper planning. Alteration of hill slopes for mining and construction of hotel, motels, resorts has increased sedimentation in creeks and estuaries. Flooding of rivers, creek erosion has become an annual phenomenon in certain parts of Ratnagiri and Sindhudurg districts. Solid waste management on beaches is a grave problem on Palghar, Kihim beach. Un-surveyed beaches for 'beach tourism' and lack of knowledge and awareness has resulted in casualties along the beaches. Ganpatipule is one such reported site for many years. Construction of Jetties and Harbors has altered the sedimentation pattern and thus the objectives of the project were defeated completely. Jetty at Miryabay in Ratnagiri is one such example. There is a need to address these issues keeping in mind sustainability of tourism in Konkan.

Tarkarli-Devbag Spit (15°57' N to 16°1'N) latitude and 73°29' to 73°31' E longitude) is a 6 km long sand spit which is about 16 km South of Malvan on Maharashtra coast (Figure 1). The spit bar is connected to mainland near Tarkarli. Southern end of bar abruptly ends at sea near Mobarwadi. The eastern edge of the spit is bordered by tidal stretch of river Karli. The western margin faces Arabian Sea. The tidal range along the bar is only 2.2 m at spring and 1.8 m at neap. The lowest tide level is only 0.6 m. The dominant wave height measured in the field is 1.6 m. Maximum surge level is 4.6 m. (Karlekar, 2002). Strom surges produced due to low atmospheric pressure and coincident with onshore south-west monsoon are frequent in the area. The pressure anomalies and storm surges cause a rise and fall of sea level by about 30 cm every year. Episodic inundation due to storm surges has been experienced since 1990 in every monsoon. The climate of the area shows a regular variation on account of south-west monsoon and fair weather. February to May is treated as pre-monsoon, June to September as monsoon and October to January as post monsoon period. Rough or very rough sea occurs during monsoon. Moderate to heavy swell waves also persists along the coast in this period. Winds on the open sea are modified when they approach the coast, mainly due to the effects of coastal configuration (Dhavalikar, 2002).

Long shore currents are north westerly from November to January with a speed of 8 to 20 cm/s. They are south easterly from February to October with speed of 30 to 40 cm/s. The prevailing winds are south westerly in monsoon and north westerly and westerly in fair weather (Karlekar, 2002).The width of the beach spit is not uniform. The spit considerably narrows to the south where it is 85 m. In the central

Table 1. M.T.D.C's resorts and hotels along Maharashtra coast

Place	District	Location in Konkan
Bordi	Thane	North
Harihareshwar	Raigad	Middle
Velneshwar	Ratnagiri	Middle
Ganpatipule	Ratnagiri	Middle
Tarkarli	Sindhudurg	South
Aronda	Sindhudurg	South

part near Vithoba Temple, it is 70 m. wider. Significant wave height in fair weather is 1 to 1.5 m. and in monsoon it is 4 to 5 m. The wave period is less than 10 seconds in fair weather which is around 6 seconds in monsoon. The tidal current velocity is less than 10 cm per second. (Naval Hydrographic report 1981). The tidal water penetrates to a distance of 30 km in Karli river. Survey of India toposheet of the year 1965-66 of 1:50000 scale shows 3 m relative height on the creek side near the southern end of the bar 4 m height is recorded just to its North. Near Tarkarli the beach height was recorded to be 6 m. The southern end of the spit was very narrow and extended well beyond Mobarwadi for a distance of 500 m towards the confluence of the creek to the sea. The toposheet also shows only six sand islands inside the creek, in the lower tidal sector of the river. The mouth of the creek is wide open without any sand accumulation blocking the entrance of the river. The western as well as the eastern bank of the river was firm and without any signs of erosion, cutting or breaching. High water line clearly demarcates the creekward boundary of the littoral terraces near Navalewadi, Dutonwadi and Karliwadi. These terraces are backed by 60 m high lateritic plateaus and hills covered by stony waste. An earlier topomap of the area in the year 1894 shows Devbag to be a small, inconspicuous settlement confined to a very small stretch of the sand spit. The southern end of the beach was a relatively wider sand formation that was curved seawards. Both the banks of the creek show a thin line of sand deposition. The tidal sector is strikingly clear without any signs of sand accumulation forms.

The IRS IC LISS III image of the area acquired in 1997 shows the smooth seaward outline of the spit and its irregular breached creekward side. The distal, southern end appears truncated and eaten up by the sea. The said islands inside are not clearly identified, although they exist in the creek. More recent satellite images clearly show sand deposition in the form of a sand plume over southern end. The entrance of the creek has shallowed by the extent of this deposition Number of configuration changes such as breaching of creekward boundary, erosion of seaward limit, frequent sand islands in the tidal sector and the overall transformation of the spit side of the creek are easily discernable. Recent Google images of the area show new morphological features and changes along the creek side of the spit bar and within the tidal sector of Karli creek.

The images of Upper and Middle part of the tidal sector show the appearance of wide, elongated sand bars and indicate shallowing of the channel. The image of the confluence area is significant in many ways as it directly shows the shape, size and orientation of newly created accumulation forms in the region of the creek entrance . All the images give a reasonably good idea about the cutting of the creek bank on both the sides. The creek water diversion due to the development of the islands, removal of bank material and its deposition in lower sector and widening of the mouth of the creek due to the separation of the southernmost tip of spit is some of the silent features that are identified from these scenes (Pisolkar, 2008).

The encroachment and erosion by sea have accelerated after the December 2004 tsunami. The problem of beach erosion at Devbag and Tarkarli is however not a recent phenomenon. It appears that the area is showing a tendency of severe breaching and erosion of shorefront continuously since 1952 (Karlekar, 2002). The earliest record of 1952 is followed by beach erosion in 1956,1980,1997,1999 and 2000 (CWPRS, 1993), (Table 2).

Recent field visits show that the encroachment from Creekside and erosion by sea have further accelerated affecting settlement from the creek as well as the seaside (Table 3). To add this growing tourism is fueling already existing geomorphic hazard. Devbag traditionally was a small fishing village upto last decade dependent on fishing for livelihood. In the last decade people sensed the opportunities for tourism sector. In the last 5 years, many people are quickly changing their livelihood from traditional

Table 2 Beach and creek erosion events at Devbag Beach

Year	Effects
1952	Beach erosion, sand spit erosion
1956	Beach and dune erosion, fall of trees on shore.
1980	Severe beach cutting and breaching, construction of anti-erosion wall.
1990	Major episode of Beach Erosion. About 60 to 70 m. A wide section of land beyond the high water level was washed away. Temporary measures for protection of Devbag settlement on the spit bar was taken by the Maharashtra Government.
1995	Breaching of anti-erosion sea wall, washing away of platforms built for drying fish. Fall of coconut gardens and casurina plantations.
1998	Beach erosion, sand spit erosion, creek side also breached mainly due to progressing tidal water and flood discharge through the creek uprooting of coconut trees.
2004	Beach erosion, sand spit erosion, especially severe in Mobarwadi section. Uprooting Casurina and coconut plantations. Since then it has almost become an annual phenomenon. Distal end of spit bar is breached severely every year and lot of morphological changes has occurred since then.

Source: Pisolkar et. al 2013

Table 3. Recent (2006-2012) Beach and creek erosion events on Devbag Beach

Year	Effects
2006	Beach Erosion, Sand spit Erosion especially severe in Mobarwadi section. Uprooting Casurina and coconut plantation and Ingression of sand affecting potable water wells, in Sadanand Tandel, Pednekar Ramesh's plot (60/7, 60/7).
2007	Construction of sea wall to protect the southern end of the Devbag spit in front of Sadanand Tandel's home, Beach and creek erosion in various sections of beach.
2008	Beach and Creek erosion. Loss of coconut, Casurina trees from varied survey no. of Devbag beach. Houses near the confluences of Karli River and the sea is becoming more and more vulnerable.
2009	The seawall is a sub ducted in various sections of beach. Beach and Creek erosion, Fiyan the cyclone hit the Devbag coast affecting various sections in village Devbag.
2010	Beach and Creek erosion in monsoon. Growing tourism, speed boats creates waves and thus breaching from creek side is no problem even in fair weather.
2011	Subduction of sea wall and the collapse of creek band affecting people and their properties. Morphodynamics at the confluence of Karli river and the sea is changing in every season, causing lot of problems to extreme southern end of the spit.

Source: Pisolkar et. al 2013

fishing to tourism resulting over 50 hotels have been constructed in the last decade in village Devbag. Rapid unplanned developments are causing accelerated alteration in beach sand dune ecosystems, land use pattern and drastic socioeconomic changes.

ENVIRONMENTAL IMPACT ON TARKARLI - DEVBAG SPIT

A study of Tarkarli - Devbag spit indicates a rapid and sudden erosion of beaches and dunes during monsoon surges. The part of the spit bar most affected is the southern end. 500 m. of beach is removed due to breaching after December 2004. The effect is clearly seen to the south and north of Mobareshwar Temple since 2004 from the river as well as from the seaside. It is clear from satellite images and confirmation from locals that already Survey No.64 & 65 are submerged under water 3000 casuarinas plantations have been stripped off due to erosion in the last 5 to 6 years. Survey Nos. 61, 56, 51 are severely affected by erosion from the seaward side (Table 4).

Survey Nos. New 55, 53, 45 just north of Mobareshwar Temple are severely affected by river inundation in monsoon (Table 5). Recently near Vithoba Temple Survey Nos. 10, 7, 3 are breaching up very quickly

Table 4. Erosion: From seaword side → South to north

Survey No. / Plot No.	House No.	Last two years' major events	History of major events in last decade (2001-2011)/Fiyan
Old 1015A→ New 67/7	904	Construction of sea-wall in Oct.-Nov.2007 has protected the plot for last 3-4 years.	2005→30 coconut trees uprooted. 2004-2000 Casurina Plantations uprooted . 5 Potable water wells diminished due to ingression of sand.
Old 1015A→ New 67/7		Construction of sea-wall in Oct.-Nov.2007 has protected the plot for last 3-4 years.	
Old 1015→ New 67/7	700	Construction of sea-wall in Oct.-Nov.2007 has protected the plot for last 3-4 years.	2006-07→7-8 coconut trees uprooted. Lost potable water well. 10-12 'Guntha' land lost.
Old 1015A→61	921	Collapsing of sea wall for about 50 m. Due to this Toilet, Bathroom, and Safety Tank collapsed. Cost approximately Rs.60,000/-. Lost two coconut plants.	
	693	Lost 10 coconut trees in 2010. Sinking of sea wall.	
Old 755→ New 55/21	692	Lost 4 Gunthas in 2008, 2009, 2010. Collapsing of house wall.	Lost 1 coconut plant at the time of Fiyan. No compensation from Government.
55	691	1-2 Gunthas lost in 2008, 2009, 2010.	Lost Langde, Bangde of approx. 22,000/-. Loss of 1 cashew, 1 coconut tree due to uprooting in Fiyan. No compensation.
55	689	Lost 6-7 Gunthas in 2008, 2009, 2010 also experienced some loss.	2 coconut trees uprooted.
	681		Lost 5-6 Gunthas in last 15 years. Fishing net lost. 5 coconut trees uprooted.
New 56	639	Lost 2 Gunthas in 2010. Due to intrusion of saline water coconut trees are shrinking.	Lost Nets at the time of Fiyan.
New 61	769A 769B	Lost 5 Casurina, 2 Coconut trees in 2011.	30 to 35 Casurina trees lost in 10 years.
	907	Water is intruding upto home.	22 Coconut trees lost in 1961 (Cyclone).
	768	Lost 1-2 Gunthas in 2009, 2010.	Laterite Bund collapsed.
	629	Subduction of sea-wall, uprooting of Casurina trees.	2 to 3 Gunthas lost in last 5 to 6 years.
51	616 A	Collapsing of Sea-wall in 2010-11.Lost 2 coconut trees in 2011.	Lost fishing net worth Rs.1000/- at the time of ' Fiyan '.
51	616 B	Collapsing of Sea-wall in 2010-11.Lost 5 coconut trees in 2011.	Intrusion of water in force.
51 Plot No.16	617	Collapsing of Sea-wall in 2011.	Lost 20 Nets at the time of Fiyan Rs.20,000/- loss.
51	619	Collapsing of Sea-wall.	Uprooting of 1 coconut tree Casurina branch collapsed on house.
7	72	Lost 2 Coconut trees erosion from sea.	Lost 7 Banana tress, Karvanaji 1 tree.

by ingression of river Karli. Significant information by locals is that the salinity of potable water in some households near and north of Vithoba Temple is increasing in the months of March, April and May.

Even though the southern part of the spit is affected severely, the northern part which was relatively less affected is also getting eroded from seaside. Protection measures constructed in front of M.T.D.C. indicates intensity of erosion has accelerated since 2009, even though it is not as grave as that in the southern part. Rapid and sudden erosion of beaches and dunes during monsoon surges is caused due to slow sea level rise. The coastal erosion due to sea level rise is so striking that there is hardly any spare space between houses situated on the beaches and water. Locals in village Devbag have opined that, anti erosion wall, up to considerable level has protected significant part of village Devbag especially from north of Vithoba Temple upto Survey No.48, even though it is not engineering structure. The loss of wall about 50-70 m. In Survey No.61, 56 is a major concern because the intensity of erosion has accelerated in last monsoon (2011). There is no protection whatsoever in the area and already locals in this survey numbers has been severely suffered last monsoon. A section of the sea wall in front of Survey No.51 has already collapsed. The area is vulnerable in next monsoon. Southern section of Survey No.61 is again protected by sea wall upto end of Survey No.61. North of Survey No.7 no sea-wall exists up to Ranchi Nala (Pisolkar et. al 2013).

Until recent many estuaries and creeks in Konkan tended towards the equilibrium condition in which sediment accumulation rate was balanced by the local sea level rise (Karlekar, 2004). Field observation, however, show that some estuaries, especially the estuaries in Sindhudurg District up to the latitudes

Table 5. Erosion: From creek side → South to north

Survey No. / Plot No.	House No.	Last Two Years' Major Events	History of Major Events in Last Decade (2001-2011) / Fiyan
Old 755→ New 55	699	Lost 1 Mango, 1 Chiku, 1 Peru tree Lost 2 Gunthas from creek.	5-6 cashew trees, 11 coconut trees uprooted 2006-2007. Lost one well 10-15 years back. Collapsing of coconut trees on house at the time of Fiyan.
Old 755→ New 55 Plot No.22	696	Anti-erosion bund collapsed in last 2 years. Constructed bund from own funding.	10-12 coconut trees uprooted in last 10 years. 1/2 Guntha lost - 2004.
Old 755→ New 55	672	60 to 70 coconut trees uprooted in last year. House collapsed due to river erosion.	
Old 75→ New 55 Plot No.5	669	5-6 coconut trees uprooted. 1 Jambhul, 1 Alphonso, 1 Mango tree lost.	3-4 Gunthas lost. 2 to 3 coconut trees uprooted. 1 collapsed on house Rs.2,400/- remuneration.
Old 755→ New 55	667	3 Cashew trees uprooted from creek side.	5 to 6 Gunthas lost. 5 to 6 trees uprooted at the time of Fiyan. Water ingressed upto well.
Old 755→ New 55	668	Toilet collapsed this year(2011). 4 Gunthas lost	
Old 753→ New 53	651	Ingression of water from creek side. 3 coconut trees uprooted.	No bund exist from Paradkar upto Khavnekar.
Old 753→ New 53 Plot No.34	636	20 to 25 coconut trees uprooted in last 10 years.	13 out of 26 Gunthas eaten up by creek.

continued on following page

Table 5. Continued

Survey No. / Plot No.	House No.	Last Two Years' Major Events	History of Major Events in Last Decade (2001-2011) / Fiyan
Survey No. / Plot No.	House No.	Last two years' major events	History of major events in last decade (2001-2011) / Fiyan
Old 753→ New 53 53/10	604	Due to ingression of creek water trees are affected	25 coconut trees uprooted in last 5-6 years.
Old 751→ New 47 Plot No.15	893	Subduction of creek band due to boating in creek. Distance between home and river 1 m. Lot of reduction in fish catch.	In 1966-67 lost house from creek side.
Old 751→ New 47 Plot No.15		Subduction of anti-erosion creek bund in last two years i.e. 2010-11. In monsoon water ingresses upto house. Vulnerable situation.	Lot of erosion at the time of Fiyan.
Old 751→ New 47 Plot No.18	821	Subduction of anti-erosion creek bund in last two years i.e. 2011-11. In monsoon water ingresses upto house. Vulnerable situation.	Roof was blown by wind in Fiyan. No remuneration.
New 10	101	Ingression of water from creek. Last 2 years ingression has increased tremendously. 2 coconut trees uprooted. In May water become saline.	Fish Net lost. Rs.20,000/- loss.
New 10	98	Ingression of water in Monsoon (Rainy season) upto house. March, April, May water becomes saline.	Uprooting of coconut trees. Collapse of bund. Constructed by own fund Rs.2,500/-.
New 7	69A	Ingression of water from creek upto house. New flooring done. Lot of luggage drained. Water becomes saline in April and May.	Lot of ingression of water at the time of Fiyan.
Old 722→ New 7	88	Ingression of water from creek upto house. New flooring done. Lot of luggage drained. Water becomes saline in April and May. 2 coconut trees uprooted. In the month of May water is not potable as well as for washing.	Two 'Udalis' trees uprooted Rs.2,000/- loss.
Survey No. / Plot No.	House No.	Last two years' major events	History of major events in last decade (2001-2011) / Fiyan
New 7	86	Ingression of water from creek upto house. New flooring done. Lot of luggage drained. Water becomes saline in April and May. 2 coconut trees uprooted. In the month of May water is not potable. As well as cannot be used for washing.	Collapse of house wall after Fiyan.
New 3	24	3 to 4 coconut trees uprooted.	About 9 out of 18 Gunthas lost in creek.

Source: Pisolkar et. al 2013

of 16°N parallel are getting silted at an alarming rate after December 2004. In addition to Karli -creek, Terekhol - creek on the border of Maharashtra and Goa can be sited as the best example of this tendency. The Karli -creek backing the beach is a bar built micro tidal, salt wedge estuary. The tidal discharge through the creek is dependent on channel dimension. Bed forms and nature of infill in the creek show that they are flood dominated. Vegetation on the sand bars helps in trapping the sediments. Due to the microtidal seasonal setting of the estuary, tidal transport of the sediments has started chocking the creek in recent years. From the discussions and interviews with locals in village Devbag, it is almost certain that due to natural factors and anthropogenic activities, there is a significant change in the morphology of Karli-creek in recent years. Some interventions in the form of construction of roads on the spit bar as well as along the creek bank, alteration of hill slopes adjacent to creeks, sand mining in creek bank, construction of bridge on river Karli and industries has affected the creek. The main channel of river Karli has shallowed considerably in the recent past (Pisolkar, 2008).

Due to the infilling of the channel by the sediments tide velocities increase, resulting in bank cutting (Woodroffe, 2002). In addition to this wave generated due to the boats used for recreational activities are enhancing the breaching problem especially in Survey Nos. 53, 55, 47 in Survey Nos. 10, 7, 3 in the recent past, inundation has increased. Fiyan the cyclone hit the coast of Maharashtra with devastating impact on 11th November, 2009. Villagers of Devbag suffered during this cyclone.

SOCIOECONOMIC ENVIRONMENTS

Up until the year 2000 villages of Devbag and Tarkarli were sleepy fishing villages with communities depend on in traditional fishing and local Agra products are their livelihood. The sudden appearance and rapid constructions for tourism industry exert great pressure upon coastal area. Village Devbag has changed drastically since then. Much of the activity revolves around tourism, with about 50 home stays or hotels and approximately the same number under construction (Figure 2). Land use is changing very rapidly with Beach sand dunes which are the cushions that takes the impact of sea waves, especially surge waves produced in monsoon are significantly altered . Many hotels and home stays are being constructed on the dune system itself. These rampant alterations are making the ecosystem vulnerable to natural forces. Waste disposal is another important aspect. Home stays and hotels, which generate waste, collect it in the pits and burn it at the end of each week. Even through care is taken to dispose waste within the area owned by hotel owner there is no mechanism by which waste generated on beaches, adjacent to big hotels.

Tourism in Tarkarli - Devbag is a seasonal activity as the main thrust is coastal tourism. The main influx of tourist is concentrated in the months of October-May. In this seasonality, period adjacent to Diwali vacation, Mid-December to first week of January and Mid-April to May is the peak season. This seasonality has its own impacts. According to news sources villages of Devbag and Tarkarli were visited by 12000 tourists in 4 days during Diwali vacations. Compare that to the population of these villages of 5000, this influx puts tremendous strain on public infrastructure and resources. In village Devbag narrow road runs through the middle of the village up to the Mobar Number of vehicles, especially in peak season have now been a concern. Parking of vehicles with limited space is creating additional pressure. Quality and availability of water are under threat in certain parts of the beach especially near Vithoba Temple. Carrying capacity has to be investigated in depth. All the facilities are underutilized in the off-season. The positive outcome of the seasonality of tourism in Tarkarli Devbag is that it has allowed the diversification of occupation of at least some percentage of the population. Economic conditions of the people are gradually improving due to this shift in occupation However, these villages historically have been fishing-oriented. Still major population depends on fishing for their livelihood. Those engaged in tourism and other primary activities, a differentiation exist.

Figure 2. Hotels in Devbag

RESOURCES, COASTAL TOURISM AND CARRYING CAPACITY

Coastal ecosystems must be protected not only for the more general functions that they perform in terms of support to human well-being but also because they provide the goods and services required for coast-based activities. Impairing them will impact adversely on the economic activities that utilize them – implicitly or explicitly. For example, the depletion of vegetation, the razing of sand dunes, and the extraction of sand that accompanies development, can result in land erosion as well as a loss of tourist attraction as the aesthetic value of the local coastline is reduced. Vegetation loss can also affect water circulation and recharging of ground water reservoirs. Coastal aquifers are doubly vulnerable: proximity to the sea means heightened vulnerability to saltwater intrusion and a public good characteristic means susceptibility to the common pool problem. In the context of coastal tourism and tourist accommodation needs, the health of coastal ground water is a concern for tourism and the host population. Both land use and land cover tends to be affected by tourism and need to be studied to the extent of change. While tourism may not be involved in initiating this process of change it can and does influence its outcomes (Naronha 2003).The concept of tourism carrying capacity is based on the general statement that any form of development within the carrying capacity of an ecosystem means sustainable utilization of tourist resources and development of ecosystem. When an area is used beyond its capacity, the quality of natural resource changes and gets destroyed. The increased inflow of either local or foreign tourists beyond the carrying capacity of the area may deteriorate the area due to in luxury hotels (Lindberg et al., 1997).

THREAT TO TARKARLI- DEVBAG SPIT

Tarkarli- Devbg Spit is a 6 km long sand spit about 16 km south of Malvan on Maharashtra coast. The spit bar is connected to mainland near Tarkarli. Southern end of the bar abruptly ends at sea near Mobarwadi. The eastern edge of the spit is bordered by a tidal stretch of the river Karli. The western margin, faces the Arabian Sea. The area has been showing a tendency of severe breaching and erosion of shorefront continuously since 1952. In fact, it appears that it has become an annual phenomenon. Tarkarli (16^{0}to16^{0}2'N latitude and 73^{0}29'to73^{0}30'E Longitude) and Devbag (15°57' to 16°01' N latitude and 73°29' to 73°31' E longitude) which were traditionally small fishing villages up to the end of the last century in Sindhudurg district of Maharashtra with a population of 1886 (2011) & 2992 (2011) respectively. People in Tarkarli and Devbag sensed the potential to develop the coast for national as well as international tourists in last two decades. Local people and government of Maharashtra is promoting coastal tourism in Tarkarli. In Devbag coastal tourism is flourishing at an alarming rate. About 50 hotels, which include 'home stays' and resorts have come up in the last one and half decade up to December, 2012. In Tarkarli village these spots are often characterized by

1. Small size, limited resources, including drinking water and heavy reliance on an external source of resources and capital.
2. Population is forever in flux. There is migration cities due to lack of opportunities for livelihood within the confines of the village.
3. Lack of connectivity with larger economic centers, resulting in isolation or higher cost of transportation.
4. Lack of social infrastructure including banks, schools or hospitals making the population vulnerable to disasters.
5. The landscape is vulnerable to forces of nature with increasing susceptibility to climate change and sea level rise.
6. The ecosystem is often very fragile. Lack of capital makes it difficult to manage.
7. Weak governance with lack of influence or connectivity with state or national level authorities and significant poverty because of lack of opportunities. Situation regarding availability of water, especially drinking water is unfortunate.

There is severe distress due to natural as well as manmade factors. Integration of water related policies on the village and state level is nowhere to be seen. The solutions are often piecemeal with scant regard for efficiency or sustainability. There is a market failure on account of both financial and technical capacity. Most of the population has personal sources of water, such as dug wells and bore wells. There is no mechanism to regulate the usage. No quality standards are available or enforced. It is difficult to execute public works projects due to the fragmented nature of land holding and lack of public space for utilities. Complete lack of regulations regarding public use of space or sharing of resources complicates the situation. Extensive integrated management solutions or regulatory interventions face daunting task of conflict resolutions arising from the lack of cooperation from the stakeholders. Water has been traditionally considered as a god given right and any notion of financial obligation towards management of water resources is promptly rejected. Existing water management systems are too unregulated and inefficient for providing sustainable future. Local government hesitates to allocate any budget towards water resources due to lack of accountability. The water quality is rapidly deteriorating where in January the salinity can be as high as 5800 ppm, yet no measures have been taken or even proposed to tackle this issue that will plague the region in years to come (Figure 3). Tarkarli and Devbag are part of the watershed, which directly connects to the Arabian Sea. For wholistic solutions the efforts are required to integrate the entire watershed along with the coast for a viable future.

Some of the proposed solutions will have to be spatially adequate within the confines of the watershed. It should include both creek and coast. Scientific study of the dynamics of the creek and sea water will have to be carried out to understand the relationship of fresh and salt water. The approach will have to be a combination of environmental, management and engineering solutions. The approach should have

Figure 3. Salinity in Devbag

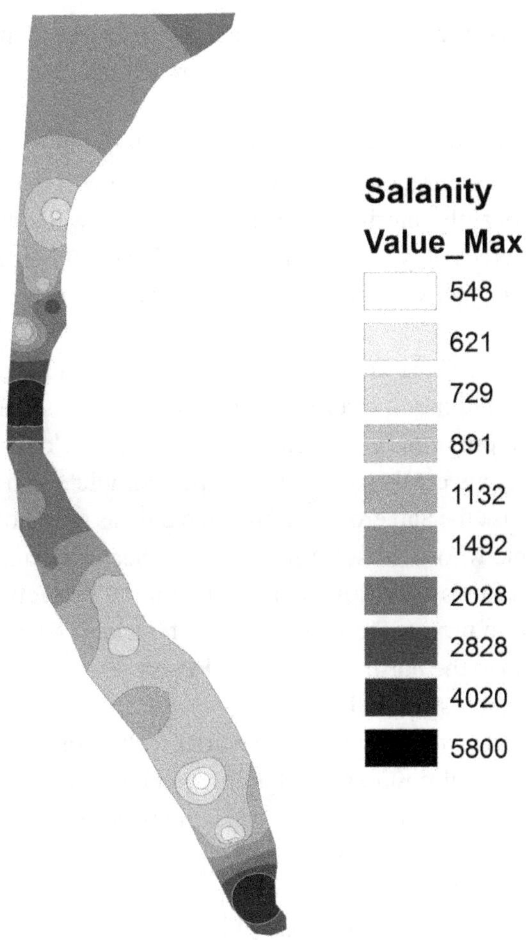

a long term view with the flexibility in making course corrections based on the outcomes. The tradeoff between the environment, management and engineering solutions must be analyzed setting community level goals.

ADAPTATION AND MITIGATION

Protection and Adaptation of Coastal Zone at Tarkarli: Devbag Spit

Tourism can, and does, impact local development and the environment. The beauty and environmental quality of tourist destinations are what attract tourist, and local people are increasingly anxious to participate in and draw the benefits of tourism while preserving their own identities, the natural environment and the historic and cultural heritage from the impact of uncontrolled tourism. As such the coast, poses a special dilemma for sustainable management because it involves a land-ocean interface; Mix of property systems, multiple users and multiple jurisdictions. Requires a special effort at governance and multidisciplinary approaches, as both the social and the natural sciences need for effective management (Noronha, 2003)

Problem at villages and Devbag and Tarkarli are threefold, -

1. As a result of natural geomorphic processes, the boundary between land and water has undergone a shift in position with time on Devbag spit. The encroachment and erosion by the Arabian Sea and Karli River have accelerated.
2. Rapid unplanned development for tourism is aggravating the problem of beach and creek erosion. Several serious issues regarding quality of tourist destination will definitely rise in near future.
3. These changes are significantly altering the water quality of wells and bore wells. Making the region rely on external sources of water mainly water tankers and bottled water. These sources are notoriously unreliable and expensive, putting the population in existential crisis.

Erosion is leaving behind several short term and long term impacts on Tarkarli - Devbag spit. A variety of interventions are needed for mitigation of erosion and adaptation to natural phenomena, integrated with development planning and program implementation in the area. Construction of anti erosion wall on the seaward side and bonds along creek side has not helped completely in protection of the spit bar at Tarkarli - Devbag, mainly because the anti erosion wall and bund has been constructed in patches, as and when demanded by local people who were suffering severely. Seawalls are a temporary solution; they are sustainable only as long as funds are available to maintain the protection continually (Woodrofle, 2002). There is an urgent need to protect the seaward side of Tarkarli - Devbag spit with good quality anti erosion wall from mobar up to the north of M.T.D.C. Resort. Provision of funds should be made by government to maintain this anti erosion wall.

The debug spit bar is bordered by the tidal sector of Karli creek. From the discussions and interviews with locals in village Devbag, it is almost certain that due to natural factors and anthropogenic activities, there is a significant change in the morphology of Karli creek in recent years. According to locals the main channel of river Karli was comparatively deeper with a depth of 7 to 10 m., and was the major navigable river upto Nerur Par till 1970-72. Present morphology to Karli creek reveals the fact that the main channel of river Karli has shallowed considerably in the recent past. A DEM model generated from the hydrographic charting (Bathymetry) for Karli creek shows that the present depth of Karli creek varied between 1.5 to 3 m from Tarkarli to the mouth of river Karli at Mobarwadi (Pisolkar 2008). Due to the infilling of the channel by sediments, tide velocities increase resulting in bank cutting (Woodrofle, 2002). The filling of the channel has decreased the strength of creek banks, which are getting eroded at a faster rate (Pisolkar, 2008).

The present researchers are of the opinion that the best solution to arrest erosion from creek side is to construct the road along the creek up to the Mobarwadi. Dredging of river Karli and material obtained from this can be utilized for this purpose. Expert guidance is needed from institutions like National Institute of Ocean Technology or National Institute of Oceanography. Conservation Architects can give the designs to maintain the aesthetic beauty along creek side as a region has quickly picked up as an important tourist destination. This upto certain extent will solve the problem of heavy traffic and mainly the creek erosion problem considerably. The tourism in the last decade has concentrated only on promoting tourism without attention to the resources, human and natural, that gave the place its charm, beauty and attractiveness. There is a lack of attention to cleanliness of beaches, to water availability and to ensuring quality tourism.

SUSTAINABLE COASTAL DEVELOPMENT

Tourism can be a seductive activity, luring the youth with easy work and money (Naronha et al., 2003). The rate at which tourism is developing on Tarkarli - Devbag spit there is a need for training, skilled local youth. There has to be an emphasis on education and professional skill formation as most of the Homestay owners have no background of providing service or running a business. Local people have skills in singing, dancing, guiding etc. Infrastructure in the sense two drama theaters also exits in village Tarkarli and Devbag. There is a need for policy to strengthen the local people's skills to become hosts for various programs of singing, drama etc. These local skills and products can provide the much needed reinforcement for the tourism industry. Tourism is a fickle industry searching for ever new locations and activities. The over reliance on "new craze" often results in unrealistic projections and over capacity. Once the raze subsidies it often leaves a lost generation in its wake. The transient nature of tourism industry must be complemented or reinforced by developing competing industry or services in order to buffer the volatility. A precautionary slant has to be adopted in policy to ensure that local communities are not affected badly in the event of such an occurrence (Naronha etc. al 2003). The role of the local Panchayats is very important. The constraints or opportunities that the Panchayat provide for the good governance need to be studied.

Vulnerability of coastal aquifers to salt-water intrusion and growing tourism demands has to be investigated in depth. There is an urgent need for parking places. Local people are providing a lot of services in the peak season. Self Help Group of women can be one meaningful solution for well being of locals in Tarkarli - Devbag. Land use change to avoid a reduction in ecosystem functionality to the loss of diversity which will affect local livelihoods. Water sports activities in river Karli are one of the main reason for reduction in fish catch, local governance to regulate the activities of tourism in more meaningful and sustainable way has to be strengthened.

Sustainable coastal tourism faces many challenges. It requires thorough understanding of consumer satisfaction, economic, health, host well being and environmental and social care (Naronha etc al 2003). To add this if the region is vulnerable to rising sea level, issues become more complex. In case of

Tarkarli - Devbag, it needs through understanding of geomorphic processes, what development plans villages should have for tourism to 'last' and harmony must be sought between the demands of tourist, the industry and the host population.

ISSUES AND CHALLENGES

People have an extremely short term view of things, especially if they are constrained by resources Strategy to create awareness of sustainable practices must be designed and implemented. Knowledge must be built and updated through long term research and policy studies. Local knowledge base of community can be incorporated in order to evoke their participation into conservation strategy.

The complexity and interlinking of issues, number of stakeholders and governance pattern had led to inadequate understanding of system among stakeholders. However, recently the Maharashtra Government has set up Maharashtra Coastal Zone management Authority (MCZMA) with the purpose of Monitoring coastal areas, feeding observations to state and national policies and act as a link between all stakeholders. Strengthening MCZMA will be the first step towards improving synergy.

Regional Centre of Expertise (RCE) is a networking model proposed by the United Nations University Institute of Advanced Studies (UNU-IAS). RCE is a dynamic network comprising of all stakeholders of an area to integrate academic research into real life issues and feed the working systems with inputs obtained from the research. Setting up RCE for sustainable coastal Development and linking it with MCZMA will connect all stakeholders.

COMPREHENSIVE PLANNING FOR FUTURE

We propose a six phase program for comprehensive planning for a sustainable future with a focus on resource management the situation with respect to water resources is likely to get dire in light of climate change predictions.

Phase 1: Survey

Comprehensive surveys are needed to do resource inventory. It is imperative that we know the status of the resources we have in order to address the deficiencies. Geomorphic surveys are needed to understand the dynamics of the changes occurring at the creek as well as on the coast. Comprehensive studies are needed to study the ecosystem and assess the threats from natural as well as anthropogenic interventions. With respect to water resources it is critical to assess the availability and quality of water. Spatio- temporal patterns need to be deciphered to address supply and demand situations. Scoio economic surveys or demographic surveys need to address the policy issues and to formulate the framework for monitoring and sharing of water resources.

Phase 2: Roadmap

After careful analysis all the stakeholders should come together to formulate a roadmap for a sustainable future. For water resources it will involve monitoring of geomorphic changes, possible environmental, engineering and management interventions. Policy on water sharing and possibility of procuring water supply from nearby reservoirs should be explored.

Phase 3: Goals

Based on the roadmap stakeholders should formulate a strategy for achieving the sustainable future. Based on the predictions on climate change, changes in demographics and economic development support strategies should be formulated. Care should be taken to incorporate multiple scenarios that might unfold. These strategies should be periodically reviewed and adjusted to better fit the unfolding scenario. The strategy thus should provide a framework but should be flexible enough to be altered

In order to achieve a good fit with an objective.

Phase 4: Planning

When a strategy is in place, roles must be given to all stakeholders so as to collectively move towards the common goal. Institutional framework must be in place along with regulations to keep the stakeholders fixed on the vision. Transparency is a must in planning process to ensure participation and accountability. National or regional authorities must be consulted to avoid any overlaps or conflicts.

Phase 5: Execution

One of the Achilles heel for successful development is the execution phase. The policies must be implemented with complete transparency and strict timelines in order to reduce the conflicts between various stakeholders. The execution often suffers due to lack of adequate coordination, lack of effective communication, leadership deficit or the skill deficit within the stakeholder community. These issues should be addressed during the planning process. One of the major drawbacks is to have unreasonable expectations while formulating the goals. These expectations need to be tempered in order to avoid early disappointments which often sabotage the execution process.

Phase 6: Monitoring & Assessment

Monitoring and assessment provide opportunity for the various stakeholders to assess the execution of the policies. It allows stakeholders to build a knowledge base, learn lessons from the experience as well as fine tune the subsequent actions. It also provides a template for future planning and policy formulation. This process will also help build institutional and legal frameworks for water resources management.

REFERENCES

Central Water and Power Research Station, Specific Note No. 3035. (1993). *Engineering Properties of rock material. Koyna Hydro Electric Project (Stage-IV)*. Pune, India: Author.

Cicin, S.B. (1993). Sustainable development and integrated coastal management. *Ocean and Coastal Management, 21*, 11-43.

Clark, J.R. (1977). *Coastal ecosystem management*. John Wiley and Sons.

Dhavalikar, L. N. (2002). Sea level scenario on Vengurla coast of Maharashtra. Transactions, Institute of Indian Geographers, 24(1-2), 63-71.

Dikshit, K.R. (1986). *Maharashtra in Maps*. Maharashtra State Board for Literature and Culture.

Group of Experts on the Scientific Aspects of Marine Pollution. (1990). *Report of the 20th session*. WHO.

Hildebrand, L.P. (1997). Introduction to the special issue on community based coastal management. *Ocean and Coastal Management, 36*, 1-9.

Hinrichsen, D. (1996, Summer). Coast in crisis. *Issues in Science and Technology, 12*(4), 39–47.

Intergovernmental Oceanographic Commision (IOC). (2015). *Scientific community proposes recommendations for integration of ocean and coasts in COP21 agreement*. UNESCO Media Service.Retrieved from http://www.unesco.org/new/en/media-services/2015

Intergovernmental Panel on Climate Change. (1994). *Preparing to meet the coastal challenges of the 21st century: Conference report*. World Coast Conference 1993, Intergovernmental Panel on Climate change, Noordwijk, Ministry of Transport, Public Works and Water Management, The Hague, Netherlands.

Karlekar, S. N. (1981). *Unpublished Ph. D. Thesis*. University of Pune.

Karlekar S.N. (2004). The impact of Coastal Landforms on the distribution of placer Minerals along Maharashtra Coast. *Indian Journal of Geomorphology, 9*(1&2), 101-114.

Kay, R., & Alder, J. (1999). *Coastal Planning and Management*. London: Routledge.

Lindberg, K., McCool, S., & Stankey, G. (1997). Rethinking carrying capacity. *Annals of Tourism Research, 24*(2), 461–465. doi:10.1016/S0160-7383(97)80018-7

Noronha, L., Lourenco, N., Lobo-Ferreira, J. P., Lleopart, A., Feoli, E., Sawkar, K., & Chachadi, A. (2003). *Managing Coastal tourism; perspectives from India and the European Union in Coastal Tourism, environment and sustainable local development. In The Energy and Resources Institute* (pp. 419–446). TERI.

Patil, Y., & Rao, P. (2015a). Industrial waste management in the era of climate change - A smart sustainable model based on utilization of active and passive biomass. In W. L. Filho (Ed.), *Handbook on Climate Change Adaptation* (pp. 1–13). Springer-Verlag Berlin Heidelberg. doi:10.1007/978-3-642-38670-1_49

Patil, Y., & Rao, P. (2015b). A Novel Biological Strategy for the Management Industrial Effluents with Simultaneous Recovery of Metal Resource. *Advanced Materials Research*, *1130*, 640–643. doi:10.4028/www.scientific.net/AMR.1130.640

Pisolkar, Y., Rawat, A., Mahankale, N., & Arora, R. (2013). *Growing Tourism on Tarkarli –Devbag Spit, its Consequences and Suggested Strategy for Sustainable Tourism, Coastal Maharashtra, India, Report 'Minor Research Project' Symbiosis Centre for Management Studies*. Pune: SCMS.

Pisolkar, Y. M. (2008). *A Geomorphic Study of Beach and Creek Erosion at Devbag, coastal Maharashtra*. (Unpublished Ph.D. Thesis). University of Pune.

Rao, P., & Patil, Y. (2015). Climate resilience in natural ecosystems in India - Technology adoption and the use of local knowledge processes and systems. In Handbook on Climate Change Adaptation. Springer-Verlag.

Rao, P., Patil, Y., & Gupte, R. (2013). Education for Sustainable Development: Trends in Indian Business Schools and Universities in a Post Liberalization Era. In Sustainability Assessment Tools in Higher Education Institutions - Mapping Trends and Good Practices Around the World. Springer International Publishing.

Rao, P., Patil, Y., Ketkar, M., Bhat, V., & Kulkarni, S. (2014). Sustainability in an educational institution: analysing the transition to paperless e-processes, an Indian case. In U. M. Azeiteiro, W. L. Filho, & S. Caeiro (Eds.), *ELearning and Education for Sustainability, Environmental Education, Communication and Sustainability Series* (Vol. 35, pp. 215–233). Peter Lang GmbH International Academic Publishers.

Salomans, W., Turner, R. K., de Lacerda, L. D., & Ramachandran, S. (1999), Perspectives on Integrated Coastal Zone Management. Springer Verlag. doi:10.1007/978-3-642-60103-3

Shrivastara, A., & Kothari, A. (2012). *Adding fuel to fire in Churning the Earth, the making of global India*. Penguin Group.

Siqueira, A. (1999). Global consumption and local processes. In *Population, consumption and environment interrelations: a tourist spot scenario*. New Delhi: Tata Energy Research Institute.

The Energy Reasearch Institute. (2000). Population consumption, and environment interrelations: a tourist spot scenario. New Delhi: Tata Energy Research Institute.

Woodrofle, C. D. (2002). *Coast: Form, process and evolution*. CambridgeUniversityPress.

World Resources Institute. (2000). *People and ecosystem: The fraying web of life*. Washington, DC: Author.

Chapter 5
Impact of Climate Change on the Retreat of Himalayan Glaciers and Its Impact on Major River Hydrology:
Himalayan Glacier Hydrology

Ram Karan Singh
King Khalid University, Saudi Arabia

ABSTRACT

Himalayan Glaciers are the largest freshwater resource on earth and the rivers originating from them are an important source of water. They significantly modify stream flow both in quantity and timing as annual basin run-off is enhanced or decreased in years of negative or positive glacier mass balance respectively. Although glacial advances and retreats are a part of its natural cyclic phenomenon, the rate of de-glaciations has accelerated in recent times due to climatic changes and global warming caused by anthropogenic activities. Some of the important glaciers of Himalayas are receding at an alarming rate, which could have dire consequences on river hydrology of the main rivers of this region namely, Indus, Ganga and Brahmaputra, initially causing floods and the paradoxically, scarcity of water later. This chapter is an attempt to summarize some of the studies on Himalayan glacier retreats and also to assess its impact on the availability of freshwater in the sub-continent.

INTRODUCTION

Glaciers represent ancient rivers of compressed snow and are the earth's largest freshwater reservoir, occupying an area that is roughly the size of South America. The earth, in its geological history, has witnessed a global climate change from being extreme cold to intense warm. This has manifested into the cyclic glacial and inter-glacial periods during the past. Geological evidences suggest that the earth has experienced glaciations during Pre-Cambrian, Perm carboniferous and in the last Pleistocene periods

DOI: 10.4018/978-1-5225-1046-8.ch005

(Ramanathan et al., 2005). There have been at least 17 major glacial advances in the last 1.6 million years alone, the most recent one being the Last Glacial Maxima, which reached its peak some 20,000 to 18,000 years ago and came to an end about 10,000 years ago (Ramanathan et al., 2005). During the peak glaciations period of the earth, 47 million sq. km was covered by glaciers, three times more than the present ice cover of the earth (Kulkarni et al., 2007). Glaciations are followed by interglacial periods, during which the glacier retreats as a result of global warming. The interglacial period typically continues for about 10,000 years before the next glaciations begin. A number of ideas were proposed to explain the repeated cycle of glaciations on the earth. The most popular idea is related to the natural variation in the earth's orbit around the sun. The orbital cycles (100000, 41000 and 22000 years) can cause 10% variation of the incoming solar radiation in the various parts of the globe (Kulkarni et al., 2007).

Natural variations in the earth's orbit are well synchronized with the atmospheric variations in methane and carbon dioxide, leading to normal cycles of glaciations (Kulkarni et al., 2007). However, the natural cycle has been largely altered due to the greenhouses effect caused by anthropogenic changes made in the environment. During the past century or so, the economic development across the globe has been coupled with increase in atmospheric pollutants that trap the solar radiation and increase the mean temperature of the earth. This has led to an increase in global temperature by 0.6±0 .2°C from the 1900 (IPCC report 2007). In addition to this, climate modeling suggests that greenhouse gases and aerosols have led to absorption of 0.85±0.15 W/sq m more energy by the earth than that emitted into space (Kulkarni et. al., 2007). This implies an increase of about 0.6°C without any change in atmospheric composition.

Therefore, in the event of such enhanced global warming, it is important to study the effects and impacts on the glaciers across the world as they represent a vital source of water for a large percentage of people all over. They also affect the land pattern and river hydrology by their repeated melting and formation. Glaciers are dynamic systems, with periodic formation due to snowfall and melting taking place depending on the season. Snowfalls in the highest part of a glacier, which is known as the accumulation area, add to its mass and as the snow slowly turns to ice, the glacier grows in weight, forcing glacial movement (Ramachandran, 2001). Further down the glacier is the ablation area where most of the melting and evaporation of glacier.

IMPACTS ACROSS THE GLOBE

Glaciers occupy a prominent portion of land mass of the earth across various continents and impact of climate change has been witnessed all throughout. A study carried out by World Wildlife Fund (WWF) on the impact of climate change on glaciers recorded conspicuous changes occurring in all glaciers and snow-covered regions of the world. Over the decades the Arctic glaciers have been observed to be shrinking, with the exception of Scandinavia and Iceland. The annual melting of ice has accelerated from100 sq km in 1980-1989 to 540 sq km in 1998. Greenland, which represents 12% of the world's ice, has been significantly thinning and ice loss around the periphery. Recent studies show that the surface melting trickles down to the glacial bed, lubricating the ice sheet and making the whole mass of ice slide towards the sea.

In Northern America, glaciers in the Rocky Mountains and Western Coastal Ranges along with Alaska have shown significant glacial melting. Two-thirds of the glaciers of Glacier National Park have disappeared since its establishment in 1910. In Banff, Jasper and Yoho National Park in the Canadian Rockies, glacial cover has receded by 25% during the past century.

The northern Andes in South America contain the largest number of glaciers in the tropical region, but their retreat has increased many folds during the 1990s. In the sub-tropical Andes, the large ice mass of North Patagonia Ice field (Chile) and South Patagonia Ice field (Chile and Argentina) have lost 4-6% of their cover till the mid-1990s.

In Europe, the majority of glaciers in the Alps have experienced significant mass loss which has been even more enhanced due to increasing temperatures during the past decade and has resulted in catastrophic events like flood and landslides. Tropical glaciers of Africa have decreased by 60-70%, with Mt. Kilimanjaro losing 80% of this ice cover in the past century. The largest mass of ice on earth Antarctica, contain 95% of the planet's freshwater, but it too has experienced melting due to increased ocean temperature. Warmer seas have also contributed to the rapid thinning and break up of ice shelves in this region.

THE HIMALYAN ARC AND GLACIER RETREAT

The Himalayan glaciers are of special importance due to the fact that they supply water to about 40% of the earth's total population and also its control on the climatic conditions of this region. Hence, any change in the ice mass across the Himalayas would spell trouble for a large section of the people inhabiting the Asian continent, especially in Nepal, China and India. The Himalayan glaciers feed water to seven of the most important rivers in Asia, namely Ganga, Indus, Brahmaputra, Salween, Mekong, Yangtze and Huang He (WWF, 2008).

The Indian Himalayan glaciers are divided broadly into three river basins: Indus, Ganga and Brahmaputra. Indus, with 3,538 glaciers has the most number of glaciers followed by Ganga basin with 1,020 glaciers and Brahmaputra with 662 glaciers (Kumar, 2005). It has been estimated that 17 percent of the Himalayas is under permanent ice cover and the principal glaciers of this region are Gangotri, Kedarnath and Chota Shigri (Kumar, 2005).

Himalayas produce considerable changes in air masses crossing the region, giving rise to a mosaic of climatic conditions which is determined by aspect, elevation and slope. During the summer monsoons,

Figure 1. Glacier distribution along the Himalayan Arc
(Source: WWF case study of Indian glaciers)

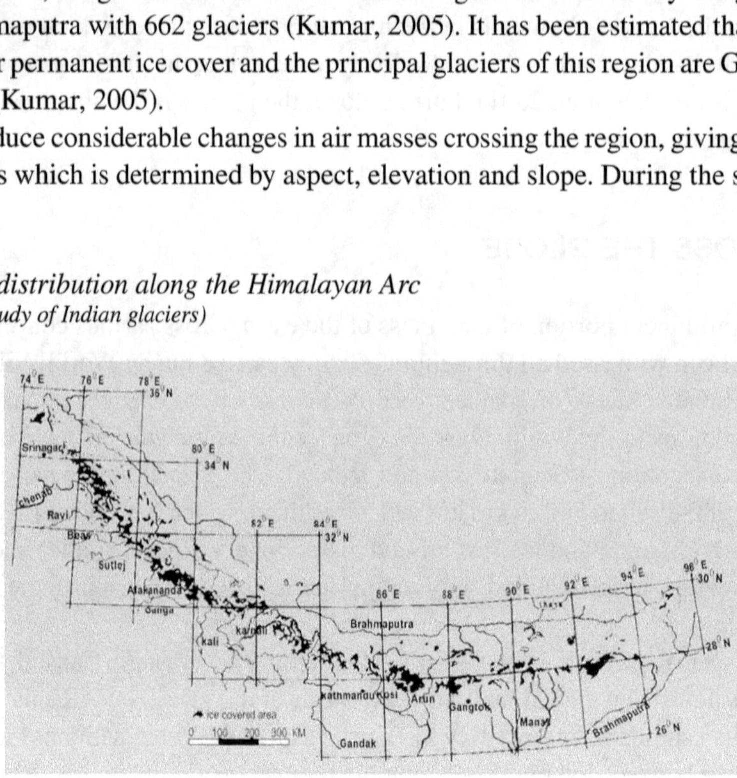

Table.1: Glacial Inventory of Indus Basin

Basins	No. of Glaciers	Glacierised area (km^2)	*Ice Volume (km^3)*
Jhelum	133	94	*3.0*
Satluj	244	420	*23.0*
Others	*3398*	*33382*	-

(WWF: Case study of Indian glaciers, 2005)

Table.2 Glacial inventory of the Ganga- Brahmaputra Basin

Basins	No. of Glaciers	Glacierised area (km^2)	Ice Volume (km^3)
Bhagirathi	238	755	67.0
Tista	449	706	40.0
Brahmaputra	161	223	10.0
Others	640	2378	-

(WWF: Case study of Indian glaciers, 2005)

which normally commences from mid-June to mid-September, the mountain range blocks the northward advancement of the monsoon causing intense rainfall on the southern slope, and dry conditions on the northern slope (Kumar, 2005). Delayed onset of the monsoon decreases precipitation along the Himalayan arc from east to north-west. The Western Himalayas get more precipitation during November to April.

The effects of climate change on the glaciers of the Himalayas have been studied for a considerable time, taking into account numerous glacio-geomorphic features including lateral morainic ridges, medial morainic ridges, mass balance, glacial lakes, crevasses and glacio-fluvial deposits (Ramanathan et. al., 2005). Monitoring of the Himalayan glaciers is normally difficult due to its rugged terrain and harsh climatic conditions. But with the advancement of science and technology, use of innovative and sophisticated tools like remote sensing has been increasingly used. A study of this region by use of remote sensing has been carried out by (Kulkarni et al., 2007) using Indian Remote Sensing (IRS) satellites IRS-P6 and IRS-1D with LISS-III and LISS-IV sensors. The glacial extent was delineated using topographic maps and GIS. Field investigations were carried out at five glaciers during the peak summer season when the glaciers are fully exposed. These include Shanue Garang Glacier in Baspa Basin, Parbati Glacier in Parbati basin and Chota Shigri, Samudra Tapu and Pastio glaciers in the Chenab basin. Snout positions were marked using GPS and by comparing the positions of snout with geomorphic features like moraines and moraine-dammed lakes, the glacier retreat was measured. It was observed that there was significant retreat of the glacier snouts during the period 1998-2003, especially in the Chota Shigri glaciers with an 800 m retreat. The overall retreat was estimated to be 21% since the middle of the last century.

Precipitation levels are another aspect associated with climate change. According to climatologists, alpine glaciers are particularly sensitive indicators of climate change and through various monsoon climate models, it has been predicted that there will be considerable change in precipitation levels in the Himalayan region due to its dependency on summer monsoon precipitation and cool summer temperatures (Kumar, 2005). As global warming continues, it will cause the zero temperature line or the snow line to shift upwards and thus, glaciers will receive more liquid than solid precipitation. This will alter the mass balance (difference in mass of ice gained and lost during a hydrological year) of the glaciers.

Therefore, higher temperature and more liquid precipitation will lead to rapid retreat of glaciers and may cause downstream flooding.

During the recent decades, considerable retreat has been observed in many of the glaciers of the Himalayas, most prominent of those being the Gangotri Glacier, Dokriani Glacier and Chota Shigri Glacier.

The Gangotri Glacier in the past had extended up to Sukhi below Jhala in the Bhagirathi valley, as indicated by the remnants of moraine at an altitude of 2300 m (Ramanathan et al., 2005). This is the oldest remnants of the glacier and it possibly represents the position of the snout during the Pleistocene period. The glacier is now rapidly receding, which is evident when Hasnain et al. compared the snout position from top of sheet map of 1985 and panchromatic satellite image from 2001. Large morainic ridges and smaller inset morainic ridges also provide evidence of major glaciations. Observations on the retreat of the Gangotri go back to 1842, when it was assessed that between 1842 and 1935 the snout of the Gangotri glacier was receding at an average of 7.3 m a year (Naithani et al., 2001). Geological Survey of India calculated that between 1935 and 1996 the snout retreated by 1147m with an average of 19m per year. This implied that the rate of retreat had more than doubled during the last century.

Dokriani glacier, located 30°49' to 30°52' N and 78°47' to 78°51' E, is a valley type glacier in the Garhwal Hiamlayas, Uttarakhand. The length of the glacier is 5.5km and the total catchment area is 15.7 km2 (Kumar, 2005). The melt-water emerging from the glacier is known as Din Gad and joins river Bhagirathi at Bhukki. The thickness of the glacier ice varies from 25-120 m between snouts to accumulation zone (Kumar, 2005). Dokriani glacier has shown a rapid frontal recession of 10%, substantial thinning at the lower elevation and reduction in volume by 20%. From 1962-1995, the glacier had receded by 550 m, with an average of 16.6 m per year (WWF, 2005). The calculated average thickness of Dokriani

Figure 2. Satellite imagery showing oraine-dammed lake and terminus of Samudra Tapu glacier, Chenab basin HP
(Source: Kulkarni et.al 2007)

Figure 3. Glacier split into four glaciers between 1962-2004
(Source: Kulkarni et.al 2007)

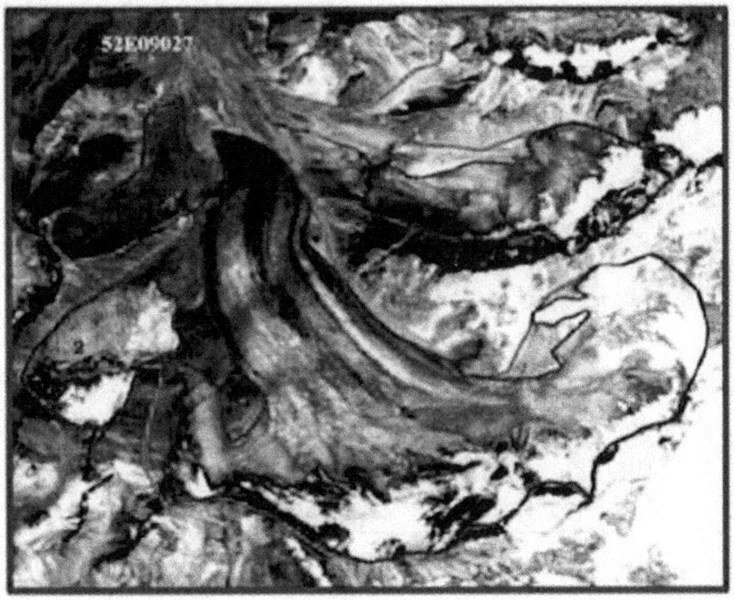

Figure 4. Retreat of the Gangotri glacier in last 220 year
(Source Jeff Kargel, USGS)

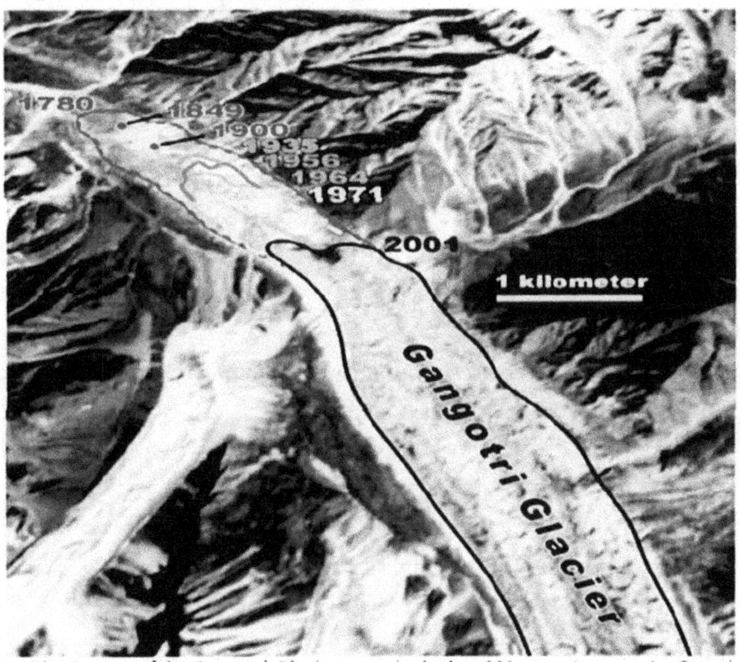

Figure 5. Drainage map of Dokriani basin
(Source WWF, 2005)

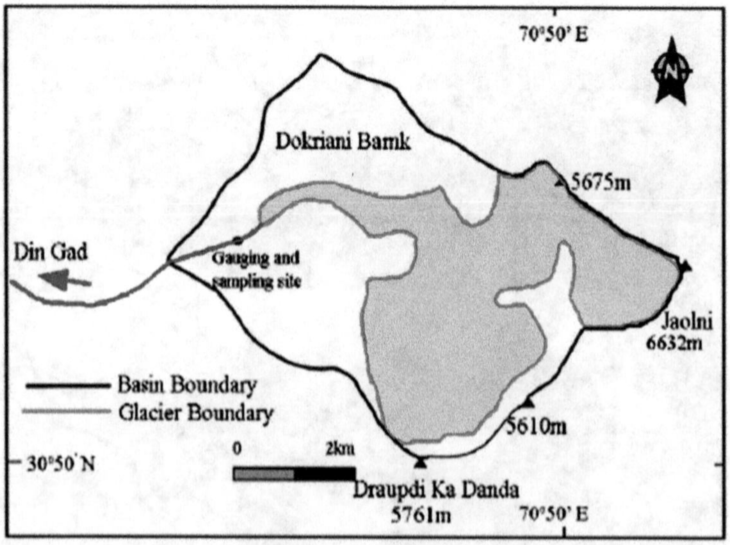

glacier was 55m in 1962 and 50 m in 1995. The average reduction in volume showed a negative and relatively increased mass balance.

Chota Shigri glaciers (32°11' - 32°17'N and 77°30' - 77°32' E is also a valley type glacier with varying altitude. It lies in the Chandra-Bhaga river basin in the Lahaul-Spiti valley of Himachal Pradesh. It falls in the monsoon-arid transition zone and is therefore considered to be an indicator of the northern limits of the intensity of the monsoons. From its snout to the accumulation zone, it extends up to a length of 9 km (Kumar, 2005). The Chota Shigri glacier drains into the river Chandra. The total drainage area of the glacier basin is about 45 km2. Four years of mass balance measurements from 2002-2006 showed an overall negative value, varying from -1.4 m water equivalent (w.e.) to a maximum of +0.1 m w.e. (Ramanathan et al., 2005). Satellite remote sensing can also be a useful tool in approximating the extent of the glacier in the last glacial maxima. Through careful image interpretation and observing various

Table 3. An overview of glaciers, glacier retreat, and subsequent impacts in Nepal, India and China

Glacier	Location	Period	Average retreat (m/Year)
Milam	Uttarakhand	1849-1957	12.5
Pindari	Uttarakhand	1845-1966	23.0
Gangotri	Uttarakhand	1935-1976	15.0
Gangotri	Uttarakhand	1985-2001	23.0
Bada Shigri	Himachal Pradesh	1890-1906	20.0
Kolhani	Jammu and Kashmir	1912-1961	16.0
Chota Shigri	Himachal Pradesh	1970-1989	7.5

(WWF, 2005)

geo-morphological features, the remnants of moraine depositions have been identified. Its presence in a distinct pattern gives an indication of the stages of glaciations encountered in the area. Through these debris cover or moraine depositions, the extent of the glacier has been traced up to the Chandra River. This implies that the present trend of retreat is a matter of serious concern.

IMPACTS OF GLACIAL RETREAT

Himalayan glaciers represent a unique reservoir that supports perennial rivers like Indus, Ganga and Brahmaputra, which are lifelines of millions of people on the Indian sub-continent. It is estimated that 33,200 km2 of the Himalaya is glaciated and glaciers occupy about 17 percent of the total mountainous area of the Himalaya, while an additional area ranging from 30-40 percent has seasonal snow cover (Kumar, 2005). Melt-water draining from this region is important in regulating the hydrology of the sub-continent. Gradual shifts in the snow line due to progressive increase in temperature and release of water from snow and ice keeps the level of water relatively constant in the Himalayan Rivers, out of which about 75 percent of discharges in the rivers occur during May-September (Kumar, 2005). The occurrences of monsoonal precipitation on the glacier surface increase albedo and reduce ice and snow melting (Colins & Hasnaian, 1995). Monsoonal solid precipitation coming to a glacier in the monsoon period is a major input in the mass balance of the glacier.

A reduction in water from the mountains would affect the economy by limiting the energy generated from hydropower plants and hampering industrial productivity. At a time when the government is promoting the use of renewable energy sources as a clean and sustainable way of meeting the nation's growing energy requirements, the possibility that the energy potential from hydropower may not be achieved has serious implications for the development plans of the whole country. Industries like food processing, mining, paper, chemical and steel production, which require a reliable supply of water, both as an input to the manufacturing process and as a medium for the dilution of effluent, would also be detrimentally affected.

The reduction in water resource that could cause the above stated problems have been revealed by a model developed by Centre for Ecology and hydrology and Department for International Development, UK known as the SAGARMATHA Project, which demonstrates that the initial period of glacier melting would result in increase in discharge by rivers causing flooding in adjacent areas (WWF, 2005). However, after a few decades the situation will reverse and the water content of rivers will start to decline permanently. This model, when run for 100 years under different climatic scenarios, shows distinct differences in the potential impacts of DE glaciation along the Himalayan arc, and within catchments. In the upper Indus, the study show initial increases of between +14 percent and +90 percent in mean flows (compared to baseline) over the first few decades of the 100 year incremental scenario runs, which are generally followed by flows decreasing between –30 percent and –90 percent of baseline (Kumar, 2005). In the headwaters of the Brahmaputra, there is a general decrease in decadal mean flows for all temperature scenarios; glaciers are few in this area and flows recede as the permanent snow cover reduces with increasing temperatures.

The impact of glaciers and snow cover on river hydrology has been studied by (Thayen et al., 2007) in a minor catchment area in the Garhwal Himalaya. Discharge of headwater streams in this region comprises of monsoons rains, glacier and snow melt. November to March records the lowest flows in these rivers and peak run-offs occur from July to August. These two months experience highest temperature, highest glacier melts and highest monsoon rains in the year. However, with the change in climatic conditions the headwater flow during these months was expected to change. The Din Gad catchment, the study area, is located in the Ganga basin in the Garhwal Himalaya. For data collection, three stations were set up to enable the monitoring at different altitudinal zones of the catchment, one each at Tela, Gujjar Hut and one very close to snout. The stations were monitored from May to October from 1998-2000 and glacier mass balance was calculated. The warmest year of the past 119 years was 1998 but the highest negative mass balance was observed in 1999, although temperature in this year was lower. This result showed that temperature and precipitation played an equally important role in glacier mass balance and run-off, and the amount of winter snow has a pre dominant role in determining the discharge volume of the glacier. The results also showed a characteristic change in the role of the glacier on stream discharge as the stream shifts further away from the glacier. The study thus showed that lowering of winter snow-cover extent and duration could reduce headwater flow drastically, and the observed 45% decline in run-off demonstrates the stress which the Himalayan region would experience in a climate change regime.

Another concern related to glacier retreat is the formation of large glacier lakes and also causing outburst floods. The formation and growth of glacier lakes is a phenomenon closely related to DE glaciation. Valley glaciers usually have glacial pond; but with the warming climate and soaring temperature, these ponds merge and grow bigger and bigger. This process is further accelerated by retreat of glaciers as they leave behind large void, depressed areas for the ponds to form. The moraine structures surrounding such depressions are generally weak and unstable, and as the water level in the pond increases beyond the carrying capacity, the moraine walls sometimes break which causes glacier lake outburst floods (GLOFs) (Kumar, 2005). A moraine dam may break due to various actions like huge waves caused due to landslide, rockslide, snow or earthquake.

A GLOF is a form of dangerous flood wave that is suddenly released from the lake and rushes downstream along the stream channel. The severity of such waves depends upon the amount of water released, amount of debris and basin characteristics of the watershed. A GLOF with high velocity represent a catastrophic impact of glacier retreats as it can have devastating consequences on the riparian communities, hydropower plants etc.

BRIDGING THE GAP

The 2007 report of the Intergovernmental Panel on Climate Change (IPCC) states that the Himalayan glaciers are retreating at a faster rate considering the present rate of global warming.(IPCC, 2007). There are very few scientists trained in this area of science and they are battling with poor resources, especially in India, Bhutan and Nepal (Padma, 2007). There is an acute shortage of scientific manpower and research gaps still exist in studies on climatology, climate and weather forecasting (UNDP, Bhutan). Financial constraints also hinder in obtaining data in these developing countries, and some of these data could be crucial in predicting events like GLOFs.

The roughness of the terrain and harsh climatic conditions also make data collection difficult in this area. Some of the Himalayan region in Bhutan and Nepal remain inaccessible for most part of the year. Although technologies like remote sensing and GIS help gather information on such inhospitable terrain, these technologies are expensive and require adequate financial support to operate.

The study of glaciers is an inter-disciplinary field, requiring collaboration of many scientific areas like glaciology, climatology, meteorology and hydrology to name a few. So information sharing and generation of awareness is the key to acquiring adequate information, which is crucial to devise a strategy to tackle this task before it turns into a catastrophe.

Between these two areas a balance is reached where snowfall equals snowmelt and the glacier is said to be in equilibrium. Whenever this equilibrium is disturbed, either by increased snowfall or by excessive melting, the glacier advances or retreats at more than its normal pace (Ramachandran, 2001).

Like a thermometer, a glacier is sensitive to the climatic variations and their advances leave behind landscape markers such as moraines, which reflect past glacial conditions and climate. In the Himalayas, for instance, the extent of the past ice masses is evident from the U-shaped valleys and in the large moraines which are up to several hundred meters in height (Ramanathan et al., 2001). Glaciers also serve as a barometer of our ever changing climate as they respond to changes in temperature and precipitation in their cycles of advances and retreats. The fluctuations of past and present glaciers serve as a valuable recorder of change within the climate regime.

CLIMATE CHANGE DEAL AT COP21 AND INDIA'S STANDING

As many as 195 countries reached a deal to save the earth from catastrophic climate change that threatens to bring about a mass extinction of species, including human beings. The deal has been termed historic because this is the first time there has been a near consensus in reaching an agreement over the grave environment challenge facing all nations equally. However, there are criticisms that the deal will put pressure on countries like India and developed nations like the US have not taken enough responsibility in dealing with the issue. Some important points of deal are as follows.

UNDERSTANDING TO DEAL WITH THE TEMPERATURE RISE

The deal is aimed at reducing the global warming which is already melting ice sheets and raising oceans threatening the lives of animals and plants. The countries that met at Paris agreed to hold the increase in the global average temperature to well below 2°C above pre-industrial levels and also make efforts to limit the temperature increase to 1.5°C. For this, the key is reducing the dependence on fossil fuels and cutting the greenhouse gas emissions. Temperatures have already increased by about 1°Celsius (1.8 degrees Fahrenheit) since pre-industrial times. In order to achieve the long-term temperature goal, the countries aim to reach global peaking of greenhouse gas emissions as soon as possible, says the agreement. It, however, recognizes that peaking of greenhouse gases emission will take longer for developing countries.

As per the agreement, developed countries should "continue taking the lead" by undertaking economy-wide absolute emission reduction targets, while developing countries should continue enhancing their mitigation efforts. They are also encouraged to move over time towards economy-wide emission reduction or limitation targets in the light of different national circumstances.

STRATEGIES TO DEAL WITH LONG TERM OBJECTIVES

For one, countries will set national targets for reducing greenhouse gas emissions every five years. More than 180 countries have already submitted targets for the first cycle beginning in 2020.The agreement requests countries that have set a target with time-frame of up to 2030 "to communicate or update by 2020 these contributions and to do so every five years thereafter".

PENALTY FOR NON-COMPLIANCE TO THE SET GOALS

The agreement has transparency rules to help encourage countries to actually do what they say they will do. That was one of the most difficult pieces to agree on, with China asking for softer requirements for developing countries. The agreement says all countries must report on their emissions and their efforts the reduce them. But it allows for some "flexibility" for developing countries that "need it."

FINANCIAL OBLIGATIONS

This is the most contentious issue: how will countries finance their shift towards greener economy? The agreement says wealthy countries should continue to offer financial support to help poor countries reduce their emissions and adapt to climate change. It also encourages other countries to pitch in on a voluntary basis. That paves the way for emerging economies such as China to contribute, even though it doesn't require them to do so. Actual dollar amounts were kept out of the agreement itself, but wealthy nations had previously pledged to provide $100 billion in climate finance by 2020. However, this seems to be a small sum. Oxfam estimates that the new deal will see developing countries facing adaptation costs of almost $800 billion a year by 2050.

For small island nations threatened by rising seas, the agreement includes a section recognizing "loss and damage" associated with climate-related disasters. The US long objected to addressing the issue in the agreement, worried that it would lead to claims of compensation for damage caused by extreme weather events. In the end, the issue was included, but a footnote specifically stated that loss and damage does not involve liability or compensation.

EXPERT OPINIONS

Environmental experts have struck a note of caution though they agreed it is a path breaking deal. They are of the opinion that it may not be the panacea to the problem of reducing use of fossil fuels but is the "starting gun" for the race towards a low-carbon future.

Delhi-based Centre for Science and Environment (CSE) has termed the climate deal as "weak and unambitious". It does not include any "meaningful" targets and has discharged developed nations from their historical responsibility, it said.

"Developing countries have got words and promise of money while the developed countries have finally got rid of their historical responsibility of causing climate change. They have no legally binding targets on finance or emissions cuts," CSE director general Sunita Narain said.

A section of the environment experts feel this is only the beginning. WWF-UK Chief Executive David Nussbaum said, "Paris is just the starting gun for the race towards a low-carbon future."

Greenpeace International echoed the sentiment. "Today the human race has joined in a common cause, but it's what happens after this conference that really matters. The Paris Agreement is only one step on long a road, and there are parts of it that frustrate and disappoint me, but it is progress. This deal alone won't dig us out the hole we're in, but it makes the sides less steep," said Greenpeace International executive director Kumi Naidoo.

Oxfam said too many governments in Paris protected their own self-interests instead of their own people at the talks. "While the richest countries have the biggest responsibility to tackle climate change, few countries stood out calling for a stronger deal," it said.

STANDING OF INDIA

The western media had projected India as a villain throwing a spanner in the works of the climate deal. Ironically, as the deal is clinched, India seems to be on the receiving end. The big win for the country is the inclusion of the phrase "common but differentiated responsibilities", recognizing the different national circumstances of developed and developing countries.

CONCLUSION

Glacial advances and retreats are natural cyclical phenomenon that occurs during the glacial and inter-glacial period. The last glacial advance has been recorded to be from 15th-19th century, which is popularly known as the Little Ice Age (LIA) (Ramanthan et al., 2005). Hence, the earth can be said to be in an inter-glacial phase now. However, the rate and extent of glacial retreat has increased considerably due to anthropogenic activities, leading to global warming and causing the climate to change. Since glaciers represent the largest freshwater reservoir on the planet, their retreat is of serious concern due to their direct impact on availability of drinking water to millions of people across the earth. Besides having an impact on the economy of a country like India, it also has negative impacts on the environment and livelihoods of people. Melting of glaciers would cause rivers to flood and the sea-level to rise, which could be dangerous for riparian communities, agriculture and hydropower industries and deltaic regions like the Sunder bans. Necessary policy measures and scientific research should be undertaken by the countries concerned. The collaboration of knowledge along with financial support from the developed nations could help in accelerating the rate of acquisition of required scientific information, so that necessary measures can be taken to address the impact of climate change on freshwater eco-systems.

REFERENCES

Arora, M., & Singh, R. D. (2007). Modeling of Melt Dominated Stream flow – A case study for a large Himalayan Basin. In Proceeding of Workshop: Glaciers in Watershed and Global Hydrology.

Cop21 Report. (2015). *The Sustainable Innovation Forum (SIF15) annual Conference of Parties (COP) Stade de France*. Retrieved on May 03,2016 from http://www.firstpost.com/world/climate-change-deal-clinched-at-cop21-heres-why-india-is-on-the-receiving-end-2544122.html

Collins, D. N., & Hasnain, S. I. (1995). Runoff and sediment transport from the glacier zed basins at the Himalayan scale. In W R Osterkamp (Eds.), *Scale on Interpretation and Management of Sediment and Water Quality,Proc. Boulder Symposium.*

IPCC. (2007). *Climate Change 2007: Impacts, Adaptations and Vulnerability*. Contribution of Working Group II to the Fourth Assessment Report of the Intergovernmental Panel on Climate Change. Retrieved on March 2, 2008 from http://www.ipcc.ch/ipccreports/ar4-wg2.htm

Kulkarini, A. V., Mathur, P., Rathore, B. P., Alex, S., Thakur, N., & Kumar, M. (2002). Effect of global warming on snow ablation pattern in the Himalaya. *Current Science*, *83*(2), 120–123.

Kumar, R. (2005). Status review of possible Impacts of Climate Change on Himalayan Glaciers, Glaciers retreat and its subsequent impacts on fresh water régime. In WWF: An Overview of Glaciers, Glacier Retreat, and Subsequent Impacts in India, Nepal and China, (pp. 29-38). WWF.

McCarthy, J. J. (2001). Climate Change 2001: Impacts, Adaptation and Vulnerability. Contribution of Working Group II to the Third Assessment Report of the Intergovernmental Panel on Climate Change.

Naithani, A. K., Nainwal, H. V., Sati, K. K., & Parsad, C. (2001). Geomorphological evidences of retreat of the Gangotri glacier and its characteristics. *Current Science*, *80*(1), 87–94.

Padma, T. V. (2007). *Monitoring Climate Change at the top of the world.* Science and Development Network. Retrieved on April 25, 2008 from cidev.net/en/features/ monitoring – climate-at- the-top-of-the-world.html.

Ramachandran, R. (2001). The receding Gangotri. *Frontline*, *18*(7).

Ramanathan, A. L., Shruti Jose, P. G., Sharma, P., & Linda, A. (2005). *Glacial Retreat of Himalayas in the Recent Past (Pleistocene to Recent Decade).* IGCP 514: field workshop.

Thayyem, R. J., Gergan, J. T., & Dobhah, D. P. (2007). Role of glaciers and snow cover on headwater river hydrology in monsoon regime – Micro-scale study of the Din Gad catchment, Garhwal Himalaya, India. *Current Science*, *80*(1), 87–94.

UNDP. (2008). *Glacial Lake Outburst Floods remains the biggest hazard of Bhutan UNDP Bhutan.* Retrieved April 8,2008, from http://www.undp.org.bt/info/pr08/ps_GLOF_8April.pdf

Vohra, C. P. (1978). Glacier Resources of the Himalaya and their Importance to Environment Studies. *Proc. Nat. Sem. on Resources Development and Environment in Himalaya Region*. DST.

Vohra, C. P. (1981). Himalayan Glaciers in the Himalaya: Aspects of Change. Oxford University Press.

WWF Report. (2008). *Going going gone: Climate Change and Global Glacier Decline*. Retrieved on March 11, 2008 from www.wwf.fi/wwf/www/uploads/pdf/glaciersraportti.pdf

KEY TERMS AND DEFINITIONS

Advance: An increase in the length of a glacier compared to a previous point in time. As ice in a glacier is always moving forward, a glacier's terminus advances when less ice is lost due to melting and/or calving than the amount of yearly advance.

Calving: The process by which pieces of ice break away from the terminus of a glacier that ends in a body of water or from the edge of a floating ice shelf that ends in the ocean. Once they enter the water, the pieces are called icebergs.

Distributaries: A tongue of glacier ice that flows away from the main trunk of the glacier. This may result from differential melting changing the gradient of part of a glacier.

Glacier: A large, perennial accumulation of ice, snow, rock, sediment and liquid water originating on land and moving down slope under the influence of its own weight and gravity; a dynamic river of ice. Glaciers are classified by their size, location, and thermal regime.

Mass Balance: A measure of the change in mass of a glacier at a certain point for a specific period of time. The balance between accumulation and ablation.

Retreat: The reduction in the length of a glacier compared to a previous point in time. As ice in a glacier is always moving forward, its terminus retreats when more ice is lost at the terminus to melting and/or calving than reaches the terminus. During retreat, ice in a glacier does not move back up the valley.

Surge: A short-lived, frequently large-scale, increase in the rate of movement of the ice within a glacier. Ice velocities may increase 10 to 100 times above normal flow rates. In some surges, the terminus of a glacier rapidly advances. Although not all glaciers surge, those that do often surge with some sort of a periodicity.

Trimline: A clear boundary line on the wall of a glacier valley that delineates the maximum recent thickness of a glacier. It may be a change in the color of the bedrock, indicating the separation of weathered from unweathered bedrock; the limit of a former lateral moraine or other sediment deposit; or the boundary between vegetated and bare bedrock.

Chapter 6

Mitigation of Climate Change Impacts through Treatment and Management of Low Quality Water for Irrigation in Pakistan

Ghulam Murtaza
University of Agriculture Faisalabad, Pakistan

Muhammad Zia-ur Rehman
University of Agriculture Faisalabad, Pakistan

Muhammad Saqib
University of Agriculture Faisalabad, Pakistan

Muhammad Naveed
University of Agriculture Faisalabad, Pakistan

Saifullah
University of Dammam, Saudi Arabia

Abdul Ghafoor
University of Agriculture Faisalabad, Pakistan

ABSTRACT

The Indus Plains of Pakistan are situated in arid to semi-arid climate where monsoon rains are erratic and mostly fall in the months of July and August. These rains are not only insufficient to grow even a single crop without artificial irrigation but also cause flood havoc very frequently that is associated with the climate change. The Indus river transports water for agriculture, industry and domestic usage within the basin and downstream. The Indus Basin is among the few basins severely affected by global warming and resulting climate change. The alteration in temporal and spatial patterns of rainfall has resulted in unexpected drought and floods. About 70 to 80% of total river flows occur in summer season due to snow melt and monsoonal rainfalls. Lack of storage reservoirs has decreased the ability to regulate flood water as well as its potential use during the drought season along with cheap hydro-electricity generation. The sedimentation in the system has limited the storage capacity of the existing three reservoirs by 28%. Consequently carry over capacity of these storage structures is only 30 days compared to 120 to 220 days in India and 900 days in Colorado Basin. Pakistan is facing shortage of good quality water due to competition among agricultural and non-agricultural sectors, this scenario will continue rather will further aggravate in future. According to the climate change scenario, the warming is reflected in the river-flow data of Pakistan, especially during the past 2-3 decades. To bridge the gap between fresh water availability and demand, ground water is being pumped to meet the irrigation requirements of crops. The

DOI: 10.4018/978-1-5225-1046-8.ch006

pumped ground water (70-80%) is brackish and could become a sustainability issue in the long run. The prolonged agricultural uses of such water will deteriorate soils, crops and human living environments. Water quality parameters usually considered include electrical conductivity (EC) for total soluble salts, and sodium adsorption ratio (SAR) and residual sodium carbonate (RSC) reflect the sodicity hazards. In order to limit or even to eliminate adverse effects of such waters, certain treatment and/or management options are considered as important pre-requisites. For bringing down high concentration of total soluble salts, dilution with good quality water is the doable practice. To decrease high SAR of irrigation water, a source of calcium is needed, dilution (with good quality water) will decrease SAR by the square root times of the dilution factor, while use of acids will be cost-intensive rather may adversely impact the soil health. For high RSC, dilution with low $CO_3^{2-}+HCO_3^-$ *water will serve the purpose, addition of Ca-salts will raise* $Ca^{2+}+Mg^{2+}$ *to bring a decrease in water RSC, while acids will neutralize* $CO_3^{2-}+HCO_3^-$ *to lower water RSC. Gypsum is the most economical and safe amendment while acids could also decrease RSC but at higher relative cost. City wastewater and seed priming in aerated gypsum solution is also presented. Such practices at small and/or large scale surely will help a lot to sustain the food security and the environment in the days to come where climate change has to be experienced round the world. Therefore, a well-coordinated program is necessary to create awareness among different sections of the society including the policy makers, general public, organizations, industrialists and farmers.*

INTRODUCTION

The Indus Plains of Pakistan are situated in arid to semi-arid climate zones where rains during monsoon are erratic and mostly fall during the months of July to September. These rains cause flood havoc very frequently that is associated with the climate change; floods during the years of 2010, 2011 and 2014 are very good examples to such climate change impacts those resulted in hundreds of human and animal deaths as well as caused billion dollar loss to infrastructures, soil quality and crops. The Indus river transports water for agriculture, industry and domestic usage within the basin and downstream. About 70 to 80% of total river flows occur in summer due to snow melt and monsoonal rains.

The effect of climate change is not limited to water availability only but it also affects crop yields and thus the food security and worsens the human living environments. The increasing soil salinity might cause additional harm in future, if less annual rainfall and higher temperatures prevailed at the current rate in future due to climate change. Due to reduction in annual rainfall, sufficient leaching of salts will not be achieved and higher temperatures will further aggravate the salt stress in regions already threatened by soil salinity (Sommer et al., 2013). The availability of water in Pakistan has decreased from 5300 m^3/year/person in 1950's to 1066 m^3/person/year in 2010 and it is estimated to be < 850 m^3 per capita by the year 2025 (WAPDA, 2011). The World Bank has included Pakistan in the list of 17 countries predicted to encounter severe water shortages by 2025. There is immense need and scope to develop additional surface water storage for drought periods since a plenty of river water is discharging into the Arabian Sea, ≈ about 34 to 37 million acre foot (MAF) water is discharged into the sea. Such policy issues have to be decided by overlooking the political interests and giving priority to national interests by all the sections of society including politicians, policy makers and techno-crates.

An additional option is ground water (unfortunately poor quality) that may supplement growing irrigation needs because of increased cropping intensity and competition for fresh water by non–agricultural sectors. At present, > 1.07×10^6 tube wells are pumping 9.05×10^6 ha–m ground water in Pakistan (Federal Bureau of Statistics, 2011-12), of which≈ 70-80% is unfit for agriculture because of high EC, SAR and/or RSC (Ghafoor et al., 2004). The irrigation with poor quality water may cause soil salinity/sodicity, poor infiltration, hard setting of soils, specific ion toxicity in plants; all combined to adversely affect the growth and economic yields of crops along with human living environments. Further to this, sub-soil drying due to draw-down of water table could be an important future concern (Wichelns and Qadir, 2015).

The ground water, storm water during floods and raw city sewage (from industry-mix) may pose serious hazards because of high EC (≥1.0 dS m^{-1}), SAR (≥10.0), RSC (≥2.5 $mmol_c$ L^{-1}), heavy metals, high Mg to Ca ratio, diseases, pathogens, detergents, azo-dyes and pesticide residues (Murtaza et al., 2010; Qadir et al., 2010). However, management or treatment depends upon many factors like crop type, plant growth stage, physical/chemical properties of soils, and reactions between water and soil solids, climatic factors, farmers' skills, drainage water quality, requirements of consumers and socio-economic conditions and acceptable decrease in produce yield as well as quality. Recently United Nations Climate Change Conference (COP21, 2015) was held in Paris ended up with the conclusions to decrease global greenhouse gases (GHGs) emission to zero level and to limit the average rise in global temperature to 2 degrees Celsius. To follow the goals and accomplish them saline water can be used for irrigation to bring more lands under cultivation. Bio saline agriculture can help in improving salt-affected soils such as cultivation of salt tolerant crops and growing of forest trees in barren lands instead of leaving them fallow. This would help in sequestering environmental CO_2 and alleviates global temperature rise. Deserts can also be made productive by installing solar energy panels. Recently, Morocco has floated a tender for 400 MW solar power project (COP21, 2015). In Pakistan a solar power project has also been installed in Bahawalpur with a capacity of 100 MW.

This chapter describes efforts related to brackish water treatment and/or management that must be critically examined to learn lessons and shape the future for safe and sustainable use of brackish water. This chapter also highlights the economic feasibility of using gypsum, acids and/or acid formers and like amendments for brackish water treatment with least disturbing the biosphere equilibrium.

Brackish Water

Brackish water is a general term for water having high EC, SAR and/or RSC. Canal water supply has become short to cope with the crop requirements because of prolonged and un-expected droughts mostly in response to climate change, silting of water reservoirs, competition between agricultural and non-agricultural demands and increased cropping intensity. As a result, pumping of ground water has increased tremendously over the years. Due to canal water shortage, number of tube wells is increasing at an alarming rate, i.e. 3000 in 1950 to > a million in 2014. Unfortunately, the pumped ground water is unsuitable for irrigation, since ≈ 70-75% tube wells were and are pumping brackish water (high EC, SAR and/or RSC) during the period of 1995 to 2014 (personal communication from Directorate of Soil Fertility Res. Inst., Lahore). Continued use of such waters for irrigation is one of the major reasons to induce soil salination/sodication along with hard setting, deterioration in produce quality/shelf life, and creating environment problems (Ghafoor et al., 2001a). In addition, sub-soil drying due to draw-down of water table is another future concern and must be addressed immediately and wisely. Indiscriminate pumping of ground water is also deteriorating the ground water quality, particularly under the decreased of ground water recharge due to lining of water courses, canals and increasing water use efficiency leading less leaching fraction; a minimum leaching fraction is the key to arrest the salt accumulation in the root zone under arid and semi-arid climates even with canal water irrigation. It is the rule of thumb that as one goes deeper into subsoil and away from rivers, ground water quality becomes poorer and poorer.

The shortage of quality irrigation water in Pakistan has attracted the international interest. Some private organizations are trying to market their products for the treatment of brackish water without proper testing to prove validity/usefulness under local agro-climatic conditions as explained in a later section.

Water Quality Criteria for Irrigation

Numerous water quality guidelines have been proposed by various scientists in Pakistan (Hussain, 1978; Sheikh, 1989; Yunus, 1977) and India (Gupta, 1990) but lack support of experimental data. The guidelines developed by the US Salinity Lab. Staff (1954) and Ayers and Westcot (1985) have been most commonly followed which more accurately predict the effect of water quality on soils and crops. Extensive experimentation is required to develop water quality guidelines keeping in view the physical and chemical properties of soils and agro-climatic conditions of Pakistan. Depending upon the degree of restriction, the three poor quality water classes (saline, saline-sodic and sodic) could be further grouped each into three homogenous subgroups (Table 1), since each subgroup needs specific management practice(s). This classification also helps to serve the purpose at regional levels.

GROUND WATER RESERVES

Previously, the quality of ground water was assessed on the basis of total soluble salts (TSS). The extent of areas possessing groundwater of different salinities within 350 feet is 40.73 MA (Ahmad &

Table 1. Grouping ground water for quality parameters

Water Quality	EC_{iw} (dS m^{-1})	SAR_{iw}	RSC ($mmol_c$ L^-)
E. Good	< 1. 5	< 1 0	< 2. 5
F. Saline			
Marginally saline	1.5-3.0	< 10	< 2.5
Saline	≥ 3.0	< 10	< 2.5
High-SAR saline	≥3.0	≥ 10	< 2.5
G. Saline-sodic			
Marginally saline-sodic	1.5-3.0	10-15	2.5-4.0
Moderately saline-sodic	1.5-4.0	10-15	≥ 4.0
Highly saline-sodic	≥ 4.0	≥ 15	≥ 4.0
H. Sodic			
Marginally sodic	0.5-3.0	10-15	2.5-4.0
Moderately sodic	1.5-3.0	≥ 15	≥ 4.0
Highly sodic	< 3.0	≥15	≥ 4.0

Source: Ghafoor et al. (2004)

Chaudhry, 1997). It may also be noted that groundwater of the reported quality does not exist within the full depth of 350 feet. Generally, quality of water in the upper depth is relatively better owing to recharge through seepage from canal water irrigation structures and useable except at certain sites. But, in general, groundwater quality deteriorates as we move deeper and away from the rivers and canals. However, highly hazardous properties of water, RSC and SAR have been neglected in the past surveys which even need scientifically more emphasis. Furthermore, EC: SAR, Ca^{2+}: Mg^{2+}, Na^{+}: K^{+} and Ca^{2+}: total cation ratios in water have seldom been considered.

TREATMENT OF BRACKISH WATER

Salinity

The EC_{iw} is an index of total soluble salts. The only treatment option to bring a decrease in total soluble salts is dilution with low salt water. Use of any chemical like gypsum, acid or acid formers will increase salt contents making the use of treated water even more hazardous. However, high EC_{iw} for initial reclamation of saline-sodic/sodic soils have proved better because of favorable effects of electrolytes on soil hydraulic properties (Ghafoor et al., 2004; Murtaza et al., 2009). Addition of organic matter to soil could help counter partially or extend the time to appear ill effects of high EC_{iw} on soils and crops (Ghafoor et al., 2002). Dilution required for high EC_{iw} can be calculated by the formula:

$EC_{BW} = [(EC_a)F_a)] + [EC_b)(F_b)]$ if $a + b = 1$, then $a = 1 - b$ or $b = 1 - a$ ..... (1), Where EC_{BW} = EC of blended water, EC_a = EC of water a, F_a = Fraction of water a, EC_b = EC of water b, F_b = Fraction of water b.

Dilution

The SAR will decrease by the square root times of the dilution factor (DF) or will increase square root times of the concentration factor (CF), i.e. Blended water SAR = Original SAR/$(DF)^{1/2}$ or Original $SAR_{iw} \times (CF)^{1/2}$

Desalination

Desalination is a process to remove excess salts and other minerals from water to make it suitable for irrigation. Sometimes table salt is obtained as a by-product. In order to sustain irrigated agriculture, construction of water desalting plants provides another option. The traditional process used in these operations is like to distillation. In this technique, systems require large amounts of energy to produce desalted water. This is possible only where low cost energy is available. Another technique (reverse osmosis) is used to remove all the types of soluble salts from brackish water. The brackish water is passed through semi-permeable membranes under pressure and concentrated brine is collected separately down the stream for safe disposal. Because of decreased volume, the concentrated brine can be used to extract some commercial-grade salts. In reverse osmosis, a pressure greater than the osmotic pressure of water is applied to saline water which causes desalted water to flow through membranes while retaining back salts. The higher the applied pressure above the osmotic pressure, the higher the rate of desalted water transports across membranes. This technology is well established and systems have been fabricated at economical costs. Such systems are in operation in USA, Saudi Arabia, China, USSR and Morocco at under government management (Ghafoor et al., 2004). This technology seems very difficult and expensive for the time being to practice in Pakistan due to energy crises and high prices.

SODICITY

Sodium Adsorption Ratio (SAR)

It is calculated using soluble ions in water or soil solution (US Salinity Lab. Staff, 1954) as:

$SAR = Na^{+}/ (Ca^{2+} + Mg^{2+}/2)^{1/2}$........(2), ions concentration in $mmol_c\ L^{-1}$ OR
$SAR = Na^{+}/ (Ca^{2+} + Mg^{2+})^{1/2}$........(2), ions concentration in $mmol\ L^{-1}$

Dilution

Like $EC_{iw,}$ mixing with low SAR or high $Ca^{2+} + Mg^{2+}$ water will decrease it by square root times of the dilution factor.

Addition of Ca-Source

Concentration of Na^{+} in irrigation water could not be decreased economically because of large volumes required for agriculture except dilution with low Na^{+} water. Concentration of Ca^{2+} can be increased through lining of water courses with gypsum stones as baffled structures/ geometry to enhance the flow velocity of water as well as to maximize the water-stone contact. This strategy is safe and economical although some problems of rodents, weeding of water courses or decreasing gypsum stone dissolution with time due to lime coating ($CaCO_3$) on gypsum stone surfaces if water has high RSC and thus may prove cost-intensive (Oster, 1994) because frequent acid flushing of stones will be essential to sustain the dissolution rate of gypsum stones in order to have effective treatment of water. The calculated amount of gypsum powder required to decrease the SAR hazards (or to counter the adverse water SAR effects) could also be soil-applied before planting the designed crop. The soil-application of gypsum considering volume of water can be calculated as:

Water has EC 3.25 dS m^{-1}, $Ca^{2+} + Mg^{2+}$ 6 $mmol_c\ L^{-1}$, Na^{+} 24 and SAR 13.87. How much 90% pure gypsum is required to lower SAR to 10 of a 10 cm-ha irrigation.

Na 24
SAR = --------------- = -------------- = 13.87
$(Ca + Mg/2)^{1/2}$ $(6/2)^{1/2}$
Additional $Ca^{2+} + Mg^{2+}$ required = 13.87 – 6.00 = 7.87 $mmol_c\ L^{-1}$
100% pure gypsum required for 10 cm-ha irrigation = $10^{6} \times 7.87\ mmol_c\ L^{-1} \times 86 \times 10^{-6}$
= 676.82 kg of 100% pure
752.20 kg required of 90% pure gypsum

Residual Sodium Carbonate (RSC)

The RSC calculations are based on the assumption of quantitative precipitation of CO_3^{2-}, HCO_3^- and Ca^{2+} ions in soils (Eaton, 1950) which could not be always true. Quantity and rate of salt precipitation is governed by the least amount of any one of these ions. Upon irrigation (dilution), the recently formed $CaCO_3$ precipitates are dissolved while reverse (re-precipitation) may be true upon concentration of soil solution through evapo-transpiration. The formula for RSC calculation is:

$RSC = (CO_3^{2-} + HCO_3^-) - (Ca^{2+} + Mg^{2+})$ -------- (3), where ions concentration in $mmol_c\ L^{-1}$.

Addition of Ca Source

This treatment could be achieved by the water course lining with gypsum stone as baffled structures/ geometry to enhance the flow velocity of water as well as to maximize the water-stone contact. The calculated amount of gypsum powder required to decrease the RSC hazards (or to counter the adverse water RSC effects) could also be soil-applied before planting the designed crop.

Neutralization of CO_3^{2-} and HCO_3^-

This could be accomplished with mineral acids (H_2SO_4, HCl, HNO_3) or acid former (S, CaS_5, FeS) but the option is and has to be highly site-specific because this treatment will simultaneously increase EC. General chemical reactions of carbonates with acids (say Na_2CO_3 and H_2SO_4) will be:

$2Na_2CO_3 + H_2SO_4 = 2NaHCO_3 + Na_2SO_4$ ------------ (4)
$2NaHCO_3 + H_2SO_4 = Na_2SO_4 + 2H_2O + 2CO_2$ ----------- (5)

Without changing its SAR, these reactions will tend to lower pH of water and the soil irrigated with such treated water for reasonably longer periods.

Dilution

Like EC_{iw}, mixing with low RSC or high $Ca^{2+} + Mg^{2+}$ water will decrease it in direct proportion of the dilution factor.

Among amendments for lowering RSC, gypsum is the most economical and safe, while acids could serve the purpose but at 5-8 times higher cost than gypsum (Ghafoor et al., 2004). The best economical solution for high SAR and/or RSC waters is soil-applied gypsum with or without farm manure or green manure before each crop considering the delta of water of crops and intensity of SAR and/or RSC problem. The blended/mixed use of canal and brackish waters is not possible in our conditions as the canal water supplies are not in the control of farmers.

High Concentration of Mg^{2+}

In some areas, irrigation water may contain high concentration of Mg^{2+} with Mg^{2+} to Ca^{2+} ratio > 1. The use of such water for irrigation and reclamation may increase exchangeable Mg^{2+} in soils (Karajeh et al., 2004). Exchangeable Mg^{2+} at excessively high levels may result in soil degradation because of its adverse effects on physical properties of soils. According to McNeal et al. (1968) soils having mixed Na^{+}–Mg^{2+} will have lower hydraulic conductivity than did soils with Na^{+}–Ca^{2+} under similar conditions. This could be due to the fact that t the hydration energy as well as hydration radius of Mg^{2+} are greater than that of Ca^{2+} (Strawn et al., 2015). Thus soil surface retains more water than where exchangeable Ca^{2+} is present, causing weakening of forces that stabilize soil structure. This, in turn, decreases the amount of energy needed to break down soil aggregates (Oster & Jayawardane, 1998). Further to this, at low levels of exchangeable sodium percentage (ESP), Mg^{2+} also exaggerates the effects of Na^{+} on clay dispersion and hydraulic conductivity (Qadir & Schubert, 2002). Oster (2001) reported that when Mg^{2+} proportion compared to Ca^{2+} increases than exchangeable Mg^{2+} resulted in greater dispersion thus causing a larger decline in hydraulic conductivity. Similarly, at a given ESP, replacing the exchangeable Ca^{2+} with Mg^{2+} decreased the infiltration rate of soils (Rengasamy et al., 1986). A recent review (dating back to the 1930s) by Smith et al. (2015) supports the general conclusion that the relative order of deleterious effects on soil hydraulic properties of the four common cations in soils was in the order of $Na^{+} > K^{+} > Mg^{2+} > Ca^{2+}$. It was concluded that Mg^{2+} was significantly less effective than Ca^{2+} in improving soil structure and water permeability.

The achieved results from the above stated brackish water quality improvements, several other untested technologies by some vested interest people are claiming. These technologies include Sulphurous Acid Generator (SAG) by Sweet Water Solution of USA, EM/BM technology from Japan, RISTECH Technology from UK, and Electro-Magnetic Membrane technology from Germany. Novel claims of these technologies are to desalinate the brackish water, convert sodium into nitrogen, change in soil texture over a period of 2-3 years, and bacteria in these recipes eat salts as well as sodium present in water or soils. Unfortunately, none of these have been demonstrated at any university or research farm. In spite of this fact, they are be-fooling the farmers at a very large scale to make money. Government should give attention immediately to stop their business and save hard earnings of farmers since farmers are drowning and try to catch a straw to save themselves.

EC_{iw} to SAR_{iw} ratio

Low EC_{iw} or EC_{e} at a given SAR_{iw} or SAR_{ss} tends to decrease soil infiltration through decreasing the thickness of Diffused Double Layer (DDL) (Ayers & Westcott, 1985; Ghafoor et al., 2004). Water having low SAR_{iw} and high EC_{iw} was found better for normal and salt-affected soils because of favorable effects of electrolytes on soil infiltration and hydraulic conductivity, while reverse was true for high SAR_{iw} with low EC_{iw}. Therefore at start of reclamation of saline-sodic or sodic soils, waters having high EC: SAR ratios were found equally useful (Ghafoor et al., 2001b) but has to switch to better quality water (low EC_{iw}, low SAR_{iw}) with the advancement of soil reclamation (Murtaza et al., 2009). It is important to consider this quality parameter for sustainable management of brackish water in order to maintain a leaching fraction as well as if used for reclaiming salt-affected soils. Some studies were conducted by authors in the Fourth Drainage Project Area (FDPA), Faisalabad using brackish water, wherein it was concluded that such waters can be used safely at the early stages of reclamation (Ghafoor et al., 2008, 2011, 2012; Murtaza et al., 2009). However, presently this parameter is not given very much importance and must be considered for future irrigation water quality guidelines.

PERMEABILITY PROBLEMS

Permeability refers to the percolation of infiltrated water through soils. Doneen (1975) suggested a formula to predict the infiltration rate which was designated as Permeability Index (PI).

$PI = [100 \{Na^{+} + (HCO_3^{-})^{1/2}\}/\{Na^{+} + Ca^{2+} + Mg^{2+}\}]$ -------- (6), where ionic concentrations in water are in $mmol_c L^{-1}$.

This relationship is quite useful in predicting the effect of water on permeability of soil and also this does not need extra determination in routine water analysis.

INFILTRATION PROBLEM

The rate at which water enters into soils is referred to as the rate of infiltration. An infiltration rate (IR) as low as 3 mm h^{-1} is considered low while a rate > 12 mm h^{-1} is high (US Salinity Lab. Staff, 1954). This parameter is affected by many factors other than water quality, like soil physical characteristics (soil texture, type of clay minerals) and chemical characteristics (exchangeable cations). The IR generally increases with increasing salinity (EC) and decreases with either decreasing EC or increasing Na^{+} concentration relative to Ca^{2+} and Mg^{2+}, i.e. SAR (Murtaza et al., 2006). Therefore, EC and SAR must be considered for a proper evaluation of the ultimate effect of water on its infiltration rate.

SPECIFIC ION EFFECTS

It is the concentration of an ion relative to total concentration which when exceeds beyond a certain limit cause specific ion effect. If Na^{+} concentration is > 60% of the total cations, water becomes hazardous for irrigation because it induces K^{+} and/or Ca^{2+} deficiency and scorching of leaf tips in addition to deteriorating physical and chemical properties of soils. It is worth mentioning that deterioration in physical properties of soils starts much earlier than specific ion effects of Na^{+} on plant growth. The concentrations of Cl^{-} and SO_4^{2-} strongly correlate with EC of irrigation water (EC_{iw}). About 5-10 $mmol_c L^{-1}$ Cl^{-} becomes harmful to sensitive plants. However, relatively wide range of sensitivity to Na^{+} and Cl^{-} effects among different plants exists which is more pronounced under sprinkler irrigation. Woody plants are more sensitive to Cl^{-} than annual crops. If Cl^{-}: SO_4^{2-} ratio becomes greater than 1: 3, the water is considered more hazardous than that where this ratio is 3: 1. The SO_4^{2-} ions are considered more harmful for roots and disturb internal metabolism of plants. These ions induce precipitation of Ca^{2+} as $CaSO_4$ which causes a rise in soil pH and SAR, and thus make the soil environment unfavorable for nutrient availability to plants. Generally waters having high EC under arid to semi-arid conditions may contain $Mg^{2+} > Ca^{2+}$. Magnesium >50% among $Ca^{2+} + Mg^{2+}$ adversely affects soils and plants. Increasing Mg^{2+} over Ca^{2+} in irrigation water increased the sodicity in soils at a given SAR_{iw} and EC_{iw} and its effects were more pronounced at higher than that at lower SAR_{iw}. Release of Ca^{2+} from the in-situ soil weathering and native lime increased with increasing Mg^{2+}: Ca^{2+} ratio in irrigation water. With an increase in Mg^{2+}: Ca^{2+} ratio and SAR of soil and/or irrigation water, the degree of dispersion increased significantly. The Ca^{2+}: Mg^{2+} ratio ≥ 1.0 is considered safe for most of the crops and soils. Generally, grasses are more sensitive to Mg^{2+} than other plants since former crops acquire high Mg^{2+} in living tissues. From previous studies, it was also noted that high Mg^{2+} water is relatively more harmful to affect rice yield as compared to wheat and cotton crops (Ghafoor et al., 1997). This aspect needs further research since a lot of ground waters contain Mg^{2+} higher than Ca^{2+} under arid regions of the world. However, the productivity of Mg^{2+} dominated soils can be improved by increasing levels of Ca^{2+} on the cation exchange complex to counter the adverse impacts of Mg^{2+}. This could be accomplished by applying sufficient amount of Ca^{2+} to soils (Vyshpolsky et al., 2008).

SEED PRIMING

Seed priming is a process where seed is hydrated for specified time periods under controlled conditions followed by re-drying that allows all the pre-germination metabolic activities but prevents radicle protrusion (Khan et al., 1992). Poor stand establishment is a widespread constraint for crop production in salt-affected soils (Harris et al., 1999; Saifullah et al., 2002). Presence of excessive soluble salts and/or exchangeable Na^+ may adversely impact the physical, chemical and nutritional properties of soil that in turn may lead to lower plant population (Qadir et al., 2014). Moreover, such plants after emergence often grow slowly and are highly susceptible to stresses such as drought, salinity, pests and diseases (Francois et al., 1986). Under such conditions farmers may choose to re-sow, although this entails severe yield penalties, increased labor and financial losses. Pre-sowing seed treatments with salts, nutrients or hormones have been advocated to enhance seed germination in saline medium (Afzal et al., 2013; Ashraf & Foolad, 2005). However, it fails in saline-sodic or sodic soils due to poor physical condition. In these soils, seeds fail to emerge out of soil due to surface crust. Gypsum helps to improve soil physical and chemical properties. In this backdrop, a three years project was initiated on salt-affected soil to monitor the effect of combined application of seed priming and soil-applied gypsum on growth and yield of wheat and rice crops using brackish water for irrigation. The results indicated that combined application of gypsum and seed treatment with saturated solution of gypsum was the best treatment in terms of soil health improvement, yield enhancement and economic output. Priming enhanced the seed germination through protein synthesis, repair of nucleic acid and membranes (Hameed et al., 2010; Fujikura & Karssen, 1995). In a recent study (Saifullah, 2012), it was concluded that sowing of wheat seed soaked for 24 hours in aerated gypsum saturated solution + gypsum at 50% of soil gypsum requirement helped better crop stand and ultimately the economic yields.

WASTEWATER

In developing countries including Pakistan, it is a common practice to discharge domestic and industrial effluent (premixed due to common water carrying channels) directly to a sewer system, a natural drain or water body, a nearby field or an internal septic tank. Unfortunately, this wastewater is not properly treated because of non-functioning and/or availability treatment plants. The sewage in drains is mostly have high EC, SAR, RSC and elevated levels of several metals, organics and associated pathogens that warrant site-specific management strategies. The wastewater used for irrigation is becoming an attractive commodity for farmers mainly because of its nutrient contents and reliability of supply (Ensink et al., 2004; Murtaza et al., 2010). It can have positive impacts on agriculture and monthly income despite producing adverse effects on soil physical and chemical properties in addition to contamination of human food chain and related health risks (Hanjra et al., 2012; Qadir et al., 2010; Vincent, 2014). Recent estimates depict (Table 2) that total volume of wastewater generated in Pakistan is about 962,335 million gallons (4.369×10^9 m^3/yr) out of which 674,009 million gallons (3.060×10^9 m^3/yr; and 5.54×10^9 m^3/yr for the year 2011) are shared by municipalities and 288,326 million gallons (1.309×10^9 m^3/yr) from industries. As with other developing countries, in Pakistan, wastewater is not subjected to any treatment and none of the cities have any biological treatment process except Islamabad and Karachi, and even these cities treat only a small proportion before disposal (Table 3). Estimates suggest that the amount of treated wastewater is not > 1% (Murtaza & Zia, 2012). The treated wastewater generally flows into open drains, and there are no provisions for its reuse for agriculture.

It is highly recommended that the wastewater should be treated at source. Different chemical and plant based technologies are available for the remediation of wastewater (Murtaza et al., 2014). One opportunity to address these concerns is the use of low cost agrowaste adsorbents for the treatment of urban wastewater for irrigation of crops, trees, forests and greenbelts (Abdolali et al., 2013). There seems week national policy and implementation on sustainable use of wastewater in developing countries including Pakistan. The problems of wastewater disposal emerge from distortions due to economy-wide policies, failure of targeted environmental policies and institutional failures. Therefore, a well-coordinated program is necessary to create awareness among different sections of the society including the policy makers, general public, organizations, industrialists and farmers to save future of nations.

CONCLUSION AND RECOMMENDATIONS

It is important to consider and adopt preventive measures immediately to avoid the disaster of soil salination and sodication in response to poor quality tube well water irrigation. The deterioration of soils adversely affects the ambient environments for all the living organisms including human beings. Thus, it seems essential and wise to reclaim the saline and/or sodic soils and adopt measures to limit the adverse effects of hazards associated with low quality irrigation waters in favor of crops/plants growth in order to keep the environment clean.

Table 2. Sector-wise wastewater production in Pakistan in the year

Sr. No.	Source	Volume	
		10^6 m^3 y^{-1}	Percent (%)
1	Industry	395	6
2	Commercial	266	5
3	Urban residential	1,628	25
4	Rural residential	3,059	48
5	Agriculture	1,036	16
	Total	6,414	100

Source: Murtaza & Zia (2012)

Table 3. Wastewater produced annually from major cities of Pakistan

City	Urban Population (1998 Census)	Total Wastewater Produced (10^6 m^3/y)	% of Total	% of Treated	Receiving water Body
Lahore	5,143,495	287	12.5	0.01	River Ravi, irrigation canals,
					vegetable farms
Faisalabad	2,008,861	129	5.6	25.6	River Ravi, River Chenab and
					vegetable farms
Gujranwala	1,132,509	71	3.1	-	SCARP drains, vegetable farms
Rawalpindi	1,409768	40	1.8	-	River Soan and vegetable
					farms
Sheikhupura	870,110	15	0.7	-	SCARP drains
Multan	1,197,384	66	2.9	-	River Chenab, irrigation canals
					and farms
Sialkot	713,552	19	0.8	-	River Ravi, irrigation canals
					and farms
Karachi	9,339,023	604	26.3	15.9	Arabian Sea
Hyderabad	1,166,894	51	2.2	34.0	River Indus, irrigation canals and SCARP drains
Peshawar	982,816	52	2.3	36.2	Kabul River
Other	19,475,588	967	41.8	0.7	-

Source: Master Plan for Urban Wastewater (Municipal and Industrial) Treatment Facilities in Pakistan. Final Report, Lahore: Engineering, Planning and Management Consultants, 2002.

The best economical solution to high SAR and/or RSC waters is to soil-apply gypsum along with farm manure or green manure before each crop, considering the delta of water of a crop and gravity of SAR and/or RSC problem. Blended/mixed use of canal and brackish waters is not possible under our conditions since canal water supplies are not at the disposal of farmers. The cyclic (alternate irrigation, or one crop with canal and next with brackish water) are already in practice (Oster & Grattan, 2002; Rhoades et al., 1992). The farmers should try to avoid irrigation with brackish water at critical crop growth stages. They should also try to use good water on good soils and poor quality water on poor soils.

There is an immense need to initiate the tube well water analyses for appropriate and site-specific advice by soil and water experts, and simultaneously it must be made mandatory and legal for tube well owners to get the water analyzed and act upon the advice. Assuming 70% pumped ground water hazardous which require addition of gypsum at 4 $mmol_c$ L^{-1} of irrigation water, 400 kg for each acre-foot and 15.6 million tons gypsum for 39 MAF tube well water annually will be required, i.e. 234 million tons quality gypsum will be essential to soil-apply in order to mitigate the adverse effects of high SAR/RSC tube well water. For a door-step supply of this much gypsum has to be ensured on credit for sustainability of irrigated agriculture and safe environment since the salt-stress land and tube well owners might not be in a position to purchase gypsum on cash payment.

The potential of virtual use of water could be exploited at regional/provincial/national levels. The idea of virtual use of water compares the amount of water embodied in a crop that can be purchased at regional/provincial/international levels with the amount of water that should be required to produce that crop natively. For example, transferring every kilogram of wheat into water stress area means to transfer about 1000 L of virtual water at a much less price than the price of the same quantity of water from the native water resources in the area itself (Qadir et al., 2003).

There is significant loss in storage capacity of water reservoirs which is increasing gradually and this loss will become about 33% of the designed capacity by the year 2020. It is strongly recommended that additional storage facility is developed, and the silt load of in-coming water into existing dams is decreased through controlling/ promoting vegetation with or without the help of engineering structures in the catchment areas. Seed priming in saturated gypsum solution for 4-8 hours did help early and enhanced germination of cereal crops which latter improved crop stand for better yield since as plants grow older, their performance in salt-stress environment (saline and/or sodic soils or irrigation waters) become promising.

The city wastewater poses threats of heavy metals. Under such situations, metals could be made less available with the soil-applied gypsum, mono-ammonium phosphate, DAP or some other economical Ca-salts. The metals could also be removed by enhancing their availability through soil-application of acids, acid formers and EDTA salts with crops not directly consumed by human beings. However, growing leafy vegetables with sewage water must be banned.

Another feasible practice for brackish water irrigation is to cultivate salt-tolerant crops requiring less irrigation, particularly to decrease the input of chemicals into soils and ground water contamination. This practice will also help decrease the problem of waterlogging. The breeders should come forward and conduct research in the fields of genomics, proteomics, biotechnology, and nanotechnology to breed salt-resistant and drought-tolerant crops (Roy et al., 2014; Seabra et al., 2014).

The allocation of irrigation water is not uniform in various provinces of Pakistan. The water allocation is more in Sindh compared to Punjab. Moreover, water allocation is higher in Southern Punjab as compared to central and upper Punjab. Improvement in water allocation is required to achieve sustainable irrigation and to avoid salinity and waterlogging.

POLICIES

To tackle the water quality problems some policies are suggested 1) give priority to increasing irrigation efficiencies, essentially with allowance for salinity/ sodicity control through effective leaching of in-coming salts, 2) move towards a demand based irrigation water supply, 3) promote water use associations (WUA) and thereafter Federation of WUA, 4) maintain the existing drainage infrastructure with site-specific emphasis on new projects, 5) strengthen the advisory service to farmers and training to extension staff.

MEASURES

Some specific measures for the safe use of brackish water on sustainable basis are 1) take incremental but specific measures towards volumetric consumption as the ultimate basis for water charges, 2) undertake studies in community management of a few distributaries on a pilot scale, 3) address problems of water table draw-down by holistic approaches, encompassing both the demand management and methods to increase water recharge, 4) promote farmer education and extension activities as well as in-service training of staff, 5) review funding, goal should be 60:40 ratio for research: administration, 6) in kind credit facility for gypsum in time and space, particularly to small and resource poor farmers, 7) make it legal for farmers to get their tube well water analyzed, mobile teams of experts be organized to collect, analyze and advice solutions for tube well water quality.

REFERENCES

Abdolali, A., Ngo, H. H., Guo, W. S., Lee, D. J., Tung, K. L., & Wang, X. C. (2014). Development and evaluation of a new multi-metal binding biosorbent. *Bioresource Technology*, *160*, 98–106. doi:10.1016/j.biortech.2013.12.038 PMID:24405652

Afzal, I., Basra, S. M. A., Cheema, M. A., Farooq, M., Jafar, M. Z., Shahid, M., & Yasmeen, A. (2013). Seed priming: A shotgun approach for alleviation of salt stress in wheat. *International Journal of Agriculture & Biology*, *15*, 1199–1203.

Ahmad, N. C., & Chaudhry, M. R. (1997). *Review of research on reclamation of salt-affected soils in Pakistan (Publication No. 175)*. Lahore, Pakistan: International Waterlogging And Salinity Research Institute, Water And Power Development Authority.

Ashraf, M., & Foolad, M. R. (2005). Pre-sowing seed treatment-a shotgun approach to improve germination growth and crop yield under saline and none-saline conditions. *Advances in Agronomy*, *88*, 223–271. doi:10.1016/S0065-2113(05)88006-X

Ayers, R. S., & Westcott, D. W. (1985). *Water quality for agriculture. Irrigation and Drainage Paper 29*. Rome, Italy: Food and Agriculture Organization of the United Nations.

COP21 (Conference of Parties 21). (2015). *United Nations Climate Change Conference*. Retrieved from http://www.scoop.it/t/cop21news

Doneen, L. D. (1975). Water quality for irrigated agriculture. In A. Poljakoff-Mayber & J. Gale (Eds.), *Plants in saline environments* (pp. 56–79). Springer-Verlag. doi:10.1007/978-3-642-80929-3_5

Eaton, F. M. (1950). Significance of carbonate in irrigation waters. *Soil Science*, *69*(2), 123–133. doi:10.1097/00010694-195002000-00004

Ensink, J. H. J., Simmons, R. W., & van der Hoek, W. (2004). Wastewater use in Pakistan: The cases of Haroonabad and Faisalabad. In C. A. Scott, N. I. Faruqui, & L. Raschid-sally (Eds.), *Wastewater use in irrigated agriculture* (pp. 91–99). Wallingford, UK: CAB International. doi:10.1079/9780851998237.0091

Federal Bureau of Statistics. (2011–12). *Pakistan statistics year book, Statistics Division*. Islamabad, Pakistan: Planning Commission Government of Pakistan.

Francois, L. E., Maas, E. V., Donovan, T. J., & Youngs, V. L. (1986). Effect of salinity on grain yield and quality, vegetative growth, and germination of semi-dwarf and durum wheat. *Agronomy Journal*, *78*(6), 1053–1058. doi:10.2134/agronj1986.00021962007800060023x

Fujikura, Y., & Karssen, C. M. (1995). Molecular studies on osmoprimed seeds of cauliflower: A partial amino acid sequence of a vigour- related protein and osmopriming-enhanced expression of putative aspartic protease. *Seed Science Research*, *5*(03), 177–181. doi:10.1017/S0960258500002804

Ghafoor, A., Gill, M. A., Hassan, A., Murtaza, G., & Qadir, M. (2001a). Gypsum: An economical amendment for amelioration of saline-sodic waters and soils, and for improving crop yields. *International Journal of Agriculture & Biology*, *3*(3), 266–275.

Ghafoor, A., Murtaza, G., Ahmad, B., & Boers, Th. M. (2008). Evaluation of amelioration treatments and economic aspects of using saline-sodic water for rice and wheat production on salt-affected soils under arid land conditions. *Irrigation and Drainage*, *57*(4), 424–434. doi:10.1002/ird.377

Ghafoor, A., Murtaza, G., Maann, A. A., Qadir, M., & Ahmad, B. (2011). Treatments and economic aspects of growing rice and wheat crops during reclamation of tile drained saline-sodic soils using brackish waters. *Irrigation and Drainage*, *60*(3), 418–426. doi:10.1002/ird.577

Ghafoor, A., Murtaza, G., Rehman, M. Z., Saifullah, , & Sabir, M. (2012). Reclamation and salt leaching efficiency of treatments for tile drained saline-sodic soil using marginal quality water for irrigating rice and wheat crops. *Land Degradation & Development*, *23*(1), 1–9. doi:10.1002/ldr.1033

Ghafoor, A., Nadeem, S. M., Hassan, A., & Sadiq, M. (2001b). Reclamation response of two different textured saline-sodic soils to EC_{iw} to SAR_{iw} ratios. *Pakistan Journal of Soil Science*, *19*, 92–99.

Ghafoor, A., Qadir, M., & Murtaza, G. (1997). Potential for reusing low quality drainage water for soil amelioration and crop production. In *Proceedings of the International Symposium on Water for the 21st Century: Demand, Supply, Development and Socio-Environmental Issues* (pp. 411-20). Lahore, Pakistan: Centre of Excellence in Water Resources Engineering, University of Engineering and Technology.

Ghafoor, A., Qadir, M., & Murtaza, G. (2002). Agriculture in the Indus Plains: Sustainability of land and water resources. *International Journal of Agriculture & Biology*, *4*(3), 429–437.

Ghafoor, A., Qadir, M., & Murtaza, G. (2004). *Salt-affected soils: Principles of management*. Lahore, Pakistan: Allied Book Centre.

Gupta, I. C. (1990). *Use of saline water in agriculture. Bombay, India*. Oxford, UK: IBH Publishing House Co, Pvt. Ltd.

Hameed, A., Afzal, I., & Iqbal, N. (2010). Seed priming and salinity induced variations in wheat (*Triticum aestivum* L.) leaf protein profile. *Seed Science & Technology*, *38*(1), 236–241. doi:10.15258/sst.2010.38.1.25

Hanjra, M. A., Blackwell, J., Carr, G., Zhang, F., & Jackson, T. M. (2012). Wastewater irrigation and environmental health: Implications for water governance and public policy. *International Journal of Hygiene and Environmental Health*, *215*(3), 255–269. doi:10.1016/j.ijheh.2011.10.003 PMID:22093903

Harris, D., Joshi, A., Khan, P. A., Gothkar, P., & Sodhi, P. S. (1999). On-farm seed priming in semi-arid agriculture: Development and evaluation in maize, rice and chickpea in India using participatory methods. *Experimental Agriculture*, *35*(1), 15–29. doi:10.1017/S0014479799001027

Hussain, G. (1978). *Determination of irrigation water quality standards*. (Unpublished doctoral dissertation). Colorado State University.

Karajeh, F., Suleimenov, M., Karimov, A., Vyshpolsky, F., Mukhamedjanov, Kh., Bekbaev, U. (2004). Technology of irrigation, water saving and improving soil fertility in Arys Turkestan canal command zone. *Kazakh Research Institute of Water Management: Taraz, 18.* (in Russian).

Khan, A. A., Maguire, J. D., Abawi, G. S., & Illas, S. (1992). Matriconditioning of vegetable seed to improve stand establishment in early field planting. *Journal of the American Society for Horticultural Science, 117*, 41–47.

McNeal, B. L., Layfield, D. A., Norvell, W. A., & Rhoades, J. D. (1968). Factors influencing hydraulic conductivity of soils in the presence of mixed-salt solutions. *Soil Science Society of America Proceedings, 32*(2), 187–190. doi:10.2136/sssaj1968.03615995003200020012x

Murtaza, G., Ghafoor, A., Owens, G., Qadir, M., & Kahlon, U. Z. (2009). Environmental and economic benefits of saline-sodic soil reclamation using low-quality water and soil amendments in conjunction with a rice-wheat cropping system. *Journal Agronomy & Crop Science, 195*(2), 124–136. doi:10.1111/j.1439-037X.2008.00350.x

Murtaza, G., Ghafoor, A., & Qadir, M. (2006). Irrigation and soil management strategies for using saline-sodic water in a cotton-wheat rotation. *Agricultural Water Management, 81*(1-2), 98–114. doi:10.1016/j.agwat.2005.03.003

Murtaza, G., Ghafoor, A., Qadir, M., Owens, G., Aziz, M. A., Zia, M. H., & Saifullah, . (2010). Disposal and use of sewage on agricultural lands in Pakistan: A review. *Pedosphere, 20*(1), 23–34. doi:10.1016/S1002-0160(09)60279-4

Murtaza, G., Murtaza, B., Niazi, N. K., & Sabir, M. (2014). Soil contaminants: sources, effects and approaches for remediation. In *Improvement of crops in the era of climatic changes* (pp. 171–196). Springer Science, Business Media. doi:10.1007/978-1-4614-8824-8_7

Murtaza, G., & Zia, M. H. (2012). *FAO Report - UNW-AIS. Individual's capacity development on the safe use of wastewater in agriculture in Pakistan.* Available at http://www.ais.unwater.org/ais/pluginfile.php/232/mod_page/content/124/pakistan_murtaza_finalc ountryreport2012.pdf

Oster, J. D. (1994). Irrigation with poor quality water. *Agricultural Water Management, 25*(3), 271–297. doi:10.1016/0378-3774(94)90064-7

Oster, J. D. (2001). *Magnesium effects on soil physical properties—Hydraulic conductivity and infiltration rate*. SLPHYCHM. Available online at http://esce.ucr.edu/oster/SLPHYCb.html

Oster, J. D., & Grattan, S. R. (2002). Drainage water reuse. *Irrigation and Drainage Systems, 16*(4), 297–310. doi:10.1023/A:1024859729505

Oster, J. D., & Jayawardane, N. S. (1998). Agricultural management of sodic soils. In M. E. Sumner & R. Naidu (Eds.), *Sodic soil: Distribution, management and environmental consequences* (pp. 126–147). New York: Oxford University Press.

Qadir, M., Ghafoor, A., Boers, Th. M., & Murtaza, G. (2003). Agricultural water management in water starved countries: Challenges and opportunities. *Agricultural Water Management, 62*(3), 165–185. doi:10.1016/S0378-3774(03)00146-X

Qadir, M., Quillérou, E., Nangia, V., Murtaza, G., Singh, M., Thomas, R. J., & Noble, A. D. et al. (2014). Economics of salt-induced land degradation and restoration. *Natural Resources Forum*, *38*(4), 282–295. doi:10.1111/1477-8947.12054

Qadir, M., & Schubert, S. (2002). Degradation processes and nutrient constraints in sodic soils. *Land Degradation & Development*, *13*(4), 275–294. doi:10.1002/ldr.504

Qadir, M., Wichelns, D., Raschid-Sally, L., McCornick, P. G., Drechsel, P., Bahri, A., & Minhas, P. S. (2010). The challenges of wastewater irrigation in developing countries. *Agricultural Water Management*, *97*(4), 561–568. doi:10.1016/j.agwat.2008.11.004

Rengasamy, P., Greene, R. S. B., & Ford, G. W. (1986). Influence of magnesium on aggregate stability in sodic red-brown earths. *Australian Journal of Soil Research*, *24*(2), 229–237. doi:10.1071/SR9860229

Rhoades, J. S., Kandiah, A., & Mashali, A. M. (1992). *The use of saline waters for crop production. Irrigation and Drainage Paper 48*. Rome, Italy: Food and Agriculture Organization of the United Nations.

Roy, S. J., Negrão, S., & Tester, M. (2014). Salt resistant crop plants. *Current Opinion in Biotechnology*, *26*, 115–124. doi:10.1016/j.copbio.2013.12.004 PMID:24679267

Saifullah, G. A., Murtaza, G., & Qadir, M. (2002). Brackish tube well water promotes growth of rice and wheat and reclamation of saline-sodic soils. *Pakistan Journal of Soil Science*, *21*, 83–88.

Saifullah. (2012). *Enhancing crop productivity on salt-affected soils through combined use of soil applied gypsum and pre-sowing seed treatments*. Final Technical Progress Report for Endowment Fund Secretariat (EFS) funded project (July 2008 – June 2011). Institute of Soil & Environmental Sciences, University of Agriculture.

Seabra, A. B., Rai, M., & Durán, N. (2014). Nano carriers for nitric oxide delivery and its potential applications in plant physiological process: A mini review. *Journal of Plant Biochemistry and Biotechnology*, *23*(1), 1–10. doi:10.1007/s13562-013-0204-z

Sheikh, I. A. (1989). Country report on problems of waterlogging and salinity in Pakistan. In *Proceeding Information Seminar on Waterlogging and Salinity Research in Some Major Problem Countries* (pp. 1-10).Lahore, Pakistan: International Waterlogging And Salinity Research Institute.

Smith, C. J., Oster, J. D., & Sposito, G. (2015). Potassium and magnesium in irrigation water quality assessment. *Agricultural Water Management*, *157*, 59–64. doi:10.1016/j.agwat.2014.09.003

Sommer, R., Glazirina, M., Yuldashev, T., Otarov, A., Ibraeva, M., Martynova, L., & de Pauw, E. et al. (2013). Impact of climate change on wheat pro-ductivity in Central Asia. *Agriculture, Ecosystems & Environment*, *178*, 78–99. doi:10.1016/j.agee.2013.06.011

Strawn, D. G., Bohn, H. L., & O'Connor, G. A. (2015). *Soil chemistry*. New York, NY: Wiley-Blackwell.

US Salinity Lab. Staff. (1954). Diagnosis and improvement of saline and alkali soils (Agriculture Handbook 60). Washington, DC: United States Department of Agriculture.

Vincent, S. (2014). Environmental health monitoring: A pragmatic approach. *International Journal of Waste Resources*, *4*(4), 164. doi:10.4172/2252-5211.1000164

Vyshpolsky, F., Qadir, M., Karimov, A., Mukhamedjanov, K., Bekbaev, U., Paroda, R., & Karajeh, F. et al. (2008). Enhancing the productivity of high-magnesium soil and water resources in central Asia through the application of phosphogypsum. *Land Degradation & Development*, *19*(1), 45–56. doi:10.1002/ldr.814

WAPDA (Water And Power Development Authority). (2011). *Hydro potential in Pakistan*. Water And Power Development Authority, Pakistan. Retrieved from http://www.wapda.gov.pk/pdf/BroHydpwrPotialApril2011.pdf

Wichelns, D., & Qadir, M. (2015). Achieving sustainable irrigation requires effective management of salts, soil salinity, and shallow groundwater. *Agricultural Water Management*, *157*, 31–38. doi:10.1016/j.agwat.2014.08.016

Yunus, M. (1977). Water quality in The Indus Plains. In *Proceedings of the Seminar on Water Management for Agriculture* (pp. 283-292). Lahore, Pakistan: Exxon Chemicals Pakistan Ltd.

Chapter 7

Characterization and Management Concerns of Water Resources around Pallikaranai Marsh, South Chennai

Avantika Bhaskar
Care Earth Trust, India

G. Babu Rao
Care Earth Trust, India

Jayshree Vencatesan
Care Earth Trust, India

ABSTRACT

Pallikaranai is one of the last remaining natural wetlands of Chennai. This marsh collects floodwater and increases groundwater levels in the region. The present study characterizes the water sources available around Pallikaranai Marsh. Groundwater was found to be the main source of water in the study area, extracted through domestic wells as well as commercially through a large number of agricultural wells. Direct surface water extraction from wetlands by private tankers was also observed in some areas. Acute water shortage and inefficient water supply by the government has led to thriving of tanker market in this area. Shrinking of the marsh and surrounding water bodies owing to construction, dumping of waste and encroachment accompanied by over-extraction of groundwater is driving this area towards extreme water crisis especially in event of climate change. Conservation of wetlands and evolving norms for sustainable water extraction of groundwater especially by commercial entities is recommended.

BACKGROUND

Hydrological cycle forms an essential part of climate system (Askew, 1987) and change in climate is associated with changes in a number of components of hydrological cycle like altered precipitation pattern, intensity, etc. (Trenberth, 2011). Wetlands perform essential hydrological functions and are known

DOI: 10.4018/978-1-5225-1046-8.ch007

to act as a sponge- 'soaking up water during wet periods and releasing it during dry periods'- signifying their importance as buffers against extreme climate change events that affect water availability (Bullock & Acreman, 2003; Maltby & Acreman, 2011). Some of the most important ecosystem services provided by wetlands are water supply and regulation, groundwater recharge and maintaining high water table, surface water storage, and removal of sediments, nutrients and other contaminants to improve water quality (Juliano & Simonovic, 1999; Zedler & Kercher, 2005). Apart from hydrological services, wetlands regulate global climate change through sequestration and long term storage of carbon dioxide from the atmosphere (Mitsch et al., 2013). Wetlands, mangroves and reefs also protect the coastal and inland communities from tropical storms and cyclones in events of extreme climate change (Turner, Oppenheimer, & Wilcove, 2009).

India sustains 16% of the world population with only 4% of the world's water resources (UNICEF, FAO and SaciWATERs, 2013). It is predicted that in future, the relative proportion of water used for irrigation will drop, whereas domestic and industrial use will increase. According to the National Water Policy, highest priority has been given to drinking water followed by irrigation, hydropower, ecology, etc. (Michael, 2009). However, with the rapid growth and expansion of metro cities, providing safe drinking water is one of the biggest challenges that the country faces today (Srinivasan, 2008).

Chennai, located on the east coast, is a water scarce city that has the lowest per capita availability of water amongst the four metro cities in India (Ruet, Saravanan, & Zerah, 2002). Although Chennai receives an average annual rainfall of 1200-1300 mm, only 5% of this actually seeps into the ground in urban areas (Janakarajan, Butterworth, Moriarty, & Batchelor, 2007). Traditionally a network of tanks (traditional rainwater storage reservoirs), ponds, temple tanks and wells formed the primary source of water in Chennai (Agarwal & Narain, 1997). Later due to increased inward migration and population pressure, these surface water sources had to be supplemented by groundwater pumped from wells in peri-urban areas (Janakarajan et al., 2007).

Chennai Metropolitan has 4 distinct systems of water supply-

1. Metro Water Board supply,
2. Municipal supply in towns adjoining Chennai city,
3. Domestic shallow wells or tube-wells, and
4. Private market including tanker trucks and retail distribution of bottled water (Anand, 2001).

Chennai Metropolitan Water Supply and Sewerage Board (also called 'Metro Water Board' or CMWSSB) obtains water from three interconnected rain-fed reservoirs, well fields and two inter basin projects (Srinivasan, 2008).

Despite the new projects, Chennai's reservoir storage capacity, mostly controlled by external agencies, remains very low (Srinivasan, 2015). As the gap between supply and demand is increasing by the day, the city is facing problems of low water supply, waste water management; fall in water table, salt water intrusion and flooding (Narain, Anand, & Banerjee, 2013).

One important source of water in Chennai Metropolitan is a large number of wetlands located in and around the city. Care Earth's assessment of the wetlands of Greater Chennai, based on a study of Survey of India topo-sheets of 1971 and composite imageries for the period 2009-11, revealed that Greater Chennai has 474 wetlands organized as complexes (Care Earth Trust, 2012). These wetland complexes perform critical ecosystem services like providing fresh water and recharging groundwater aquifers. However it is estimated that more than half of these wetlands have been converted for various uses, encroached upon or used as dumping sites (Gupta & Nair, 2011).

Adding to the water woes is the changing climate which is one of the major factors that threatens water security (Schewe et al., 2014). The freshwater resources have been projected to be strongly affected by climate change (Bates, Kundzewicz, Wu, & Palutikof, 2008). The impact of climate change is expected to be more severe in developing countries and IPCC has predicted increased water stress in India (Mertz, Halsnæs, Olesen, & Rasmussen, 2009). Climate change can affect water resources in terms of changes in precipitation and runoff patterns, rise in sea level and subsequent altered land use and water demands (Frederick & Major, 1997). Projected impact of climate change on water resources suggest that water supply may either increase or decrease in water stressed countries in the near future. This can be accompanied by water quality degradation, increase in salt water intrusion into coastal aquifer, and an increase in flood and drought frequency (Erwin, 2009). Extreme and intense rainfall events associated with climate change resulting in increased runoff will not alleviate water stress but would rather lead to enhanced flooding if the excess water is not stored (Arnell, 2004).

Pallikaranai Marsh

Historically, a large part of South Chennai was a floodplain comprising of Pallikaranai Marsh, smaller satellite wetlands and tracts of pasture land (Vencatesan, 2007). Pallikaranai, one of the last remaining natural wetlands in the city collects flood water and increases ground water levels in this landscape (Care Earth, 2002, 2005; Drescher et al., 2007; Joshi & Kale, 2013). As Chennai is bound by Bay of Bengal in east and Andhra Pradesh in north, much of the recent expansion of the city has been towards south covering areas like Velachery, Madipakkam, Thoraipakkam, Pallikaranai, etc. swallowing several agricultural villages and wetlands. These areas constitute the peri-urban region of Chennai city. These transitional areas around the city not only provide land and water resources for urban expansion but also serve as a sink for urban wastes (Narain, 2010). Further, these areas in Chennai have been witnessing population escalation as well as rise in water demand (Appasamy, 1995). The recent expansion of the city with a spurt in Information Technology industry, residential complexes and large number of informal settlements has resulted in problems of encroachment of the land and wetlands, pollution and unsustainable groundwater abstraction. This has led to fall in water table, flooding, water contamination, salinity intrusion, health problems, etc. The present study focuses on characterizing and assessing the current status of the water resources located around Pallikaranai Marsh in South Chennai. This study was supported and funded by the Tamil Nadu Forest Department.

ASSESSMENT OF WATER RESOURCES

The study was conducted around Pallikaranai Marsh (12.949371°N latitude and 80.218184°E longitude) in Chennai Metropolitan Area. Pallikaranai is largely a freshwater marsh located nearly 20 km South of Chennai city. It drains an area of about 250 km^2 through two outlets - Okkium Maduvu and Kovalam creek and falls into Bay of Bengal (Vencatesan, Daniels, Jayaseelan, & Karthick, 2014). Over the years, the size of the marsh has reduced from 6000 ha in 1900s to the present 593 ha. Pallikaranai is one of the 94 wetlands identified under National Wetland Conservation and Management Programme of the Government of India.

A large part of South Chennai was historically a floodplain comprising of the Pallikaranai Marsh, smaller satellite wetlands and tracts of pasture land and forests. These small wetlands formed an important source of irrigation for the area. Presently, the occurrence of the freshwater aquifer running parallel to the coast has significantly contributed to the expansion of the city in this direction (Vencatesan et al., 2014). Over the past 50 years large parts of the marsh have been reclaimed into terrestrial habitats and converted to residential colonies. The assessment of water resources was carried out within 5 km radius from the centre of the marsh mainly covering residential areas of Perungudi, Thoraipakkam, Karapakkam, Velachery, Medavakkam, Madipakkam-Puzhudivakkam and Pallikaranai (Figure 1).

Systematic field surveys were carried out from January 2013-2014, in addition to the ongoing longitudinal surveys of Care Earth Trust on Pallikaranai Marsh for the past 13 years. The survey focused on the 5 km radius area from the center of the marsh to assess and characterize the water resources. Surveys were undertaken by walking along the highway roads, inside the colony streets and barren fields. Total area of 37.608 km^2 was covered within 5 km radius. Water sources were identified in the field and categorized on the basis of the nature of the source as surface and groundwater. These sources have been further classified as Primary and Secondary source. 'Primary source' defines the direct source of water for the consumer while 'Secondary source' are the sources that supply water to the consumer via a supply chain and storage structure. In this case the municipal water tanks and village tanks have been classified as secondary sources as they are filled by water tankers which draw water from groundwater sources. Another secondary source is packaged drinking water plants which process groundwater to be supplied as water cans, packets, etc. to the consumer. However the actual number of groundwater sources has not been accounted for these two observations. All the shops selling water in the form of packets/cans/bottles have been classified as 'small scale water source' in the 'Others' secondary source category as the source (whether ground or surface) for these packaged units could not be ascertained in the present study. Accordingly, water sources are categorized as wetlands (including lakes), small ponds (relatively smaller water bodies), temple tanks, well sources (private dug well and agricultural wells), manmade

Figure 1. Map showing the study area within the 5 km radius of Pallikaranai Marsh

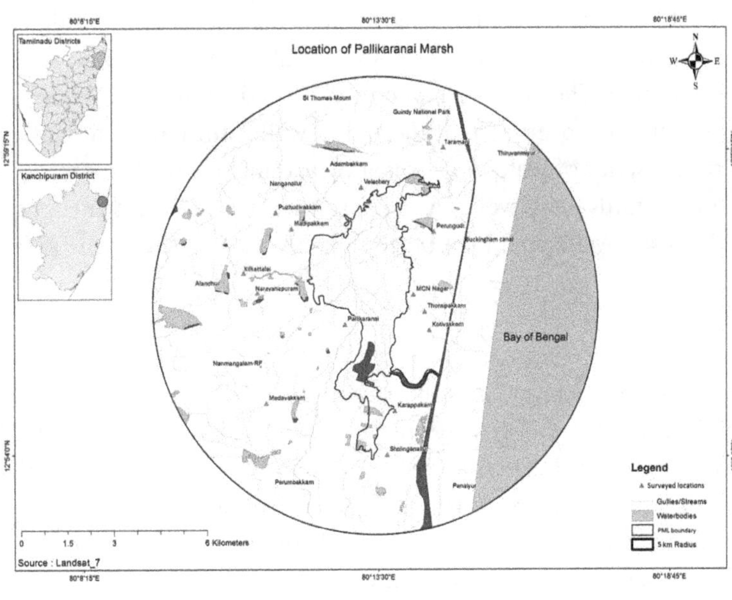

water tanks (village tanks), and municipal water tanks (Sintex tanks erected by the government). The geographical position of source (latitude and longitude) was marked using GARMIN GPS. To quantify the fresh water usage around Pallikaranai Marsh, data was obtained by interviewing people at lorry tanker filling stations. Data collected included the number of water tankers being filled every day, cost of filling water tanker at filling stations and cost of delivery at final point.

On the basis of field survey, eleven water resources were identified around Pallikaranai Marsh. Table 1 classifies the water sources around Pallikaranai Marsh providing details about their location and their use. The important water sources in the region are described in the following section.

Groundwater

Groundwater is the dominant source of water in the study area available to the consumers through municipal water tanks and manmade/village tanks. Municipal tanks supply water to large slum areas located on the northern side of the Pallikaranai Marsh whereas private residential buildings, apartments and industries depend mainly on pump sets and private water tankers. More than hundred municipal water tanks were observed in all the areas around the marsh. Water supply to these tanks is under the control of CMWSSB. Maximum number of municipal tanks was observed on northern side of the marsh in areas like Velachery, Perungudi and Taramani followed by Thoraipakkam, and MCN Nagar located on the eastern side of the marsh. These tanks are filled by water tankers which extract water from agricultural and private wells present around Pallikaranai Marsh. Apart from this, nearly forty man-made water tanks/village tanks were observed mainly on eastern and western side, in Thoraipakkam, Kilkattalai, Puzhudivakkam and Madipakkam areas. These are tanks constructed in each village filled by groundwater extracted using pump sets. The water from these tanks is supplied through an underground pipeline system to the households.

Wells are one of the most important water sources in the area used for extraction of freshwater. About thirty agricultural wells located mainly on the western side of the marsh in Madipakkam, Kilkattalai and Narayanapuram were observed to be used for commercial extraction of water by both private and municipal tankers. The availability of well sources is higher in the west and south as some years back these areas were completely under agriculture, and the farming activities depended on seasonal wetlands and agricultural wells. However, as a result of commercial water extraction, farming activities in the area have almost discontinued. The area has given way to residential houses, commercial buildings and educational institutions. It was observed that these dug wells still have a remarkable storage of water possibly because of the presence of water bodies nearby and inflow of groundwater. Apart from this, the seasonal rainfall also safeguards these wells. As a household survey for water supply was not conducted, the extent of use of domestic wells could not be completely ascertained in the present study.

Surface Water

Nearly 30 wetlands, 4 temple ponds and 13 small ponds were observed around Pallikaranai Marsh. The wetlands are mainly concentrated in the western and northern side of the Marsh. The seasonally filled wetlands around the marsh include Lakes of Velachery, Adambakkam, Perungudi, Kilkattalai, Muvarasapattu, Madipakkam, Puzhudivakkam, Nesavallar Nagar and Narayanapuram. These water bodies are used to a small extent for domestic purposes like washing, bathing, cleaning of vehicles, by livestock, etc. Few tankers were also observed pumping surface water from Kilkattalai Lake nearly 3 km away from

Table 1. Classification of water sources present around Pallikaranai Marsh, their number in each direction and use

Source	Primary Source	Secondary Source	No. of Sources in Each Direction Around Pallikaranai Marsh					% of Source	Use
			East	West	North	South	Total		
GW	Hand Pump				2		2	0.6	Domestic
		Manmade water tank (village tank)	16	15	5	8	44	14.2	Domestic
		Municipal water tank	33	20	52	16	121	39	Domestic
		Packaged drinking water plant	1	1			2	0.6	Drinking, cooking
	Pump set (Bore well)		2			1	3	1.0	Domestic
	Well (domestic)			5		1	6	1.9	Domestic
	Well (Agricultural or Private well)			19	1	10	30	9.7	Commercial (tanker)
SW	Small pond		1	6	4	2	13	4.2	Bathing, washing
	Temple pond			2	2		4	1.3	Bathing
	Wetlands		6	10	8	5	29	9.4	Domestic/cleaning vehicles, livestock
		Surface water pumping		1			1	0.3	Commercial extraction by private tankers
Others		Small scale water source	9	21	20	6	56	18.1	Drinking

Pallikaranai Marsh. Narayanapuram Lake in the west has three open wells which supply fresh water to surrounding areas through underground pipeline system. The water in the lake was relatively clean and is used by local people for bathing, washing and fishing. However signs of pollution were also observed in different parts of this lake.

The number of small ponds is higher along the western side of the marsh. These water bodies are seasonally filled and provide fresh water, fodder, fuel, and fishery resources for local people. As a cultural tradition in Tamil Nadu, old temples generally house ponds in the vicinity, and 4 temple ponds (temple tank) were observed in the entire 5 km radius of the study site which serve as a vital habitat for aquatic flora and fauna, though not used for domestic purposes.

Tanker Market

The water vending by private tankers depends on groundwater drawn from agricultural and private wells situated in areas like Perungudi, Madipakkam, Medavakkam, Velachery and Thoraipakkam. Every day, on average 700-800 tankers are filled at source (Figure 2). The tankers with an average capacity of 9,000 liters per vehicle provide water every day to the low-income areas which are not connected with the piped supply system. These tankers transfer water to storage tanks made of concrete or plastic (Sintex tanks) installed by the CMWSSB. The cost of water tanker filling is ₹150-190 (USD 2-3) but the cost of delivery is ₹700-800 (USD 10-12), nearly 3-4 times the filling cost. The delivery cost varies depending upon the season and distance to the delivery point. The main consumers of water provided by tankers include domestic households, industries, hotels, academic institutions and offices across the city.

Two water processing plants were observed in the vicinity of the marsh in Medavakkam area. In addition, more than fifty shops selling packaged drinking water were recorded around the marsh. Though the actual source of this water supply could not be ascertained in the present study, the water processing plants generally draw groundwater.

Figure 3 shows the various sources of water available around Pallikaranai Marsh.

MAJOR CONCERNS

The expansion of Chennai city has led to growth of peri-urban areas. As a consequence, the water bodies in peri-urban areas are in the process of being reclaimed to build urban infrastructure in addition to extraction of water for supply to both urban and peri-urban areas (Narain et al., 2013). Flooding and over-extraction of groundwater are known to affect the water security of urban and peri-urban areas (Janakarajan, 2009). The area around Pallikaranai Marsh which was originally a low lying floodplain is now a composite of industrial, residential, institutional and commercial establishments, and remnant natural habitats. In this peri-urban area, the water security is governed by the analogous processes of urbanization and climate change (Narain, 2010). On the basis of the present study, the following possible causes of water stress in the study area have been identified.

Figure 2. A long queue of water tankers at the periphery of Pallikaranai Marsh

Figure 3. Various sources of water around Pallikaranai Marsh, A. Piped water supply through village tanks, B. Tanker water supply, C. Wetlands, and D. Domestic wells

Over-Extraction of Groundwater

Chennai experiences acute water scarcity especially during low rainfall years. During these times groundwater plays an essential role in meeting the water demand (Janakarajan, Zehra, & Llorente, 2006). Growth of the city in the absence of adequate supply infrastructure has resulted in unsustainable extraction of groundwater (Srinivasan, Seto, Emerson, & Gorelick, 2013). The present study showed that nearly thirty agricultural wells located mainly on the western side of the marsh were used for extraction of groundwater by both private and municipal tankers. In addition to over-extraction, the increase in impervious surface owing to rapid urbanization has reduced the area for groundwater recharge. The impacts of groundwater depletion can be varied, the most direct and important being the loss of base-flow. This can further result in drying up of wells, increased cost of pumping and well infrastructure, salt-water intrusion, reduction in biodiversity, etc. (Ponce, 2006). Rapid urbanization together with disappearance of traditional network of tanks and ponds has resulted in over-exploitation of groundwater resources in Chennai. Chennai city during the past two decades has depended heavily on private and agricultural wells for water supply impacting groundwater table, changing the land-use pattern and affecting the livelihood of the farmers (Ruet, Gambiez, & Lacour, 2007). According to Janakarajan et al. (2007) nearly 80% of Chennai groundwater has already depleted and further exploitation may result in enhanced sea water intrusion.

Degradation of Wetlands

Wetlands provide various ecosystem services of value to the society like preventing floods, filtering nutrients and pollutants thereby improving water quality, groundwater buffering, and serving as an important habitat for flora and fauna (Gren, Folke, Turner, & Batemen, 1994). Coastal cities similar to Chennai are one of the most rapidly urbanizing places and rapid urbanization affects the hydrology, geomorphology and ecology leading to drastic changes in the urban wetlands (Ehrenfeld, 2000). More than half of the wetlands in Chennai have already been converted to other uses (Chandramohan & Bharthi, 2009). A large part of South Chennai where the present study was conducted was historically a floodplain comprising of Pallikaranai Marsh and other smaller satellite wetlands. In our survey 29 large and 13 small water bodies were identified around the Pallikaranai Marsh. The historical images of Pallikaranai Marsh and surrounding areas show that a large number of water bodies drained into the marsh mainly during seasonal rains creating an effective drainage system (Care Earth, 2002 & 2005). However, over the years, in addition to Pallikaranai Marsh, the surrounding water bodies including Adambakkam, Kilkattalai, Narayanapuram Lake, Velachery Lake and other similar wetlands have shrunk drastically. This is attributed to rapid urban development, road laying and encroachment resulting in blocking of water channels in addition to draining of domestic sewage into the marsh. An inner ring road laid at the center of the Adambakkam Lake has completely separated it from the marsh while the Thoraipakkam radial road constructed in the center of the Kilkattalai Lake has divided the lake into two parts altering the water flow. As these wetlands belonging to a single system of water bodies, aid in crucial groundwater percolation, their disappearance will significantly impact water sustainability in this region (Chandramohan & Bharthi, 2009). Further, a large number of studies have shown that degradation of wetlands often results in loss of water retention and enhanced risk of flooding (Brody, Highfield, Ryu, & Spanel-Weber, 2007).

Informal and Unregulated Water Market

According to Janakarajan (2013) piped water supply accounts for only one fifth of the total water demand in Chennai while Ruet et al. (2002), state that at least 60% of water used in Chennai comes from sources other than public system. The limitations in groundwater availability and unreliable piped supply have led to an increase in demand for water tanker supply. This private water market is largely informal as it is not monitored by a regulatory agency in terms of quantity of water extracted, quality as well the economic value of water in the system. This informal commercial water sector consists of bulk suppliers like tankers and small scale suppliers including cans, bottles, sachets, etc. (Venkatachalam, 2012). Present study showed that an average of 700-800 private tankers extract groundwater from agricultural and private wells around Pallikaranai Marsh. Tanker water is nearly 3 to 50 times more expensive than municipal piped water supply (Srinivasan, Gorelick, & Goulder, 2010). Further these informal tanker markets have led to drastic decline in groundwater in peri-urban areas affecting agriculture and livelihood of farmers (Londhe et al., 2004). In order to regulate groundwater extraction and illegal transportation of water the Chennai Metropolitan Area Ground Water (Regulation) Act was enacted in 1987, foreseeing the registration of existing wells, regulations for sinking new wells, issuing licenses to extract water for non-domestic use and issue of licenses for transportation through goods vehicle (Ruet et al., 2002). However this Act is grossly violated as both private companies and Metro Water Board are responsible for overexploitation of groundwater in many peri-urban areas apart from the private lorry and tankers that are operating in the Metropolitan Area without proper licenses (Janakarajan et al., 2006). In addition to water tankers, more than fifty shops were found to sell packaged drinking water in the study area. These small scale suppliers provide value added service in terms of treated water thereby setting up a strong consumer base and involving large money transactions. Studies have shown that informal water market serves both the high and low income groups in Chennai ensuring good quality water supply (Venkatachalam, 2012).

Anticipated Water Stress Due to Climate Change

Extreme and frequent weather events as a result of climate change are expected to alter the quantity as well as quality of available water resources in the future (Bahri, 2012). Water availability is the most critical and direct impacts of climate change in urban and peri-urban areas. As the population rises in these regions, the water demand goes up while the water becomes limiting as rainfall patterns and groundwater levels change (UN-Habitat, 2011). It has been projected that climate change will affect water supply primarily through flood damage and increased water treatment costs. On the other hand, prolonged periods of drought will affect surface water supply and enhance the vulnerability of groundwater systems (Bahri, 2012). It has been predicted that in India, climate change would affect water resources in terms of decline of the glaciers in the Himalayas, increase in both drought-like situations and flood events, change in groundwater quality and groundwater recharge, and increased saline intrusion of coastal and island aquifers due to rising sea levels (INCCA Report, 2010). In Chennai, a city located on the east coast, the future climate hazards include cyclones and floods resulting in drinking water shortage (Tanner, Mitchell, Polack, & Guenther, 2009). Further studies project occurrence of more extreme temperatures as well as rainfall events on the east coast of India (Dash & Hunt, 2007). An increase of flood risk has been observed across the last two decades over the eastern coast (Guhathakurta, Sreejith, & Menon, 2011).

Climate change affects precipitation variability, resulting in frequent floods and drought (Miller, Rhodes, & MacDonnell, 1997). These two major possible scenarios arising due to climate change affecting water resource availability are discussed below.

Flooding

Just as the government delegations from various countries including India were committing to limiting climate change at COP21, Paris; catastrophic flooding affected hundreds of people causing loss of life and property in and around Chennai. Even though such events are unpredictable and cannot be directly linked to climate change, climate change with its generally warmer sea water causes increase in number of storms and flooding, indicating that El Niño has a part to play in the sequence of extreme weather events (Gross, 2016). Climate change can affect the intensity and frequency of floods as a result of extreme rainfall events and such events are projected to become more frequent over a large number of regions (Bates et al., 2008). In addition to this, urbanization increases impervious surfaces and loss of natural drainage resulting in floods. Chennai receives an average annual rainfall of 1300 mm, most of which falls during North-East monsoon in the months of October through December (Figure 4). Meteorological data analyses shows that though there is a year-to-year variability, there are been no significant upwards or downwards trend during the last 200 years. However, heavy precipitation events in the last few decades - 1985, 1996, 1998, 2005, 2008, 2010 and 2015 have resulted in catastrophic floods in Chennai ((Walsh, Glaser, & Militzer, 1999; Gupta & Nair, 2010). Pallikaranai Marsh in southern Chennai is an important area for the storage of storm water runoff from the surrounding areas, allowing the overflow to drain into the sea. Historically the seasonally flooded marsh was used for supporting agriculture as evidenced by the absence of dug out wells in the landscape until the last five decades. While agriculture declined over the years, the marsh continued to collect the overrun of the surrounding wetlands in south Chennai. The reason for rise in floods especially around the marsh is not only because of heavy rainfall events but also due to extensive urbanization of the surrounding areas which poses a physical obstruction to runoff to the marsh. Lack of adequate storm water drainage, garbage dumping in the marsh and

Figure 4. Monthly rainfall pattern in Chennai over a period of 5 years. Source of data: Customized Rainfall Information System (CRIS), Hydromet Division, India Meteorological Department

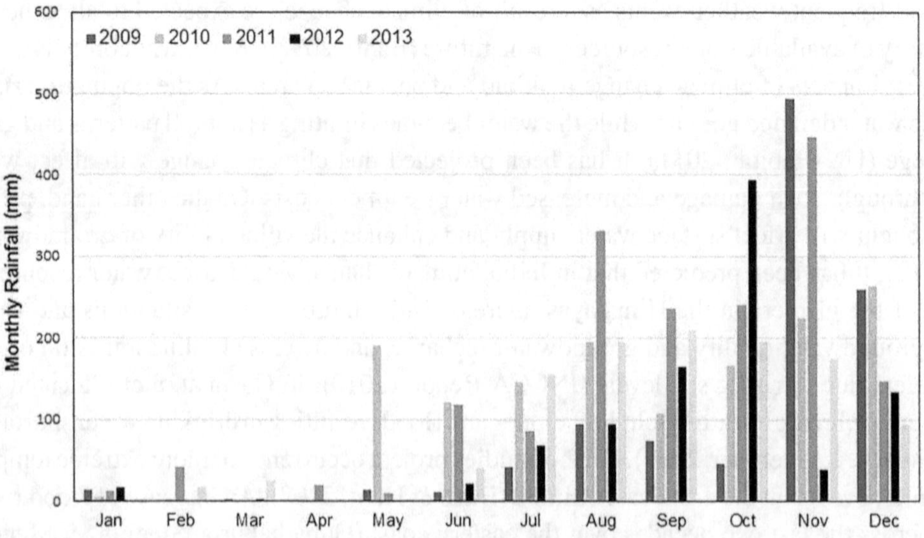

encroachment of the wetlands also compound the problem. As a result, the natural drainage is highly obliterated leading to frequent floods in the area. Figure 5 shows one of the flooding events in an area surrounding Pallikaranai Marsh. A favorable microclimate and assured water availability has made the area prime spot for large scale real estate development. This however is adding to the water stress in the region. Figure 6 shows change in the extent of the Pallikaranai Marsh for the time period between 1970 and 2010. Degradation of the marsh over the years has severely curtailed its water holding capacity and studies have shown that events of above normal precipitation would cause flooding of the residential

Figure 5. A flooding event in south Chennai

Figure 6. Maps showing water spread area in Pallikaranai Marsh (1970-2010) (Source: Vencatesan et al., 2014)

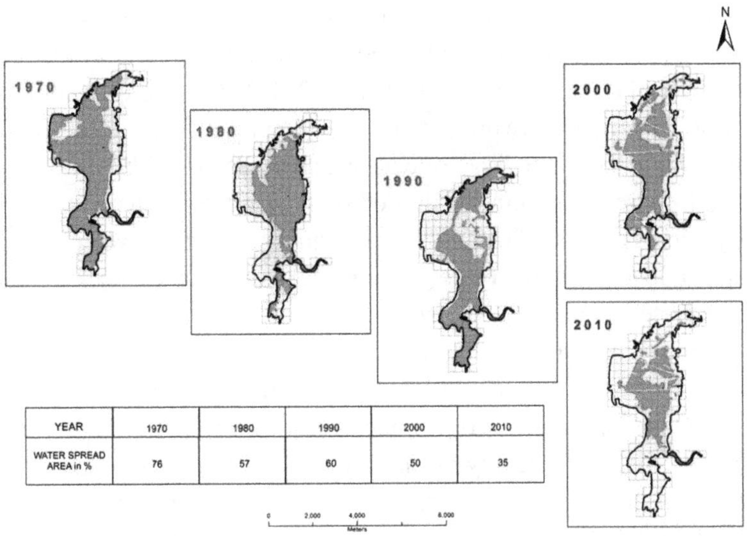

YEAR	1970	1980	1990	2000	2010
WATER SPREAD AREA in %	76	57	60	50	35

and commercial spaces within this landscape (Apfelbacher, 2008; Glaser et al., 2008; Vencatesan et al., 2014). More than 5,550 hectares of wetlands in the area has been developed into commercial real estate. Large scale encroachment of the wetlands around the marsh has also resulted in increased incidents of flooding. The density of habitation is maximum on the northern and north-western portions of the marsh which are impacted intensively during flooding (Vencatesan et al., 2014). Recent heavy rains in Chennai resulted in disastrous flood largely attributed to ignoring hydrological concerns during urban planning.

Wetlands are known to play a significant role in modifying the hydrological cycle and either reduce or delay floods (Bullock & Acreman, 2003). Nearly 80% of the studies suggest that floodplain wetlands reduce flooding by reducing both the magnitude and frequency of peak flow events (Acreman & Holden, 2013). Wetlands reduce the amount and speed of rainwater runoff and improve surface water quality in urban areas (Jha, Bloch, & Lamond, 2012). Alteration of wetlands is known to reduce the natural water retention and enhance flooding risk (Brody et al., 2007). Flooding can also contaminate water supplies in addition to rendering areas inhabitable and imposing additional costs on storm water drainage (Bahri, 2012). Worldwide, encroachment into of low lying areas closer to the water bodies has increased the damage potential due to floods. Further, globally, an increase in the number of inland flooding together with a rise in economic losses has been reported (Kron & Berz, 2007; Kundzewicz et al., 2014). Wetland restoration and conservation in Hubei Province, China and New Orleans, USA helped withstand storm surges and reduced the volume of rainwater runoff in these urban cities, resulting in improved resilience to flooding (Renaud, Sudmeier-Rieux, & Estrella, 2013). Further, a study by Walters & Babbar-Sebens (2016) in Midwestern United States, showed that presence of wetlands provided steady peak flow reductions across all climate change scenarios.

Drought

Decreased precipitation, enhanced evapotranspiration and reduced soil moisture have been defined as the main causes for drought. Climate change is predicted to impact water availability by altering both the frequency and severity of droughts (McDonald et al., 2011). According to IPCC Assessment Report (2007), the areas affected by drought are predicted to expand in addition to rising sea levels in the coastal areas. Growing demand from the expanding population and excessive withdrawals can thereby aggravate the problem of drought. Increase in vulnerability of shallow groundwater systems is also projected in case of extended dry periods (Bahri, 2012). The importance of groundwater as a perennial source of freshwater is projected to increase under more frequent and intense extreme climate events like drought. Wetlands have been reported to play an important role in groundwater recharge through both rain-fed discharge and focused recharge (Taylor et al., 2013).

A study by Dash and Hunt (2007) has shown that the maximum and minimum temperature have been increasing by 0.6°C and 0.4°C respectively on the eastern coast of India. Chennai city lying in this region faces acute water scarcity especially during low rainfall years. During such period groundwater plays an essential role in meeting the water demand and the south Chennai aquifer is an important source of groundwater during drought years. A study by Seifelislam (2013), analyzing the effective recharge amount over the period from 2001 to 2013 showed that there has been a decrease in the groundwater level of the aquifer recharged by Pallikaranai Marsh. The reduced recharge capacity is mainly attributed to decrease in the recharge area owing to large-sale construction.

Study by van der Kamp and Hayashi (1998) has shown that the wetlands in the semi-arid northern Prairie region are important sites of groundwater recharge. Their study also showed that drainage of wetland lowers the water table in the vicinity of the wetland thereby reducing the yield of surrounding shallow wells. Various other studies have recorded that the recharge rates of wetlands vary from 1 to 45 mm per year (Hayashi, van der Kamp, & Rudolph, 1998). Study by Acharya and Barbier (2000) showed that the Hadejia-Nguru wetlands in northern Nigeria formed by floodwaters of the region help in re-charging of the aquifers which are used for domestic consumption and irrigation during the dry season. Another study in Okavango delta in northern Botswana showed that 80 to 90% of seasonal floodwater recharges the groundwater (McCarthy, 2006).

Events of low rainfall, drying up of reservoirs and reduced groundwater recharge makes Chennai highly vulnerable to multiyear drought (Srinivasan et al., 2013). The south Chennai aquifer is important, especially during the drought years for meeting the water requirements. However it is speculated that increasing urbanization of the area will lead to over-extraction of groundwater resulting in sea water intrusion.

CONCLUSION

Although climate change scenarios predict high risk to wetlands, intact wetlands would play a key role in climate change adaptation. Wetlands act as sinks for organic carbon in addition to storing water to buffer runoff and recharging groundwater table, thereby, positively impacting water availability (Junk et al., 2013). The water sources identified in the present study can suffice the supply of water to the area but not without proper maintenance and restoration. It was found that wells around the marsh retained water supply even during peak summer months owing to groundwater recharge by Pallikaranai and surrounding wetlands. The marsh also plays an effective role in collecting floodwaters and overflows from nearby wetlands in the surrounding area. Therefore Pallikaranai Marsh together with the surrounding wetlands is vital to Chennai city as a means of adapting to climate change. Hence it is imperative to maintain and restore these wetlands and tanks for ensuring sustainable supply of water to surrounding areas. It is suggested that institutions like CAP (Conservation Authority of Pallikaranai Marsh), CMWSSB, Central Groundwater Board, Tamil Nadu Pollution Control Board, etc. should undertake steps to conserve these wetlands, preventing further construction, reclamation, pollution and road laying. Further, in view of the rampant violation of the Groundwater Act, an attempt should be made to regularize the informal tanker market. Protocols for extraction of groundwater especially by commercial entities should be in place and unsustainable extraction of water from bore wells inside the wetlands should also be monitored. These few steps would be in line to facilitate the transformation of Chennai from water starved to a water sustainable city.

REFERENCES

Acharya, G., & Barbier, E. B. (2000). Valuing groundwater recharge through agricultural production in the Hadejia-Nguru wetlands in northern Nigeria. *Agricultural Economics*, *22*(3), 247–259. doi:10.1111/j.1574-0862.2000.tb00073.x

Acreman, M., & Holden, J. (2013). How wetlands affect floods. *Wetlands*, *33*(5), 773–786. doi:10.1007/s13157-013-0473-2

Agarwal, A., & Narain, S. (1997). *Dying Wisdom. Rise, Fall and Potential of India's Traditional Water Harvesting System.* Faridabad, India: Thomson Press Ltd.

Anand, P. B. (2001). *Water scarcity in Chennai, India: Institutions, entitlements and aspects of inequality in access*. WIDER Discussion Papers // World Institute for Development Economics (UNU-WIDER), No. 2001/140. Retrieved February 15, 2015, from http://hdl.handle.net/10419/53004

Apfelbacher, J. (2008). *Land Use/Cover Change in the Context of Global Change around the Pallikaranai Marshland.* (Unpublished bachelor's thesis). Department of Physical Geography, University of Freiburg, Germany.

Appasamy, P. (1995). *Water allocation and management in Madras. Report funded by Overseas Development Institute*. Chennai: Madras Institute of Development Studies.

Arnell, N. W. (2004). Climate change and global water resources: SRES emissions and socio-economic scenarios. *Global Environmental Change*, *14*(1), 31–52. doi:10.1016/j.gloenvcha.2003.10.006

Askew, A. J. (1987). Climate change and water resources. In S. I. Solomon, M. Beran, & W. Hogg (Eds.), *The influence of climate change and climatic variability on the hydrologic regime and water resources* (Vol. 168, pp. 421–340). Wallingford, UK: IAHS Publishers.

Bahri, A. (2012). Integrated urban water management. *Global Water Partnership Technical Committee Background Papers*, (16). Retrieved February 13, 2015, from http://www.gwp.org/Global/The%20Challenge/Resource%20material/GWP_TEC16.pdf

Bates, B., Kundzewicz, Z. W., Wu, S., & Palutikof, J. P. (Eds.). (2008). *Climate change and water. Technical Paper of the Intergovernmental Panel on Climate Change, IPCC Secretariat, Geneva*. Geneva, Switzerland: IPCC.

Brody, S. D., Highfield, W. E., Ryu, H. C., & Spanel-Weber, L. (2007). Examining the relationship between wetland alteration and watershed flooding in Texas and Florida. *Natural Hazards*, *40*(2), 413–428. doi:10.1007/s11069-006-9003-3

Bullock, A., & Acreman, M. (2003). The role of wetlands in the hydrological cycle. *Hydrology and Earth System Sciences Discussions*, *7*(3), 358–389. doi:10.5194/hess-7-358-2003

Care Earth. (2002). Conservation of Urban Wetlands- Pallikaranai Marsh, A Baseline study. Report supported by Tamil Nadu Pollution Control Board.

Care Earth. (2005). Current Status of the Pallikaranai Marsh and Recommendations for Protection and Restoration. Report submitted to the Tamil Nadu Pollution Control Board.

Care Earth Trust. (2012). *Identification of wetlands for protection, restoration and conservation in Greater Chennai. Report submitted to the Chennai Metropolitan Development Authority*. Chennai, India: Government of Tamil Nadu.

Chandramohan, B. P., & Bharati, D. (2009). Role of Public Governance in the Conservation of Urban Wetland System: A Study of Pallikaranai Marsh. In *Proceedings of the Indian Society for Ecological Economics (INSEE), 5th Biennial Conference, Ahmedabad, India. Customized Rainfall Information System (CRIS)*. Hydromet Division, India Meteorological Department. Retrieved February 16, 2015 from http://hydro.imd.gov.in/

Dash, S. K., & Hunt, J. C. R. (2007). Variability of climate change in India. *Current Science*, *93*(6), 782–788.

Drescher, A., Glaser, R., Pfeiffer, C., Vencatesan, J., Schliermann-Kraus, E., Glaser, S., & Dostal, P. (2007). *Risk assessment of extreme precipitation in the coastal areas of Chennai as an element of catastrophe prevention. In 8. Forum DKKV/CEDIM: Disaster Reduction in Climate Change 15./16.10.2007*. Germany: Karlsruhe University.

Ehrenfeld, J. G. (2000). Evaluating wetlands within an urban context. *Ecological Engineering*, *15*(3), 253–265. doi:10.1016/S0925-8574(00)00080-X

Erwin, K. L. (2009). Wetlands and global climate change: The role of wetland restoration in a changing world. *Wetlands Ecology and Management*, *17*(1), 71–84. doi:10.1007/s11273-008-9119-1

Frederick, K. D., & Major, D. C. (1997). Climate change and water resources. *Climatic Change*, *37*(1), 7–23. doi:10.1023/A:1005336924908

Glaser, S., Glaser, R., Drescher, A., Pfeiffer, C., Schliermann-Kraus, E., Lechner, M., & Vencatesan, J. (2008). *Geo-communication for risk assessment and catastrophe prevention of flood events in the coastal areas of Chennai*. iEMSs 2008: International Congress on Environmental Modelling and Software Integrating Sciences and Information Technology for Environmental Assessment and Decision Making 4th Biennial Meeting of iEMSs. Retrieved from http://www.iemss.org/iemss2008/index.php?n=Main.Proceedings

Gren, M., Folke, C., Turner, K., & Batemen, I. (1994). Primary and secondary values of wetland ecosystems. *Environmental and Resource Economics*, *4*(1), 55–74. doi:10.1007/BF00691932

Gross, M. (2016). World under water. *Current Biology*, *26*(2), R47–R50. doi:10.1016/j.cub.2016.01.001 PMID:27213204

Guhathakurta, P., Sreejith, O. P., & Menon, P. A. (2011). Impact of climate change on extreme rainfall events and flood risk in India. *Journal of Earth System Science*, *120*(3), 359–373. doi:10.1007/s12040-011-0082-5

Gupta, A. K., & Nair, S. S. (2010). Flood risk and context of land-uses: Chennai city case. *Journal of Geography and Regional Planning*, *3*(12), 365–372. Available online at http://www.academicjournals.org/JGRP

Gupta, A. K., & Nair, S. S. (2011). Urban floods in Bangalore and Chennai: Risk management challenges and lessons for sustainable urban ecology. *Current Science (Bangalore)*, *100*(11), 1638–1645.

Hayashi, M., van der Kamp, G., & Rudolph, D. L. (1998). Water and solute transfer between a prairie wetland and adjacent uplands, 1. Water balance. *Journal of Hydrology (Amsterdam)*, *207*(1), 42–55. doi:10.1016/S0022-1694(98)00098-5

INCCA Report, .Indian Network for Climate Change Assessment. (2010). *Climate Change and India: A 4x4 Assessment A Sectoral and Regional Analysis for 2030s*. Ministry of Environment & Forests, Government of India.

IPCC (Intergovernmental Panel on Climate Change). (2007). *Synthesis Report*. Contribution of Working Groups I, II and III to the Fourth Assessment Report of the Intergovernmental Panel on Climate Change. Geneva, Switzerland: IPCC.

Janakarajan, S. (2009). Urbanization and peri-urbanization: Aggressive competition and unresolved conflicts. The case of Chennai city in India. *South Asian Water Studies*, *1*(1), 51–76.

Janakarajan, S. (2013). *Optimising Local Water Resources and Availability in Urban and Peri-urban Chennai* [PDF document]. Retrieved from cseindia.org/userfiles/Janakarajan.pdf

Janakarajan, S., Butterworth, J., Moriarty, P., & Batchelor, C. (2007). Strengthened city marginalised peri-urban villages: stakeholder dialogues for inclusive urbanisation in Chennai, India. In J. Butterworth, R. Ducrot, N. Faysse, & S. Janakarajan (Eds.), *Peri-urban water conflicts: Supporting dialogue and negotiation* (pp. 51–75). Delft, Netherlands: IRC International Water and Sanitation Centre.

Janakarajan, S., Zehra, M. H., & Llorente, M. (2006). *Urban Water Conflicts in Indian Cities: Man-made Scarcity as a Critical Factor*. MIDS Working Paper No.196. Chennai, India: Madras Institute of Development Studies.

Jha, A. K., Bloch, R., & Lamond, J. (Eds.). (2012). *Cities and flooding: a guide to integrated urban flood risk management for the 21st century*. World Bank Publications. Retrieved from http://hdl.handle.net/10986/2241

Joshi, V. U., & Kale, V. S. (2013). Environmental Conflicts in Coastal Metropolitan Cities in India: Case Studies of Mumbai and Chennai Metropolitan Regions. In A. Z. Khan, L. X. Quynh, E. Corijn & F. Canters (Eds.), *Environmental Conflicts in Coastal Urban Areas: Towards a Strategic Assessment Framework for Sustainable Development*. Retrieved from http://digilab-epub.uniroma1.it/index.php/SECOA/article/download/103/92

Juliano, K., & Simonovic, S. P. (1999). *The impact of wetlands on flood control in the Red River Valley of Manitoba*. University of Manitoba. Retrieved February 10, 2015, from http://www.ijc.org/rel/pdf/wetlandsman.pdf

Junk, W. J., An, S., Finlayson, C. M., Gopal, B., Květ, J., Mitchell, S. A., & Robarts, R. D. et al. (2013). Current state of knowledge regarding the worlds wetlands and their future under global climate change: A synthesis. *Aquatic Sciences*, *75*(1), 151–167. doi:10.1007/s00027-012-0278-z

Kron, W., & Berz, G. (2007). Flood disasters and climate change: trends and options—a (re-)insurer's view. In J. L. Lozán, H. Grabl, P. Hupfer, L. Menzel, & C. D. Schönwiese (Eds.), *Global change: enough water for all?* (pp. 268–273). University of Hamburg, Hamburg.

Kundzewicz, Z. W., Kanae, S., Seneviratne, S. I., Handmer, J., Nicholls, N., Peduzzi, P., & Muir-Wood, R. et al. (2014). Flood risk and climate change: Global and regional perspectives. *Hydrological Sciences Journal*, *59*(1), 1–28. doi:10.1080/02626667.2013.857411

Londhe, A., Talati, J., Singh, L. K., Vilayasseril, M., Dhaunta, S., Rawlley, B., . . . Mathew, R. P. (2004). *Urban-hinterland water transactions: a scoping study of six class I Indian cities*. IWMI Tata Water Policy Programme Working Paper.

Maltby, E., & Acreman, M. C. (2011). Ecosystem services of wetlands: Pathfinder for a new paradigm. *Hydrological Sciences Journal*, *56*(8), 1341–1359. doi:10.1080/02626667.2011.631014

McCarthy, T. S. (2006). Groundwater in the wetlands of the Okavango Delta, Botswana, and its contribution to the structure and function of the ecosystem. *Journal of Hydrology (Amsterdam)*, *320*(3), 264–282. doi:10.1016/j.jhydrol.2005.07.045

McDonald, R. I., Green, P., Balk, D., Fekete, B. M., Revenga, C., Todd, M., & Montgomery, M. (2011). Urban growth, climate change, and freshwater availability. *Proceedings of the National Academy of Sciences of the United States of America*, *108*(15), 6312–6317. doi:10.1073/pnas.1011615108 PMID:21444797

Mertz, O., Halsnæs, K., Olesen, J. E., & Rasmussen, K. (2009). Adaptation to climate change in developing countries. *Environmental Management*, *43*(5), 743–752. doi:10.1007/s00267-008-9259-3 PMID:19184576

Michael, A. M. (2009). *Irrigation: theory and practice*. New Delhi, India: Vikas Publishing House Pvt Ltd.

Miller, K. A., Rhodes, S. L., & MacDonnell, L. J. (1997). Water allocation in a changing climate: Institutions and adaptation. *Climatic Change*, *35*(2), 157–177. doi:10.1023/A:1005300529862

Mitsch, W. J., Bernal, B., Nahlik, A. M., Mander, Ü., Zhang, L., Anderson, C. J., & Brix, H. et al. (2013). Wetlands, carbon, and climate change. *Landscape Ecology*, *28*(4), 583–597. doi:10.1007/s10980-012-9758-8

Narain, V. (2010). *Peri urban water security in a context of urbanization and climate change: A review of concepts and relationships*. Peri Urban Water Security Discussion Paper Series, Paper No. 1, SaciWATERs. Retrieved February 16, 2015, from http://saciwaters.org/periurban/newsletter/pdfs/discussion_papper_1.pdf

Narain, V., Anand, P., & Banerjee, P. (2013). *Periurbanization in India: A review of the literature and evidence*. Report for the project – Rural to Urban Transitions and the Peri-urban Interface. SaciWATERs. India. Retrieved February 16, 2015, from https://saciwaters.org/east-west-center/pdf/status-paper.pdf

Ponce, V. M. (2006). *Groundwater Utilization and Sustainability*. Retrieved February 9, 2015, from Lecture Notes Online Web site: http://groundwater.sdsu.edu/

Renaud, F. G., Sudmeier-Rieux, K., & Estrella, M. (Eds.). (2013). *The role of ecosystems in disaster risk reduction*. Tokyo, Japan: United Nations University Press.

Ruet, J., Gambiez, M., & Lacour, E. (2007). Private appropriation of resource: Impact of peri-urban farmers selling water to Chennai Metropolitan Water Board. *Cities (London, England)*, *24*(2), 110–121. doi:10.1016/j.cities.2006.10.001

Ruet, J., Saravanan, V. S., & Zérah, M. H. (2002). *The Water and Sanitation Scenario in Indian Metropolitan Cities: Resource and Management in Delhi, Calcutta, Chennai, Mumbai.* CSH Occasional Paper No.6. New Delhi: Centre de Sciences Humaines.

Schewe, J., Heinke, J., Gerten, D., Haddeland, I., Arnell, N. W., Clark, D. B., & Kabat, P. et al. (2014). Multimodel assessment of water scarcity under climate change. *Proceedings of the National Academy of Sciences of the United States of America*, *111*(9), 3245–3250. doi:10.1073/pnas.1222460110 PMID:24344289

Seifelislam, A. (2013). *A rapid assessment of the impact of wetland's degradation on groundwater resources in the fast growing city Chennai. (Unpublished Internship-Report for the European Program Chance2Sustain)*. Pondicherry, India: French Institute of Pondicherry.

Srinivasan, V. (2008). *An integrated framework for analysis of water supply strategies in a developing city: Chennai, India* (Doctoral dissertation). Available from ProQuest Dissertations and Theses database. (UMI No. 3332994)

Srinivasan, V. (2015). Reimagining the past–use of counterfactual trajectories in socio-hydrological modelling: The case of Chennai, India. *Hydrology and Earth System Sciences*, *19*(2), 785–801. doi:10.5194/hess-19-785-2015

Srinivasan, V., Gorelick, S. M., & Goulder, L. (2010). Factors determining informal tanker water markets in Chennai, India. *Water International*, *35*(3), 254–269. doi:10.1080/02508060.2010.487931

Srinivasan, V., Seto, K. C., Emerson, R., & Gorelick, S. M. (2013). The impact of urbanization on water vulnerability: A coupled human–environment system approach for Chennai, India. *Global Environmental Change*, *23*(1), 229–239. doi:10.1016/j.gloenvcha.2012.10.002

Tanner, T., Mitchell, T., Polack, E., & Guenther, B. (2009). Urban governance for adaptation: assessing climate change resilience in Ten Asian Cities. *IDS Working Papers*, (315), 01-47.

Taylor, R. G., Scanlon, B., Döll, P., Rodell, M., Van Beek, R., Wada, Y., & Konikow, L. et al. (2013). Ground water and climate change. *Nature Climate Change*, *3*(4), 322–329. doi:10.1038/nclimate1744

Trenberth, K. E. (2011). Changes in precipitation with climate change. *Climate Research*, *47*(1), 123–138. doi:10.3354/cr00953

Turner, W. R., Oppenheimer, M., & Wilcove, D. S. (2009). A force to fight global warming. *Nature*, *462*(7271), 278–279. doi:10.1038/462278a PMID:19924191

UN-Habitat. (2011). *Cities and climate change: global report on human settlements 2011*. London: Earthscan.

UNICEF, FAO, & SaciWATERs. (2013). *Water in India: Situation and Prospects, 11*. Retrieved February 9, 2015, from www.unicef.org/india/Final_Report.pdf

van der Kamp, G., & Hayashi, M. (1998). The groundwater recharge function of small wetlands in the semi-arid northern prairies. *Great Plains Research: A Journal of Natural and Social Sciences*. Retrieved February 16, 2015, from http://digitalcommons.unl.edu/greatplainsresearch/366

Vencatesan, J. (2007). Protecting wetlands. *Current Science-Bangalore*, *93*(3), 288–290.

Vencatesan, J., Daniels, R. J. R., Jayaseelan, S. S., & Karthick, M. N. (2014). *Comprehensive Management Plan for Pallikaranai Marsh. Technical Report submitted to the Conservation Authority for Pallikaranai Marsh*. Tamil Nadu Forest Department.

Venkatachalam, L. (2012). *Role of Informal Water Markets in Urban Water Supply: A Household Survey Based Case Study of Chennai City*. Working Paper, Institute for Water Policy, National University of Singapore.

Walsh, R. P. D., Glaser, R., & Militzer, S. (1999). The climate of Madras during the eighteenth century. *International Journal of Climatology*, *19*(9), 1025–1047. doi:10.1002/(SICI)1097-0088(199907)19:9<1025::AID-JOC402>3.0.CO;2-F

Walters, K. M., & Babbar-Sebens, M. (2016). Using climate change scenarios to evaluate future effectiveness of potential wetlands in mitigating high flows in a Midwestern US watershed. *Ecological Engineering*, *89*, 80–102. doi:10.1016/j.ecoleng.2016.01.014

Zedler, J. B., & Kercher, S. (2005). Wetland resources: Status, trends, ecosystem services, and restorability. *Annual Review of Environment and Resources*, *30*(1), 39–74. doi:10.1146/annurev.energy.30.050504.144248

KEY TERMS AND DEFINITIONS

Aquifer: An underground layer of rock saturated with water through which water can be extracted.

Ecosystem services: These are services provided by a particular ecosystem for the benefit of human beings.

Floodplain: It is a low lying area near a stream or river which is at a risk of flooding.

GW: Groundwater.

Marsh: Marsh is a type of wetland which is either frequently or continually inundated, dominated by emergent vegetation. It can range from freshwater to brackish to saltwater.

Peri-urban areas: Areas around urban centres having both urban and rural characteristics.

SW: Surface water

Urbanization: It is the increase in population in urban areas resulting in social, economic and environmental transformations.

Wetland: Land which is either temporarily or permanently saturated with water, characterized by unique soil properties and vegetation.

Chapter 8
Achieving Climate Smart Agriculture with a Sustainable Use of Water:
A Conceptual Framework for Sustaining the Use of Water for Agriculture in the Era of Climate Change

Sneha Kumari
Symbiosis International University, India

Yogesh Patil
Symbiosis International University, India

ABSTRACT

With time there has been an unpredictable climate change affecting the requirement of water for agriculture. Survival of agriculture has become a matter of concern with an efficient management of water. The aim of the chapter is to design a conceptual framework in sustaining water for agriculture in the era of climate change. The authors in the present chapter have used secondary data from previous research work and critically analysed the cases on water management for agriculture. The chapter discusses on practices like drip irrigation, new adaptations, mitigation technologies, vapour pressure, agro-ecological zone model and other water management strategies and the agricultural practices which though increases the yield but is leaving an adverse impact on climate. The chapter designs a conceptual framework to sustain water for agricultural use in the era of climate change and discusses it. This must be dealt through in order to attain sustainability.

DOI: 10.4018/978-1-5225-1046-8.ch008

INTRODUCTION

Agriculture is the soul of every country. With time there has been an unpredictable climate change affecting the requirement of water for agriculture (Branca et al., 2011). Survival of agriculture has become a matter of concern with an efficient management of water (Vanclay, 2004). With the growing population there has been rapid cutting down of forests for socio-economic development. This has resulted into strong buildings and roads which has disturbed the climate. Looking into the past fifty years one can find that there is a drastic change in climate and has severely influenced water resources and agriculture (Sakhuja, 2008). The change in climate is responsible for an increase or decrease in availability of water which has an influence on the biodiversity hotspots, ecosystems and agriculture. Water availability has declined since 1947 from 5150 cubic metre to 2200 cubic metre in 2000. According to the estimations, the country will be water stressed by 2017 when water availability will be only 1600 cubic metres (Kenyon, 2004). The World Bank has reported about Ethiopian agriculture which is dependent on rainfall. As a result of climate change the country is facing severe problems of land degradation and soil erosion due to heavy runoff and deforestation. Morison (1996), in his study has stated that scientists working on climate change are of the opinion that gases leading to climate change will go on increasing which will have a direct effect on water resources and an indirect effect on agriculture. There is a strong link between agriculture, climate change and water which can be clearly seen today (Adger, 2005). One can find a link of the three factors which affects one another directly or indirectly. Looking into the past fifty years one can find that there has been a drastic change in climate and has severely influenced water resources and agriculture (Sakhuja, 2008). Human activities are responsible for the change in climate one or the other way (Singh, 2012). The change in climate is responsible for an increase or decrease in availability of water which has an influence on the biodiversity hotspots, ecosystems and agriculture. This has affected the crop yield leading to food shortage and malnutrition problems.

Several techniques have been evolved for an efficient and integrated water use in agriculture. The authors in the present chapter have used secondary data from previous research work and critically analysing case studies on supportive climate smart agricultural development with watershed management. It's high time now that one has to be prepared for the future and examine all the domains economic, social, institutional and biophysical in order to manage the risks and shocks.. Today the major issue of concern is the evil impacts of climate change on agriculture and the evil impacts of agricultural and water practices to climate change (FAO, 2016). This must be dealt through in order to attain sustainability. The reason for increasing global warming is not just pollution but also the agricultural practices which though increases the yield but is leaving an adverse impact on climate. In one or the other way the water availability has been a matter of concern for agriculture.

Minimum wastage of water can lead to a sustainable water utilisation for agriculture. With time there has been found a vast variation in climate. The reasons for this are rise in population, cutting down of trees, construction of tall buildings, industries, rise in pollution and unsustainable practices followed in the country. These causes have led to change the climate into too hot, humid and followed by unpredictable rainfall. The ultimate sufferers of this change are agriculture whose lands have shifted into buildings and industries and the remaining land is left to be swept away by heavy rainfall or becomes a victim of drought conditions. The objectives of the chapter is to discuss the impact of water availability in agriculture in the era of climate change, design a framework in sustaining water for agriculture in this era of climate change and to discuss on practices like drip irrigation, new adaptations, mitigation technologies, vapour pressure, agro ecological zone model and other water management practices.

INTERLINKAGE BETWEEN AGRICULTURE, CLIMATE CHANGE AND WATER

There has been a gradual rise in temperature, precipitation, humidity, rainfall, evaporation and radiation which indicates that climate has changed a lot so far. Due to change in climate agriculture has been suffered from drastic soil erosion, floods and droughts. These natural calamities have led the agriculture productivity to decline. Water scarcity in agriculture has created a problem for the farmers and it has been realised that there is a strong interlink age among the three factors namely agriculture, water and climate change (Rao & Patil, 2015). Agriculture has been affected by the climate change but it has been found that the unsustainable practices followed in agriculture have also led to climate change indirectly. The applications of chemical fertilisers and pudding practices followed in agricultural field have led to emissions of greenhouse gases affecting climate which ultimately leads to a disturbance in water management. Thus a cycle can be found inter-linking water, agriculture and climate which have been shown in Figure 1.

The scientists working on climate change are of the opinion that gases leading to climate change will go on increasing which will have a direct effect on water resources and an indirect effect on agriculture. Rise in temperature increases the rate of photosynthesis process reducing the life of the crops. For an every rise of 2°C the life of agriculture will fall by two weeks. According to Adam et al. (2001) climate changes strongly affect the yield of the agricultural crops. High temperatures leads to low yield and high yield variances while high precipitation leads to high yield and reduces the variability

As per the study conducted it has been found that agricultural practices in India contributed to emission of 28% of greenhouse gases, 78% methane and nitrous oxides emissions leading to climate change (http://www.ritimo.org/Climate-Change-and-Agriculture-in-India-Avenues-for-Mitigation#) . Carbon dioxide generated through burning fossil fuels, crop residues, Methane generation from inundated paddy fields, spread of numerous pesticides and chemical fertilisers has resulted into emission of greenhouse gases. Puddling in paddy fields is a common practice which has led to reduce the infiltration rate of water in the soil emitting methane and nitrous oxide. The global warming potential of nitrous oxide has been found to be 296 times that of carbon dioxide. Large water reservoirs constructed as water resource

Figure 1. Interlinkage between water, agriculture and climate

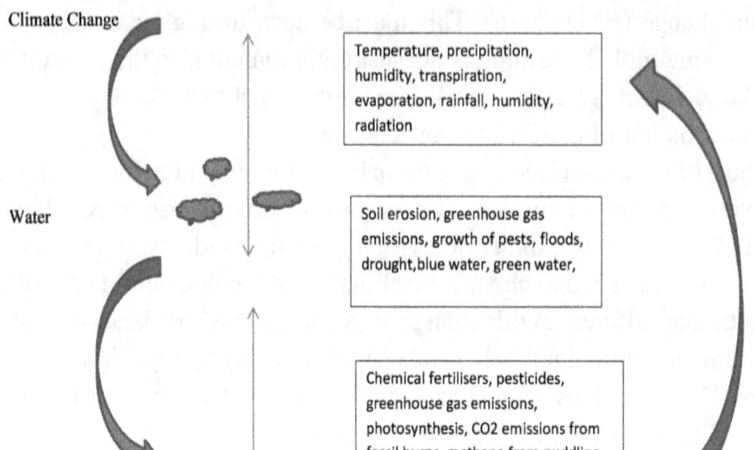

contribute to18.7% of emissions of greenhouse gases (Ramanjaneyulu & Kuruganti, 2009). Water is also a source of greenhouse gas emissions affecting climate. Table 1 shows the share of water resources which emits greenhouse gases.

Several water storage canals, panels and tanks had been constructed in the past for irrigation. With time such ponds and tanks have been filled up in order to construct roads and buildings. The need of such canals was felt at times when there is heavy rainfall leading to floods which results in the accumulation of water in the fields spoiling the agriculture crops. Kharif crops are sown in the month of July which requires a good rainfall. Due to climate change there is spatial and temporal variations in rainfall which ultimately affects the crops. Lack of erratic rainfall results in lowering down the underground water table, thereby reduces the carbon content of the soil and making it easier for the pests to attack the crops. Delay in the monsoon season results in low rainfall, less water storage and availability of water for irrigation. This has resulted into shift in the sowing dates of the kharif crops and rescheduling of water and nutrients. Delay in sowing of kharif crops results in late harvesting of crops and again late sowing of rabi crops. The unpredictable rainfall in rabi season leaves an impact on critical grain crops like wheat, barley, oats and maize. Low availability of water and high temperatures results in low productivity of crops. According to CSREES (2005), in the era of climate change water shortage has become problematic for agriculture. Blue water is the indicator of rivers, lakes, ponds and aquifers which is the main source of irrigation for crops. Rain water constitutes of oxides of nitrous and sulphur (acid rain) or constituents of salt with some trace elements of pollutants which again deteriorates the crops or affects the nutritional content of the crops. Fipps (2007) in his study has reported that the salinity can be destroyed by irrigation which maintains the fertility of the soil. Therefore water if used sustainably can make the soil fertile or if not managed in time can make the agricultural land.

IMPACT OF CLIMATE CHANGE ON AGRICULTURE

India is a diverse country with diversity in climate in every region of the state and so in the entire world. The climate of a region is influenced by the latitudes and longitudes a particular region is located at. Since there is variation in climate therefore the pattern of agricultural crops grown varies from place to place. The alterations in climate largely influence the availability of food (Joshia & Chaturvedib, 2013). The emission of greenhouse gases and other harmful gases leads to deterioration of the quality of food and affects climate which again leads an adverse impact on agriculture. Increase in carbon dioxide concentration in the atmosphere leads to rise in the sea level and also global warming. Due to change in climate the demand for water is expected to rise by 2050 to 11 percent (Arnell, 1999). Table 2 further

Table 1. Emissions of methane in million tonnes from several sources of water storage

Source of Water storage	Emission of Methane (Million Tonnes)
Large dams	33.5
Reservoirs	1.1
Spillways	13.2
Turbines of hydropower	19.2

Source: Ramanjaneyulu & Kuruganti (2009), Sustaining Agriculture in the era of climate change in India

Table 2. Adverse impact of climate on agriculture

Sr.no.	Adverse Impacts of Climate on Agriculture	Reference
1	For every rise in temperature above 29 degree Celsius there is a fall in the yield of corn buy half a percent	Burke, 2012
2.	The effect of temperature with the trends of yields was observed in U.S and it has been observed that warmer temperatures have a positive relation with decline in the yields.	Lobell, 2003
3.	Due to climate change since last 30 years, the floods and heavy rainfall have eroded several fertile soils and there has been variation in the amount of soil eroded with time	Hornbeck, 2012
4.	It has been observed that higher temperatures causes less damage to the crop yield than warm temperatures.	Burke, 2012
5.	With change in climate there has been variation in the availability of water leading to frequent floods and droughts.	Singh, 2013
6.	Increase in temperature reduces the productivity of rice by 4.17%	Upadhyay, (2012)
7.	High temperatures at night has an adverse effect on productivity of rice. For every 1 degree rise in temperature there is a decline of 10 percent in the yield	Thornton et al., 2012
8.	Yam and cassava crops are drought resistant crops but with changing climate they are easily infected by pests, insects and diseases	-
9.	With increase in rainfall the dry crop called potato is also severely affected by pest and insects.	-
10.	Global warming results in growth of weeds	
11.	Climate change results in 20 to 30 percent reduction in the productivity of grains	Darwin, 1995
12.	Increase in temperature with carbon dioxide accumulation promotes the growth of pests and insects	Ziska et al, 2010

explains the adverse impacts of climate on agriculture and it can be concluded that change in climatic conditions has an adverse impact on agriculture yield.

IMPACT OF CLIMATE CHANGE ON GROUND WATER SUPPLY

In the coming future with increasing temperatures, the glaciers and snow will melt down into surface water and result into scarcity of water. During such scenario utilisation of ground water from domestic, industrial and agricultural purpose will be the key. Groundwater can be defined as water stored in terms of water flow or subsurface water in soil, aquifer or deep zones (Green, 2011). Groundwater comprises of one third portion of total fresh water which is used for industries, agricultural activities and domestic use. 40% of the water supply from industry, 50% for domestic use and 20% of water supply is made to irrigation from the ground water (Zektser, 2004). Most of the ground water is utilized for drinking as well as in rural areas. Groundwater is utilized when the surface water goes beyond the limit to fulfil the requirements. During the periods of low or scarce rainfall conditions these groundwater resources acts as the main source of water supply. Therefore this groundwater is the rich source to overcome the need of water in case of climate change. Climate and land cover are largely responsible for the availability of groundwater resource. Extreme high temperatures with no rainfall results in the lower level of ground water supply for domestic as well as agricultural purposes since the withdraw of water exceeds the recharge of water. In the coming future irrigation will require the maximum use of groundwater as the rise in global warming will make the land dry and groundwater will be scarce (Taylor, 2013). Moreover harnessing groundwater will require tremendous amount of energy inputs in terms of electric motor pumps. Currently, 70% of such electricity generated in developing countries is from fossil fuels.

ACHIEVING CLIMATE SMART AGRICULTURE WITH A SUSTAINABLE USE OF WATER

Agriculture sector with the upcoming researches have led to adoption of new practices and technologies for making agriculture as a climate smart agriculture with a sustainable use of water. Climate change and availability of food is the major issue in today's world. The protection of productivity of agriculture is also essential for the local people since seventy- five percent of the rural people residing in rural areas depends upon agriculture for their livelihood (FAO, 2016). The impacts of climate on agriculture are basically due to changing temperatures and rainfall which affects the productivity of crops. There are a number of domestic practices in agriculture which can meet the challenges of climate. Sustainable land management is one of the practices which results into low emission of greenhouse gases to the environment and improves soil water retention. Encouragement of agro forestry results in reducing soil and water erosion, maintaining water requirement and reduces the variation in the crop yield (Ajayi et al., 2006). Agro-forestry, land management and grazing land management are some of the control measures to sustain the land from climatic effects. Conservation of forests and agriculture helps to reduce the erosion of soil during heavy rainfall or floods. Several organic materials and planting materials like seed, fruits, aggregate fruits, bulbs and tubers are dug into the soil in order to increase the moisture capacity of the soil and maintain the fertility of the soil.

The cropping pattern in agriculture is also a step to make it smart climate agriculture. Plantation of cover crops and rotational crops helps to reduce the growth of weeds in the farm which are harmful for the growth of agricultural crops as they intake the nutrients and water which was to be taken by the agricultural crops. The growth of weeds not only takes away the nutrients from the crops resulting into poor yield but also results in unsustainable water practice. Thus the cover crops and rotational crops must be promoted in order to sustain water for agriculture use. Similarly alley cropping practice also helps to increase the organic matter of the soil reducing the soil erosion and helps in efficient utilization of water (Blanco, 2008). The residues left after the crop cycle can be deposited in the soils which increase the moisture content of soil as discussed above. Reduction of weeds and pest control done during a particular stage of the crop is helpful for the farmers not only to increase the yield but also saves water use. Promotion of terraces in areas affected by heavy rainfall is also an effective measure for making agriculture a smart climate one. Gully and riled erosion cause tremendous damage to the soil therefore it is recommended to grow agricultural crops on the slopes in order to conserve soil and water. Grazing in the regions which are affected by heavy rainfall must be minimized. This process helps in reducing soil erosion and conserve water from wastage.

Due to improper practices of agriculture, agriculture sector has been the major contributor of greenhouse gas and carbon dioxide gas emissions. About 50% of nitrogen oxides and 47% of methane gas are emitted from the improper practices of agriculture. Therefore, in order to reduce the hazardous gases to be explored in the environment sustainable practices in agriculture should be followed (Godfray et al., 2011). Technological developments, farm production practices, farm financial practices and government programs are some of the key factors which makes agricultural practices sustainable (Smit & Skinner, 2002). Technologies like improved land practices, organic matter, vermin-composting, fertilizer application, tillage and storage of seeds have improved the results from agriculture. Several climate adapted seeds have been developed for survival of agricultural crops in the field. Agro forestry are the best measures to adapt the climatic change. The practice like tree pruning helps in reducing the transpiration rate and sustains water (Mbow et al., 2014). Figure 2 and 3 depicts some of the measures for improving the water availability for agriculture.

CONCEPTUAL FRAMEWORK FOR SUSTAINING THE USE OF WATER FOR AGRICULTURE IN THE ERA OF CLIMATE CHANGE

Several researches and study has been conducted for sustaining the use of water for agriculture in the era of climate change. Through extensive literature following items have been identified which has been used to develop a conceptual framework for sustaining the use of water for agriculture in the era of climate change. These items are water requirement of different crops and water use efficiency formula, adaptations, control of loss of water, policy, and application of technologies and use of treated water. These constructs of the conceptual framework has been explained in detail in Figure 4.

WATER REQUIREMENT OF DIFFERENT AGRICULTURAL CROPS AND WATER USE EFFICIENCY FORMULA

The requirement of water for different crops is different. Therefore the quantity of water given to a crop must be given as per its requirement for optimum yield of the crops and also for the sustainable use of water. There are several factors which determine the requirement of water for a particular crop. Climate is one of the major factors which determine the use of water by each crop and the changing climate in today's era has led to variation in the utilisation of water for different crops. A crop grown in a hot and

Figure 2. Hollow dugs to store rain water for agriculture purposes

Figure 3. Pruning leads to reduce the transpiration rate

Figure 4. Conceptual Framework for sustaining the use of water for agriculture in the era of climate change

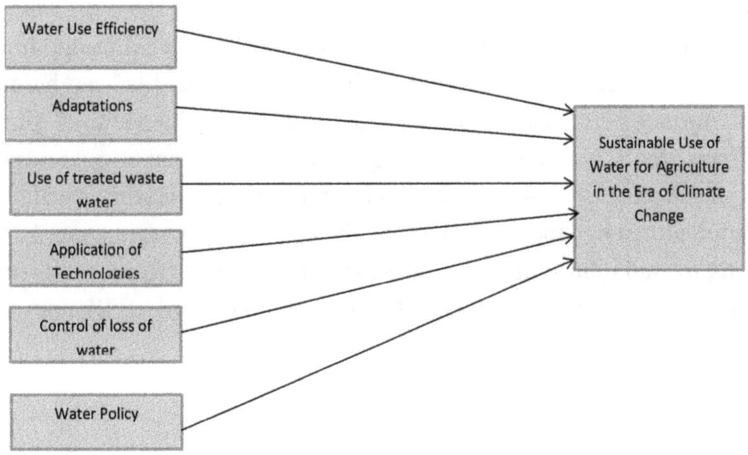

humid climate requires more water than that grown in a cooler climate. For further analysis it is essential to target a single crop and notice its water requirement in different climate conditions. The concern to sustain water availability with changing climate has been discussed below. Another factor which determines the utilisation of water is the type of the crop. A single crop has several kind of varieties and every variety needs for water requirement are different. Therefore, it is recommended to grow those variety crops which require less water in the area resistant with drought and heat. The duration of crops also is a factor for concern in the era of climate change. Few crops are short duration crops and complete their cycle with the season. Such crops can be grown in order to overcome the challenges of climate.

The requirement of water can be calculated for different crops by the following measures:

Equation 1:

$$\mathrm{ET}crop = \mathrm{ET}o \times \mathrm{K}c$$

where,
ET crop = Crop Evapotranspiration or Crop Water Need (mm/Day)
ET_0 = Reference Evapotranspiration (mm/Day)
K_c = Crop Factor
Both ET crop and ET_0 are expressed in the same unit: usually in mm/day (as an average for a period of one month) or in mm/month
Source: http://www.fao.org/docrep/s2022e/s2022e07.htm

The crop factor or the crop coefficient varies with the development of crops in the initial stage, mid-season stage and in the late season stage. This can be calculated by the pan evaporation method. Under this method a pan is placed in the field with water as marked and left for 24 hours. Observation is made after 24 hours with the level of water in the pan. The decrease in the level of water gives the evapotranspiration rate occurring from the pan and the evapotranspiration rate is calculated using a formula as evapotranspiration rate = decrease in the depth of pan * coefficient of the pan. The similar study can be done with the crops which can survive in water for 24 hours and left for observation. The evapotranspiration rate for the crop will be the change in the reduction of the depth of water before the crop was placed in the pan and the change in the depth of water after the crop was placed into the pan. This measure of evapotranspiration comes out with its limitation that water depth may change in such a small unit that it can be difficult to measure. For this a second measure can be made for calculating the evapotranspiration rate of crop which is known as Blanney Cridle Method. Under this method the average maximum temperatures is calculated as the sum of all the maximum temperatures divided by the number of days in a month. Similarly the average minimum temperature is calculated by the sum of all the minimum temperatures in a month divided by the number of days in a month. Taking the average of maximum temperature and minimum temperatures as calculated by the above formula gives the mean temperature which is used in the Blanney Cridle formula named after H.F Blanney and W.D Criddle as

Equation 2:

$$\mathrm{ET}o = p\left(0.46\mathrm{T}mean + 8\right)$$

where,
ET_o = Reference crop evapotranspiration (mm/day) as an average for a period of 1 month
T mean= mean daily temperature (°C)
p = mean daily percentage of annual daytime hours
Source: http://www.fao.org/docrep/s2022e/s2022e07.htm
The water requirement for crops can also be calculated as per the Penman–Monteith equation (FAO, 2016)

Equation 3:

$$\mathrm{ET}o = \frac{0.408\Delta\left(Rn - G\right) + \gamma\frac{900}{\mathrm{T}+273}U2\left(es - ea\right)}{\Delta + \gamma\left(1 + 0.34U2\right)}$$

where
ET_o reference evapotranspiration [mm day^{-1}],
R_n net radiation at the crop surface [MJ m^{-2} day^{-1}],
G soil heat flux density [MJ m^{-2} day^{-1}],
T mean daily air temperature at 2 m height [°C],
u_2 wind speed at 2 m height [m s^{-1}],
e_s saturation vapour pressure [kPa],
e_a actual vapour pressure [kPa],
e_s - e_a saturation vapour pressure deficit [kPa],
D slope vapour pressure curve [kPa °C^{-1}],
g psychrometric constant [kPa °C^{-1}].

Water use efficiency in agriculture was then taken as yield per unit of evapotranspiration rate. In agronomy this term has been defined as the amount of crop yield produced divided by the amount of water used to produce that yield. This irrigation efficiency can be can be improved by agronomic practices which includes plant spacing, tillage practice, conservation of water, crop residues and introduction of improved crop varieties, engineering, management and institutional practices (Wallace & Batchelor 1997).

Howell, (2001) proposed water use efficiency in the field using a formula as:

Equation 4:

$$WUE = \frac{\left(HI \times DM\right)}{\left\{T\left(1 - WC\right)\left[1 + \frac{\mathrm{E}}{\left(P + I + SW - D - Q - E\right)}\right]\right\}}$$

where

HI = Harvest Index (Dry yield per unit dry matter)
DM= Dry Matter in gm^{-2}
T = Transpiration in mm
WC = Standard Water Content used to express the economic yield
E = Soil Water Evaporation in mm
P = Precipitation in mm
I = Irrigation in mm
SW = Soil Water Depletion from the root zone in mm
D = Deep percolation below the root zone in mm
Q = Surface runoff in mm

This equation given by Howell, (2001) undertook all the agronomic and engineering practices to enhance water use efficiency. This can be an effective step towards a sustainable use of water in agriculture.

Table 3 above shows the water required for different crops. This requirement can be taken as guidance for growing crops under severe climatic conditions. Crops like rice, sugarcane, banana, alfalfa and cotton requires a good amount of water for their growth. This means that the areas which are drought should not go for cultivation of these crops. These crops should only be grown where there is adequate supply of water. However in drought areas the farmers can grow crops like tomato, potato, wheat and onion.

The figure compares the requirement of water of different crops with different climatic condition as shown in Table 3 with the different climatic conditions. This shows that the water requirement of every crop varies with temperature, humidity, wind speed and rainfall.

ADAPTATION OF AGRICULTURE TO CLIMATE CHANGE

Since long agriculture is known to be affected by the climate change and efforts are being made to adapt the agriculture systems to climate change (Samuelson, 1947) and there are several reports for the adaptations to be carried in agriculture to meet the challenges of climate. With time several adaptations came into existence in agriculture. There can be long run or short run adaptations towards climate as studied through the case studies which are explained in the Table 3. Farmers have shifted the cultivation date of their agricultural crops as per the climate predictions. This has made their work easy from sowing the seeds to harvesting the crops.

Table 3. Amount of water required in the production of the crops

Sr. No.	Crops	Water Requirement in mm
1.	Rice	900-2500
2	Wheat	450-650
3	Sorghum	450-650
4	Maize	500-800
5	Sugarcane	1500-2500
6	Groundnut	500-700
7	Cotton	700-1300
8	Soybean	450-700
9	Tobacco	400-600
10	Tomato	600-800
11	Potato	500-700
12	Onion	350-550
13	Grape	500-1200
14	Ragi	400-450
15	Gingelly	350-400
16	Pineapple	700-1000
17	Citrus	900-1200
18	Banana	1200-2200
19	Peas	350-500
20	Cabbage	380-500
21	Castor	500
22	Sunflower	350-500
23	Chillies	500
24	Alfalfa	800-1600
25	Melon	400-600
26	Cotton	700-1300
27	Sunflower	600-1000
28	Sugar beet	550-750
29	Tomato	400-800
30	Pepper	600-900
31	Peanut	500-700
32	Onion	350-500
33	Barley	450-600
34	Oats	450-600

Source: agropedia.iitk.ac.in/node/4977

Figure 5. Climatic factors influence crop water needs (http://www.fao.org/docrep/s2022e/s2022e07.htm)

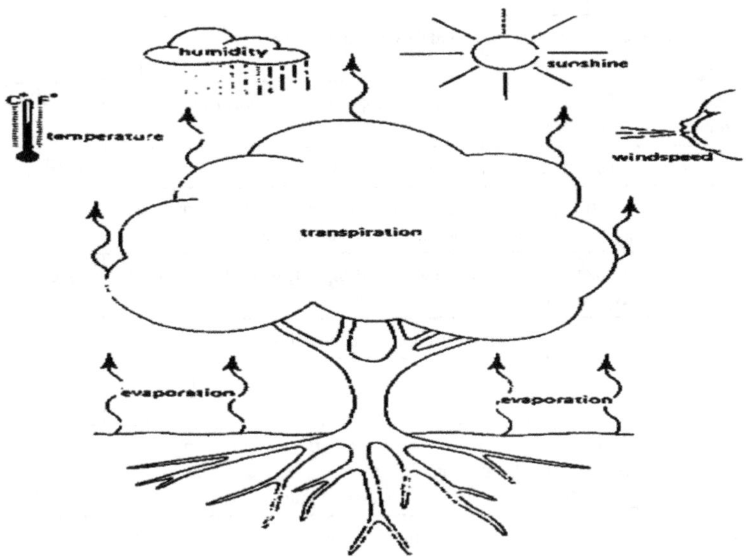

CONTROL OF LOSS OF WATER

The unsustainable use of water leads to huge loss in agriculture ultimately giving rise to water scarcity. Water in the basin when goes for irrigation, in the process it undergoes several losses. The basis loss which a stored water undergoes are mainly evaporation which leads to loss of water, some of the water is percolated by the surface and some are drainage water contaminated with many pollutants and chemicals which make the water good for nothing. Lot of water applied for irrigation leads to water logging and salinisation while the water in closed system are polluted to such an extent that it cannot be reused. Water loss is mainly made by the unsustainable use of water. The cause of unsustainable water is mainly variations in the availability of water and variations for demand of water (Gleick, 1998).

The loss of water depends upon the intensity of rainfall that is the amount of rain water which falls in a given period of time. The intense of rainfall is responsible for the runoff damages. When rain water continues to flow it directly goes into the soil where the infiltration rate of the soil helps to absorb the water up-to a limit. When water exceeds to a particular limit the soil puddles into the rain water and then continues to run-off. The infiltration rate of different soil is different and varies with the structure and texture of the soil. The infiltration capacity of every soil depends upon the moisture content of the soil and the porous characteristics of the soil. The amount of water loss also depends upon the type of vegetation cover in the farm. The vegetation cover helps to maintain the infiltration rate of the soil and reduces the run-off from the soil. The slope of a land also decides the amount of water to be lost. A steep slope leads to more water runoff than a gentle slope. Therefore it is not recommended to have steep slopes in order to conserve water. The run-off from rain water can be calculated as the percentage of runoff * rainfall depth

Table 4. Several long run and short run adaptation of agriculture towards climate change

Sr. No.	Adaptations of Agriculture Towards Climate Change	Reference
1	Irrigation facilities made to agriculture by ground water pumps can be a short run adaptation in a drought prone area	Fisher et al, 2012
2	Shift in the time periods of crops can save them from the climate disaster	Burke, 2012
3	Alteration in planting date, irrigation technologies, variation in crops and management of inputs supplied to the field can fight with the challenges of climate change	Burke, 2012
4	Variations in the management practices carried on land or shifting forests to pastures or cropland can be a good adaptive measure for agriculture	Lobella, 2006
5	Switching to different seed varieties acts as a measure for climate adaptations.	Burke, 2012
6	Water can be recycled and reused for agricultural needs. If needed heavy fines must be charged for returning polluted water	National Water Policy, 2012
7	In 2011 it has been found that productivity of rice was affected by the climate change in Assam. As an adaptation strategy it has been found that growing new rice varieties adaptable to climate change and shift in cultivation dates is a good practice.	Upadhyay, (2012)
8	Production of two or more crops in a narrow strip commonly known as strip intercropping is an adaptation for sustaining water need with change in climate	Candelaria, (2012)
9	Removal of older leaves commonly called as leaf pruning is a practice to reduce transpiration rate. Reduction in transpiration rate leads to reduced water requirement	-
10	Minimum tillage leads to less water and soil loss	-
11	Mulching can be done in order to conserve soil. This helps to retain the moisture in the soil	-
12	Flat and sloping lands helps to overcome the challenges of climate change	-
13	Drought tolerant crops like gram and legumes are a good source of nitrogen fixation in soil and also leads to survival in the era of climate change	-
14	Increase in small and large irrigation lands leads to expansion of the agricultural land	Bosello et al. (2013)
15	Non-irrigation practices should be followed by irrigation practices	Bosello et al. (2013)
16	Reduce tree cutting	-
17	Ban on grazing	-
18	Construction of soil nullah bunds	-
19	Consumption of water by drip irrigation practice	-
20	Restrictions on water intensive crops	-
21	Promotion of sprinkle irrigation	-
22	Restrictions on use of bore wells for irrigation	-
23	Prohibition of water intensive crops like sugarcane and banana especially in drought prone areas	-
24	Plantation of bio-fuel crop	Thornton & Cramer (2012)
25	Construction of pools	Thornton & Cramer (2012)
26	Use of pits, dams and ridges	Thornton & Cramer (2012)
27	Organic farming leads to reduce the need of irrigation by 30-50%	-
28	Organic soils have good water holding capacity and drainage which reduces the risk of drought and soil erosion	-

POLICY

There has been competition for need for water in agriculture, industries and cities. With climate change water has become the most challenging parameter in agriculture in today's era. Ground water is being polluted by the surface water which has been affected by industrial wastes. These water logging, desalinisation water or polluted water is deteriorating the productivity of irrigated lands. This leads to the wastage of around 60% of water for irrigation. The irrigation potential has been found to increase from 19.5 million hectare since 1947 to 106 million hectares in 2010 (Singh, 2013). National water policy in 2012 clearly states that water budgeting and water auditing will be made as a mandatory practice for regulation of water use efficiency. Water tariffs and water charges will be taken for sustaining water in the coming years. Agriculture and water related issues are approached through partial and general equilibrium (Ponce et al., 2012). Partial equilibrium considers the equilibrium conditions for one sector that is comparative view of prices in one with all across the markets. Computable equilibrium models are used by the general equilibrium approach which needs to solve supply, demand, prices and other parameters for several markets. With the rise in population there has been shortage of agricultural food which needs water as a demand to be fulfilled. Berritella et al. (2007) were the first to raise the issues related to water. With such an important aspect of water, water intensity coefficient was defined by FAOSTAT and AQUASTAT as the quantity of water required for a unit. The total water requirement can be defined as water supplied by means of irrigation commonly called as blue water and water supplied in agriculture commonly known as green water.

Water has never been charged but looking at the Computable equilibrium model in the coming years if water supply cannot meet the demands of water requirement then water has to be charged in terms of rupee to meet the problems of water scarcity. Strzepek et al., (2002) in this aspect defined the term water price elasticity. It means that in order to realise the value of water, price elasticity is a good term. Water price elasticity has been introduced with a concept that as the price of water will increase, the use of water will reduce. Berritella et al., (2005) through his research analysed that in order to have a good flow of water the availability of water should be decreased. He introduced four parameters in order to avoid water scarcity. These were sustainable water supply, pricing of water, liberalisation of trade and china water transfer. It explains that water should be charged at amount per cubic metre of water used and removal of all the trade barriers. Later on Berritella et al. (2006) introduced tax on water supply as water tax being charged differently for agriculture and industry sector. Due to the arrival of water tax the amount of water demand reduced in regions around the world. Berritella et al., (2007) again studied the impacts of water scarcity. The water supplied should be restricted in order to save water. The author again came up with few parameters like water price elasticity, shortage of water, water property rights and specific water to save water use. The efficient use of water in agriculture has been defined by Calzadilla et al., (2008) as an improved irrigation management. He defined irrigation efficiency as the ratio of irrigation water utilised by the crop to the water applied. This helps in improving efficiency of irrigation in water stressed regions. The sustainable use of water in agriculture can be obtained by suppressing the ground water exploitation.

Dudu & Chumi (2008) proposed the Computable General Equilibrium model constituting 20 sectors comprising 12 in agriculture, 4 agricultural related sectors and 4 non agriculture related sectors. The model constitutes of land, labour, water, capital and other inputs. The five elements are essential for productivity of agriculture sector. Another computable general equilibrium model proposed by Gomezet et al. (2004) liked water with trade liberalisation and the two sectors which produced water traditional and desalinisation charged different costs. They divided agriculture sector into irrigated agriculture and rainfed agriculture constituting land, labour, capital, water and sea water as the factors of production. Strzepek et al. (2008) proposed a computable general equilibrium model to analyse that the land water relationship is very crucial. He analysed the effect of construction of aswan dam for farmers. These dams acted as the best source for water conservation during summers which otherwise would lead to drought conditions or floods. Water requirement from these dams for agriculture differs from season to season. Therefore water should be charged differently for different seasons. Inspite of the fact that every state can frame their water policies it has been felt mandatory to involve the national frameworks for water

APPLICATION OF TECHNOLOGIES

There is a need for innovation in promoting technologies for reducing water wastage and it's application for the survival of agricultural crops. Meijers (2014) has discussed about six technologies which are used for saving water for agricultural growth in Table 5.

Dinar and Yunan (1996) studied different irrigation technologies for application of water for agriculture use. He studied Tradition Irrigation using furrows and bunds, sprinkler irrigation, solid set spriklers above canopy, solid set spriklers below canopy, low volume micro sprinklers and drip irrigation. The Figure 3 explains the irrigation technology used in agriculture for a sustainable use of water.

USE OF TREATED WATER

Whenever there is a scarcity of good quality water it is essential to treat the wastewater and apply in agriculture. Typical domestic waste water constitutes of total solids, chloride, dissolved solids, suspended

Table 5. Technologies used for agricultural growth

Technology					
Ondavia	**Fruition Science**	**Grove Labs**	**Precision Hawk**	**Farm Hack**	**Bitoponics**
It performs laboratory test to identify harmful toxic elements in water for agricultural growth.	This is software to monitor weather conditions, humidity and wind speed for agriculture crops.	This helps in allowing web connected data for hydroponics and greenhouse growing operations	This is an airborne platform which allows the agriculture sector to gather remote sensing data	This consists of tools for developing resilient sustainable agriculture.	It is a very essential device which automates web services and monitors the garden health and sends the data to cloud in real time which controls light and water pumps for the garden growth. This process helps in making any changes or taking actions as needed.

Figure 6. Irrigation technologies for sustainable use of water

solids, nitrogen, phosphorous, alkalinity, grease and biochemical's (Pescod, 1992). Municipal waste water also consists of various inorganic substances and industrial waste water constitutes of several toxic elements. Many ions which are beneficial for the crops at low concentration becomes toxic at high concentration therefore for utilizing treated wastewater for agriculture use it is essential to evaluate the water quality in terms of physical parameters like temperature, electric conductivity, hardness, sediments, Total dissolve solids and chemical properties like acidity, basicity and presence of the chemical elements (Pescod, 1992).

CONCLUSION

Water flow in agriculture must be efficient for a sustainable country. The availability of water is dependent on several factors. Natural factors like climate change have influenced the availability of water to a great extent. The rise in population has led to overuse of water. For a sustainable water use the major issue of concern should be the rising population and the innovative technologies (Gleick, 1998). The unlimited population has led to destruction of agricultural land and overuse of water affecting the climate and sustainability. Similarly the innovative technologies have led to involvement of huge water supply as an input and results into several polluting agents which again have an impact on climate. The use of water should be made giving priority to human being and the environment. The basis requirement of water for human being is the key criteria to look into wastage of water. Secondly is the environment which requires water for the ecosystem. This introduced a term called water use efficiency. Water use efficiency is defined as the crop yield obtained per unit of water use. In other words water consumed by the crop is utilised as carbohydrates formed in photosynthesis and water per unit transpiration occurring in a crop. This gave the concept of water use efficiency for irrigation (Bos, 1980).

The world's more than 80% of water supply are consumed by the agricultural field (Howell 2001). There are several irrigation technologies applied in the farm which helps in enhancing the water use efficiency and reduces the water loss. Water can be utilised efficiently by increasing the water use efficiency that is yield per unit of evapotranspiration, reducing water from being polluted, reducing wastage of water and allocation of water from lower value to a higher value (Seckler, 1996)

Water quality standards are one of the important aspects for sustainable water. Water quality standard should be developed for several aspects like agriculture, industries or consumption. There should be an upper limit to water use. Several adaptation strategies have been introduced for overcoming the challenges of climate change in agriculture making a sustainable use of water. Water storage must be done through participatory action research (Agrawal, 2009). Rainwater harvesting is one of the active practices that must be followed in every home to make the supply of water uniform. Rain water harvesting is defined as collecting and storing the rain water (Myers, 1975) and gathering it to fulfil the needs of uncertain climate in need instead of wasting it. Rain water harvesting depends on several factors namely rainfall intensity, storage capacity and infiltration rate. Rain water harvesting largely depends upon the quantity of water which can be stored under a given climatic condition in a particular area. The runoff efficiency for harvesting water largely depends upon the surface, management of vegetation and chemicals that reduce the infiltration rate and increases the runoff (Boers, 1982). This rain water harvesting must be done in every district of the country so that water can be utilised for the agricultural crops at times of need very efficiently. Rain water must also be collected at an individual level by gathering the water from the roof top and collecting it for storage. The cost effective measure must be applied collects a specific volume of water from an area of a roof which gives the maximum capacity of water storage (Liaw, 2004).

The biggest limitations found in climate smart agriculture is that farmers do not accept the fact that climate are changing and adaptations are strictly required in agriculture streams (Burke, 2012). In recent past there has been drastic change in climate and the biggest limitation to overcome the challenges of climate change is that farmers practice very limited adaptations in agriculture. The unanswered question is whether they lack the capability to practice any adaptations or they do not feel any need for adaptations.

The application of water use efficiency is difficult since it involves many management factors like fertility of soil, pest management, soil-water content, planting date, spacing and other land practices (Howell, 2001). Since the water use efficiency is defined as the yield per unit of water. The term yield is not only dependent on water. It also depends on other factors as mentioned above. Efforts have been made to calculate the minimum water required for the ecosystem but it is quite difficult since the ecosystem comprises of the natural ecosystem and the ecosystem and the ecosystem which is managed by humans. Groundwater is very difficult to monitor and access. Pumping of ground water becomes difficult, cost effective and time consuming. Government and private bodies should fund projects for promoting irrigation in agriculture.

FURTHER RESEARCH DIRECTIONS

Despite the emerging interest of sustainable use of water there is still a dearth of literature with regards to the sustainable use of water for agriculture in the era of climate change. Majority of the research has explored the issues such as water problems, poor agriculture growth and farmer's suicide due to water unavailability, however the derivation of insights from the ocean of researches requires experience and knowledge of the way to make these information reach to farmers. In this era of climate change, although researchers have recognised the need of sustainable use of water for agriculture, however research to address the implications of this knowledge still exists.

REFERENCES

Adams, Chen, McCarl, & Schimmelpfennig. (2001). Climate variability and climate change: Implications for agriculture. In The Long-Term Economics of Climate Change: Beyond a Doubling of Greenhouse Gas Concentrations. Emerald Group Publishing Limited.

Agrawal, A., & Perrin, N. (2009). Climate adaptation, local institutions and rural livelihoods. In *Adapting to climate change: thresholds, values, governance*, (pp. 350-367). Retrieved from agropedia.iitk.ac.in/node/4977

Ajayi, O. C., Akinnifesi, F. K., Sileshi, G., & Kanjipite, W. (2009). Labour inputs and financial profitability of conventional and agroforestry-based soil fertility management practices in Zambia. *Agrekon,* (3), 276-292.

Arnell, N. W. (1999). Climate change and global water resources. *Global Environmental Change, 9,* S31–S49. doi:10.1016/S0959-3780(99)00017-5

Berrittella, M., Hoekstra, A. Y., Rehdanz, K., Roson, R., & Tol, R. S. (2007). The economic impact of restricted water supply: A computable general equilibrium analysis. *Water Research, 41*(8), 1799–1813. doi:10.1016/j.watres.2007.01.010 PMID:17343892

Berrittella, M., Hoekstra, Y., Rehdanz, K., Roson, R., & Tol, R. (2005). Virtual water trade in a general equilibrium analysis. In *GTAP Conference Paper* (Vol. 1715).

Berrittella, M., Rehdanz, K., & Tol, R. S. J. (2006). *The economic impact of the South-North water transfer project in China: A computable general equilibrium analysis.* FEEM Working Papers, No. 2006,154.

Blanco, H., & Lal, R. (2008). Soil and water conservation. Principles of soil conservation and management. Springer.

Boers, T. M., & Ben-Asher, J. (1982). A review of rainwater harvesting. *Agricultural Water Management, 5*(2), 145–158. doi:10.1016/0378-3774(82)90003-8

Bos, M. G. (1980). Irrigation efficiencies at crop production level. *International Commission on Irrigation and Drainage Bulletin, 29*(2), 18–25.

Bosello, F., Campagnolo, L., & Eboli, F. (2013). *Climate change and adaptation: the case of Nigerian agriculture.* Working Paper. Retrieved from https://www.econstor.eu/bitstream/10419/72963/1/745144 42X.pdf

Branca, G., McCarthy, N., Lipper, L., & Jolejole, M. C. (2011). Climate-smart agriculture: a synthesis of empirical evidence of food security and mitigation benefits from improved cropland management. *Mitigation of Climate Change in Agriculture Series, 3.*

Bruinsma, J. (2009, June). The resource outlook to 2050. In Expert meeting on how to feed the world in (Vol. 2050, pp. 1-33).

Burke, M., & Emerick, K. (2012). *Adaptation to climate change: Evidence from US agriculture.* Academic Press.

Calzadilla, A., Rehdanz, K., & Tol, R. S. (2008). *Water scarcity and the impact of improved irrigation management: A CGE analysis* (No. 1436). Kiel Working Paper.

Calzadilla, A., Rehdanz, K., & Tol, R. S. (2010). The economic impact of more sustainable water use in agriculture: A computable general equilibrium analysis. *Journal of Hydrology (Amsterdam)*, *384*(3), 292–305. doi:10.1016/j.jhydrol.2009.12.012

Candelaria, A. P. (2012). Mitigating Climate Risks Utilizing Good Practices Options in Upland Ecosystems of Bicol Region, Philippines. *OIDA International Journal of Sustainable Development*, *4*(03), 97–108.

Cerri, C. E. P., Sparovek, G., Bernoux, M., Easterling, W. E., Melillo, J. M., & Cerri, C. C. (2007). Tropical agriculture and global warming: Impacts and mitigation options. *Scientia Agricola*, *64*(1), 83–99. doi:10.1590/S0103-90162007000100013

CSREES. (2005). *CREES Agricultural Water Security White Paper, Cooperative State Research, Education and Extension Service*. Washington, DC: CSREES.

Darwin, R., Tsigas, M., Lewandrowski, J., & Raneses, A. (1995). *World agriculture and climate change: economic adaptations*. Agricultural Economic Report No.703. Natural Resources and Environmental Division, Economic Research Service. U.S. Department of Agriculture.

Diao, X., & Roe, T. (2003). Can a water market avert the double-whammy of trade reform and lead to a win–win outcome? *Journal of Environmental Economics and Management*, *45*(3), 708–723. doi:10.1016/S0095-0696(02)00019-0

Dinar, A., & Yaron, D. (1992). Adoption and abandonment of irrigation technologies. *Agricultural Economics*, *6*(4), 315–332. doi:10.1016/0169-5150(92)90008-M

Döll, P. (2009). Vulnerability to the impact of climate change on renewable groundwater resources: A global-scale assessment. *Environmental Research Letters*, *4*(3), 035006. doi:10.1088/1748-9326/4/3/035006

Dossiers, A. (2010). Climate Change and Agriculture in India: Avenues for Mitigation. *India and Climate Change*. Retrieved January 25, 2016, from http://www.ritimo.org/Climate-Change-and-Agriculture-in-India-Avenues-for-Mitigation#

Duda, H., & Chumi, S. (2008). *Economics of irrigation water management: A literature survey with focus on partial and general equilibrium models*. World Bank Policy Research Working Paper Series.

FAO. (2016a). *Climate-Smart Agriculture*. Retrieved January 25, 2016, From http://www.fao.org/climate-smart-agriculture/en/

FAO. (2016b). *Irrigation Water Management: Irrigation water needs. Climate-Smart Agriculture*. Retrieved January 25, 2016, From http://www.fao.org/docrep/s2022e/s2022e07.htm

Fipps, G. (2007). *Irrigation water quality standards and salinity management strategies*. Available at: http://lubbock.tamu.edu/irrigate/documents/2074410-B1667.pdf

Fisher, A. C., Hanemann, W. M., Roberts, M. J., & Schlenker, W. (2012). The economic impacts of climate change: Evidence from agricultural output and random fluctuations in weather. *The American Economic Review*, *102*(7), 3749–3760. doi:10.1257/aer.102.7.3749

Gleick, P. H. (1998). Water in crisis: Paths to sustainable water use. *Ecological Applications*, *8*(3), 571–579. doi:10.1890/1051-0761(1998)008[0571:WICPTS]2.0.CO;2

Godfray, H. C. J., Pretty, J., Thomas, S. M., Warham, E. J., & Beddington, J. R. (2011). Linking policy on climate and food. *Science*, *331*(6020), 1013–1014. doi:10.1126/science.1202899 PMID:21273449

Gomez, C. M., Tirado, D., & Rey-Maquieira, J. (2004). Water exchanges versus water works: Insights from a computable general equilibrium model for the Balearic Islands. *Water Resources Research*, *40*(10), n/a. doi:10.1029/2004WR003235

Green, T. R., Taniguchi, M., Kooi, H., Gurdak, J. J., Allen, D. M., Hiscock, K. M., & Aureli, A. et al. (2011). Beneath the surface of global change: Impacts of climate change on groundwater. *Journal of Hydrology (Amsterdam)*, *405*(3), 532–560. doi:10.1016/j.jhydrol.2011.05.002

Hornbeck, R. (2012). The Enduring Impact of the American Dust Bowl: Short and Long-Run Adjustments to Environmental Catastrophe. *The American Economic Review*, *102*(4), 1477–1507. doi:10.1257/aer.102.4.1477

Howell, T. A. (2001). Enhancing water use efficiency in irrigated agriculture. *Agronomy Journal*, *93*(2), 281–289. doi:10.2134/agronj2001.932281x

Joshia, K., & Chaturvedib, P. (2013). Impact of climate change on agriculture. *Octa Journal of Environmental Research, 1*(1).

Kalra, N., Chander, S., Pathak, H., Aggarwal, P. K., Gupta, N. C., Sehgal, M., & Chakraborty, D. (2007). Impacts of climate change on agriculture. *Outlook on Agriculture*, *36*(2), 109–118. doi:10.5367/000000007781159903

Kenyon, Pollett, & Wills-Johnson. (2000). *Australia Sustainable Water Management Practices: Lessons*. Academic Press.

Liaw, C. H., & Tsai, Y. L. (2004). Optimum storage volume of rooftop rain water harvesting systems for domestic use. *Journal of the American Water Resources Association*, *40*(4), 901–912. doi:10.1111/j.1752-1688.2004.tb01054.x

Lobell, D. B., & Asner, G. P. (2003). Climate and management contributions to recent trends in US agricultural yields. *Science*, *299*(5609), 1032–1032. doi:10.1126/science.1077838 PMID:12586935

Lobell, D. B., Bala, G., & Duffy, P. B. (2006). Biogeophysical impacts of cropland management changes on climate. *Geophysical Research Letters*, *33*(6), L06708. doi:10.1029/2005GL025492

Mann, W., Lipper, L., Tennigkeit, T., McCarthy, N., Branca, G., & Paustian, K. (2009). *Food Security and Agricultural Mitigation in Developing Countries: Options for Capturing Synergies*. Rome: FAO.

Mbow, C., Smith, P., Skole, D., Duguma, L., & Bustamante, M. (2014). Achieving mitigation and adaptation to climate change through sustainable agroforestry practices in Africa. *Current Opinion in Environmental Sustainability*, *6*, 8–14. doi:10.1016/j.cosust.2013.09.002

McCarthy, N., Lipper, L., & Branca, G. (2011). *Climate-smart agriculture: smallholder adoption and implications for climate change adaptation and mitigation. Mitigation of Climate Change in Agriculture.* Working Paper, 3.

Meehl, G. A., Arblaster, J. M., & Branstator, G. (2012). Mechanisms contributing to the warming hole and the consequent US east–west differential of heat extremes. *Journal of Climate, 25*(18), 6394–6408. doi:10.1175/JCLI-D-11-00655.1

Meijers, N. (2014). *Water-Saving Agricultural Technologies*. Retrieved January 25, 2016 From http://foodtechconnect.com/2014/03/21/6-water-saving-agricultural-technologies/

Meijers, N. (2014). The platform for good food innovation. *Water-Saving Agricultural Technologies*. Retrieved January 25, 2016, From http://foodtechconnect.com/2014/03/21/6-water-saving-agricultural-technologies/

Morison, J. I. L., & Morison, J. I. L. (1996). Climate change and crop growth. *Environmental Management and Health, 7*(2), 24–27. doi:10.1108/09566169610112980

Myers, L. E. (1975). Water harvesting 2000 B.C. to 1974 A.D. *Proc. Water Harvesting Syrup*. Phoenix, AZ: USDA. Retrieved January 25, 2016, From http://www.fao.org/docrep/s2022e/s2022e07.htm

National Water Policy. (2012). *Government of India*. Retrieved January 25, 2016 http://wrmin.nic.in/writereaddata/NationalWaterPolicy/NWP2012Eng6495132651.pdf

Neil Adger, W., Arnell, N. W., & Tompkins, E. L. (2005). Successful adaptation to climate change across scales. *Global Environmental Change, 15*(2), 77–86. doi:10.1016/j.gloenvcha.2004.12.005

Pescod, M. B. (1992). *Wastewater Treatment and Use in Agriculture*. FAO Irrigation and Drainage Paper 47. Food and Agriculture Organization (FAO) of the United Nations.

Ponce, R., Bosello, F., & Giupponi, C. (2012). Integrating Water Resources into Computable General Equilibrium Models – A Survey. Climate Change and Sustainable Development Series.

Ramanjaneyulu, G. V., & Kuruganti, K. (2009). *Sustaining Agriculture in the era of Climate Change in India*. Centre for Sustainable Agriculture.

Rao, P., & Patil, Y. (2015). Climate Resilience in Natural Ecosystems in India: Technology Adoption and the Use of Local Knowledge Processes and Systems. In Handbook of Climate Change Adaptation (pp. 2063-2077). Springer Berlin Heidelberg.

Sakhuja, N. (2008). Hiware Bazaar - A Village with 54 Millionaires. *Down to Earth, 16*(20080131).

Samuelson, P. A. (1947). *Foundations of Economic Analysis*. Harvard University Press.

Seckler, D. W. (1996). *The new era of water resources management: from" dry" to" wet" water savings* (Vol. 1). IWMI.

Singh, K. M., Singh, R. K. P., Meena, M. S., & Kumar, A. (2013). *Water Policy in India: A Review*. Available at SSRN 2226877

Singh, R. K. P., & Singh, K. M. (2012). Climate Change, Agriculture and ICT: An Exploratory Analysis. *ICT for Agricultural Development Under Changing Climate*, 17-28.

Smit, B., & Skinner, M. W. (2002). Adaptation options in agriculture to climate change: A typology. *Mitigation and Adaptation Strategies for Global Change*, *7*(1), 85–114. doi:10.1023/A:1015862228270

Strzepek, K. M., Yohe, G. W., Tol, R. S., & Rosegrant, M. W. (2008). The value of the high Aswan Dam to the Egyptian economy. *Ecological Economics*, *66*(1), 117–126. doi:10.1016/j.ecolecon.2007.08.019

Tarawali, G., Manyong, V. M., Carsky, R. J., Vissoh, P. V., Osei-Bonsu, P., & Galiba, M. (1999). Adoption of improved fallows in West Africa: Lessons from mucuna and stylo case studies. *Agroforestry Systems*, *47*(1-3), 93–122. doi:10.1023/A:1006270122255

Taylor, R. G., Scanlon, B., Döll, P., Rodell, M., Van Beek, R., Wada, Y., & Treidel, H. et al. (2013). Ground water and climate change. *Nature Climate Change*, *3*(4), 322–329. doi:10.1038/nclimate1744

Thornton, P. K., & Cramer, L. (2012). *Impacts of climate change on the agricultural and aquatic systems and natural resources within the CGIAR's mandate*. Academic Press.

Upadhyay, J. (2012). *Climate Change and its Impact on Rice Productivity in Assam*. Available at SSRN 2172183

Vanclay, F. (2004). Social principles for agricultural extension to assist in the promotion of natural resource management. *Animal Production Science*, *44*(3), 213–222. doi:10.1071/EA02139

Wallace, J. S., & Batchelor, C. H. (1997). Managing water resources for crop production. *Philosophical Transactions of the Royal Society of London. Series B, Biological Sciences*, *352*(1356), 937–947. doi:10.1098/rstb.1997.0073

Wiegandt, E. (2008). Framing the study of mountain water resources: an introduction. In E. Wiegandt (Ed.), *Advances in Global Change Research 31: Mountains: Sources of Water, Sources of Knowledge* (pp. 3–13). Dordrecht: Springer. doi:10.1007/978-1-4020-6748-8_1

Wikipedia. (n.d.). *Energy policy of India*. Retrieved January 25, 2016, From https://en.wikipedia.org/wiki/Energy_policy_of_India

Zektser, I. S., & Everett, L. G. (Eds.). (2004). *Groundwater Resources of the World and Their Use*. Retrieved from http://unesdoc.unesco.org/images/0013/001344/134433e.pdf

Ziska, L. H., Blumenthal, D. M., Runion, G. B., Hunt, E. R., & Diaz-Soltero, H. (2010). Invasive Species and Climate Change: An Agronomic Perspective. *Climatic Change*, *105*(1-2), 13–42. doi:10.1007/s10584-010-9879-5

Chapter 9
Entangled Systems at the Energy-Water-Food Nexus:
Challenges and Opportunities

Joseph Nyangon
University of Delaware, USA

Nabeel Alabbas
University of Delaware, USA

Lawrence Agbemabiese
University of Delaware, USA

ABSTRACT

This chapter assesses energy, water, and food resource systems based on their inter- and intra-sectoral imperatives of large scale development investments at the institutional level (including private and public activities) and how to achieve security of resource supplies. It identifies key interrelated processes, practices, and factors that underpin integrated resource management (IRM) and their attendant benefits. Applying the E4 framework concerned with energy, economy, environment, and equity to identify the main threats to these systems, the chapter evaluates their institutional, political, economic, cultural and behavioral components, and characterizes the forces that drive each of them at different governance scales. The chapter is guided by political economy, economic, and sociological theories that suggest that institutional structures affect economic factors and processes (i.e. production, distribution, and consumption processes). A case study of energy, water, and food (EWF) conflicting sectoral imperatives in Delaware is discussed in detail to better understand how these policy and institutional processes occur, which forms they take, and in which ways they define the quality and quantity of EWF resource systems in the State. In order to verify these parameters, the analysis considered the advantages of a sectorally balanced, E4 framework, in particular to evaluate the valency and magnitude of cross-sectoral connections, balance competing needs, and identify policy options that address various trade-offs.

DOI: 10.4018/978-1-5225-1046-8.ch009

INTRODUCTION

The energy-water-food nexus is essential for sustainable development and conceptually relevant to mitigating the risk of unintended consequences of large-scale sector-specific investments and negative trade-offs. Energy is required to produce, transport and distribute food as well as to extract, pump, lift, collect, treat, and transport water (Halstead et al., 2014). Water is required in energy generation and in the cultivation of food crops. Likewise, food is required to support the world's growing population that both generates and relies on water and energy services (Belden et al., 2008). This highlights the interlocking between water, energy and food resource systems (Figures 1 & 2).

Furthermore, policymakers and researchers in the United States, China, Spain, and Australia duly recognize the important role of a nexus approach as opposed to static experiments in deepening our understanding on "how the occurrence, valency and magnitude of sectoral connections emerge and are altered as a consequence of single sector interventions in a water–food–energy nexus" (Smajgl et al., 2016). For instance, in the United States, a proposal to create specific institutions (such as an Energy-Water Architecture Council to foster data collection, reporting, and technology innovation) to administer and research water, food, and energy ''provisioning" regulatory and planning regimes and promote optimal cross-sectoral coordination was included in the Energy and Water Research Integration Act of 2012 (GAO, 2012). However, the bill was never enacted.

REALIZING A NEW PARADIGM

The Nexus Framework

To date, water, energy, and food resource systems are still largely organized, studied, and prescribed independently. However, decisions, actions, choices, and preferences in each of the three domains fun-

Figure 1. An integrated model showing the complexities between energy, water, and food systems (Belden et al., 2008)

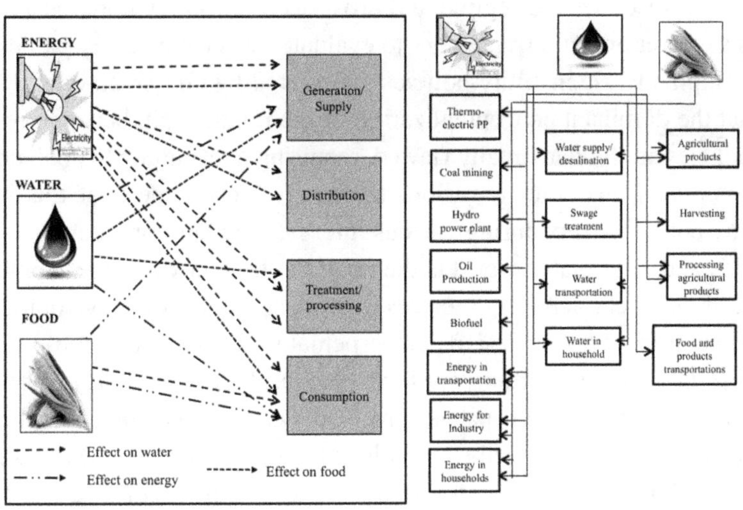

Figure 2. Value chain of water, energy and food sectors and their interactions (World Economic Forum [WEF], 2011)

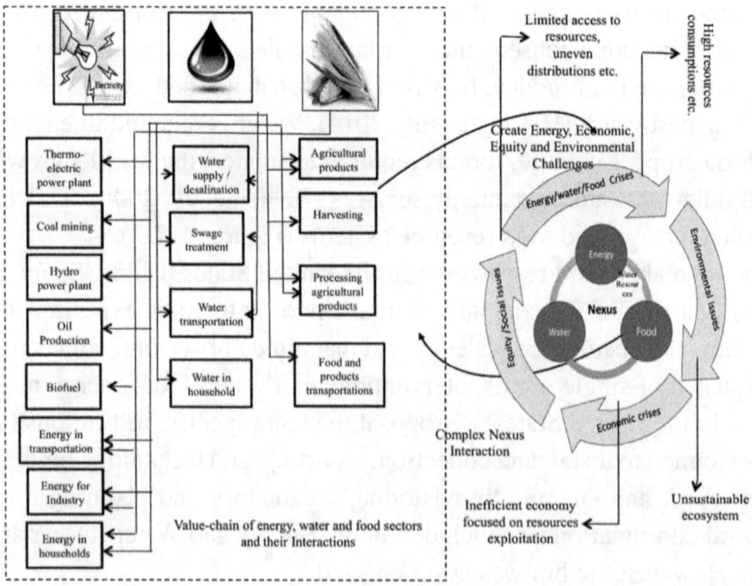

damentally affect the others, often negatively (Stockholm Environmental Institute [SEI], 2011). The "Nexus approach" is required to effectively analyze these interacting resource systems. So what does a nexus approach entail? In this chapter, a nexus approach refers to multidisciplinary analysis of the relationship between water, energy, and food, in order to help ease trade-offs in production, distribution, and consumption while at the same time developing, integrating, and promoting synergies across these different sectors (Halstead et al., 2014). The synergistic and systems approach applied here is consistent with inter alia, Smajgl et al. 2016 and Foran (2015) (as cited in Smajgl et al., 2016) summarized as: "incorporating social and political context, essential for effective cross sectoral negotiations, can be achieved by reconceptualizing the Nexus as the cumulative effects of development projects coupled with an appraisal of prevailing water, food and energy "provisioning" regulatory and planning regimes" (p. 534).

Developing efficient resource use regime and cross-sectoral policy coherence based on the nexus approach demands a clearer analytic framework to evaluate the sustainability paradigms in the energy-water-food relationship and sustainability policies articulated by Glassman et al. (2011). Glassman et al. (2011) argue that the dominant conceptualization of the relationship between resource exploitation and the attendant policies are fundamentally flawed, requiring a reframing to "balance competing needs and identify policy options that address various trade-offs." The benefit of this integrated approach emphasized by many policy makers during various international environmental governance fora, in an increasingly complex and interrelated world is the potential for a nexus-wide, ripple effects of policies in the management of resources across the sectors [European Report on Development (ERD), 2012].

Priorities set in institutional policy design can perpetuate existing inequalities—allowing greater opportunities and access for those with certain privileges (social, class, ethnic, etc.) to access these assets thus limiting equitable participation by all the parties involved. On the other hand, adapting and repositioning existing institutional arrangements effectively could intervene to modify these distribution disparities and prevent resource collapse (Smajgl et al., 2016). However, this requires a clear understanding and use of terms such as 'institutions', 'behaviors', and 'beliefs' (Hadfield and Weingast, 2014). Embracing relational ontologies that see agencies as "distributed and emergent" and as part of unfolding action nets that emerge around issues and events (Garud et al., 2010) constitute institutional adaptation phenomenon. Viewed from this perspective, the presence of differing ontologies of the relationships between agency and structure result in confusion and constraints of time, resources and institutional adaptation needed to address the complexities of water, energy, and food resources, and who bears responsibility for the decisions. To fully characterize water, energy, and food resource systems and the nexus approach, it is imperative to further explore these interactions by examining the institutional linkages between public and private activities. The nexus core (Figure 3) consists of drivers critical for the energy-water-food linkage as it applies to the E4 framework dynamics and cross-sector feedbacks.

Benefits of Integrated Conceptualization of Energy-Water-Food Nexus

The benefit of integrating EWF resource systems on the economy, the environment, energy, and equity (E4) are fourfold. First, it promotes water, energy, and food security by advancing explicit cross-sectorial perspective and response options that supersede traditional sectorial approaches which continue to prioritize isolated investment in one specific sector (WEF, 2011). Second, it supports equitable, inclusive and sustainable growth (ERD, 2012). Third, it fosters a resilient environment (Termeer et al., 2011). And finally, it maintains ecosystem integrity (Norgaard, 2010). In this regard, conservation efforts in water and energy use improve the E4 balance, by enhancing sustainability, building synergies, reducing unintended side-effects and negative sectorial trade-offs, and improving governance across sectors to enable deliberative, legitimate change in social systems (Future Earth, 2013; Gorddard et al., 2016; SEI, 2011; Wang et al., 2006)

This highlights the need to critically assess trade-offs such as between the cost of investing in water technologies now and the potential future return on investments as well as the return on the environ-

Figure 3. Typical interactions in the water, energy and food cross-sectoral connections (Smajgl et al., 2016; SEI, 2011)

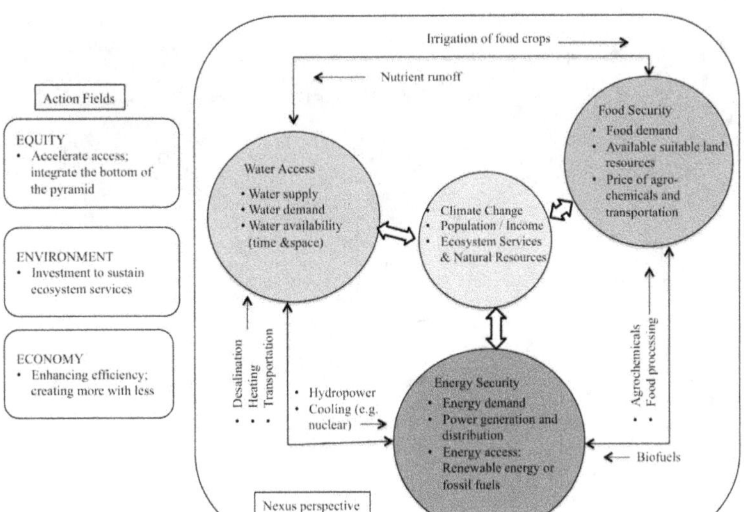

ment because these benefits depend on each other and are interlinked. The nexus approach thus provides a dynamic, integrated way of tackling these complex issues in energy and water, and soon hopefully food, globally. Moreover, to meet the growing water demands of energy and food production, historical, social, and political trajectories that give rise to contemporary energy-water-food planning and regulatory regimes should be integrated into policymaking decisions (Foran, 2015). Incorporating social and political context, necessary for effective cross-sectorial negotiations and agency cooperation between public entities, between private institutions, and between public and private institutions, therefore requires conceptualizing the energy-water-food nexus as the cumulative effects of "systems" of multiple and dynamic inter-linkages rather than "chains" of causal, linear linkages (Smajgl et al., 2016). In Figure 4 we conceptualize the additional benefits of a nexus perspective by expanding the E4 framework to include institutional linkages between public and private entities. We therefore depart from a largely conceptual, abstract domain to actual implementation by 'transmitting' ripple effects throughout the EWF nexus system.

1. Environment / Food

Reduced water waste and improvements in water efficiency will lower the amount of energy that is required to extract, pump, lift, collect, and transport water. This reduction in energy use will translate in reduced carbon emissions from power generation. Advances in agricultural productivity and focused management of water development through risk management instruments offer opportunities for improving water conservation and avoiding substantial breakdowns between nexus sectors (Leese & Meisch, 2015). This is largely due to the positive side-effects of safeguarding ecosystem services—provisioning, regulating, habitat or supporting, and cultural—and ecosystem functions such as nutrient cycling, bioturbation, plant growth enhancement, secondary seed dispersal, trophic regulation and pollination

Figure 4. Water, energy, and food integration benefits (Smajgl et al., 2016; Wang et al., 2006)

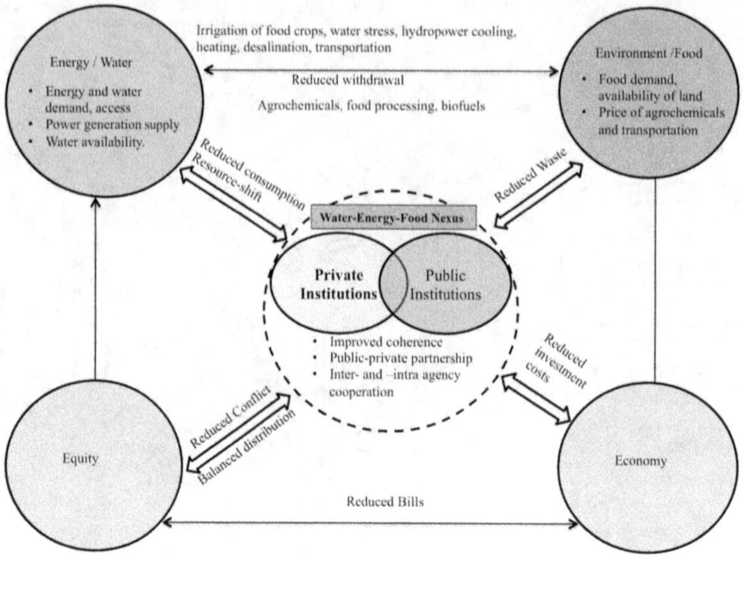

that support the coordinated and sustainable development of nature, economy and society (Changshun et al., 2015).

As ecosystem services and functions become scarce due to population growth, impact of human development, and effects of climate change, their demand and economic value increase resulting in changes in composition, transfer and consumption pattern of these services and functions (Nichols et al., 2008). Serious ecological and environmental problems such as ecological degradation, environmental pollution and biodiversity loss, largely associated with the destruction of natural capital such as forests, water, marine and coastal resources, as well as erosion of soils and pollution of air, threaten food production systems. For instance, water use efficiency varies depending on crop type, and bioenergy offers opportunities for new types of crop production that use water more efficiently (Organisation for Economic Co-operation and Development [OECD], 2014). Water conservation measures also have a direct benefit to the environment because of reduced rate of water extraction, enhancing ecological integrity (Andrews-Speed et al., 2015)

2. Economy

Considering the variety of characteristics common to energy, water, and food sectors, as far as national security interest is concerned, efficiency improvements in these sectors can bring long-term economic benefits, such as lowering costly investment in supply-side facilities and reducing tensions between upstream and downstream users. Reduced power and water utility bills for customers, lowered costs of modernizing or upgrading the energy-water infrastructure, and implementing adaptive institutional management and arrangements such as in fishing stock management, could help mitigate unproductive land and inefficient irrigation systems (Belden et al., 2008; Wang et al., 2006).

3. Energy / Water

In the U.S. nearly 13 gallons of water is required to produce every gallon of gasoline that fuels millions of vehicles (Story, 2014). Approximately 5 million gallons of water is required for the hydraulic fracturing of a shale gas horizontal well (this amount varies with well depths), requiring states to develop urgent strategies to reduce water usage in hydraulic fracturing and to implement cost-effective water recycling technology (Moniz et al., 2011). Studies on hydraulic fracturing in the Marcellus and Barnett shale plays showed that life cycle water for unconventional shale gas is more water intensive per unit of energy (4–10 gal/10^6 British thermal units, Btu) than conventional gas production (~3 gal/ 10^6 Btu) (Clark et al., 2013). Additionally, U.S. power plants combined withdraw nearly 200 billion gallons of water a day, with the majority of these withdrawals efficiently returned to waterways after being used for cooling (Story, 2014). According to the U.S. Department of Energy, about $4 billion is spent annually for energy costs to operate water and wastewater utilities (Ibid: 22). In this regard, reduction in energy consumption is urgently required and can be achieved at both the upstream and downstream levels:

1. Through reduction in energy use for surface and ground water withdrawal and wastewater treatment and discharge by water utilities, and
2. The use of efficient home appliances such as dishwashers, laundry machines, etc.

In the 2014 report entitled, "Water-Energy Nexus: Challenges and Opportunities," the U.S. Department of Energy outlined an integrated challenge / opportunity space around the water-energy nexus including: investment in smart water management to pioneering a smart water grid; improving pump efficiencies; developing new patented technologies to reduce electricity and added chemicals in wastewater treatment; investing in renewable energy sources; and developing new innovative solutions for water use and recycling (Department of Energy [DOE], 2014). From a technology perspective, reducing water pressure reduces leakage, which in turn translates in reduced energy consumption for collecting, treating and transporting water. Investing in water recycling also has the potential to optimize opportunity space around the water-energy nexus. Over 90% of treated wastewater in the United States is not recycled (Story, 2014). In the current era of climate change and taking into consideration the problems associated with conventional technologies of wastewater treatment, research efforts are being also accomplished towards developing innovative solutions for resource conservation with simultaneous management of wastewater (Itankar & Patil, 2015; Patil, 2012; Patil & Rao, 2015). This presents another tremendous opportunity for onsite water treatment options, wastewater reuse for non-portable applications, and development of new local and regional water treatment facilities needed to reduce the cost of energy incurred in producing and transporting the water.

4. Equity

Investing in a balanced EWF nexus paradigm, resource conservation, and efficiency improvements help in optimizing their allocation between competing users during times of scarcity such as climate change-induced energy supply disruptions, especially droughts, food shortages, and electricity blackouts and brownouts (Gorddard et al. 2016; Wang et al, 2006). Moreover, climate change produces new concerns about procedural fairness and equity such the rules and expectations regarding compensation for damage, highlighting the need for inclusive, deliberative process-oriented approaches in the decision processes, particularly ethical and equity propositions (Abel et al., 2011). In this regard, reformulating decision contexts in order to optimize resource allocation could help to reduce conflicts over riparian rights, food scarcity, and energy consumption. The implication of resource induced conflict is profound because of its potential to change social relationships and perceptions of mutual rights and responsibilities between individuals, social groups and the state as well as the possibility to alter the perceived legitimacy of institutions and their obligations over resources (Belden et al., 2008)

THE MAIN THREATS TO ENERGY, WATER AND FOOD SUPPLIES IN DELAWARE: THE E4 ANALYTICAL PERSPECTIVE

Institutional Arrangements in Water and Food Sectors

To examine the main threats to the quality and quantity of EWF supplies in Delaware, as well as the occurrence, valency, and magnitude of sectoral connections in the energy–water–food nexus, it is vital to analyze how these resource systems are managed, commodified, traded and consumed in the state as a common-pool resource (Ostrom, 2010) The paradigms, processes, and practicalities that define the quality and quantity of these resources emerge at different scales, from the individual scale (private) to the state and national scales (public) (Halstead et al., 2014) Within the Delaware state government, increased confidence in the role of institutions charged with managing water and food supplies may stimulate a movement toward deregulation of these resources. Consequently, the emerging institutional arrangements managing these resources can either be a centralized system, or a decentralized one or a combination of both—i.e. regulated or deregulated. In this regard, policy and management decisions on energy, water and food related issues could come from different state departments and elected officers responsible for planning decisions (e.g. agriculture, health, water, etc.), or be the exclusive purview of a single department of energy and water. Furthermore, the emerging organizational structure can either be a top-down, or a bottom-up principally led by local governance and administration (DOE, 2014).

However, the complexity of energy-water-food resource systems precludes top-down management and policy panaceas, thus requires 'polycentric' structures of actions (Byrne et al., 2015). The Paris Agreement, which has ushered in a new policy commitment to ramp-up climate mitigation and adaptation worldwide by containing global temperatures to "well below 2°C above pre-industrial levels" focused on applying "polycentric strategies" based on efforts outlined in the national pledges on climate action called ("intended nationally determined contributions" or INDCs) (Taminiau & Byrne, 2015). The Agreement is fundamentally a revolution in the Conference of the Parties (COP) process. It commits all nations—developed and developing—to curb emissions and tracks their performance over time. To realize just and sustainable climate actions, and in accordance with Decisions 1/CP.19 and 1/CP.20, Parties to the United Nations Framework Convention on Climate Change (UNFCCC) submitted their climate plans based on their national circumstances, rather than focusing on dividing up that responsibility among nations. Additionally, while procedural aspects of the agreement (e.g. communication of nationally determined contributions which is to be housed in a public registry and maintained by the Secretariat) are legally binding, substantive elements (e.g. their content and targets) are not.

Because INDCs are bottom up, countries determine their targets, which they communicate to the UNFCCC Secretariat. As a result, the Agreement promises a flexible, balanced and hybrid approach between a bottom-up system of national pledges-and-review (mandated through a set of transparency measures every five years after 2023) and a top-down, rules-based system for compliance and transparency framework. The polycentric approach lets the process of setting emissions cuts play out within each country. The model "can be considered 'ecological' seeking to enhance institutional 'fit' with the complexity of Earth's social-ecological systems" (Taminiau & Byrne, 2015). The agreement thus could be summed up as an outcome and a push in that direction. It replaces a "non-linear, uncertain, and unpredictable character of environmental degradation" with this "pledge and review" strategy organized through "polycentric networks of creative innovation and leadership" (Taminiau, 2015).

Institutional change demands effective coordination by officers responsible for planning decisions at different levels. Polycentric strategies can be "considered 'ecological', seeking to enhance institutional 'fit' with the complexity of Earth's social-ecological systems" (Taminiau & Byrne, 2015). The agreement thus could be summed up as an outcome and a push in that direction. It replaces a "non-linear, uncertain, and unpredictable character of environmental degradation" with this "pledge and review" strategy organized through "polycentric networks of creative innovation and leadership" (Taminiau, 2015). Equally, realizing a new investment paradigm (i.e. decentralized and community driven solutions) in the energy, water, and food sectors requires mobilizing 'polycentric financing strategies' by engaging the private sector, capital markets, cities, and transnational and subnational authorities. Against this backdrop, Table 1 provides a brief overview of some institutions in the water, energy, and food sectors in the Delaware State.

Table 1. Major institutions in the water, energy and food sectors

Institution	Function	Sector
The Delaware Public Service Commission (DPSC)	Sets rates on water and wastewater services to cover the cost of water collection, treatment, testing, and delivery.	Water
The Delaware Department of Natural Resources and Environmental Control (DNREC)	• DNREC Division of water oversees pollution regulation for both surface water and groundwater discharges. • The U.S. Environmental Protection Agency has delegated authority to DNREC to administer permits under the National Pollutant Discharge Elimination System (NPDES) pursuant to Section 402 of the Clean Water Act. • DNREC also monitors water quality in the state. DNREC Division of Energy and Climate administers programs to avoid the adverse impacts of energy use on environment, health, and economy.	Water and energy
Public Service Commission (PSC)	Regulates the distribution of electricity in the state.	Energy
Delaware River Basin Commission (DRBC)	Manages water quality protection, supply allocation, permitting, conservation, planning, drought management, flood control, and recreation in four basin states (Delaware, New Jersey, Pennsylvania, and New York).	Water, energy, food.
Delaware Agricultural Lands Preservation Program	• Landowners who place their lands into Agricultural Preservation Districts agree to not develop their lands for at least 10 years, devoting the land only to agriculture and related uses. • The owners receive tax benefits, right-to-farm protection, and an opportunity to sell an easement to the state that keeps the land free from development permanently.	Food
Delaware Agricultural Forestland Preservation Program	• The Program protects forest lands through perpetual conservation easements. • The program currently receives a $1 million appropriation annually.	Food
Delaware Office of Food Protection	• Ensures the regulatory foundation of programs in retail food protection and safety includes which current, science-based requirements. • Ensures that complaints and outbreaks associated with the food consumption are appropriately investigated. • Ensures that effective compliance and enforcement procedures for food establishments, production and processing are promulgated and are used appropriately to reduce the risk of foodborne illness.	Food
The United States Department of Agriculture Farm Service Agency (FSA)	• Oversees a number of voluntary conservation-related programs which address farming and ranching related conservation issues including: o Drinking water protection o Reducing soil erosion o Wildlife habitat preservation o Preservation and restoration of forests and wetlands.	Food
Federal agencies	Federal Energy Regulatory Commission (FERC) for regulating prices and even licensing hydropower plants; USDA Natural Resources Conservation Service, USDA (Farm Services Agency, Forest Service), Environmental Protection Agency (EPA) with water conservation programs.	Water/ Energy/ Food
Delaware's Sustainable Energy Utility (SEU)	A unique non-profit organization offering a one-stop resource through its Energize Delaware initiative to help residents and businesses save money through clean energy and efficiency.	Energy
Delaware's Renewable Portfolio Standard (RPS).	Eligible renewable energy technologies include: geothermal electric, solar thermal electric, solar photovoltaics, fuel cells using non-renewable fuels, landfill gas, wind, anaerobic digestion, fuel cells using renewable fuels, biomass, hydroelectric, tidal, wave, ocean thermal. RPS is set at 25% by compliance year 2025-2026 (Delaware State Senate, 2016).	Energy

Exogenous Factors

The success of these institutions in achieving their missions depends on a number of considerations: the availability of the energy, water, and food resources; the occurrence of external shocks; and the synchronicity with other trends or events occurring within the system (Graedel et al., 2014) Therefore, understanding the interactions between these exogenous variables, which shape either the quality or quantity of the three resources in the Delaware state, or a combination of the two is fundamental to the nexus approach. The following development strategies were identified for this book chapter: Energy (e.g. hydropower, concentrated solar thermal and concentrated solar power (CSP) systems, energy crops); water (e.g. water diversion); and food (e.g. irrigation projects, fisheries management, and corn production).

These interactions are shaped and influenced by a trifecta of variables, including: (i) resource and/or biophysical conditions, (ii) rules and norms, and (iii) community attributes that influence consumption patterns of the respective resources (Ostrom, 2005). Resource or biophysical conditions including: the size of the resource (actual); the magnitude of the resource (potential); and its location (Macknick et al., 2011) explain the pattern and rate of consumption and whether the resource is evenly distributed or geographically dispersed within the state.

Within the state, rules and norms that determine the quality and quantity of the resources (i.e. energy, water and food supplies), include government regulations (i.e., water and food policies at the City, County and State levels); established precepts, moral behavior, values or norms that guide behavior on use of water and food resources (e.g., decisions such as use of hydraulic fracturing and its potential impact in contaminating underground water resources); state instructions (i.e., existing strategies on improving and sustaining quality and quantity of water and food supplies); and principles that guide these sectors (e.g., water and food resources are location specific) (Hanlon et al., 2013). Finally, community attributes include culture and values that define different energy-water-food choices on what is perceived to meet the set standards of quality and the quantity consumed thereof; the State's homogeneity and preferences for certain foods and/or brands of treated water over the other; the level of commonality in water-food options, the composition and size of the community using different resources; and the equity and ethical implications that exist in accessing these resources in the state. Figure 5 displays the interactions of these exogenous variables at the institutional level in the E4 framework.

The three variables—resource conditions, community attributes and rules and norms—as illustrated in figure 4, taken together present a coherent longitudinal orientation of the constraints on water, energy, and food resources, with conceptual integration and dialectical rationality as their cornerstones (FAO, 2011; Halstead et al., 2014; Ostrom, 2005; SEI, 2011). For instance, rules and norms shape and define community attributes, and in turn define resource conditions and consumption patterns. Figure 6 provides a summarized schematic framework of such interactions at the institutional scale.

Key tenets of this interaction, driven by self-organizing emergent processes embodied by the E4 framework towards long-term sustainable energy-water-and-food system, energy resiliency, long-term sustainable economic growth, and equitable distribution of the resources, however, have rested on essentially narrow models of society. Consequently, conceptual frameworks and theoretical positioning of the linkages between lifestyle choices and the resultant macro-behavior have yielded inadequate sustainable resource insights even as demand for these resources has grown. As a result, Delaware State faces constant threats to maintain sustainable quality and quantity of energy, water, and food supplies.

Figure 5. An illustration of how exogenous variables influence quality and quantity of energy-water-food supplies at public and private institutional levels in Delaware State (Ostrom, 2005)

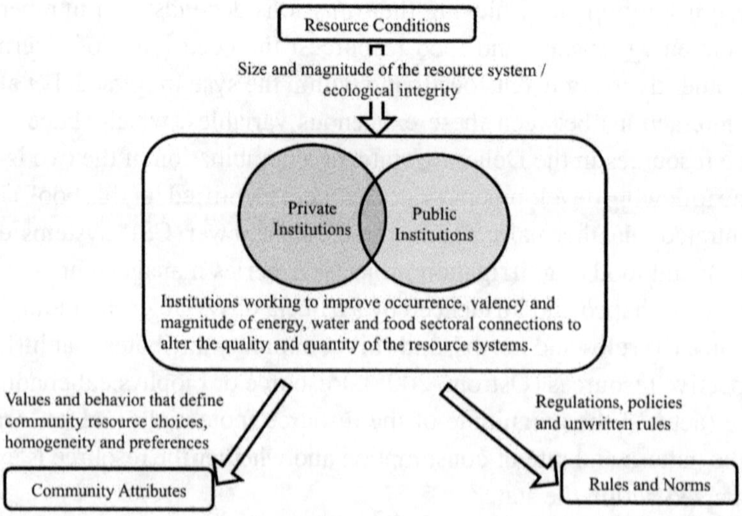

Figure 6. A simple framework for institutional analysis (Ostrom, 2005)

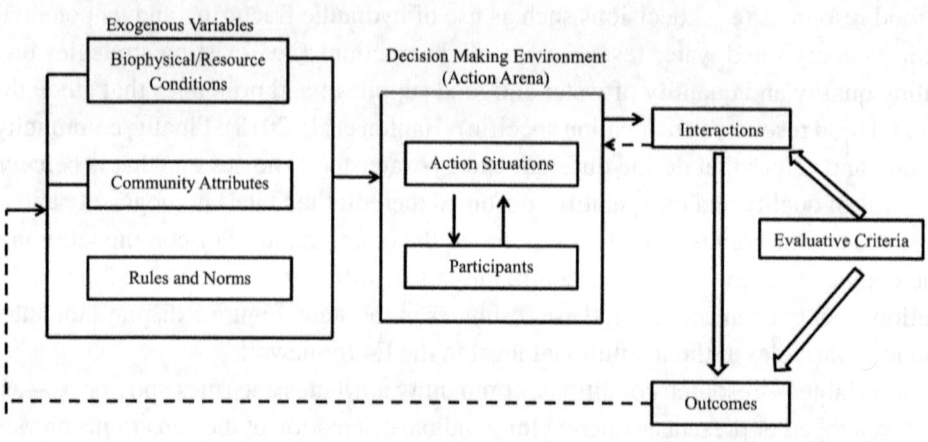

Table 2 provides a summary of different categories of threats with respect to these resources.

Investments in the EWF systems in Delaware remain inadequate sub-optimal by individual sectors. Statewide connectivity implies that investment factors (or drivers) interact between nexus sectors, transmitting the effects of magnitude and valency of these interactions and feedbacks from one part of the State to another and to other sectors. According to the Nature Conservancy (2014), "94 percent of the State's [Delaware] creeks and rivers fail to meet fish and wildlife needs under the Clean Water Act, 86 percent are unfit for swimming and 30 rivers carry warnings against eating fish caught in them" (p. 4). In March 2014, Delaware Governor Jack Markell proposed an increase of $800 million in water sector investment to clean Delaware's waterways, curb stormwater runoff and flooding, and protect drinking water supplies (Montgomery & Murray, 2014) Although the cost of the program was to be met through a statewide tax that would cost homeowners $45 a year, once implemented the governor's proposals

Table 2. Threats to quality and quantity of energy, water and food supplies in Delaware

Categories of Threats	Threats
Political and Institutional risks	Lack of coordination and skilled personnel to develop, design, finance, build, operate, and maintain water and food related projects and programs.
	Reliance on non-renewable energy sources to collect, treat and transport water and food supplies.
	Uncoordinated research and development (R&D) in water and food sectors.
Economic and Financial risks	Unclear signals sent to potential investors in water and food sectors which may lead to high risk perception by investors and heightened uncertainty.
	High upfront cost of constructing water infrastructure projects and lack of structured investment options and innovative financial mechanisms in the water and/or food sectors.
	Lack of one-stop investment repository of water and food resources in the state.
	Lack of up-to-date best practices in investment decision making in water and food sectors.
Regulatory risks	Existing regulatory structures that favor large scale commercial food companies and not small food producers and processors.
	Lack of a centralized database (for all county, city and state focused) of water and food related regulations and stages in their implementation.
	Lack of a strong targeted regulatory presence that focuses on both policy and administration of water and food organizations.
	Conflicting research findings on genetically modified organisms (GMOs) food crops; biotechnology and balanced intellectual property rights thus creating uncertainty on food security in the state; concerns and implications of hydraulic fracturing on underground water resources.
Technical risks	Preference for pipeline extensions and centralized water treatment facilities by water utilities.
	Connectivity, perishability, availability, and quantity concerns related to food distribution and transportation.
	Lack of clear quality standards for home water purification and related accessories.
Socio-cultural risks	Little involvement and participation of consumers in influencing water and food related decisions and related services.
	Decision making precludes local participation.
	Poor information diffusion on the use and benefits of quality water and food to producers and consumers.

would have improved the State's capacity to secure its fresh waters, improve its coastal resilience, and broaden the constituency for water conservation (Jerome, 2016). However, this proposal was rescinded as other sectors were prioritized for State financial investments. "Everyone wants a cleaner environment, but no one wants to pay for it," the governor observed (Bittle, 2016).

Like most states, financing water infrastructure through debt securitization is still very expensive. The high cost of debt directly affects investments in water utilities and related sectors including food and food processing companies that depend on availability of uninterrupted water supplies.

SOLUTIONS AND RECOMMENDATIONS FOR DELAWARE

The application of the E4 framework to study the main threats to the quality and quantity of energy, water and food resources in Delaware represent one aspect of this analysis. However, it is imperative to note that these threats do not operate in a vacuum. Economic, institutional, political, regulatory, and sociotechnical innovations that nurture sustainable trajectories in order to respond effectively to the threats identified above also provide insightful synergistic perspectives. Therefore, employing a robust integrated framework that clearly outlines the interactions between exogenous variables (i.e. resource conditions, rules and norms, and community attributes) and their influence on water and food resource systems in Delaware is a prudent strategic response. Major gaps and needs between the desired resource quality and quantity, and public and private institutional interactions, however, remain wanting and in need of policy and institutional reorientation.

Policy Options

Achieving massive reforms that are associated with huge investments in the EWF sectors undoubtedly require socioeconomic restructuring and additional monetary and fiscal resources to stimulate sustainability transformation. This requires a strong State government policymaking regime (i.e. political will) as well as multi-stakeholder participation especially in the public and private sectors to prioritize implementation of policy options that incentivize promising sociotechnical innovations. Leaving management of water, energy, and food resource systems completely to the whims of market factors often do not effectively support inclusive and sustainable transformation (OECD, 2014). Thus, policy options that translate nonmarket concerns such as sustainable watershed funding, water, and food pollution control, water quality assessment and treatment, avoidance of population migration peaks due to change in access to the resources, and identifying the ensemble of EWF system criticalities and understanding them as intervention points instead of a singular focus on one sector are primary strategies. However, government policies alone are not adequate. In this regard, rather than relying completely on government policies in order to effectively address current and future threats, effective solutions to energy-water-food resource challenges should be bold and integrative and should not be constrained by economic and political undertones (as a means to embody the E4 framework (SEI, 2011).

Second, given that both federal and state water resource appropriations have been on the decline, Delaware must identify other sources of revenue to fund its huge capital-intensive water pollution control program in the Delaware Basin. For instance, potential sources of revenue include the user/polluter pays principle by addressing the negative externalities from upstream water pollution that can impair downstream consumers. Successful user/polluter pays approaches have been implemented by New York City (to protect its Catskill/Delaware water estuaries in upstate New York), San Francisco, and Boston to promote cost-effectiveness of resource use, innovation, and alignment of financial benefits with pollution reduction (Hanak, 2011). Other funding sources include the use of water quality trading by deploying successful pollution control principles as successfully demonstrated by the Clean Air Act to reduce the threat of SO_2 and acid rain from atmospheric emissions (DOE, 2014).

Third, there is need to re-evaluate the value of water because water prices charged currently do not reflect the full opportunity cost of utilizing the resource. Non-pricing tools such as landscaping ordinances and restrictions (e.g., limits on the planting of lawns and use of outdoor watering), public education, plumbing and appliance standards, tiered rate structures, and rebates to encourage new technology adoption can encourage conservation (Hanak, 2011). The current water cost structure whereby consumers pay for water at its average cost when the resource is abundant and not based on its overall scarcity value has contributed to undervaluing water resources. As a result, investments in the water sector have remained small as federal, state, and local governments continue to underinvest in water resources and water pollution control programs. In addition, solutions to the financial challenges confronting interrelated sectors such as the food sector may lie in effectively implementing river basin management strategies whereby beneficiaries and users of the river resources bear some of the costs of the restoration.

Fourth, with regards to the three domain sectors, local level planning should be prioritized because it incorporates community attributes—their choices and their preferences about the kinds of changes in food supply, distribution, processing and transportation they want implemented. Further, three policy perspectives that legitimize the role of local institutions in the three domain sectors in Delaware have been offered:

- Analytical-operative, which considers the local scale the best means to establish diversified and effective regulatory policies, and in obtaining contextual knowledge of food, water, energy, environmental, and public health issues.
- Policy perspective, which requires bottom-up endogenous development of decentralized energy, water and food management strategies in which the local dimension is prioritized through avenues such as Farmers Markets, Farm to Table, etc. This perspective favors inclusive decentralized knowledge at the local level as the most suitable scale at which to introduce the nexus perspective as it is effective in closing the cycles of material flows and improving food production system efficiency, thereby improving resource conditions (Bagliani et al., 2010)
- Finally, the third policy perspective supports the local dimension approach due to its potential to drive variants of small signal transformations by 'transmitting' ripple effects statewide throughout the nexus system, through better articulation of the regulated energy, water, and food functions; improved coherence and understanding of sectoral linkages; and enhanced investment readiness in the domain sectors.

However, like the water and energy sectors, management of food resource systems in Delaware State can hardly be described as bottom-up driven. Its history has been heavily top-down and the complexity of this high conceptualization has tended to be ineffective.

INSTITUTIONAL MEASURES AND STAKEHOLDER ENGAGEMENT

As discussed in the E4 framework, successful realization of an integrated policy regime in energy, water and food resource systems requires commitment by all stakeholders to help deliver broad economic, equity, energy security and environmental benefits. In Delaware, it is evident that various stakeholders including the federal and state governments, private firms and non-profits all support opportunities that seek to mitigate policy, finance, regulatory, and institutional threats to these resources. However, as revealed in Table 1, multiple, interlocking missions often hinder progress towards an integrated energy-water-food perspective rather than segregated series of liability. Overcoming these threats thus requires multiple, mutually reinforcing actions that are often beyond the scope of the current institutional capacity.

The central aim, here, is to provide institutional measures and deliberative stakeholder engagements that hasten cross-sectoral coordination in areas where progressed has slowed or even regressed by:

- Creating a conducive environment to private investment in renewable energy technology deployment, water and food security;

- Developing adequate infrastructure options for adaption needed to scale up and deploy sustainable energy-water-food solutions with the least resource intensive options;
- Ensuring stricter requirements for more efficient energy use, and water and food conservation by government agencies;
- Implementing utility reforms to enhance electricity grid reliability and sustainable service delivery, values-based water management strategies, and rules-based changes in food administration including legislative reform; and
- Strengthening local financial institutions by positioning their product rollout and investment in the energy, water, and food services towards a nexus paradigm.

A water-centric analysis of the institutional challenges to the quality and quantity of supplies sustained by the Delaware River Basin reveals that management and administration of water resources in the State is confronted by complex institutional and governance concerns that require explicit cross-sectoral perspective and urgent response options (Cody & Carter, 2009). As a result, this can put the various state governments managing the basin in dispute with their upstream or downstream neighbors leading to interstate conflicts. To respond to these institutional and governance challenges, the state should implement a sectorally balanced, dynamic nexus approach, such as:

- Moderating competing water uses for energy generation and agricultural food production between upstream and downstream stakeholders;
- Balancing and adapting institutional arrangements at the federal, state, city, and local levels, for example for fishing, hydropower development and irrigation;
- Promoting dynamic processes that reveal a set of decision-making elements that are important in 'transmitting' ripple effects throughout the energy-water-food nexus system to advance multidisciplinary knowledge networks in the fields of science and policy in these domains.
- Developing mechanisms to develop cost effective solutions, pricing structures, and rate designs of energy, water and food resources; and
- Finally, implementing market-based instruments to control nonpoint sources of water pollution—the runoff from agricultural fields, mining, urban streets, timber harvesting, gardens, and construction sites.

Kneese and Bower (1984) raised three fundamental questions on the state of Delaware Rivers, which we still have not answered: "Do we want good water quality 90%, 95%, or 98% of the time? How can we achieve a desired level of good water quality at least cost? What are the best institutional arrangements for managing water quality in [the Delaware] river basins? These questions underscore why water-centric policy approaches have yielded relatively minor additional dividends. Hence, the need for a variant of integrated approach in designing and assessing investment across multiple sectors and scales. Identifying policy, finance and market interventions such as energy, water, and food conservation campaigns for monetary incentives to change consumptive behavior and habits as well as undertaking better assessment of the impacts of energy choices on water availability for agricultural purposes. Furthermore, state governments must undertake regular complementary analysis of their freshwater withdrawal for agriculture and quality assessment by energy type (e.g., breakdowns by technology including aggregate data and details by industry sector such as energy, agriculture, manufacturing, and domestic use).

FUTURE RESEARCH DIRECTIONS

Rules and norms, social structures, and values, which are internalized by institutions in order to sustain existing social arrangements at different governance scales, could perpetuate distribution disparities in energy, water, and food resources (Gorddard et al. 2016). Understanding institutional structures and operational and implementation mechanisms through which these structures perpetuate or create exclusion is thus the first and foremost step in achieving a comprehensive energy-water-food nexus perspective. The chapter examined the EWF resource systems in Delaware at the institution level given underlying factors implicit therein at private and public levels. However, further research is required to explore how complexity of socio–ecological systems, economic preferences, and EWF resource distributions in turn influence institutional structures, the occurrence, valency, and magnitude of sectoral connections that alter the quality and quantity of the resource systems.

CONCLUSION

Given the complexity of modern energy, water, and food planning and management as common-pool systems, cross-sectoral coordination and investment strategies in these sectors based on the nexus architecture remain a protracted process. There is neither no guaranteed proven system for successful management nor established paradigms with guaranteed end results, and any proven policy interventions and institutional strategies will require regular reviews and adjustments to make them relevant to the prevailing market, as the change and impacts of climate change on water resources information becomes evident and population grows. In this regard, formulation of desirable goals for these sectors may seem a hopeless endeavor in a fast changing and unpredictable modern society, where serendipitous political and economic changes have become ubiquitous. However, the nexus paradigm for energy, water, and food sectors examined in these asset-intensive sectors can happen very fast, as has happened in the telecoms sector in other parts of the world especially Africa, Asia, and Latin America. Yet in the water, food, and energy sectors, adding a 'wicked' problem such as climate change to this mix further complicates management of policy, investments and market criticalities needed to produce more sustainable outcomes. For example, uncoordinated management of these three domains, can make the sustainable transition process takes much longer because of the time required to transform sociotechnical systems and realize the return on investments. But one thing is clear. Through innovative financing mechanisms such as user/polluter pays and energy/water trading schemes ("cap and trade" systems), unlocked by market competition and fair pricing, integrated and coordinated management of these vital resources while meeting social and ecological needs and promoting economic development, significant incentives and opportunities can be optimized, hastened and realized in the energy-water-food nexus approach.

REFERENCES

Abel, N., Gorddard, R., Harman, B., Leitch, A., Langridge, J., Ryan, A., & Heyenga, S. (2011). Sea level rise, coastal development and planned retreat: Analytical framework, governance principles and an Australian case study. *Environmental Science & Policy*, *14*(3), 279–288. doi:10.1016/j.envsci.2010.12.002

Andrews-Speed, P., Bleischwitz, R., Boersma, T., Johnson, C., Kemp, G., & Vandeveer, S. D. (2015). *Want, waste or war? The global resource nexus and the struggle for land, energy, food, water and minerals*. New York, NY: Routledge.

Bagliani, M., Dansero, E., & Puttilli, M. (2010). Territory and energy sustainability: The challenge of renewable energy sources. *Journal of Environmental Planning and Management*, *53*(4), 457–472. doi:10.1080/09640561003694336

Belden, A., Cole, P., Conte, H., deMooy, J., Jose, S., Karki, J., … Alleng, G. (2008). Integrated policy and planning for water and energy. Newark, DE: Center for Energy and Environmental Policy, University of Delaware.

Bittle, M. (2016). *Funds needed to clean Delaware's water*. Retrieved from http://delawarestatenews.net/government/funds-needed-to-clean-delawares-water/

Bizkova, L., Dimple, R., Darren, S., Venema, H., & MacCandless, M. (2013). *The Water–Energy–Food Security Nexus: Towards a Practical Planning and Decision-Support Framework for Landscape Investment and Risk Management. International Institute for Sustainable Development*. Winnipeg, Canada: IISD.

Byrne, J., Taminiau, J., Kim, K. N., Seo, J., & Lee, J. (2015). A solar city strategy applied to six municipalities: Integrating market, finance, and policy factors for infrastructure-scale photovoltaic development in Amsterdam, London, Munich, New York, Seoul, and Tokyo.Wiley Interdisciplinary Reviews: Energy and Environment, 4-5.

Changshun, Z., Chunlan, L., & Na, L. (2015). Mechanisms for Realizing the Economic Values of Ecosystem Services. *Journal of Resources and Ecology*, *6*(6), 412–419. doi:10.5814/j.issn.1674-764x.2015.06.009

Clark, C. E., Horner, R. M., & Harto, C. B. (2013). Life cycle water consumption for shale gas and conventional natural gas. *Environmental Science & Technology*, *47*(20), 11829–11836. doi:10.1021/es4013855 PMID:24004382

Cody, B., & Carter, N. (2009). *35 years of water policy: The 1973 national water commission and present challenges*. Congressional Research Service Report.

Delaware State Senate. (2016). *An Act to Amend Title 26 of the Delaware Code Relating to Delaware's Renewable Energy Portfolio Standards and Delaware-Manufactured Fuel Cells*. Senate Bill No. 124. Retrieved from www.legis.delaware.gov/LIS/lis146.nsf/vwLegislation/SB+124/$file/legis.html

Department of Energy. (2014). *Water-energy nexus: Challenges and opportunities*. Washington, DC: U.S. Department of Energy (DOE).

Earth, F. (2013). Future Earth Initial Design. Paris: Future Earth Interim Secretariat. International Council for Science. Retrieved from http://www.futureearth. org/sites/default/files/Future-Earth-Design-Report_web.pdf

European Report on Development (ERD). (2012). *Confronting Scarcity: Managing Water, Energy and Land for Inclusive and Sustainable Growth.* Brussels, Belgium: European Commission, Overseas Development Institute (ODI), European Centre for Development Policy Management (ECDPM), German Development Institute (GDI/DIE).

FAO. (2011). The state of the world's land and water resources for food and agriculture (SOLAW) – managing systems at risk. Food and Agriculture Organization of the United Nations.

Foran, T. (2015). Node and regime: Interdisciplinary analysis of water–energy–food Nexus in the Mekong region. *Water Alternatives*, *8*(1), 655–674.

GAO. (2012). Energy-Water Nexus: Coordinated federal approach needed to better manage energy and water tradeoffs (Congressional Research Service Report, GAO-12-880). Washington, DC: GAO.

Garud, R., Kumaraswamy, A., & Karnøe, P. (2010). Path dependence or path creation? *Journal of Management Studies*, *47*(4), 760–774. doi:10.1111/j.1467-6486.2009.00914.x

Glassman, D., Wucker, M., Isaacman, T., & Champilou, C. (2011). *The water-energy nexus: Adding water to the energy agenda. (World Policy Paper).* New York: World Policy Institute.

Gorddard, R., Colloff, M. J., Wise, R. M., Ware, D., & Dunlop, M. (2016). Values, rules and knowledge: Adaptation as change in the decision context. *Environmental Science & Policy*, *57*, 60–69. doi:10.1016/j.envsci.2015.12.004

Graedel, T., Elshkaki, A., & van der Voet, E. (2014). Entangled circles: Energy and its resource connections. In G. J. Kramer & B. Vermeer (Eds.), *The colours of energy*. Amsterdam: Shell.

Hadfield, G. K., & Weingast, B. R. (2014). Microfoundations of the rule of law. *Annual Review of Political Science*, *17*(1), 21–42. doi:10.1146/annurev-polisci-100711-135226

Halstead, M., Kober, T., & van der Zwaan, B. (2014). *Understanding the energy-water nexus No. ECN-E--14-046*. Netherlands: ECN.

Hanak, E., Lund, J., Dinar, A., Gray, B., Howitt, R., Mount, J., & Thompson, B. et al. (2011). *Managing California's Water*. Public Policy Institute of California.

Hanlon, P., Madel, R., Olson-Sawyer, K., Rabin, K., & Rose, J. (2013). *Food, water and energy: Know the nexus*. GRACE Communications Foundation.

IEA. (2010). *World energy outlook 2010*. Paris: OECD/ International Energy Agency.

Itankar, N., & Patil, Y. (2015). Biosorptive Removal of Hexavalent Chromium from Aqueous Industrial Solutions by A Novel Wood Apple Shell Powder Biomass. *Advanced Materials Research*, *1130*, 689–692. doi:10.4028/www.scientific.net/AMR.1130.689

Jerome, S. (2014). *Potential Water Tax Could Pay For Infrastructure Updates In Delaware*. Retrieved from http://www.wateronline.com/doc/potential-water-tax-could-pay-for-infrastructure-updates-in-delaware-0001

Kneese, A., & Bower, B. (1984). *Managing water quality: Economics, technology, institutions*. Washington, DC: Resources for the Future.

Leese, M., & Meisch, S. (2015). Securitising sustainability? Questioning the 'water, energy and food-security Nexus'. *Water Alternatives*, *8*(1), 695–709.

Macknick, J., Newmark, R., Heath, G., & Hallett, K. (2011). *A review of operational water consumption and withdrawal factors for electricity generating Technologies. National Renewable Energy Laboratory*. NREL.

Moniz, E. J., Jacoby, H. D., Meggs, A. J. M., Armstrong, R. C., Cohn, D. R., Connors, S. R., … Yang, Y. (2011). The Future of Natural Gas. Massachusetts Institute of Technology.

Montgomery, J., & Murray, M. (2014). *Markell seeks new tax to improve water.* Retrieved from http://www.delawareonline.com/story/news/2014/03/04/markell-water-quality-tax/6008877/

National Academy of Sciences. (n.d.). *Our Energy System*. Retrieved from http://needtoknow.nas.edu/energy/interactive/our-energy-system/

Nature Conservancy. (2014). *Partners turn collective focus to Delaware's freshwater resources*. Retrieved from http://www.nature.org/ourinitiatives/regions/northamerica/unitedstates/delaware/delaware-acorns-newsletter-spring-2014.pdf

Nichols, E., Spector, S., Louzada, J., Larsen, T., Amezquita, S., & Favila, M. E. (2008). Ecological functions and ecosystem services provided by Scarabaeinae dung beetles. *Biological Conservation*, *141*(6), 1461–1474. doi:10.1016/j.biocon.2008.04.011

Norgaard, R. B. (2010). Ecosystem services: From eye-opening metaphor to complexity blinder. *Ecological Economics*, *69*(6), 1219–1227. doi:10.1016/j.ecolecon.2009.11.009

Organisation for Economic Co-operation and Development. (2014). *Climate change, water and agriculture: Towards resilient agricultural and water systems.* Working paper. Organisation for Economic Co-operation and Development (OECD).

Ostrom, E. (2005). *Understanding institutional diversity* (1st ed.). Princeton, NJ: Princeton University Press.

Ostrom, E. (2010). Polycentric systems for coping with collective action and global environmental change. *Global Environmental Change*, *20*(4), 550–557. doi:10.1016/j.gloenvcha.2010.07.004

Patil, Y., & Rao, P. (2015). Industrial waste management in the era of climate change - A smart sustainable model based on utilization of active and passive biomass. In W. L. Filho (Ed.), *Handbook on Climate Change Adaptation* (pp. 2079–2092). Springer-Verlag Berlin Heidelberg. doi:10.1007/978-3-642-38670-1_49

Patil Yogesh, B. (2012). Development of an innovative low-cost industrial waste treatment technology for resource conservation - A case study with gold-cyanide emanated from SMEs. *Procedia: Social and Behavioral Sciences*, *37*, 379–388. doi:10.1016/j.sbspro.2012.03.303

Siddiqi, A., & Anadon, L. (2011). The water-energy nexus in Middle East and North Africa. *Energy Policy*, *39*(8), 4529–4540. doi:10.1016/j.enpol.2011.04.023

Smajgl, A., Ward, J., & Pluschke, L. (2016). The water–food–energy Nexus – Realising a new paradigm. *Journal of Hydrology (Amsterdam)*, *533*, 533–540. doi:10.1016/j.jhydrol.2015.12.033

Stockholm Environmental Institute. (2011). *Understanding the nexus. Background paper for the Bonn 2011 nexus conference - the water, energy and Food security nexus: Solutions for the green economy.* Working paper. Stockholm Environmental Institute (SEI).

Story, S. (2014). The water energy dependency: Unlocking the smart cities. *Energy Biz*, *11*, 20–22.

Taminiau, J. (2015). *A paradigm analysis of ecological sustainability*. (Unpublished doctoral dissertation). University of Delaware, Newark, DE.

Taminiau, J., & Byrne, J. (2015, December). *A Polycentric Response to the Climate Change Challenge Relying on Creativity, Innovation, and Leadership.* Paper presented at the 21st Conference of the Parties (COP 21) to United Nations Convention on Climate Change (UNFCCC), Paris, France.

Termeer, C., Dewulf, A., van Rijswick, H., van Buuren, A., Huitema, D., Meijerink, S., & Wiering, M. et al. (2011). The regional governance of climate adaptation: A framework for developing legitimate, effective, and resilient governance arrangements. *Climate Law*, *2*, 159–179.

Vogt, K., Patel-Weynand, T., Shelton, M., Vogt, D., Gordon, J., Mukumoto, C., Suntana A.S., & Roads, P.A. (2010). *Sustainability Unpacked Food, Energy and Water for Resilient Environments and Societies.* London: Eaethscan.

Wang, Y. D., Smith, W., Byrne, J., Scozzafava, M., & Song, J. (2006). Freshwater management in industrialized urban areas: The role of water conservation. In V. Grover (Ed.), *Water: Global common and global problems*. Velma Grover. doi:10.1201/b11005-27

World Economic Forum. (2011). Water Security: Water–Food–Energy–Climate Nexus. *The World Economic Forum Water Initiative*. Island Press.

KEY TERMS AND DEFINITIONS

Community Attributes: These include influence of culture and values that define how choices in resource use are made i.e. cultural values and behavior that define different energy, water, and food resource choices, homogeneity, and preferences for a given resource over the other.

Complexity of Socio-Ecological Systems: These include occurrence, valency and magnitude of sectoral connections in the water–food–energy nexus.

E4 Framework: Conceptualizes the benefits of integration from economic, energy, equity and environmental perspectives.

Rules and Norms: Determine the quality and quantity of the resource (e.g. water and food supplies) and include *inter alia*: regulations, established precepts or moral behavior, values, and principles that guide the resource sectors.

APPENDIX

Hybrid Sankey Diagram of Interconnected Water and Energy Flows

Energy and water are interconnected through various sectors of the U.S. economy. Large quantities of water are used for cooling energy systems, while significant quantities of energy are required to pump and heat water. Water is also used in small but important ways in fuels production. Figure 7 displays a hybrid Sankey diagram of energy flow for the U.S. produced by Lawrence Livermore National Laboratory for 2014.

Figure 7. Hybrid Sankey diagram of energy flow in U.S. (National Academy of Sciences, 2014)

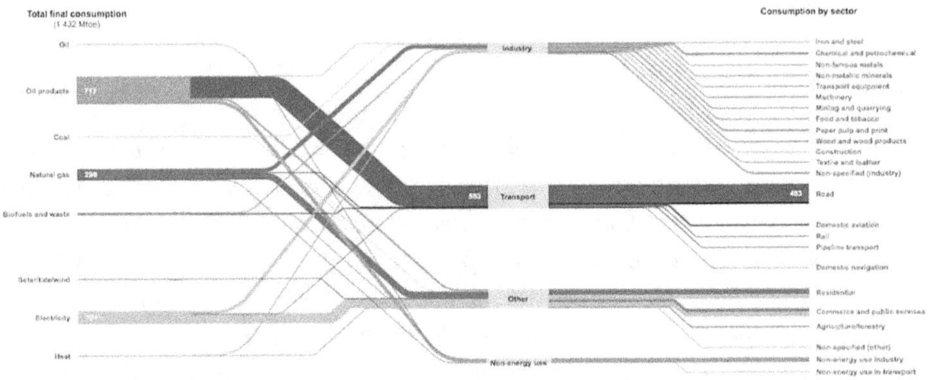

Chapter 10
Inter Linkages of Water, Climate, and Agriculture

Sunil Londhe
World Agroforestry Centre (ICRAF), India

ABSTRACT

Climate is the primary determinant of agricultural productivity and evidence shows possibility of shifts in earth's climate. Concern over the potential effects of long-term climatic change on agriculture has been raised over the past decade. Change in the climatic conditions on the globe created threat to the availability water for agriculture production. The present chapter is an attempt to distil what is known about the likely effects of climate change on water availability to agriculture for food security and nutrition in coming decades. Apart from few exceptions, the likely impacts of climate change on agriculture water resources in the future are not understood in any great depth. There are many uncertainties as to how changes in various environmental parameters will interact with the availability of water and further agriculture production. The future consequences of water resources on agriculture are discussed and summarized. Possible mitigation and adaptations to changing water availability for agriculture are also discusses.

INTRODUCTION

The world population, which took more than 50,000 years to reach the first billion, has just surpassed 7 billion. Even if fertility continues to decline at the world level and with it population growth rates, the United Nations projects that the world population could reach 9.3 billion by 2050 and surpass 10 billion by the end of the century (United Nations, 2011). If fertility were to be higher than in that projection, the population may surpass 10 billion by 2050 and may be several billions higher by 2100.

To feed a growing world population, we have no option but to intensify agriculture and crop production. Further, numerous factors shape and drive the agricultural sector and climate is the primary determinant of agricultural productivity. Given its inherent link to natural resources, agricultural production is also at the compassion of uncertainties driven by climate variation, including extreme events such as flooding and drought. The fundamental role of agriculture in human welfare, concern has been expressed

DOI: 10.4018/978-1-5225-1046-8.ch010

by federal agencies and others regarding the potential effects of climate change on agricultural productivity. Interest in this issue has motivated an extensive of research on climate change and its impact on agriculture production over the past decade.

There may be increasing threat to agriculture production due to climate change which is now largely accepted as a real, pressing and truly global problem. There is also increasingly aware that the risks of climate change are so great, that ignoring or delaying in addressing them would be far more costly than not doing so. Climate change is now global problem for the agriculture production and food security on the globe.

The long-term climatic risk to agricultural assets and agricultural production may be linked to availability of water which is known with great uncertainty. There may be threat to risk of loss of rural livelihoods and income due to insufficient access to drinking water. The risk of alarms for agricultural irrigation and reduced agricultural productivity, particularly for farmers and pastoralists in semi-arid regions. There may be risk of loss of terrestrial and inland water ecosystems, biodiversity, and the ecosystem goods, functions, and services they provide for human being and their livelihoods. In general agriculture production and food security may face serious consequences due to inadequate water resources. Apart from this increasing demand for water from urban, industry, etc. has exaggerated the problem in many folds. From the available literature overall demand and supply relationship with linkage to risk for overall development is compiled and presented in Figure 1.

It is well known fact that agriculture production is dependent on set of climatic conditions. Climatic resources are the deciding factor for successful cultivation of any crop and which cannot be manipulated by the human beings. The availability of water for irrigation and the source of the water both are climate dependent factors. Both shortage and excess of water will interfere the agriculture production to greater loss. The latest reports (FAO, 2013) of statistics of utilization of world land says that thirty percent of the earth's land is used for crops and pastures and seventy percent of all abstracted freshwater is directed towards irrigation to produce the food that people and livestock need for a stable food supply. This large-scale utilization of land and water resources is increasingly threatening environments. Further-

Figure 1. Water demand and supply relationship

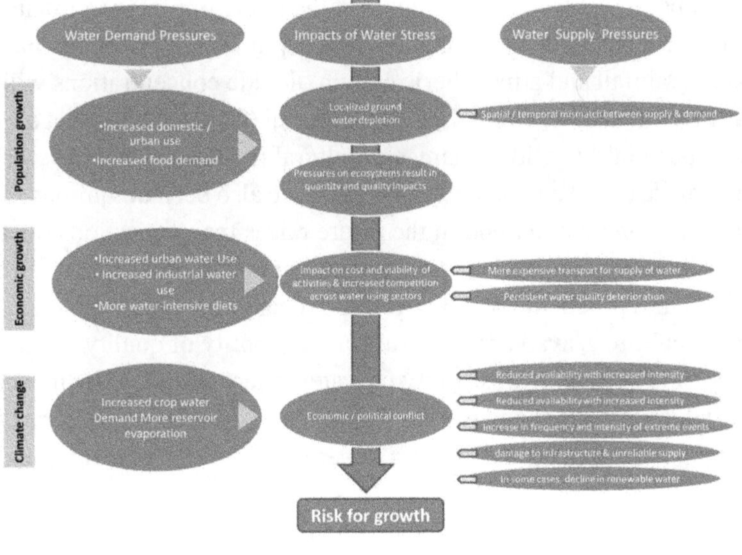

more, farming is important because it provides the livelihood of hundreds of millions of people. Water resources are important for agriculture production in all the regions of the globe. Agriculture system is vulnerable because of too hot and dry climate, limited and variable water supply, low and degraded soil quality and lack of adaptive capacity.

The chapter "Inter linkages of water, climate, and agriculture" discuss about the probable impacts of climate change on availability of water for agriculture. The document encompasses not only irrigation, but also other forms of water control intended to optimize growing conditions for crops, livestock and pasture. Change in climatic conditions like precipitation, temperature, Carbon dioxide (CO_2), etc. may be having adverse effect on global agriculture production. Hence the core of the document concerns adaptive and mitigation options and activities that can contribute to maintaining global food security and supporting farmers' livelihoods in changing water availability.

Keeping in the view the challenges of availability of quality water for agriculture in changing climate, the chapter focuses on the impact of climate change on agriculture, knowledge on the relationship between climate change, water and food security and how agriculture is able to adapt such climate change. The chapter also explores future availability of water and what will be the impending impact of availability and what mechanisms can be implemented to mitigate the resulting impact?

In order to meet the elevated flow of agricultural production further research and extension activity need to be focused keeping in mind the changing climatic situations. Hence, at the end, the chapter discuss about the latest trends in the water and agriculture research for changing climatic situations. However, the impact on agriculture and its ability for adaption may vary with different parts of world.

The purpose of the chapter is to enrich the current state of knowledge of readers on the relationship between water availability, climate change and food security and nutrition to provide an evidence base discussion. The chapter provides an analysis of empirical evidence results to highlight the relationship between water for agricultural production in climate change situations and food security.

Background

Survival of human and livestock on the globe is highly dependent on the agriculture. Agriculture and allied activities are highly climate dependent and change in the climatic parameters may severely affect food security on the globe. Apart from few exceptions, the likely impacts of climate change on agricultural sector in the future are not understood in any great depth. There are many uncertainties as to how changes in temperature, rainfall and atmospheric carbon dioxide concentrations will interact in relation to agricultural productivity (Londhe, 2016). Lobell & Gourdji (2012) noticed that climate trends over the past few decades have been fairly rapid in many agricultural regions around the world, and increases in atmospheric carbon dioxide (CO_2) and ozone (O_3) levels have also been ubiquitous. The virtual certainty that climate and CO_2 will continue to trend in the future raises many questions related to food security and health. One of the important research question mingling in agriculture research environment across globe is weather agricultural production will affect due to changing climatic conditions. One of the most important input for agriculture is availability of sufficient quantity of quality water. Change in climatic condition may result some regions with scarcity of water where as others with excess water. Both the situations are not cozy for agriculture and will having serious impacts on global agricultural production and livelihood of rural poor and marginal farmers in various agro-ecological regions.

In addition to water many factors will shape global food security over the next few decades which includes increasing human population, agricultural productivity, income growth and distribution, dietary preferences and disease incidence. Apart from agriculture, there will be increased demand for land and water resources for nonagricultural uses also. In this situation, maintaining increased flow of agricultural production is important to meet the need of growing population and new food habits. The main question of interest here is: how important will climate change in the contest of quality and timely water availability in shaping future crop yields and other agricultural production at the global scale. This question helps to set the challenge of climate change adaptation in context.

Apart from these important climatic parameters like atmospheric concentrations of carbon dioxide (CO_2) and Ozone (O_3), and changes in temperature and precipitation conditions will be having direct or indirect impact on water availability and agriculture production. It may be expected that due to climate change there is possibility of change in climatic regimes and shift in seasons. This will create need based on the water availability to change in cropping patterns in some of the regions.

The vast majority of the Earth's water resources are salt water, with only 2.5% being fresh water. Approximately 70% of the fresh water available on the planet is frozen in the icecaps of Antarctica and Greenland leaving the remaining 30% (equal to only 0.7% of total water resources worldwide) available for consumption. From this remaining 0.7%, roughly 87% is allocated to agricultural purposes (IPCC, 2007). According to the Comprehensive Assessment of Water Management in Agriculture, one in three people are already facing water shortages (CA, 2007). Further FAO and UN Water (Anonymous, 2007) reported that around 1.2 billion people, or almost one-fifth of the world's population, live in areas of physical scarcity, while another 1.6 billion people, or almost one quarter of the world's population, live in the developing country that lacks the necessary infrastructure to take water from rivers and aquifers which is known as an economic water shortage.

Projections of climate change are inherently uncertain, due to the natural variability in the climate system, imperfect ability to model the atmosphere's response to any given emissions scenario. Taken in to consideration the recommendations of Intergovernmental Panel on Climate Change (Christensen et al., 2013) with identified model agreement on future changes, the global monsoon, aggregated over all monsoon systems, is likely to strengthen in the 21st century with increase in its area and intensity, while the monsoon circulation weakens. Monsoon onset dates are likely to become earlier or not to change much and monsoon retreat dates are likely to delay, resulting in lengthening of the monsoon season in many regions. The Predictions of Intergovernmental Panel on Climate Change may further increase the complexity of food security many folds. Again these uncertainties are compounded by the paucity and unreliability of basic information related to agricultural production.

Water resource availability may be altered by changed rainfall patterns and increased rates of evaporation. Changes in rainfall will also increase variability in groundwater recharge and river flow, thus may affect all water sources. For many of the world's poorest people, rainfall variability may be a major impediment to their livelihoods. The inability to predict and manage rainfall, and consequent runoff, variability is a key contributing factor to their food insecurity and poverty. Frequent periods with too much water are followed by periods with too little and intermittent water scarcity is often a direct consequence of rainfall variability.

National economies, highly dependent on rain-fed agricultural production, are exceedingly vulnerable to fluctuations in rainfall. Although there are opposing views, there is little evidence that water scarcity by itself is a major factor limiting economic growth in most countries (Barbier, 2004). However, in contrast, rainfall variability has been shown to be a significant factor that has an effect on economic growth (Brown & Lall, 2006).

Under these circumstances, water storage can, by safeguarding domestic supplies and supporting crops and/or livestock during dry periods, significantly increase agricultural and economic productivity and enhance people's well-being. Sector-wise utilization and per capita water consumption is given in Figure 2. Water storage can also contribute to electricity generation and providing water supply to commercial and industrial enterprises. Consequently, it has an important role to play in poverty reduction, sustainable development and adaptation to climate change. Yet, despite greater rainfall variability than many other places, per capita water storage is lower in Africa and Asia than elsewhere in the World. Lack of water storage infrastructure is posited by some as a major constraint to economic development in many developing countries (Grey & Sadoff, 2006).

Physical water storage, for the future needs to rethink water storage in a future of rapidly rising population and increasing uncertainty related to climate change, and for better planning and management

large dams are often the first thing that comes to mind when "water storage" is mentioned. Mainly because of their considerable financial requirements, as well as the political opportunities that they represent, large dams (defined as those greater than 15 meters (m) high or with storage capacity exceeding 3 million cubic meters (Mm^3) for heights between 5 and 15 m (ICOLD, 2003) have often been the principal focus of water storage efforts in recent decades. Just under 50% of the 50,000 large dams constructed globally - since the 1950s - have been built to support irrigation.

Many others have been built to provide hydropower and domestic and industrial water supply as well as to provide flood protection. In many cases a single dam supports many of these functions (ICOLD,

Figure 2. Per capita water storage in cubic meters (m^3) in human-made reservoirs by continent
Source: White, 2005
**For a more accurate representation of this figure, please see the electronic version.*

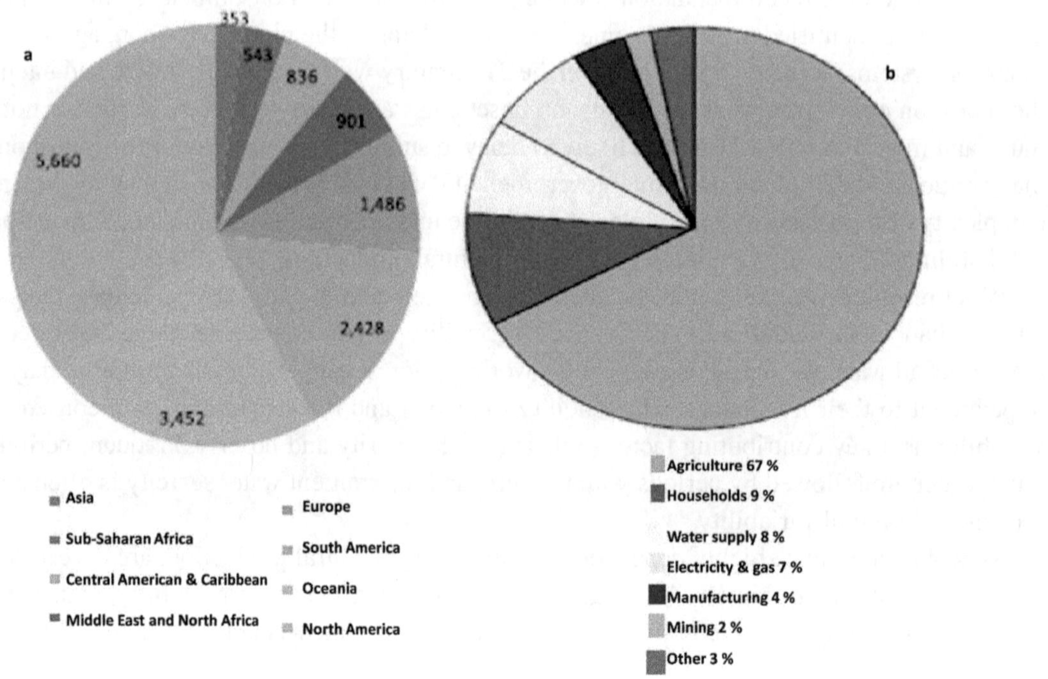

For a more accurate representation of this figure, please see the electronic version.

2003). Numerous large dams have brought significant social and economic benefits. The broad links between infrastructure development (including dams), increased agricultural productivity and economic growth have been documented (Hanjra et al., 2009; Hussain & Hanjra, 2004). However, having a high per capita storage in large reservoirs is no guarantee of national economic development.

In common with all human development, large dams also have costs. Still water storage is important because in changing climatic situations, the water requirement of agriculture may be more and the stored water in the large dams may be having the potential for fulfilling the water scarcity. From the available literature it is not very clear that the large dams will fully satisfy the future need completely in climate change situations.

Agricultural and Water Management

Increasing agriculture production with population growth is the most important issue in almost all over the globe. In order to achieve this objective, the immediate solution to the farmer is enhancing the inputs to the agriculture which may not always proper for environmental health. One of the study (Fuglie, 2012) analyzed that how much of the growth in output is due to increased resources, and how much of it is due to improved productivity? After nearly four decades, of primarily resource-driven growth, a dramatic shift to productivity-driven increases in global agricultural output began around the early 1990s (Figure 3). Between 1961 and 2009, total resources and inputs grew about 60 percent as fast as growth in total agricultural output, implying that improvement in Total Factor Productivity (TFP) accounted for only 40 percent of total output growth. But TFP's contribution to output rose over time, and between

Figure 3. Sources of growth in global agricultural production
Source: Fuglie, 2012

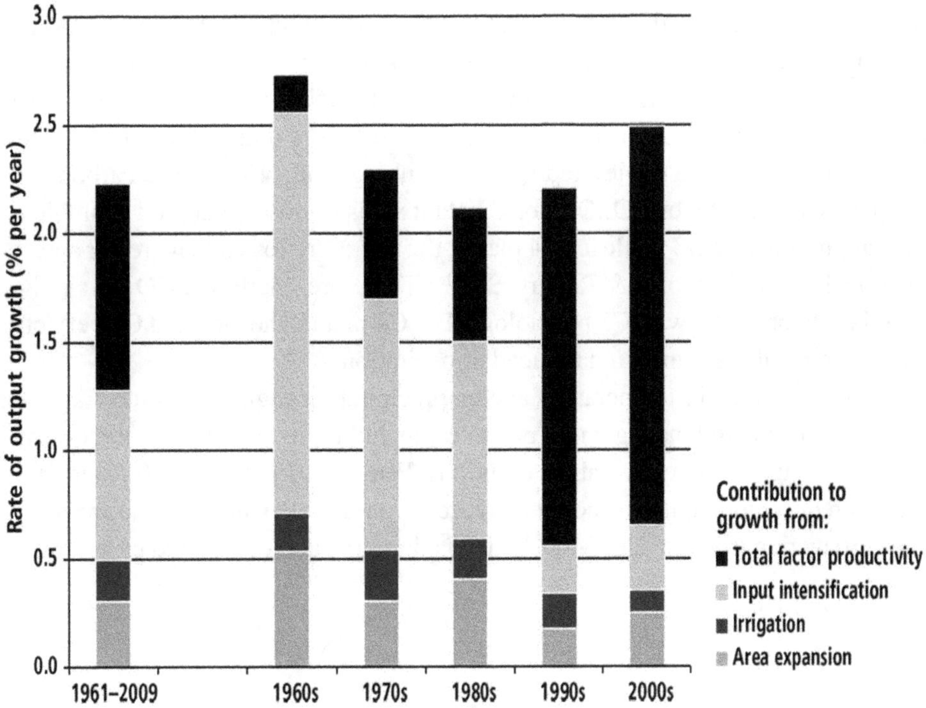

2001 and 2009 it accounted for about 75 percent of the growth in global agricultural production. The contribution of natural resources (including land and water) to output growth has decreased gradually over time while that of input intensification (including the amount of labor, capital, and materials per hectare of land) has fallen sharply.

Water management in agriculture production is one of the important parameter for high and sustainable yields. Water requirement for growth and development of every crop is different. Some crops like paddy requires more water for the growth as compared with others like sorghum and bajara. Crop stress at critical growth stages my severely reduce yield of crops or completely failure of the crop. The water requirement also vary with region as well as season. It depends on the other environmental parameters like temperature, CO_2 etc. along with type of soils they grow. Rain-fed agriculture can be severely affected by long dry spells and uneven distribution of rainfall. Hence water management in agriculture depends on many parameters. Some of the environmental factors which influence crop water requirement are described briefly in below paragraphs.

C3 and C4 Plants

Most agricultural plants are categorized by their photosynthetic mechanisms that control the chemical processes in their glucose manufacture from CO_2 and H_2O (water) as C3 and C4 species because of their photosynthetic pathways. Other plants are called CAM that stands for Crassulacean Acid Metabolism after the plant family in which it was first found (Crassulaceae) and because the CO_2 is stored in the form of an acid before use in photosynthesis. Common C3 species include wheat, cotton, soybean, and most legumes like alfalfa while common C4 crop species include sorghum, corn, and sugarcane. Some grass species are either C3 or C4 types. C3 plants fix atmospheric CO_2 directly onto 5 carbon sugar RuBP (ribulose bisphosphate) and thus into glucose. C4 plants first fix atmospheric CO_2 into 4-carbon acids in the mesophyll of the leaf and decarboxylate the 4-carbon acids in the bundle sheath cells where the CO_2 is then fixed by RuBP carboxylase (all of this takes place during the day). CAM plants first fix atmospheric CO_2 into malic acid and other 4C-acids at night. During the day, malic acid is decarboxylated and the CO_2 released is then fixed by rubisco (all of this takes place in the same cell). Generally, the C4 photosynthetic pathway is considered more water efficient than C3 species. However, C3 species typically are more sensitive to elevated CO_2 (Rosenberg et al., 1988). The carbon-fixing efficacy of Rubisco depends on the ratio of CO_2:O_2. For C3 plants, this is closely coupled to ambient conditions, and efficacy is approximately 2/3 while for C4 plants, the CO_2:O_2ratio is much greater and carboxylation efficacy is nearly 100% (Ainsworth & Rogers, 2007). Therefore, increased CO_2 in air should directly increase assimilation for plants with C3 physiology. For C4 plants, the elevated CO_2 effects are indirect due to increased stomatal resistance and reduced transpiration.

Climate change (changes in temperature and/or precipitation regimes) would likely lead growers to change crops, cultivars, and management practices, including irrigation, to mitigate any adverse effects or to take advantage of more favorable conditions. Peterson & Keller (1990) suggested that higher temperatures and reduced precipitation could increase crop water demand in some areas and prompt the development of irrigation in regions previously devoted to dry-land or rain-fed cropping.

CLIMATE CHANGE IMPACTS ON AGRICULTURAL WATER MANAGEMENT

Agriculture encompasses various activities including raising crops, raring livestock, poultry, fisheries etc. Availability of water for successful agriculture production is very important. As discussed above each crop water requirement is specific which is based on soil and various climatic parameters. Cropping pattern of any region is established over period of time based on available soil and climatic situation and farmer's knowledge base. Scientific evidences shows that due to global warming the climatic parameters are changing which may be having direct or indirect impact on global agriculture production.

Plant systems, and hence crop yields, are influenced by many environmental factors, and these factors, such as moisture and temperature, may act either synergistically or antagonistically with other factors in determining yields (Waggoner, 1983). In most land regions the frequency of warm days and warm nights may likely increase in the next decades, while that of cold days and cold nights will decrease. The impact of various climatic parameters on global agriculture from available literature is summarized in Table 1.The below discussion is trying to analyze the impact of climate change on water availability to agriculture and its production.

Precipitation

Precipitation that regulates life on Earth and soil moisture is essential to sustain plant growth and biodiversity flourishes where water is abundant. Precipitation also has a bearing on local climate (Priestley, 1966). Variation in the precipitation may be having large impact on agriculture production on the globe. The general pattern of wet-get-wetter (also referred to as 'rich-get-richer', e.g., Held & Soden 2006; Chou et al., 2009; Allan et al., 2010) and dry-get-drier has been confirmed, although with deviations in some dry regions at present that are projected to become wetter by some models.

It has been demonstrated that the wet-get-wetter pattern implies an enhanced seasonal precipitation range between wet and dry seasons in the tropics, and enhanced inter-hemispheric precipitation gradients (Chou et al., 2007).

Intergovernmental Panel on Climate Change in AR5 report (Kirtman et al., 2013) presented results of Coupled Model Intercomparison Project (CMIP5) for projections of near-time changes in precipitation (Figure 4). It is clear from the Figure 4 that the basic pattern of wet regions tending to get wetter and dry regions tending to get dryer is apparent. There may be some regional deviations in the precipitation pattern. There may be possibility of the uncertainty in the substantial spread in the magnitude of projected change based on the simulation models used. The Figure 4 also highlight the large amplitude of the natural internal variability of mean precipitation. On regional scales, mean projected changes are almost everywhere smaller than the estimated standard deviation of natural internal variability. For zonal means and at high latitudes, the projected changes relative to the recent past exceed the estimated standard deviation of internal variability. The Figure 4 also depicts that the overall, zonal mean precipitation will very likely increase in high and some of the mid latitudes, and will more likely than not decrease in the subtropics. In case of the regional scales precipitation changes may be influenced by anthropogenic aerosol emissions and may be strongly influenced by natural internal variability.

Table 1. Climate change fact and impact on crop production

Parameter	Fact	Effect
Temperature	Global Mean Surface Temperature	Has increased since the late 19th century.
	Comparison with past	Each of the past three decades has been successively warmer at the earth's surface than all the previous decades in the instrumental record, and the first decade of the 21st century has been the warmest.
	Global Mean Temperatures	Will continue to rise over the 21st century if Greenhouse Gas (GHG) emissions continue unabated.
	Impact on different regions	Temperature change will not be regionally uniform.
	Temperature extremes	In most places, there will be more hot and fewer cold temperature extremes.
	Global temperatures rise.	Mean sea level pressure is projected to decrease in high latitudes and increase in the mid-latitudes.
Water Cycle	Precipitation	In the long term, global precipitation will increase with increased global mean surface temperature.
	Average precipitation in warmer world	Will exhibit substantial spatial variation. Some regions will experience increases, other regions will experience decreases and yet others will not experience significant changes at all.
	Annual surface evaporation	Projected to increase as global temperatures rise over most of the ocean and is projected to increase over land.
	Humidity	Global near surface and tropospheric air specific humidity have increased since the 1970s.
Extreme Events	Day and night temperature	It is very likely that the numbers of cold days and nights have decreased and the numbers of warm days and nights have increased globally since about 1950.
	Warm days and warm nights	In most land regions the frequency will likely increase in the next decades.
	Cold days and cold nights	In most land regions the frequency will likely decrease in the next decades.
	Precipitation	Since about 1950 the number of heavy precipitation events over land has increased in more regions than it has decreased.
Sea Level	The rate of sea level rise	Increased from the early 19th century to the early 20th century, and increased further over the 20th century.
GHG	CO_2	Increased by 40% from 278 ppm about 1750 to 390.5 ppm in 2011.
	CH_4	Increased by 150% from 722 ppb about 1750 to 1803 ppb in 2011.
	N_2O	Increased by 20% from 271 ppb about 1750 to 324.2 ppb in 2011.
Land Use Change	Radiative Forcing (RF)	There is robust evidence that anthropogenic land use change has increased the land surface albedo, which leads to Radiative Forcing (RF) of -0.15 ± 0.10 W m^{-2}.
	Zonal mean precipitation	Will very likely increase in high and some of the mid latitudes, and will more likely than not decrease in the subtropics.
	Specific humidity	Increases in near-surface over land are very likely.
	Evaporation	Increases in over land are likely in many regions.
Earth's Energy Budget	Independent estimates of effective RF of the climate system	The observed heat storage, and surface warming combine to give an energy budget for the Earth that is closed within uncertainties (high confidence), and is consistent with the likely range of climate sensitivity.

continued on next page

Table 1. Continued

Parameter	Fact	Effect
Monsoon Systems	Future changes	The global monsoon, aggregated over all monsoon systems, is likely to strengthen in the 21st century with increases in its area and intensity.
	Monsoon circulation	The global monsoon circulation weakens.
	Monsoon onset and retreat dates	Onset likely to become earlier or not to change much and monsoon retreat dates are likely to delay, resulting in lengthening of the monsoon season in many regions.
	Annual precipitation change	Follows a 'warmer-get- warmer'pattern, increasing where warming of sea surface temperature exceeds the tropical mean and vice versa.
Cyclones	Projections	It is likely that the global frequency of occurrence of tropical cyclones will either decrease or remain essentially unchanged, concurrent with a likely increase in both global mean tropical cyclone maximum wind speed and precipitation rates.
	Regional projections	The future influence of climate change on tropical cyclones is likely to vary by region.
Crop Yield	Crop yield	In developing countries, climate change will cause yield declines for the most important crops. South Asia will be particularly hard hit.
	Irrigated crop	Will have varying effects on yields across regions, but yields for all crops in South Asia will experience large declines.
	Higher maximum temperatures, more hot days and heat waves over nearly land areas	Increased risk of damage to a number of crops
	Increased summer drying over most mid-latitude continental interiors and associated risk of drought	Decreased water resource quantity and crop yields.
	Intensified droughts and floods associated with El niño events in many different regions	Decreased agricultural and rangeland productivity in drought- and flood-prone regions.
	Increased intensity of mid-latitude storms	Increased damage to coastal ecosystems.
	Food price	Will result in additional price increases for the most important agricultural crops–rice, wheat, maize, and soybeans.
	Fodder price	Higher feed prices will result in higher meat prices. As a result, will reduce the growth in meat consumption slightly.
	Calorie	Calorie availability in 2050 will not only be lower than in the no-climate-change scenario-it will actually decline relative to 2000 levels throughout the developing world.
	Calorie and child nutrition	By 2050, the decline in calorie availability will increase child malnutrition by 20 percent relative to a world with no climate change. Climate change will eliminate much of the improvement in child malnourishment levels that would occur with no climate change.
Livestock	Higher maximum temperatures, more hot days and heat waves over nearly land areas	Increased heat stress in livestock and wildlife

Source: Hartmann et al., 2013, Collins et al. 2013, Rhein et al. 2013, Myhre et al., 2013, Christensen et al., 2013, Nelson et al. 2009 & IPCC, 2001

Figure 4. Multi-model projections of changes in annual and zonal mean (a) precipitation (%) and (b) precipitation minus evaporation (mm day–1) for the period 2016–2035 relative to 1986–2005 under RCP4.5. The light blue denotes the 5 to 95% range, the dark blue the 17 to 83% range of model spread. The grey indicates the 1σrange of natural variability derived from the pre-industrial control runs
Source: Kirtman et al., 2013

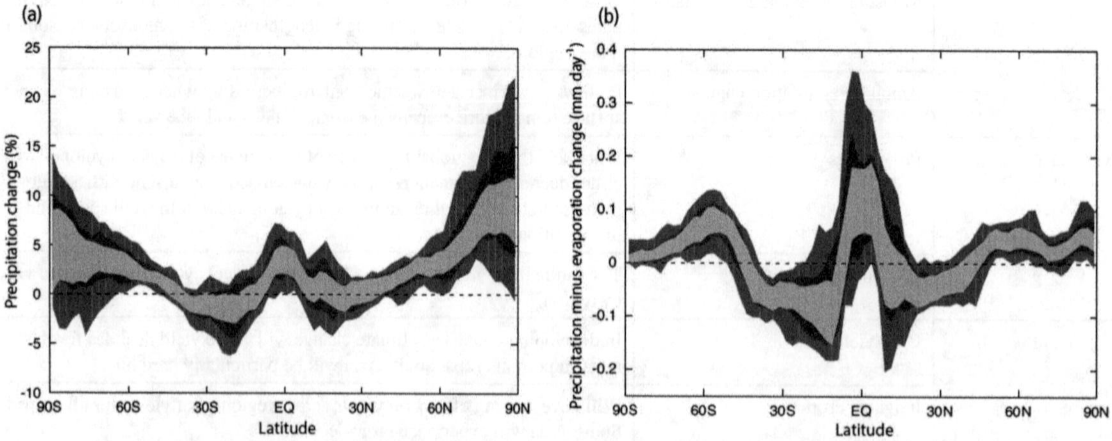

Figure 4 Time series of relative change relative to 1986–2005 in precipitation averaged over land grid points over the globe in October to March. Thin lines denote one ensemble member per model, thick lines the CMIP5 multi-model mean. On the right-hand side the 5th, 25th, 50th (median), 75th and 95th percentiles of the distribution of 20-year mean changes are given for 2081–2100 in the four RCP scenarios.

In response to GHG forcing, dry land areas tend to show a reduction of evaporation and often precipitation, accompanied by a drying of the soil and an increase of surface temperature, in response to decreases in latent heat fluxes from the surface (e.g., Fischer et al., 2007; Seneviratne et al., 2010).

Evapotranspiration

The main source of water for agriculture production is through precipitation and the loss of water in the form of evaporation depends on the climatic conditions of the region. It has long been recognized (Priestley, 1966) that the availability of surface water regulates evaporation and local temperature. Where there is abundant surface water and plant growth, the local evaporation constrains temperature and in hot climates temperatures are very much higher in the absence of surface evaporation. So it is necessary to have proper balance between precipitation and evaporation for agriculture production.

The evaporation of surface water and condensation of atmospheric water vapor to form clouds and precipitation, is an essential component of energy flow through the climate system. This phenomenon may be explained by the Clausius-Clapeyron relationship which is the rate of increase of saturated vapor pressure with temperature and is important in regulating many of the processes of the hydrological cycle. The global atmospheric water content is constrained by the Clausius–Clapeyron equation to increase at around 7% K^{-1}; however, both the global precipitation and evaporation in global warming simulations increase at 1 to 3% K^{-1} (Lambert & Webb, 2008; Lu & Cai, 2009).

Consideration of hydrological drought conditions employs a range of different dryness indicators, such as soil moisture or other drought indices that integrate precipitation and evaporation effects (Seneviratne et al. 2012). In order to provide an indication of future changes of water availability, figure 4 (b) presents zonal mean changes in precipitation minus evaporation (P - E) from Common Management Information Protocol CMIP5 (Kirtman et al., 2013). As in the case of precipitation (Figure 4), the uncertainty is dominated by model differences as opposed to natural variability (compare blue versus grey shading). The results are consistent with above discussion on the wet-get-wetter and dry-get-drier pattern. In the high latitudes and the tropics, most of the models project zonal-mean increases in P – E, which over land would need to be compensated by increases in runoff. In contrast, zonal mean projected changes in the subtropics are negative, indicating decreases in water availability.

Evapotranspiration is the combined effect of evaporation and transpiration, and it is a key part of the hydrologic cycle. Evaporation is simply the phase change from liquid water to water vapor as temperature increases. Transpiration involves the uptake of water in soil by plants, the transport of water through the plant, and its eventual evaporation from plant leaves and other surfaces. Evapotranspiration is a natural cooling process because heat at the earth's surface is captured by water droplets, which, through evaporation, disperse the heat out of the atmosphere and into space.

Change in CO_2 Levels

The prevailing view among researchers is that global climate change may prove beneficial to many farmers at least in the short term. The logic is straightforward: Plants need atmospheric carbon dioxide to produce food, and by emitting more CO_2 into the air will cause some crops and trees to grow bigger and faster.

The latest analysis of observations from the Global Atmosphere Watch (GAW) Programme (WMO 2014) shows that the globally averaged mole fractions of Carbon dioxide (CO_2), Methane (CH_4) and Nitrous Oxide (N_2O) reached new highs in 2013, with CO_2 at 396.0±0.1 ppm, CH_4 at 1824±2 ppb and N_2O at 325.9±0.1 ppb (Table 2).

The change in atmospheric CO_2 levels will have impact on the vegetation. Changes in evapotranspiration over land are influenced not only by the response to Radiative Forcing (RF), but also by the vegetation response to elevated CO_2 concentrations. Physiological effects of CO_2 may involve both the stomatal response, which acts to restrict transpiration (Cao et al. 2009; Field et al. 1995; Hungate et al. 2002; 2010; Lammertsma et al., 2011), and an increase in plant growth and leaf area, which acts to increase evapotranspiration (Bounoua et al. 2010; El Nadi, 1974).

Table 2. Global annual mean abundances (2013) and trends of key greenhouse gases. Units are dry-air mole fractions, and uncertainties are 68% confidence limits

	CO_2	CH_4	N_2O
Global abundance in 2013	396.0 ± 0.1 ppm	1824 ± 2 ppb	325.9 ± 0.1 ppb
2013 abundance relative to year 1750[a]	142%	253%	121%
2012–2013 absolute increase	2.9 ppm	6 ppb	0.8 ppb
2012–2013 relative increase	0.74%	0.33%	0.25%
Mean annual absolute increase during last10 years	2.07 ppm/yr	3.8 ppb/yr	0.82 ppb/yr

[a]Assuming a pre-industrial mole fraction of 278 ppm for CO_2, 722 ppb for CH_4 and 270 ppb for N_2O. Stations used for the analyses numbered 124 for CO_2, 121 for CH_4 and 33 for N_2O.

[Source: WMO (2014)]

PROSPECTS FOR ADAPTATION MITIGATION

Intergovernmental Panel on Climate Change (IPCC, 2001), suggested that adaptation is gaining attention as an expected answer to the challenges posed by climate change. The increasingly uncertain climatic conditions to which actors are exposed may becoming a constraint for their well-being. Given the range of warming predicted by the scientific community, regional and local variation in impacts on the agricultural production may likely to be high. A critical challenge for farmers in developing countries is how to cope with increased variability and uncertainty in current and future rainfall patterns, a phenomenon which is affecting rain-fed and irrigated farming alike. Further, the concern with climate change is sensitive given the linkage of the agricultural sector to the poverty. It may be anticipated that adverse impacts on the agricultural sector will exacerbate the incidence of rural poverty. Hence, improved water resource management will be vital to sustaining crop productivity in the face of climate variability. Food insecurity may continue to be a serious issue in coming decades. Climate change may significantly increase production risk and rural vulnerability, particularly in regions that already suffering from chronic soil and water resource scarcity, high exposure to climatic extremes including droughts, flooding, poverty and hunger.

In order to sink with the climatic variability and retain enhanced flow of food production, some measures in the form of adaptation and mitigation will be required. Adaptation to climate impacts in the agricultural sector is not a new phenomenon. According to Rosenzweig & Liverman (1992) and Rosenberg (1992) natural systems have continuously been adapting autonomously, or in accordance with a plan, to a changing environment throughout history.

Adaptation may be defined as a process or action of adjusting to different circumstances or conditions, in this case as a result of a changing climate. Availability of water for agriculture depends on many climatic parameters like precipitation, temperature, atmospheric $CO_{2,}$ etc. Hence judicious management of water resource for sustainable agriculture development may require proper adoption to the changing climatic parameters. The adaptations includes several steps starting from micro level adaptation as farm production adjustments such as diversification and intensification of crop and livestock production; changing land use and irrigation; and altering the timing of operations. The adaptations can also include numerous market responses that have emerged as potentially effective adaptation measures to climate change, development of crop and flood insurance schemes, innovative investment opportunities in crop shares and futures, credit schemes, and income diversification opportunities. The options of adaptations may include government responses and institutional changes, pricing policy adjustments and technological developments. Hence, adaptations can be broadly grouped in to Short-Term and Long-Term adaptations (Figure 5).

Short-Term Adaptations

The short- term adaptations to the climate change may be having benefits to the farmers for shorter period of time and can be implemented immediately. Existing literature suggests that such types of adaptations are more appropriate to address short-term climatic concerns. These adaptations may include the following.

Figure 5. Apaptations in agriculture

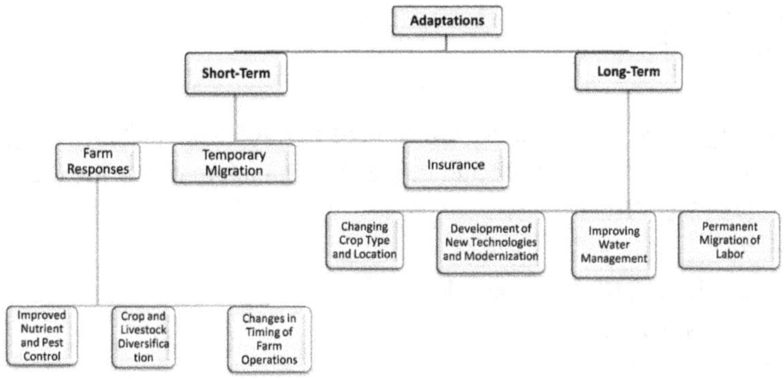

Farm Responses

Among the most important and direct current adaptations to climate variability are some of the farm level responses. The farm response adaptations can include the crop and livestock diversification as well as changes in timings of various farm operations. Improved nutrient and pest control management can also be part of adaptation in farm response.

A. Crop and Livestock Diversification and Changes in Timing of Farm operations

Crop diversification refers to the addition of new crops or cropping systems to agricultural production on a particular farm taking into account the different returns from value-added crops with complementary marketing opportunities. Diversification may be a key factor in reducing risk and means of coping with an uncertain climate and further unpredicted water for agriculture production. Crop is the major activity worldwide for livelihood followed by livestock production. Diversification of crop and livestock varieties, having the potential to increase productivity in the face of temperature and moisture stresses may be helpful. Diversity in seed genetic structure and composition may be an effective defense against numerous factors, including water stress, disease and pest outbreak and importantly climate hazards.

It is also predicted, there may be shift of cropping seasons and adjusting to such situations options include changes in the timing and intensity of production. Delcourt & Van kooten (1995), Brklacich et al. (2000), de Loe et al. (1999) have given following options for addressing impacts on yields and soils from climate change.

- Changing land-use practices such as the location of crop and livestock production.
- Rotating or shifting production between crops and livestock.
- Shifting production away from marginal areas can help reduce soil erosion and improve moisture and nutrient retention.
- Altering the intensity of fertilizer and pesticide application as well as capital and labor inputs.
- Adjusting the cropping sequence, including changing the timing of sowing, planting, spraying, and harvesting, to take advantage of the changing duration of growing seasons and associated heat and moisture levels.

- Altering the time at which fields are sowed or planted can also help farmers regulate the length of the growing season.
- Changing the timing of irrigation.

Many researchers suggested various adaptation measures for livestock and rangeland management. Following are few suggestions for livestock and rangeland management outlined by Baker and Viglizzo (1998), Chiotti et al. (1997), IPCC (1996).

- Change in grazing timing, duration, and location.
- Varying supplemental feeding.
- Changing the location of watering points.
- Altering the breeding management program.
- Changes in rangeland management practices.
- Modifying operation production strategies as well as changing market strategies.
- In temperate climatic areas, planned adaptation measures in livestock management that are advocated include the use of vegetative barriers or snow fences to increase soil moisture, or windbreaks to protect soil from erosion.
- In warmer climates, adverse climatic conditions such as heat stress can be moderated by the adoption of appropriate technology such as the use of sprinklers in livestock buildings or feedlots
- Adjusting livestock stocking rates.
- Implementing feed conservation techniques and fodder banks to moderate the consequences for animal production during periods of poor crops.
- Increasing native rangeland vegetation or plant-adapted species.

Working with various adoptions there may be numerous constraints farmers and livestock holders may encounter which make these farm level adaptations difficult. The short term adaptations are not costless. For example (Skees et al. 1999) the most significant problem to overcome is that diversification is costly in terms of the income opportunities that farmers forego (that is, switching crop varieties can be expensive, making crop diversification typically less profitable than specialization). In addition to this traditions can often by difficult to overcome and will dictate local practices. In such situation, if a local region has a long and rich tradition of planting a particular crop variety, the change to newer and more suitable varieties can be hard.

B. Improved Nutrient and Pest Control Management

Increased CO_2 levels and higher temperatures are likely to create stress in crops and livestock and may induce a need for more plant protection. Changes in the application of various chemicals may further enhance the problem. For the climate change situation alternative production techniques and crops, as well as locations, that are resistant to water stress, infestations and other risks may also be relied upon as effective response strategies. Following are a range of management practices have emerged that can assist farmers adapt to loss of soil moisture and organic carbon contents and increased soil erosion as a result of changing climate situations (Erda 1996 and Parry 2000).

- Improved nutrient management techniques to maintain soil fertility and prevent erosion.
- For the increased frequency of droughts, farmers can further adapt by changing the selection of crops.
- Introduction of management techniques that conserve soil moisture, such as reduced or no tillage, in order to maintain soil organic carbon contents can result in improved soil structure and fertility.
- Increasing production per unit of evapotranspiration with the use of new and improved varieties.
- Reducing water use in land preparation as well as loss (through seepage and percolation) during the crop growth period.
- Adoption of efficient water use methods.
- The diffusion of appropriate technology to enhance greater water use efficiency like drip-irrigation.

C. Temporary Migration

Temporary migration is also called circular migration in agriculture. It includes seasonal migration, where workers undertake off-farm or non-farm activities for some part of the year, and returning back during harvest time. The temporary migration is the movement of labor from one agricultural area to another area, or across sectors. This also includes migration between and within urban and rural to ensure security in livelihood.

It is not much clear from the available literature to what extent climate change per se can be attributed as the primary factor in the decision making process of households engaged in agriculture on whether or not to migrate. But in case of adverse climate the livelihood of poor and marginal household families may be difficult. However, in such situation the temporary migration to the other agriculture or non-agriculture areas for the period of adverse condition may be an option of livelihood. The option of such types of migration for livelihood may be useful for extreme events in climate change.

D. Insurance

There may be much uncertainty about the agriculture production as successful crop yield depends on the availability of water and climate of the region. So the households engaged in agriculture may face problem for the livelihood in case of crop failure and inherently need insurance mechanisms to cope with income risks. The types of risks faced by the agricultural sector are outlined below (Moreddu 2000).

- Production risks due to weather variation, crop disease and various other causalities,
- Ecological risks from climate change, pollution, and natural resource management,
- Market risks, which depend on input and output price variability, and
- Regulatory or institutional risks due to state intervention in agriculture.

Furthermore, households with extremely low incomes may be at highly risk and will limits their financial ability and willingness to adopt new technologies that can maintain and enhance crop productivity. In such situation, the insurance will help farmers to reduce the financial risk when they adopt for new technology or adaptation. The insurance may encourage the farmers to with stand in small climatic variations by reducing financial risk of livelihood in case of failure of the crops.

Long-Term Adaptations

Efforts are necessary to withstand and minimize the adverse impacts of short term climate variability and uneven water availability for agriculture. The measures will be necessary to reduce vulnerability to anticipated future impacts of climate change. So the combination of adaptations both at farm and policy level will be necessary for long term adaptation for future water related consequences scarcity due to climate change. Following paragraphs describes a range of adaptation options as most relevant in water availability and overall climate change.

The interaction of water for agriculture production is governed by climatic situation. However, given that warming is largely a concern of the future, long term adaptation must also be in the future. Kurukulasuriya and Rosenthal (2003) also reported that addressing climate change- a long term phenomena, should entail a comprehensive long term response strategy at the national or local level but will require a dynamic approach. Some of the important long term strategies for adaptation of availability of water for agriculture and climate change are discussed below.

A. Changing Crop Type and Location

The main purpose of changing the crop types or location of cultivation of crops is to cultivate stress resistant crops on its suitable location. The potential options include switching to more robust varieties that are better suited to the changing environment.

Agricultural analyses of future climates indicate also that crops will move poleward with warming (Mendelsohn and Neumann 1999). The extent of this migration depends upon the severity of the warming. This types of adaptation require a number of underlying prerequisites are in place. Obviously, the scope for shifting production to new lands, particularly in developing countries where, may be limited because of population pressures and the availability of cultivable land. There may be another constrain that farmers may not be willing and ability to move. Further, land use regulations on agricultural production are governed by country laws and may hinder such adaptations in some of the countries. In such situation, crop rotations that may not be optimal in interest of available water for agriculture production and a changed environment may be persisted. This may consequence in severe losses in the long term.

Appropriate land reform policies that establish or strengthen property rights in changing demand of water for agriculture need to be enforced. It is also necessary that appropriate measures that boost farmer's financial ability to undertake the necessary adaptation may be necessary. These financial measures may include improving access to credit and banking facilities in rural areas. Furthermore, investment in diffusion of access to sufficient and timely irrigation is important. Appropriate institutional support to promote the dissemination of knowledge through dedicated extension network may play important role in such situations.

There is good amount of debate in agriculture research that the new crops are equally (if not, more) profitable to the farmer. There is also serious concern that changing crop types will not automatically maintain previous levels of food production as well as or nutritional quality levels. While the national objective may be to grow crops that will fulfill food security with minimum water requirement.

B. Development of New Technologies and Modernization

Agriculture research on the globe has crossed many milestones and has many success stories. Research and technological innovation in crop and animal productivity is important to achieve food security in changing environmental conditions. This will enable farmers to cope with various climatic conditions and may be fundamental to the growth and development of agriculture. There are two basic types of technological options (Smithers and Blay-Palmer 2001), important for agriculture viz. mechanical and biological. Mechanical innovations include irrigation, conservation tillage, and integrated drainage systems. On the other hand, biological options also have an important role in enabling cropping systems to adapt to a wide range of climatic conditions. Investment in crop breeding, the promotion of climate-resistant varieties that offer improved resistance to changing diseases and insects, breeding of heat- and drought-resistant crop varieties, the use of traditional varieties bred for storm and drought resistance, and investment in seed banks may be necessary for success in overcoming vulnerability to climate.

Further, advances in science and biotechnology offer much powerful tools that hold promise to overcome the challenges posed by scarcity of water resources and threats posed to the agricultural production. However, it is not very clear that how much role the advocacy of biotechnology will play as adaptation in climate change scenario. In addition to this the issue of affordability of new technology to poor farmers may be serious concern for effective climate change. Several innovative concepts and technologies like geographical information systems, software-based weather systems, low-cost and decentralized technologies, industrial symbiosis (Rao & Patil, 2015) and as well as innovative waste treatment with simultaneous resource conservation method have also been reported (Patil & Rao, 2015).

C. Improving Water Management

Availability of water is most important for successful agriculture production. Improved water resource management will be vital to sustaining crop productivity levels in the face of both climate variability and longer-term change. In areas that are currently dependent primarily on rain-fed agriculture, the conjunctive use of surface and ground water resources will play an increasingly important role in enabling farmers to adapt to fast changing climatic conditions. However, it is also clear that in the face of rising domestic and industrial demand, additional efforts are necessary to ensure efficient management of water resources. With climate change and variability increasing pressure on available water resources, improved water management is one of the most important long-term adaptation options that countries must pursue. According to recent estimates, irrigation efficiency in developing countries is extremely low. The average efficiency of irrigated agriculture in developing countries is very low.

Preserving water resources using suitable storage may play a crucial role in availability of water in changing climate. However, if the planning of water storage is not improved, it may likely that many investments may fail to deliver intended benefits. McCartney et al. (2013) reported that, national economies, highly dependent on rainfed agricultural production, are exceedingly vulnerable to both intra-annual and inter-annual fluctuations in rainfall and hydrology. In addition to this traditional adaptation techniques may also be effective to deal with shortages in water.

Apart from water storages there may be challenges with regard to water use and availability that also need to be overcome. Increased demand for water by municipal and industrial sectors may further become serious issue in future. Un-sustained and over exploited irrigation may lead to groundwater depletion, soil salinization and water-logging. In some regions, numerous challenges need to be overcome in order to increase water supply. As emphasized by Dinar (2000), these challenges include financial crises; low-cost recovery of the investment in the water system; the role of political parties, electoral systems and interest groups. Dinar argues for improved forecasting procedures, simulation models, and improved data monitoring systems. Other necessary preconditions include overcoming economic constraints. The improvements in irrigation and other high-efficiency water conservation technology require major, long-term and costly investments. Furthermore, undertaking requisite institutional reforms that hinder the pursuit of effective water resource management strategies may also be helpful. However, given the uncertainty surrounding forecasts of regional changes in precipitation, it has not yet been proven that making water adaptations in advance of climate changes is in fact very practical.

D. Permanent Migration of Labor

Encompasses permanent migration in the form of the movement of migrants into new economic areas. The possible reasons for migration may be possibly due to policies or permanent changes in their previous environment. The agriculture labor migration may be from poorer agricultural areas in one region to lowlands in other regions. Desanker (2002) stresses, long-lasting climate pressures, such as prolonged drought, which increase the vulnerability of migratory groups to climate change (by limiting the scope of areas to move to), can be disastrous. Short-term migrants may be forced into becoming more permanent migrants, resulting in dire consequences such as pressures on land and water resources.

The adaptations outlined in the above sections may not fully address the concerns of permanent labor migration with short term climate variability may not be adequate in the face of new and long lasting climate conditions. Reilly et al. (2003) reported that even insurance programs will not be sufficient if productivity of land becomes unviable and may in fact act as a deterrent to change despite market signals that indicate otherwise. If unmanaged or uncontrolled, large-scale migration of this type may have consequence in significant impacts on the environmental resource bases as well as indigenous societies.

In such complex situations, a clearly defined system of property rights and enforcement may be necessary to prevent potential stress. The establishment of an appropriate system of property rights both at the individual or community level may be part of the likely solution. Well placed training and extension services may be act as other solutions that would be needed based on the situations.

E. Adaptations Irrespective of the Temporal Dimension of Climate Impacts

Non-climatic parameters are also equally important for agriculture production on sustainable basis. Suitable policies aimed at growing the resiliency of the agriculture sector to other, including non-climatic factors may be necessary and may help in improving capacity to cope with both climate variability and climate change. Further, Smit and Olga (2001) emphasized on solving immediate concerns facing domestic agriculture sectors do not delay the formulation and implementation of efficient responses to promote long term sustainability issues.

Such effective adaptation may also depend considerably on primary local environmental, institutional, and socio-economic conditions. Both domestic, and regional cooperation in science, resource management, and development may be extremely important. Some of the major economic and institutional issues may likely to be beneficial irrespective of the nature of the climate change are as follows.

- Investment and accumulation of capital.

- Reform of pricing schemes, development of open markets, and other reforms.
- Adoption of new technologies.

F. Promotion of Trade

Suitable market and appropriate return to the agriculture produce is the key to success for the farmers. Suitable market for agriculture produce is important for both buyers and consumers. Here suitable trade may play an important role during periods of variable climatic. There may be possibility that agricultural trade may moderate impacts by enabling farmers in regions less adversely affected to sell their produce in areas more severely affected by climate change. The role of trade policy may important because of repercussions on the prospects for adaptation. Further, regional and international trade may lead to improvements in access to international markets, which in turn may help a country diversify and reduce of risk of food security from water scarcity and climate change.

G. Extension Services

Extension services have played a key role in promoting agricultural productivity in developing countries. Their role in promoting various adaptations to climate change may be very important. Traditionally, extension services have generally been in the purview of services provided by government, given that agricultural research is typically a public good. However, private and non-governmental agencies or the formation of research cooperatives do play a significant role in some countries. Crucially, as Evenson (1997) notes, the economic contribution of extension services is governed by location-specific factors. In this regard, numerous programs have been found to be ineffective given the underperformance of agents, design limitations, and management failures. Overall, efficient, farmer friendly extension policies may coup farmer's efforts to maintain sustainable flow of agriculture production in uneven availability of water in changing climate.

H. Diversification of Income-Earning and Employment Opportunities

Seasonal effects and climatic uncertainty that characterize the agricultural sector effectively mean that diversification of income and employment opportunities may be an important adaptation strategy for households in the agriculture sector. In dryland areas, traditional practices to help cope with drought include the accrual of a surplus in a superior year, in the form of cash or assets (for example, cattle) for use in poorer years (Burton 2001). While measures such as crop storage, sales, and household savings can and do offer relief from temporary or seasonality effects, risk and market imperfections that abound in rural settings render diversification into off-farm opportunities necessary to reduce income instability (Alderman and Paxson 1992). Consequently, policies that provide the opportunities to pursue alternative livelihood options need to be encouraged.

I. Dissemination of Climate Data

A reason in literature frequently cited for not adapting in time to climatic impacts is the lack of reliable climate monitoring and forecasting data. Barnett (2001) argues that an increase in the availability of information to understand the biophysical and social environment is necessary. The timely dissemination of climate forecasting information and early warning to farmers including information on risks can strengthen the ability of farmers to cope and optimize the management of hydrological variability and change. Monitoring data and indicators of change are also necessary across all sectors in society, not just policy makers. Along with information on risks, the information on suitable mitigation measures may be helpful to farmers.

J. Institutional Planning and Implementation

Insufficient institutional and decision making structures to support long-term planning in governments in developing countries has long been recognized to be a problem in pursuing general development objectives. At World Bank report underscored the finding that in some countries, such as Bangladesh, planning for climate change is not even mandatory, an outcome of planning agencies' being "formed not by law but by administrative resolution" (Kurukulasuriya and Rosenthal 2003). Long term institutional planning related to climatic variability and suitable implementations are important.

Mitigation

Mitigation is human intervention to reduce the extent of climate change. It includes strategies to reduce greenhouse gas sources and emissions, and enhancing greenhouse gas sinks. A number of mitigation strategies in the agriculture and forestry sectors have been identified as useful in achieving the goal of stabilization of atmospheric concentrations between 450-550 ppm CO_2. These include reduced deforestation and degradation of tropical forests (REDD), sustainable forest management (SFM) and forest restoration (FR), including afforestation and reforestation (A/R). In agriculture, they involve reduction of non-CO_2 gases through improved crop and livestock management and agroforestry practices, enhanced soil carbon sequestration in agricultural soils via reduced tillage and soil biomass restoration (Table 3).

Table 3. Mitigation potential in agriculture and forestry in 2030

Sr. No.	2030 Reductions	G tonnes CO_2 e yr^{-1}
1	Global	15-25
2	Agriculture	1.5-5.0
3	Methane, N_2O	0.3-1.5
4	Agroforestry	0.5-2
5	Agricultural soils	0.5-1.5
6	Forest	2.5-12
7	REDD	1-4
8	Sustainable forest management (SFM) SFM	1-5
9	Forest Restoration (FR)	0.5-3
10	Bioenergy	0.1-1.0
	Total	4-18

Source: IPCC (2007)

Global reductions in 2030 correspond to those needed to achieve stabilization of atmospheric concentrations between 450-550 ppm CO_2, under a mid-range IPCC SRES.

In case of long term sustainable availability of water for agriculture can be achieved by investing in wider best practices. These practices can include the following.

I. Reducing Emissions of Green House Gases

The greenhouse gases include CO_2, methane and nitrous oxide and keep green gases under the limit, by adopting the effective ways. The practices that will help to reduce Carbon dioxide (CO_2) levels may include land conversion, reduced deforestation, improved ways to manage wildfires and alternatives to the burning of crop residues. The Methane and Nitrous Oxide in the environment can be managed using practices like improved nutrition for ruminant livestock, efficient management of livestock waste, and efficient management of irrigation water on rice fields, efficient applications of nitrogen fertilizer on cultivated fields and reclamation of treated municipal wastewater for aquifer recharge and irrigation.

The mitigation practice may also include reduction of emissions from commercial fishing operations and more efficient energy use by forest dwellers, commercial agriculture and agro-industries.

II. Sequestering Carbon

Sequestering Carbon is fixing the excess of Carbon dioxide in to ecosystem. The some of the means for such sequestering CO_2 includes improved management of soil organic matter, conservation agriculture involving permanent organic soil cover with minimum mechanical soil disturbance and appropriate crop rotation. The Carbon sequestering can also be achieved by reducing use of fossil fuel, improving management of pastures and sustainable grazing practices on natural grasslands. In the agriculture introduction of integrated agro-forestry systems, proper use of degraded and marginal lands and planting of carbon sink trees will also help in sequestering Carbon in the ecosystem.

FUTURE RESEARCH DIRECTIONS

The increased demand for other uses coupled with recurrent drought and climatic changes in countries of limited water resources, is producing unprecedented pressure for reducing the share of fresh water used in irrigation. Many countries give priority of water allocation to the domestic sector followed by tourism and industry, and what is left is allocated to agriculture. At the same time that agriculture is asked to give water to other uses, the increasing population demand requires increase in food production. This creates a conflict that may be resolved and by examining different options of water uses for sustainable agriculture. Increasing the efficient use of water is a key non-structural approach to water resources management. The agriculture water use efficiency and water productivity is very important as they are largely inefficient in many countries due to poor distribution systems and excess irrigation. The average agricultural water use efficiency in developing countries is very poor. Water planners and decision makers as well as researchers are facing with difficulties and challenges. Some options for such constraint may be helpful to sustain agriculture in changing climate in the various regions on the globe. The probable solution to this conflict requires greater understanding and systematic research. The future agriculture production depends on water resources and their allocation and management in agriculture. The domains of future research may include: improving water use efficiency, reducing crop consumption of water, irrigation with reclaimed water, practicing deficit irrigation and irrigation with desalinated water. The research vision for the next few years could include the few set of actions *viz.* have more efficient use and allocation for water use in irrigation; improved water productivity by introduction of new management measures such as deficit irrigation; and introduction of high yielding low water demanding varieties.

Desalination of brackish and sea water can offer limitless fresh water that can be used for agriculture. There is a need to find cheap methods of desalination such as the use of solar and renewable energy. Therefore, researchers should consider these needs in setting up their research priorities.

CONCLUSION

The world's climate is changing, and the changes will have an enormous impact on people, ecosystems, and energy use on the globe. There is threat that climate change will impact food availability, access, and utilization due to change in the climatic parameters like temperature, precipitation, CO_2.etc. The change in environmental parameters may change availability of water for agriculture. The demand of growing population for food security is crucial and uneven availability of water due to climate change will exaggerate the problem to greater extent. The impact of non-availability of sufficient and timely water for agriculture will lead to more poverty and degradation of land resources. Apart from this there is also increasing requirement of water for urban and industries. So increasing agriculture production, sharing water for nonagricultural uses and uneven availability of water due to climate change may exaggerate the problem to the greater extent in the future. Though researchers and water managers are trying hard for the solution and the complexity of the issue will be more serious in coming decades.

The aggregate impact of climate change on the water resources of the globe is not fully understood. The changing climate is playing central role in availability of water for agriculture production and individual climatic parameters and their interactions are equally important. There is an emerging consensus that changes in water resources can have detrimental impacts on the food security, in the absence of adaptation. Most of the impacts are not easy to quantify because they depend on a range of assumptions. The available quantitative scientific studies suggest that the wet regions will be more wet and dry will be more dry in coming decades which is serious concern for food security. This may reduce crop yields and the land suitable for agricultural production or change in cropping pattern and planting and harvesting schedules. Institutional development will have to play a greater role through polices supported by efficient extension programmes for mitigation of the future impact of climate change. Overall, the climate change may be having positive as well as negative effect on the water resources, agriculture production and food security of the world. This seems a huge challenge on the globe and require proper mitigations and adaptations measures.

REFERENCES

Ainsworth, E. A., & Rogers, A. (2007). The response of photosynthesis and stomatal conductance to rising (CO_2): Mechanisms and environmental interactions. *Plant, Cell & Environment*, *30*(3), 258–270. doi:10.1111/j.1365-3040.2007.01641.x PMID:17263773

Alderman, H., & Paxson, C. H. (1992). *Do the Poor Insure? A Synthesis of the Literature on Risk and Consumption in Developing Countries*. World Bank Policy Research Working Paper WPS 1008. Retrieved February 10, 2015, from https://www.princeton.edu/rpds/papers/Alderman_Paxson_Do_the_Poor_Insure_1994.pdf

Allan, R. P., Soden, B. J., John, V. O., Ingram, W., & Good, P. (2010). Current changes in tropical precipitation. *Environmental Research Letters*, *5*(2), 025205. doi:10.1088/1748-9326/5/2/025205

Anonymous. (2007). *Coping with water scarcity challenge of the twenty - first century*. Retrieved February 05, 2015, from http://www.fao.org/nr/water/docs/escarcity.pdf

Baker, B., & Viglizzo, E. F. (1998). Rangeland and Livestock. In J. F. Feenstra, I. Burton, J. B. Smith, & R. S. J. Tol (Eds.), *Handbook on methods for climate change impact Assessment and Adaptation Strategies*. Amsterdam: UNEP.

Barbier, E. B. (2004). Water and economic growth. *The Economic Record*, *80*(248), 1–16. doi:10.1111/j.1475-4932.2004.00121.x

Barnett, J. (2001). Adapting to Climate Change in Pacific Island Countries: The Problem of Uncertainty. *World Development*, *29*(6), 977–993. doi:10.1016/S0305-750X(01)00022-5

Bounoua, L., Hall, F. G., Sellers, P. J., Kumar, A., Collatz, G. J., Tucker, C. J., & Imhoff, M. L. (2010). Quantifying the negative feedback of vegetation to greenhouse warming: A modeling approach. *Geophysical Research Letters*, *27*, L23701.

Brklacich, M., Bryant, C., Veenhof, B., & Beauchesne, A. (2000). Agricultural Adaptation to Climatic Change: A comparative assessment of two types of farming in central Canada. In H. Millward, K. Beesley, B. Ilbery, & L. Harrington (Eds.), *Agriculturaland Environmental Sustainability in the New Countryside*. Winnipeg: Hignell Printing Limited.

Brown, C., & Lall, U. (2009). Water and economic development: The role of variability and a framework for resilience. *Natural Resources Forum*, *30*(4), 306–317. doi:10.1111/j.1477-8947.2006.00118.x

Burton, I. (2001). *Vulnerability and Adaptation to Climate Change in the Drylands*. Washington, DC: The Global Drylands Partnership.

CA (Comprehensive Assessment of Water Management in Agriculture). (2007). *Water for food, water for life: A comprehensive assessment of water management in agriculture*. Sri Lanka: IWMI.

Cao, L., Bala, G., Caldeira, K., Nemani, R., & Ban-Weiss, G. (2009). Climate response to physiological forcing of carbon dioxide simulated by the coupled Community Atmosphere Model (CAM3.1) and Community Land Model (CLM3.0). *Geophysical Research Letters*, *36*(10), L10402. doi:10.1029/2009GL037724

Cao, L., Bala, G., Caldeira, K., Nemani, R., & Ban-Weiss, G. (2010). Importance of carbon dioxide physiological forcing to future climate change. *Proceedings of the National Academy of Sciences of the United States of America, 107*(21), 9513–9518. doi:10.1073/pnas.0913000107 PMID:20445083

Chiotti, Q., Johnston, T., Smit, B., & Ebel, B. (1997). Agricultural Response to Climate Change: A preliminary Investigation of Farm-level Adaptation in Southern Alberta. In B. Ilbery, Q. Chiotti, & T. Rickard (Eds.), *Agricultural Restructuring and Sustainability: A Geographical Perspective*. CAB International.

Chou, C., Neelin, J. D., Chen, C. A., & Tu, J. Y. (2009). Evaluating the rich-get-richer mechanism in tropical precipitation change under global warming. *Journal of Climate, 22*(8), 1982–2005. doi:10.1175/2008JCLI2471.1

Chou, C., Tu, J. Y., & Tan, P. H. (2007). Asymmetry of tropical precipitation change under global warming. *Geophysical Research Letters, 34*(17), L17708. doi:10.1029/2007GL030327

Christensen, J. H., Krishna Kumar, K., Aldrian, E., An, S. I., Cavalcanti, I. F. A., de Castro, M., & Zhou, T. et al. (2013). Climate Phenomena and their Relevance for Future Regional Climate Change. In T. F. Stocker, D. Qin, G. K. Plattner, M. Tignor, S. K. Allen, J. Boschung, & P. M. Midgley et al. (Eds.), *Climate Change 2013: The Physical Science Basis. Contribution of Working Group I to the Fifth Assessment Report of the Intergovernmental Panel on Climate Change*. Cambridge, UK: Cambridge University Press.

de Loë, R. C., Kreutzwiser, R., & Mararu, L. (1999). *Climate Change and the Canadian Water Sector: Impacts and Adaptation*. Guelph: Natural Resources Canada. Retrieved March 01, 2016, from http://www.uoguelph.ca/~c-ciarn/documents/World_Bank_Paper.pdf

Desanker, P. V. (2002). Impact of climate change on Africa. Center for African Development Solutions.

Dinar, A. (2000). *The Political Economy of Water Pricing Reforms*. Washington, DC: World Bank.

El Nadi, A. H. (1974). The significance of leaf area in evapotranspiration. *Annals of Botany, 38*(3), 607–611.

Erda, L. (1996). Agricultural Vulnerability and Adaptation to Global Warming in China. *Water, Air, and Soil Pollution, 92*(1-2), 63–73.

Evenson, R. (1997). Rice Varietal Improvement and International Exchange of Rice Germplasm. In P. L. Pingali & M. Hossain (Eds.), *Impact of Rice Research*. Manila: International Rice Research Institute.

FAO. (2013). *FAO Statistical Yearbook 2013. World Food and Agriculture*. Rome: Food and Agriculture Organization of the United Nations.

Field, C. B., Jackson, R. B., & Mooney, H. A. (1995). Stomatal responses to increased CO_2-Implications from the plant to the global-scale. *Plant, Cell & Environment, 18*(10), 1214–1225. doi:10.1111/j.1365-3040.1995.tb00630.x

Fischer, E. M., Seneviratne, S. I., Vidale, P. L., Luthi, D., & Schar, C. (2007). Soil moisture- atmosphere interactions during the 2003 European summer heat wave. *Journal of Climate, 20*(20), 5081–5099. doi:10.1175/JCLI4288.1

Fuglie, K. (2012). Productivity growth and technology capital in the global agricultural economy. In K. Fuglie, S. L. Wang, & V. Eldon Ball (Eds.), Productivity growth in agriculture: An international perspective. Oxfordshire, UK: CAB International.

Grey, D., & Sadoff, C. (2006). *Water for growth and development.* Thematic documents of the IV World Water Forum. Mexico City: Comision Nacional del Agua.

Hanjra, M. A., Ferede, T., & Gutta, D. G. (2009). Pathways to breaking the poverty trap in Ethiopia: Investments in agricultural water, education and markets. *Agricultural Water Management, 96*(11), 1596–1604. doi:10.1016/j.agwat.2009.06.008

Hartmann, D. L., Klein Tank, A. M. G., Rusticucci, M., Alexander, L. V., Brönnimann, S., Charabi, Y., & Zhai, P. M. et al. (2013). Observations:Atmosphere and Surface. In T. F. Stocker, D. Qin, G. K. Plattner, M. Tignor, S. K. Allen, J. Boschung, & P. M. Midgley et al. (Eds.), *Climate Change 2013: The Physical Science Basis. Contribution of Working Group I to the Fifth Assessment Report of the Intergovernmental Panel on Climate Change*. Cambridge, UK: Cambridge UniversityPress.

Held, I., & Soden, B. (2006). Robust responses of the hydrological cycle to global warming. *Journal of Climate, 19*(21), 5686–5699. doi:10.1175/JCLI3990.1

Hungate, B. A., Reichstein, M., Dijkstra, P., Johnson, D., Hymus, G., Tenhumen, J. D., & Drale, B. G. et al. (2002). Evapotranspiration and soil water content in a scrub-oak woodland under carbon dioxide enrichment. *Global Change Biology, 8*(3), 289–298. doi:10.1046/j.1365-2486.2002.00468.x

Hussain, I., & Hanjra, M. A. (2004). Irrigation and poverty alleviation: *Review of the empirical evidence. Irrigation and Drainage, 53*(1), 1–15. doi:10.1002/ird.114

ICOLD (International Commission on Large Dams). (2003). *World Register of Dams, 2003.* Paris, France: International Commission on Large Dams.

IPCC. (1996). *Impacts, Adaptations, and Mitigation of Climate Change: Scientific-Technical Analyses - Contribution of Working Group II to the IPCC Second Assessment Report.* Cambridge, UK: Cambridge University Press.

IPCC. (2001). Impacts, Adaptation and Vulnerability. In J. J. McCarthy, O. F. Canziani, N. A. Leary, D. J. Dokken, & K. S. White (Eds.), *Contribution of Working Group II to the Third Assessment Report of the Intergovernmental Panel on Climate Change*. Cambridge, UK: Cambridge University Press.

IPCC. (2007). Climate Change 2007: *Mitigation of Climate Change.* In B. Metz, O. R. Davidson, P. R. Bosch, R. Dave, & L. A. Meyer (Eds.), *Contribution of Working Group III to the Fourth Assessment Report of the Intergovernmental Panel on Climate Change, 2007.* Cambridge University Press.

Kirtman, B., Power, S. B., Adedoyin, J. A., Boer, G. J., Bojariu, R., Camilloni, I., & Wang, H. J. et al. (2013). Near-term Climate Change: Projections and Predictability. In T. F. Stocker, D. Qin, G. K. Plattner, M. Tignor, S. K. Allen, J. Boschung, & P. M. Midgley et al. (Eds.), *Climate Change 2013: The Physical Science Basis. Contribution of Working Group I to the Fifth Assessment Report of the Intergovernmental Panel on Climate Change*. Cambridge, UK: Cambridge University Press.

Kurukulasuriya, P., & Rosenthal, S. (2003). *Climate change and agriculture: a review of impacts and adaptations. Climate Change Series Paper No. 91*. Washington, DC: World Bank.

Lambert, F. H., & Webb, M. J. (2008). Dependency of global mean precipitation on surface temperature. *Geophysical Research Letters*, *35*(16), L23803. doi:10.1029/2008GL034838

Lammertsma, E. I., de Boer, H. J., Dekker, S. C., Dilcher, D. L., & Lotter, A. F. (2011). Global CO_2 rise leads to reduced maximum stomatal conductance in Florida vegetation. *Proc. Acad. Sci. U.S.A.*, *108*(10), 4035–4040. doi:10.1073/pnas.1100371108 PMID:21330552

Lobell, D. & Gourdji, S. (2012). The influence of climate change on global crop productivity. *Plant Physiology*. .112.20829810.1104/pp

Londhe, S. L. (2016). Climate Change and Agriculture: Impacts, Adoption, and Mitigation. In *Handbook of Research on Climate Change Impact on Health and Environmental Sustainability*. IGI Global. doi:10.4018/978-1-4666-8814-8.ch019

Lu, J., & Cai, M. (2009). Stabilization of the atmospheric boundary layer and the muted global hydrological cycle response to global warming. *Journal of Hydrometeorology*, *10*(1), 347–352. doi:10.1175/2008JHM1058.1

McCartney, M., Rebelo, L. M., Xenarios, S., & Smakhtin, V. (2013). *Agricultural water storage in an era of climate change: assessing need and effectiveness in Africa*. Colombo, Sri Lanka: International Water Management Institute (IWMI). Retrieved February 21, 2015, from http://www.iwmi.cgiar.org/Publications/IWMI_Research_Reports/PDF/PUB152/RR152.pdf

Mendelsohn, R., & Neumann, J. E. (1999). *The Economic Impact of Climate Change on the Economy of the United States*. Cambridge, UK: Cambridge University Press. doi:10.1017/CBO9780511573149

Moreddu, C. (2000). Overview of Farm Household Strategies and Government Intervention. In Income Risk Management in Agriculture. Paris: OECD.

Myhre, G., Shindell, D., Bréon, F. M., Collins, W., Fuglestvedt, J., Huang, J., & Zhang, H. et al. (2013). Anthropogenic and Natural Radiative Forcing. In T. F. Stocker, D. Qin, G. K. Plattner, M. Tignor, S. K. Allen, J. Boschung, & P. M. Midgley et al. (Eds.), *Climate Change 2013: The Physical Science Basis. Contribution of Working Group I to the Fifth Assessment Report of the Intergovernmental Panel on Climate Change*. Cambridge, UK: Cambridge University Press.

Nelson, G. C., Rosegrant, M. W., Koo, J., Robertson, R., Sulser, T., Zhu, T., . . . Lee, D. (2009). *Food Policy Report- Climate Change, Impact on Agriculture and Costs of Adaptation*. International Food Policy Research Institute. Retrieved February 19, 2015, from http://www.ifpri.org/sites/default/files/publications/pr21.pdf

Parry, M. L. (2000). *Assessment of Potential Effects and Adaptations for Climate Change in Europe. The Europe Acacia Project*. University of East Anglia.

Patil, Y., & Rao, P. (2015). Industrial waste management in the era of climate change - A smart sustainable model based on utilization of active and passive biomass. In W. L. Filho (Ed.), *Handbook on Climate Change Adaptation* (pp. 2079–2092). Springer-Verlag Berlin Heidelberg. doi:10.1007/978-3-642-38670-1_49

Peterson, D. F., & Keller, A. A. (1990). Effects of climate change on U.S. irrigation. *Journal of Irrigation and Drainage Engineering*, *116*(2), 194–210. doi:10.1061/(ASCE)0733-9437(1990)116:2(194)

Priestley, C. H. B. (1966). The limitation of temperature by evaporation in hot climates. *Agricultural Meteorology*, *3*(3-4), 241–246. doi:10.1016/0002-1571(66)90031-8

Rao, P., & Patil, Y. (2015). Climate Resilience in Natural Ecosystems in India: Technology Adoption and the Use of Local Knowledge Processes and Systems. In Handbook of Climate Change Adaptation. Springer Berlin Heidelberg.

Reilly, J., Tubiello, F., McCarl, B., Abler, D., Darwin, R., Fuglie, K., & Rosenzweig, C. et al. (2003). U.S. Agriculture and Climate Change: New Results. *Climatic Change*, *57*(1/2), 43–69. doi:10.1023/A:1022103315424

Rhein, M., Rintoul, S. R., Aoki, S., Campos, E., Chambers, D., Feely, R. A., & Wang, F. et al. (2013). Observations: Ocean. In T. F. Stocker, D. Qin, G. K. Plattner, M. Tignor, S. K. Allen, J. Boschung, & P. M. Midgley et al. (Eds.), *Climate Change 2013: The Physical Science Basis. Contribution of Working Group I to the Fifth Assessment Report of the Intergovernmental Panel on Climate Change*. Cambridge, UK: Cambridge University Press.

Rosenberg, N. J. (1992). Adaptation of Agriculture to Climate Change. *Climatic Change*, *21*(4), 385–405. doi:10.1007/BF00141378

Rosenberg, N. J., Kimball, B., Martin, P., & Cooper, C. (1988). Climate change, CO_2 enrichment and evapotranspiration. In P. E. Waggoner (Ed.), *Climate and Water: Climate Change, Climatic Variability, and the Planning and Management of U.S. Water Resources*. New York, NY: John Wiley and Sons.

Rosenzweig, C., & Liverman, D. (1992). Predicted Effects of Climate Change on Agriculture: A Comparison of Temperate and Tropical Regions. In S. K. Majumdar (Ed.), *Global Climate Change: Implications, Challenges, and Mitigation Measures.* Guelph: Natural Resources Canada.

Seneviratne, S. I., Corti, T., Davin, E. L., Hirschi, M., Jaeger, E. B., Lehner, I., & Teuling, A. J. et al. (2010). Investigating soil moisture-climate interactions in a changing climate: A review. *Earth-Science Reviews*, *99*(3-4), 125–161. doi:10.1016/j.earscirev.2010.02.004

Seneviratne, S. I., Nicholls, N., Easterling, D., Goodess, C. M., Kanae, S., Kossin, J., . . . Zhang, X. (2012). Changes in climate extremes and their impacts on the natural physical environment. In A Special Report of Working Groups I and II of the Intergovernmental Panel on Climate Change (IPCC SREX Report). doi:10.1017/CBO9781139177245.006

Skees, J. R., Hazell, P., & Miranda, M. (1999). *New Approaches to Crop Insurance in Developing Countries.* EPTD Discussion Paper No. 55. International Food Policy Research Institute.

Smit, B., & Olga, P. (2001). Adaptation to Climate Change in the Context of Sustainable Development and Equity. In J. J. McCarthy, O. F. Canzianni, N. A. Leary, D. J. Dokken, & K. S. White (Eds.), *Impacts, Adaptation, and Vulnerability - Contribution of Working Group II to the Third Assessment Report of the Intergovernmental Panel on Climate Change*. Cambridge University Press.

Smithers, J., & Blay-Palmer, A. (2001). Technology Innovation as a Strategy for Climate Adaptation in Agriculture. *Applied Geography (Sevenoaks, England)*, *21*(2), 175–197. doi:10.1016/S0143-6228(01)00004-2

United Nations. (2011). *Seven billion and growing: the role of population policy in achieving sustainability.* Technical Paper No. 2011/3. Retrieved Feb 26, 2015, from http://www.un.org/esa/population/publications/technicalpapers/TP2011-3_SevenBillionandGrowing.pdf

Waggoner, P. E. (1983). *Agriculture and a climate changed by more carbon dioxide. Changing climate.* Washington, DC: National Academy Press.

White, W. R. (2005). *World water storage in man-made reservoirs: A review of current knowledge.* Marlow, UK: Foundation for Water Research.

WMO. (2014). WMO Greenhouse. *Gas Bulletin, 10*. Retrieved February 18, 2015, from http://www.wmo.int/pages/prog/arep/gaw/ghg/documents/GHG_Bulletin_10_Nov2014_EN.pdf

KEY TERMS AND DEFINITIONS

Adaptation: Adaptation is the process by which living organism changes to become better suited to survive in their environment. Adaptation can be a physical or genetic trait that helps an organism to be better suited to survive in the environment.

Agricultural Productivity: Agricultural productivity is the ratio of agricultural outputs to agricultural inputs. The agriculture inputs may include water, chemicals, fertilizers etc. used to grow crops to achieve desired yield as output.

Climate Change: Climate change refers to any significant change in the measures of climate parameters lasting for an extended period of time. Climate change includes major changes in temperature, precipitation, or wind patterns, among other effects, that prevailing in a region over several decades or longer.

Climate: It can be described as average temperature and precipitation over a period of time. Climate can be defined as an area's long-term weather patterns. The useful elements for describing climate include the type and the timing of precipitation, amount of sunshine, average wind speeds and directions, number of days above freezing, weather extremes, and local geography.

Mitigation: Mitigation in the context of climate change is an intervention intended to reduce adverse human influence on the climate system which includes strategies to reduce greenhouse gas sources, emissions and enhance greenhouse gas sinks.

Resilience: Resilience is the capacity of organism or system to withstand stress and catastrophe. In case of climate change the capacity of crop and livestock to withstand and sustain in the changing environment is the resilience.

Water Use Efficiency (WUE): Water use efficiency is the amount of water uptake by the plant and the amount of water used for metabolism and transpiration by the plant. WUE is simply ratio of water used in plant metabolism to water lost by the plant through transpiration.

Chapter 11
Role of Water-Energy-Waste Inter-Relatedness to Drive Sustainability amid Climate Concerns

Salil K. Sen
Philippines Asian Institute of Management, Philippines

Junya Pookayaporn
Sripatum University, Thailand

ABSTRACT

The Sustainability value parity framework is proposed to deepen the understanding of the importance of 'inter-relatedness' of water-energy-waste with the goal of balancing water usage, aligning energy intensity and optimizing waste utilization. Generic waste that is burgeoning is a deterrent to the practice of sustainability that aligns water, energy, infrastructure, health, food, and lifestyle (Sachs, 2007). This chapter delineates the gap between globalization at the macro-level and global citizenry at the grassroots-base and posits a value bridge assessed by appropriate thresholds of water - energy - waste. The emergent need to strengthen climate resilience and to usher into the sustainable pathway of climate-proofed development needs tuning of processes, lifestyle, hazardous substances and consumption. Climate change manifests as an over-arching risk that is strewn with unpredictability, multiple dimensions, uncertainties, spikes, imbalances leading to inequity.

INTRODUCTION

The Asia Pacific is straddled with rapid growth vis-a-vis emergent need for sustainability practice that aligns development with value parity on water - energy - waste. The Sustainability value parity has the ability to balance water usage, align energy intensity and optimize waste utilization. 2050 scenarios on climate change, food, security, ecosystems, bio-diversity and ecological productivity is grim. The growth aspirations are emphatic across Asia and the Pacific, which is in tandem with other global growth nodes.

DOI: 10.4018/978-1-5225-1046-8.ch011

In this milieu, the root-core attributes of sustainability value creation would substantively depend on how water - energy - waste parity is maintained. Extended Exergy Analysis is an emerging methodology that inter-weaves technology with economics on one part, as well as exergetic balance of labor and environmental remediation expenditures (Golberg, 2015). Climate change is "unequivocal" with it's impact on water evident from warming oceans and water bodies, rise in water levels, precipitation trends display strong variability, water bodies have absorbed over 30 percent of emitted anthropogenic carbon dioxide intensifying acidification that is detrimental to agriculture and may trigger food insecurity. Climate unpredictability, disaster recurrence and greenhouse proliferation, triggers severe warnings to sustainability-critical resources like water – energy – land – air (Giovannucci et. al., 2012). Sectors, such as, food, infrastructure, health, transportation and telecom production faces multiple limiting factors or value-parity thresholds for waste that needs to be dynamically harnessed with respect to water – energy sustainability.

This crisis has potential to manifest as Sustainable Development Goal (SDG) – compliant innovations on technology, climate proof financing, disaster resilient habitats. Water – energy – waste sustainability are key enablers for inclusion, environmental sustainability, and regional integration in the Asia Pacific (Thomas, 2015). Economic, societal and environmental outcome evaluation would lead to enhanced outcomes with sustainable results by climate-resilient and inclusive infrastructure pathways. Generic waste, has stockpiled that are attributable to processes, lifestyle, hazardous substances and use-and-throw lifestyle. Risks and uncertainties proliferate, businesses are no longer usual with spikes, imbalances, inequities, non-uniformities and unpredictability becoming the norm (Miller, 2015). With 2010 ~ 2014 being one of the most climate variable years on record since 1998, climate concerns are ever increasing. Unpredictable extreme weather devastations tend to pose as the predominant and overarching challenge. All other developments would be meaningless without the ecology thriving along with growth of economy.

Motivation

Is it feasible to align economy, ecology and society through the practice of sustainability? The key action-essence is to reach a dynamic balance. The diverse perspectives of the key actors or stakeholders, such as the processors, the private sector, the service providers, the community, the typical locale-specific issues, need to be kept aligned sustainably. The policies, management structure, enforcement, adoption, and innovation needs to respond quickly to the inertial imbalances, inequities and non-uniformities. This chapter proposes a framework to align the governance with resource usage that lead to dynamic equilibrium termed as water - energy - waste parity.

Gaps

Aim of this chapter is to broad-base the impacts of climate change beyond the primary focus on water supply, usage, and management to the inter-relatedness among water - energy - waste. This extended purview would help better formulate and understand the multiple benefits that may be derived from climate resilience with waste-to-energy opportunities, water-waste nexus and water - energy inter-dependency. Policy makers, processors/service providers, grassroots community and other stakeholders can derive and align their respective objectives through this analysis of inter-relatedness. Key benefits anticipated are aligned sustainable development, budget sharing in public - private - community format, better livability and quality of life, dynamically correlating demand and supply sides on manufacturing, agriculture and services and creating a state of readiness for energy delivery with attention to impacts on environment and society.

Research Objectives

The framework interlocks water - energy - waste perspectives to offer following benefits:

1. Why services, including the corporate sector, should invest in water - energy - waste parity.
2. How proactive adoption of the larger perspective of water - energy - waste can leading to the development, diffusion, usage and disposal of climate benign products, services and lifestyle.
3. How to create a value pressure onto policy makers for the need for more incentives for climate benign water - energy - waste parity approaches to derive sustainable value.

Research Framework: Water-Energy-Waste Value Parity for Sustainability

The concept of parity is an important to create the value proposition for energy - water - waste. Value parity comes into play when the price or value of an asset is directly linked to another asset category. In case of convertibles and options, value parity is the price when convertible security equals the value of the under-lying stock and when an option is trading at its intrinsic or parity value.

The issue on value parity as applied to water – energy – waste can be cited from literature on how value parity may drive private public partnerships. Parity issues such as, expansion of private water withdrawal in developing countries (Hall & Lobina, 2008), effects on distribution equity resulting in sub-quantum financing, governance issues in pricing (Kessides, 2004), demonstrate no significant efficiency enhancement between public sector-led and private – public partnered water – energy – waste services.

There is a substantive equivalence of deploying the value parity concept here on the inter-relatedness of energy - water - waste seen through the creation of value "lens". Water, health, infrastructure, food, lifestyle and energy are the climate sensitive sectors that pose as key sustainability challenges. Waste is a common generic entity that affects all. Waste may unleash negative or positive 'spin of influence' depending on how waste wealth is harnessed, as well as what is the attitude to waste, termed as empathy. Ecology and economy is aligned by empathy, leading to the sustainability citizenry of policy, people and services.

Path-breaking policy initiatives and significant influx of capital are being invested in health, water and related climate sensitive areas. Complex sustainability solutions rely on mission-relevant, transformational science and engineering development. Innovative transformations are taking shape in fossil to nuclear energy, reactive to proactive environmental management, waste to energy in civil, hazardous and military footprint.

In this context, there is a gap to be fulfilled. Corresponding advances in communication must occur to disseminate and sensitize the grassroots level the empathy that creates sensitivity, understanding and attitude to accept and adopt new sustainability citizenry values.

Curbing waste on water is an opportunity that has potential to serve as a driver of change: The explosive growth in debris (Table 1) and waste impacts the developed economies as severely as the developing regions on our planet and affects resources like fresh water, energy and land. For instance, the component indicators of water including resource availability and water quality are correlated (Jarraud, 2010). In the dynamic interplay of water, waste and energy with policy, people and services, attitude is prime and

Table 1. 2010 ~2014 was the most climate variable period generating debris/waste

Debris/Waste Created by	Economy Effect (US $)	People	Policy Region	Services (Estimated)
Debris from natural disasters	109 billion losses	89% of 207 million affected worldwide	Asia accounted for 89%	373 cases
Earthquake tsunami	38 billion losses	222500 dead (estimated)	Central Am (Chile, Haiti)	222500 Dead
Floods	9.5 billion losses	High losses	SE Asia (Pakistan)	Industry, Agriculture Damage
Floods Landslides	18 billion losses	High losses	East Asia (China)	Infrastructure, Energy damage
Heat-Wave	Very High Losses	55,736 dead	East E/A (Russia)	Health, Food Affected

alignment is supreme. Driving factors like climate policy, energy alternatives, demographic changes, food and water security and foreign direct investments become key at local, national and global levels (Ringler, et al., 2010). Waste is inseparably linked with quality of water, land and air at the resources level and with policy, people and services at the attitudinal mind-space levels. In general, attitude to waste receives ignorance or nonchalance, relegating waste to the end of material cycles and mind sets. Amidst globally burgeoning waste and debris, attitude innovations on 'curb waste on water' create sustainability sensitivity. Such orientations at the grassroots level would create 'sustainability needs' and usher appropriate products, services and entrepreneurship (Table 1).

Sustainable Development Goals Aligned to Water - Energy - Waste sustainability

The implement-ability needs of the post-CoP21 Sustainable Development Goals (SDGs) sets the priorities on water – energy – waste at the elemental level. The SDGs permeate into our value parity framework that puts waste at focus.

The reader would agree that the combination of water and energy are positive drivers of sustainability, while waste, is de facto, the negative attribute of sustainability. It is absolutely necessary to develop, maintain and sustain waste management below the critical threshold, in order to ensure implement-ability of the Sustainable Development Goals (SDGs).

Sustainable Development Goal (SDG) clusters are conceptualized and linked with impacts on water- energy-waste (author's own). Threshold Cluster, zero poverty (SDG1), null hunger (SDG2); 'Minus Carbon' Cluster, Sanitation and process (SDG6); Affordability, safe-guards:

Energy ~ minus ~ carbon processes (SDG7); Climate proofing (SDG13); Life below water (SDG14); Life on land (SDG15).

Measurement, reporting, verification systems on energy optimization and waste minimization in water sanitation, process usage and potable segment (including agriculture/irrigation water) internationally negotiated global mean temperature targets.

Dynamics and proactivity are key to reach parity of the root-core attributes of sustainable value creation from water-energy-waste. The framework assigns typicality of locales, distinctive characteristics of their habitats, optimization of scale to remain in the water - energy - waste parity threshold, as well as innovative options for connectivity of these resources with other regions. The issue of inter-relatedness is key. Instead of seeking respective water, energy, waste solutions to climate-triggered sustainability challenges, innovative options open up when inter-relatedness assumes as primary value driver.

Climate Resilient Developmental Pathways

The economics and business metrics often relegate waste/debris to low priority levels. Footprint impacts on ecology, incessant proliferation of bottom-line focused growth and consumption splurge has exponentially increased waste generation. Due to frequent climate catastrophes, debris is created overnight. The key challenge is to probe feasible modalities to strike a balance of sustainability of resources, water-energy-waste.

Climate Concerns

Progressively reducing greenhouse gases is a precondition for harnessing climate change (Pachauri et al., 2014). A total anthropogenic GHG emission has incessantly spiraled during 1970 to 2010 with the decade of 2000 to 2010 proving to be contributing the highest. About 78 percent of the global warming emissions are attributable to traditional methods of energy production using fossil fuels that power industrial processes in the span of last 40 years. Looking at the baseline situation, the energy sector would emanate direct Carbon Dioxide that portends to hike two to three times by 2050. This is a key premise of considering water and waste/debris as the other core domains would be affected. The clear message is that along with measurable and consistent improvements of energy intensity should be coupled with water usage and waste utilization (Figure 1).

Water Alarms

Changes in ocean temperature, hike in sea levels, diminution of snow/ice cover, high variability of precipitation, are fast becoming a norm. This sets water alarms that intensely affect energy production (among other usages of water), as very large volumes of clean water is necessary for energy production. The vast expanse of ocean water absorbing about 30 percent of the emitted anthropogenic carbon dioxide, exacerbates the water factor as there is distinct increase in acidification.

Waste: When Appropriately Managed Creates Value

In a recent report compiled by the Earth Engineering Center at the Columbia University based on the 2013 Survey of Waste Management in the United States presented the opportunity in terms of waste-to-energy for municipal solid wastes (MSW) and non-recycled plastics (NRP). This study indicated alarmingly high requirement of land surface to accommodate burgeoning landfills. A whopping 6100 acres that is equivalent to nearly 4000 standard-sized Olympic stadiums.

The 2013 EEC Survey reported that in 2011 the United States generated 389 million tons of municipal solid waste (MSW). What is of consequence is that the recycled component is 87.8 million tons, with compost contributing to 24.6 million tons. The waste-to-energy (WTE) utilized 29.5 million tons that otherwise would have gone to clutter valuable space as landfills.

Figure 1. Impact on water - energy - waste from greenhouse gas emissions
(Source IPCC 2014)

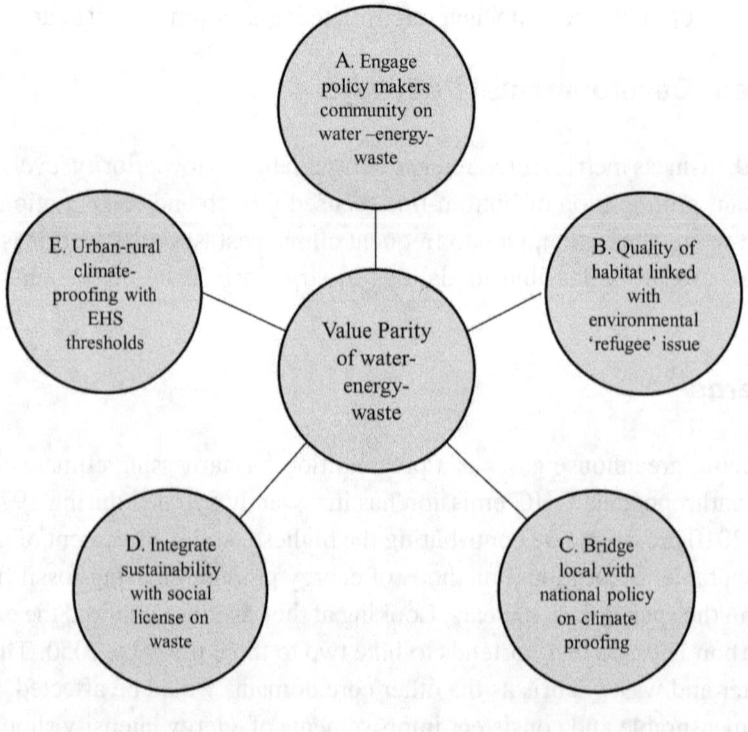

Waste usually is a strewn with recyclable plastics, such as polypropylene (PP) rigid plastics, polyethylene terephthalate (PET) as well as high-density polyethylene (HDPE) bottles that are exponentially piling-up from the consumer-driver lifestyle resulting in films, bags, and wraps. However, there is one component in the waste-mix, the non-recycled plastics (NRP), which are not economical to recycle. This provides the vital link of waste - energy - water, as waste gets converted to energy, with valuable de-contamination impact on water.

Significantly, plastics constitute over 11% (39.3 million tons) of the total waste category. The opportunity for energy - water - waste interplay is to provide climate resilience that includes disaster preparedness, is the fact that waste/debris could be used as alternative fuel for cement production (Themelis, 2014). The rate of recycling plus energy recovery has potential to alter the economics of sustainability as non-recycled plastics (NRP) as well as automotive shredder residue (ASR) is becoming feasible to be factored in.

Revisiting Current Literature

Water, health, infrastructure, food, lifestyle and energy are the climate sensitive sectors that pose as key sustainability challenges (Sachs, 2015). Waste is a common generic entity that affects all. Waste may unleash negative or positive 'spin of influence' depending on how waste wealth is harnessed, as well as what is the attitude to waste, termed as empathy. Ecology and economy is aligned by empathy, leading to the sustainability citizenry of policy, people and services.

Path-breaking policy initiatives and significant influx of capital are being invested in health (Koplan, 2002), water (Lei, 2004) and related climate sensitive areas (Frison et. al. 2004). Complex sustainability solutions rely on mission-relevant, transformational science and engineering development (Brown et. al. 2006). Innovative transformations are taking shape in fossil to nuclear energy, reactive to proactive environmental management, waste to energy in civil, hazardous and military footprint. The following table outlines sets of literature progression, namely, positive neo-liberal, neutral-transformative and negative political, linked with globalization and global citizenship (Fiss & Hirsch, 2008) (Table 2).

Given this purview, the inter-relatedness of water - energy and waste can help bridge the perspective gap between globalization at the macro level with global citizenry, at the grassroots level. Timely progression in communication must occur to disseminate and sensitize the grassroots on the "globalization from below" as is referred in Table 2. The positive and negative issues of market driven (for positive neo-liberal), macro dependencies (for neutral transformative) and democracy & rights (for negative political) aspects are distinctive to the grassroots globalization qualities. Water-energy-waste has the "bandwidth" to remain relevant at the functional level for globalization pathway, as also for the motivation through common vision for the global citizenship approach. As land use, water quality and other climate and GHG emissions can be harnessed when water - energy - waste can align the macro-level globalization propositions along with grassroots-level global citizenry proposition with empathy, sensitivity, understanding and attitude percolating to the bottom of the pyramid.

Table 2. Value parity of energy - water - waste linkage with globalization and global citizenship

Globalisation			Global Citizenship	
Perspective	**(Stance / Positive and Negative Aspects)**		**Perspective**	**Qualities**
Positive Neoliberal	Economics Market driven	(+) rational Functional (-) hegemonic	Instrumental Utility function 'Rational'	Special skill sets and language abilities; Competitive; Strategic; market-oriented Elitist; reproduce hegemony "Globalisation from above"
References: Fiss & Hirsch, 2005; Kotz, 2002; Rawls, 1971; Schultz, 2007; Schattle, 2008				
Neutral Transformative	Globalisation as matter of fact Recognises macro dependencies	(+) self critical; recognizes neoliberal and radical perspectives (-) neutralization; homogenization; flattering of the world	Recognizes the positive and negative effects of globalization Cosmopolitan vision	Proactive approach to advert negative effects of globalization through engagement and participation Motivated by common vision and higher goals Pluralistic Self-critical
References: Fiss & Hirsch, 2005; Beck, 2002; Harvey, 2005; Schultz, 2007; Schattle, 2008				
Negative Political	Radical power, democracy and rights	(+) challenges hegemony; strong ethical positions (-) emotional; "irrational"	Democracy and human rights Nationalistic	Assertive in maintaining social structure Actionsl aimed at forcing radical changes "Globalization from below"
References: Fiss & Hirsch, 2005; Held, 1995; Starr, 2000; Schultz, 2007; Schattle, 2008				

Value Proposition 1: Why Services, Including the Corporate Sector, Should Invest in Water - Energy - Waste Parity

Locales with high sustainability value on an appropriate scale, with positive water - energy - waste parity indicate that waste has the potential to be the turning point of inflexion for sustainable development. Investor friendly, bankable benefits of this are, for instance, waste to energy, de-mining, recovery of metals, freed spaces traditionally occupied by landfills and more.

This framework on inter-relatedness can augment/supplement independent initiatives on water, energy and waste management. The climate dividend can be proactively derived from water - energy - waste parity-led lifestyle, creation of new demand for innovative energy usage, water segmentation, private sector involvement in investing in hitherto low-priority sectors such as waste.

Value proposition 1: Sustainability is at the intersection of competitiveness and alignment. The value-based competencies for better governance with respect to water - energy - waste creates equity in the sustainable term for policy makers, service providers including the corporate sector. This may be termed as the Sustainability-centricism of the value parity framework, that portrays good governance, societal inclusiveness and environmental care. Poor performance have often resulted in spectacular failures, affecting not just shareholders but key stakeholders such as, society, environment and quality of growth. The linkage of water - energy - waste with policy - process - communications is of prime concern globally linking emerging markets to developed economies. Sustainability-centricism imparts resilience to risks and enables collaboration on trans-boundary generic issues such as, water - energy - waste.

This value proposition is justifies issues arising from contemporary business practices, policy dilemmas, environmental sustainability and societal concerns. The aim is to blend with policy-level globalization dimension to the grassroots-level global citizenry aspiration with proper aligned emphasis on water - energy - waste.

Value Proposition 2: How Proactive Adoption of the Larger Perspective of Water - Energy - Waste Can Leading to the Development, Diffusion, Usage and Disposal of Climate Benign Products, Services and Lifestyle

The objective is to attempt to view the global sustainability spectrum across individual, societal, business, policy issues spanning countries, regions and trans-boundary locales. The keywords are corporate societal responsibility, value added, resilience to risks, sustainability practice, governance, inclusiveness, trans-boundary perspectives, inter-disciplinarity, quality of growth, sustainable production and consumption.

Value proposition 2: The concept of sustainability spectrum is posited wherein a sustainability citizenry scale offers differential diagnoses across locales and their habitats on how aligned the resources, such as, water, waste and energy, are with the citizenry parameters like policy, people and services. The innovative approach incorporates the dynamic alignment represented by the 'spin of influence' of the sustainability parameters, such as, water, waste and energy with citizenry parameters of policy, people and services. This approach interlocks ecology and economics with empathy to provide indicators to align perspectives to create climate benign products, services and lifestyle.

Value Proposition 3: How to Create a Value Pressure Onto Policy Makers for the Need for More Incentives for Climate Benign Water - Energy - Waste Parity Approaches to Derive Sustainable Value

Dynamically aligning resources (water - energy - waste) with citizenry (policy - services - people) is to be superimposed on an appropriately scaled measure for sustainability value . The two sets, resources and citizenry needs to be aligned dynamically to portray an aligned index termed as the sustainability scale. In this approach, it is assumed that each parameter of the resources set, as well as the citizenry set, have orthogonal orientation representing respective perspectives that need to be aligned.

Value proposition 3: The sustainability scale measures water - energy - waste tipping point. The combined dynamic resultant is superimposed on a sustainability scale. While in a state of dynamic sustainability citizenry equilibrium, the range of variability is limited between a sustainability benign range. When the variability is uncontrolled, waste excess cause adverse impacts on water - energy balance and corrective action is called for. The pitfalls of excess debris can be monitored, estimated and budgeted by the sustainability scale. This is fed-in policy formulation, service/manufacturing offerings and grassroots societal readiness. This innovative approach enables differential diagnosis to align the varying perspectives of investors, service providers and users and is versatile across climates, cultures, habitats and time.

Climate Resilient Development Pathways can Sustain with Water-Waste "Ownership" at Community Level. Waste is Critical at Appropriate Thresholds of Water - Energy Systems

The value parity framework envisages an inter-linked chain on service offering, manufactured modules serving food, health, infrastructure, telecom and transport. At core are water – energy – waste aligned in a sustainable proportion. These are critical for climate proofing, disaster resilience, as well as creating attractive results-based returns to private sector investors. This schematic on water and energy balance, waste serves as a critical determinant. An appropriate threshold of tolerance on waste injects the sustainability to a habitat. Generic waste takes various forms: as construction debris, hazardous waste, process left-overs, energy waste, contamination of water as waste, poor air quality as a form of waste, water salinity as a manifestation of waste, as such others.

Each habitat, at each season, at any point on temporal and spatial scale, is governed by a waste threshold. Below the threshold, the habitat is sustainable, provided, continuous replenishments in terms of corrective measures are deployed. This ushers sustainability. This leads to positive and virtuous cycles on private financing, community level entrepreneurship, disaster preparedness, enhanced and equitable livelihoods, and more. For instance, in the Gulf Cooperation Council countries, implementation of renewable energy infrastructure creates opportunity to reduce water withdrawals for power generation to the extent of 20% by 2030 (Adnan, 2015).

Challenges in Preparing Data for the Water - Energy - Waste *Parity* Framework

Waste, as such, is shunned, ignored and treated with apathy. The economics and business metrics often relegate waste to low priority levels. Footprint impacts on ecology, incessant proliferation of bottom-line focused growth and consumption splurge has exponentially increased waste generation. The key issue is to strike a balance of sustainability of resources, like water, waste and energy. A key challenge is to prepare the data set for this model. Through the application of Hierarchical Bayes (HB) Regression (logit or linear regression) methodology, it is possible to iterate beta estimates of waste/debris at locales. The betas could be estimated at locales at the first level of the HB iteration. In the proposed framework model, the predictor variables of water, energy, exert 'influence' on the entity of waste. The outcome is the sustainability value index that provides a differential diagnostic measure of overall alignment for a specified time horizon for a designated habitat.

An initial qualitative estimate of betas of the predictor variables is estimated for each parameter as an initial starting impactor for curbing waste. At the initial level, probabilities of achieving sustainability aligned outcomes (both in magnitude and direction) are determined based on multinomial logit or linear regression model. At the next hierarchical level the initial betas are considered together as a multivariate normal distribution comprised of a sustainability vectors of means and a magnitude matrix of co-variances. After interaction, a new estimate of the water - energy - waste *parity* is determined by iteration of individual betas as well as the mean and co-variances of the distribution of betas. Habitat specific data on water - energy - waste are estimated from cases to generate current estimates. This process converges to correct estimates of the sustainability value driver. The process may be refined with dynamic feedback processes.

The outcomes of water - energy - waste *parity* estimated by the iteration in Hierarchical Bayes (referred above) is essentially a vector of means and a magnitude matrix of co-variances. For instance, the policy level incentives to 'curb waste on water' generate the momentum to adopt the waste-neutral processes. This would lead to investments backed by appropriate policy incentives for probing water - energy - waste innovations. This in turn creates products and services that appeals to the sensitivity of society, who re-look at their own attitude and habits towards waste. The water - energy - waste aligned framework aggregates and gets connected to other regional locales with similar characteristics. The aggregated framework gathers *momentum* to navigate the climate challenges and tends to reach a dynamic equilibrium creating sustainable value.

Rather than the end point, waste can be the turning point of inflexion for sustainable development. Waste to energy, recovery of metals, freed spaces traditionally occupied by landfills help to reduce ecological footprint. Awareness lacks at grassroots consumer levels to the fact that global warming gases are added at 1.3 tonnes per tonne of waste as they end up in landfills (Themelis, 2003). Additionally, incessant extreme climate catastrophes create debris in a matter of seconds that impacts livelihoods, health, infrastructure and resources. Practical sustainable solutions to waste management for the next few generations will be composting, recycling and energy recovery technologies (Themelis, 2007). The quest for a sustainable world builds on waste to energy with communications driving such waste citizenry.

Discussion on Why Water - Energy - Waste Would be a Key Determinant for Sustainability to in The Quest of Climate Resilience?

The water and energy demand in the developing and transition economies are poised to increase significantly. Side-by-side the generation of waste would exacerbate, fueled by the consumption driven lifestyle with use-and-throw plug-in type accessories. This calls for value parity, and is the key essence for sustaining growth amid need for climate resilience (Figure 2).

Figure 2. Configuring the water - energy - waste value parity framework for sustainability and climate proofing

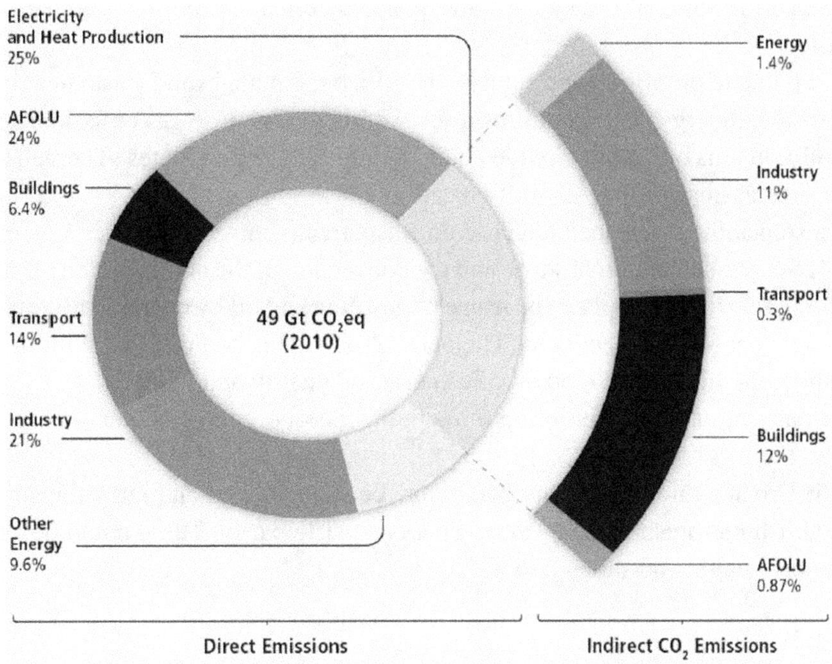

Energy - Water Inter-Relatedness

Fossil fuel based energy systems as well as nuclear based energy systems need steam for driving turbines that generate electricity. The steam is subsequently cooled. For instance in 'once-through' cooling circuits, assive quantities of water is needed, which is usually withdrawn, and post-usage, dumped to source, usually at higher temperature, that is detrimental. The one-pass water circuits harm aquatic habitats and inflict 'thermal shock' as the waste water is at higher temperature. Wind and solar energy can improve the sustainability, justifying the threshold value parity of water – energy, determined by waste as the critical sustainability enabler (Lerner, 2016)

A: Engage policy-makers with community through energy - water linkages that have good environmental value parity:
Instance 1: Renewable energy technology, such as 4 kilowatt solar micro off-grid helps to increase access to energy, thereby increasing economic activities to augment source of livelihood. A typical solar micro off-grid constitutes an labyrinth of twenty 200 watts solar photovoltaic panels arranged in five arrays with four panels per array. About 1.6 kilowatt is allocated for water management in the community borehole project facility, while the remaining 2.4 kilowatt is distributed for electrification.

B: Environmental 'refugee' issue linked with quality of the locale habitat:
Instance 2: Landfills with municipal solid waste that also contaminate water, is a trigger for environmental 'refugees'. Sustainable waste management of municipal solid waste (MSW) can positively impact locale habitat. A ton of MSW treated in waste-to-energy (WTE) plants provides

compelling reason to utilize landfill debris to WTE units. This underscores the water - energy - waste value parity potential.

C: Bridge Local with National Policy for convergence on climate proofing with focus on water - energy – waste:
Instance 3: Climate proofing opportunity is to utilize municipal solid waste in a timely manner to generate waste to energy. This could result in abating greenhouse gas emissions to the tune of at least 123 million tons of carbon dioxide equivalent for the United States (Themelis, 2015). This is equivalent to emissions of over 23 million cars.

D: Integrate sustainability with social license on waste treatment:
Instance 4: Source-separation of waste and the conversion of the non-recycled plastics to synthetic oil (through pyrolysis) can replace the usage of gasoline and oil by over 5.7 billion gallons and 136 million barrels per year respectively (Themelis, 2015). This could fuel 8.9 million cars, thereby re-establishing the implementation of value parity on energy and waste.

E: Enhance urban - rural climate-proofing through the incorporation of water - energy -waste thresholds:
Instance 5: Urban - rural climate proofing may be standardized and rationalized with the provision of establishing thresholds on water - energy - waste parity. These could be critical to assess environmental, health, and safety issues.

CONCLUSION

Water-Energy-Waste Nexus in Context to Climate Change

The inter-relatedness, alignment, self-correcting, dynamic equilibrium are synonymous to the concept of nexus (Biggs, 2015) that is manifested here as value parity. Value in terms of *bank-ability* to private sector is created when resource-intensive growth model paves way for resource productivity. For instance, USD 2.9 tri would be saved by 2030 through resource productivity (Dobbs et. al., 2011). Waste has the potential as a new resource may be translated to value-added, that would in turn make the proposition bankable.

Water ~ energy ~ waste inter-relatedness provides good degree of resilience to climate change, as waste exacerbates contamination that deteriorates water, needing more energy to treat. Fossil fuel based energy emits greenhouse gases, thereby contributing to climate change. Public ~ private ~ community partnerships foster awareness, promotes entrepreneurship, and drives motivation to utilize waste, curb water usage and clean-up energy. This creates shared value (Giovannucci et.al., 2012). Smart models for the management of industrial waste in the era of climate change is also being researched out world across as a part of adaptation to the climate change (Leal Filho, 2015; Patil & Rao, 2015, Rao & Patil, 2015; Saini et al., 2012).

Worsening air quality is responsible for curbing life-spans of over 660 million lives, according to a recent research (Greenstone & Sudarshan, 2015), that attributes a country's overt dependence on fossil fuel based energy systems. Though such energy quantum is necessary to drive economic growth, megapolises are turning to be strewn with polluted air, contaminated water, all attributable to burgeoning waste/debris. The issues on sustainability for climate resilience with it's implications for water - energy - waste, is further illustrated in Table 3.

This underscores the express need to incorporate health consequences of water - energy - waste into traditional measures of growth such as GDP. Specifically, the study points out, life expectancy is curbed by a minimum of 3.2 years for every 100 micrograms of PM2.5 (fine particulate matter) beyond safe levels. PM2.5 particles are a cause of serious concern as they penetrate deep into lungs.

Climate Change and Water Management Viewed with Respect to Energy and Waste

Human influence on climate system is now assessed with 95 percent certainty, wherein people-factor has exacerbated observed global warming since the mid-20th century. Such hikes have been heightened over recent decades. Such influence manifested via atmospheric warming, shifts in water cycle, water-stress due to dwindling snow and ice, mean sea level rise, and sudden variants in extreme climate. Water management in rural settings are adversely impacted by waste effluents, inferior energy-fuel standards, rampant garbage burning, heavy dependence on diesel for irrigation in pockets not served by power grid.

Future Pathways

To Manage The Planet For 10 Billion People: Sustainability Driven Energy - Water - Waste Value Parity

Sustainability driven energy - water - waste value parity is a pas the desired destination. Risks and uncertainties are plenty, like currency imbalances, energy inequities, and policy non-uniformities. With 2010 being one of the ten warmest years1 on record since 1998, climate concerns are ever increasing and unpredictable extreme weather devastations tend to pose as the predominant and overarching challenge. All other developments would be meaningless without the ecology thriving along with growth of economy.

Value parity establishes the history in relation to the environment for future growth (Sachs, 2015). Patterns of growth in the past have inflicted intense strains on our planet and generated severe risks to lives and livelihoods. Globally economies are on the causeway of growth. Asia propels the momentum.

Table 3. Issues on sustainability for climate resilience: Implications for water - energy – waste

Issues on Sustainability for climate resilience	Water	Energy	Waste
Atmospheric CO_2	Oceanic Acidification	Increase in average temperatures would need more energy for air-conditioning	Corresponding increase in methane (CH_4) and nitrous oxide (N_2O) affects landfills
Groundwater availability	Flow of groundwater through porous substrate would be affected due to flux (Water, Heat, Charge, Solute) (Hafemeister, 2014)	When Gradient of flux of flow (Water, Heat, Charge, Solute) would alter the water for energy balance, energy needed to treat water would be higher.	Source separation of non-recycled plastic (NRP) from smidge, would provide option to use for waste to energy
Carbon Flux	Greenhouse gas emissions' impact from ships	Energy Efficiency	High-Nutrient, Low-chlorophyll (HNLC) for ocean fertilization (Warner and Schofield, 2012)
Carbon sinks	Ship traffic energy and environment models	Renewable energy	Water-Energy-waste threshold for climate proofing

Sustainability driven citizenry has the potential to harness climate benign quality of life aspirations. Reality checks emanating from climate change concerns such as energy alternatives, health robustness, and water assurance necessitate the deployment of sustainability drivers through creative communications. The mandate 'Curb waste on Water' at the grassroots level up can aggregate to create groundswell of Sustainability driven Citizenry or simply 'Sustainability citizenry'.

REFERENCES

Adnan, A. (2015). *Renewable Energy in the Water, Energy and Food Nexus*. IRENA.

Beck, U. (2002). The cosmopolitan society and its enemies. *Theory, Culture & Society*, *19*(1-2), 17–44. doi:10.1177/026327640201900101

Biggs, E., Bruce, E., Boruff, B., Duncan, J., Horsley, J., Pauli, N., & Imanariet, Y. et al. (2015). Sustainable development and the water–energy–food nexus: A perspective on livelihoods. *Environmental Science & Policy*, *54*, 389–397. doi:10.1016/j.envsci.2015.08.002

Brown, D., Livermore, L., Bell, J., Estep, D., Gropp, W., Hendrickson, B., …Wright, M. (2008). *Using models of complex systems to inform policy makers*. US Department of Energy.

Frison, E. A., Cherfas, J., Eyzauirre, P. B., & Johns, T. (2004). *Biodiversity, nutrition and health: Making a difference to hunger and conservation in the developing world*. Keynote Address to the Seventh Meeting of the Conference of the Parties to the Convention on Biological Diversity (COP 7).

Giovannucci, D., Scherr, S., Nierenberg, D., Hebebrand, C., Shapiro, J., Milder, J., & Wheeler, K. (2012). *Food and Agriculture: the future of sustainability, Sustainable Development in the 21st century (SD21)*. United Nations Department of Economic and Social Affairs, Division for Sustainable Development. Retrieved on 20 Feb 2016, from https://sustainabledevelopment.un.org/content/documents/1443sd21brief.pdf

Golberg, A. (2015). Environmental Exergonomics: Sustainability analysis of Energy systems considering impacts on eco-systems services. *Proceedings of the International Conference on Efficiency, cost, optimization, simulation and environmental impact of energy systems*.

Greenstone, M. (2015). *Global-post on air quality*. Energy Policy Institute at Chicago.

Hafemeister, D. (2014). *Physics of Societal Issues: Calculations on National Security, Environment and Energy*. Springer. doi:10.1007/978-1-4614-9272-6

Hall, D., & Lobina, E. (2008). *Water Privatisation*. Public Services International Research Unit (PSIRU). Accessed on 18 Feb 2016 http://gala.gre.ac.uk/1704/1/PSIRU_Report_(9820)_-_2008-04-W-over.pdf

Harvey, D. (2005). *A brief history of neoliberalism*. OUP Oxford.

Held, D. (1995). *Democracy and the global order: From the modern state to cosmopolitan governance*. Stanford University Press.

Holt, D. (2003). What becomes an icon most? *Harvard Business Review*, *81*(3), 43–49. PMID:12632803

Jarraud, M. (2010). *Report of the Global Atmospheric Watch January 2011*. United Nations World Meteorolical Organization.

Kallidaikurichi, S., & Rao, B. (n.d.). Index of drinking water adequacy for the Asian economies. Water Policy, 12(S1), 135–154.

Kessides, I. (2004). *Reforming Infrastructure: Privatization, Regulation, and Competition*. World Bank Policy Research Report, co-publication the World Bank and Oxford University Press.

Koplan, J. (2002, March). Chronic Disease: The Role of Behavior and the Environment. *Health Affairs*, *21*(2), 179–184. PMID:11900157

Kotz, D. M. (2002). Globalization and neoliberalism. *Rethinking Marxism*, *14*(2), 64–79. doi:10.1080/089356902101242189

Leal Filho, W. (Ed.). (2015). *Handbook of Climate Change Adaptation*. Berlin: Springer Berlin Heidelberg. doi:10.1007/978-3-642-38670-1

Lei, C. (2011). *Investments in China in water resources*. Report of the Ministry of Water Resources.

Lerner, H. (2016). *The Water Footprint of Energy*. Accessed on 29 Feb 2016 http://www.gracelinks.org/196/the-water-footprint-of-energy

Miller, A. (2015). *Creating Value-Added weather & climate services through innovative PPPs, C4 Eco Solutions*. United Nations Development Program. Accessed on 22 Feb 2016 http://www.adaptation-undp.org/resources/planning-meeting-presentations/cirda-ppp-workshop-presentation-alan-miller

Patil, Y., & Rao, P. (2015). Industrial waste management in the era of climate change - A smart sustainable model based on utilization of active and passive biomass. In W. L. Filho (Ed.), *Handbook on Climate Change Adaptation* (pp. 2079–2092). Springer-Verlag Berlin Heidelberg. doi:10.1007/978-3-642-38670-1_49

Rao, P., & Patil, Y. (2015). Climate Resilience in Natural Ecosystems in India: Technology Adoption and the Use of Local Knowledge Processes and Systems. In Handbook of Climate Change Adaptation. Springer Berlin Heidelberg.

Rawls, J. (1971). A Theory of Justice. Cambridge, MA: Harvard University.

Ringler, C., Bryan, E., Biswas, A., & Cline, S. (2010). Water and Food Security Under Global Change. *Water Resources Development and Management*, *1*, 3–15. doi:10.1007/978-3-642-04615-5_1

Sachs, J. (2015). *The Age of Sustainable Development*. Columbia University Press. doi:10.7312/sach17314

Samir, S., Prakash, R., & Yogesh, P. (2012). City based analysis of MSW to energy generation in India, calculation of state-wise potential and tariff comparison with EU. *Procedia: Social and Behavioral Sciences*, *37*, 407–416. doi:10.1016/j.sbspro.2012.03.306

Schattle, H. (2008). Education for global citizenship: Illustrations of ideological pluralism and adaptation. *Journal of Political Ideologies*, *13*(1), 73–94. doi:10.1080/13569310701822263

Schultz, W. (2007). Reward. *Scholarpedia*, *2*(3), 1652. doi:10.4249/scholarpedia.1652

Sciubba, S. (2003, November). Extended exergy accounting applied to energy recovery from waste: The concept of total recycling, Strategies of Waste Energy Usage. *Energy*, *28*(13), 1315–1334. doi:10.1016/S0360-5442(03)00111-7

Starr, A. (2000). *Naming the enemy: Anti-corporate movements confront globalization*. Zed Books.

Themelis, N. (2003). *An overview of the global waste-to-energy industry*. Waste Management World.

Themelis, N., & Mussche, C. (2014). *Energy and economic value of municipal solid waste (MSW), including non-recycled plastics (NRP), currently land-filled in the fifty states*. Earth Engineering Center.

Themelis, N., & Ulloa, P. (2007, June). Methane generation in landfills. *Renewable Energy*, *32*(7), 1243–1257. doi:10.1016/j.renene.2006.04.020

Thomas, V. (2015). *Better results in a dynamic world*. Asian Development Bank. Accessed on 30 Nov 2015 http://www.adb.org/publications/inspiring-adbs-future-better-results-dynamic-region

Warner, R., & Schofield, C. (2012). *Climate Change and the Oceans*. Edward Elgar. doi:10.4337/9781781006160

Chapter 12
Congo Basin's Shrinking Watersheds:
Potential Consequences on Local Communities

Bila-Isia Inogwabini
Saint Pierre Canisius Institute of Agriculture and Veterinary Sciences (ISAV), Congo & Swedish University of Agricultural Sciences, Sweden

ABSTRACT

Rainfall time series data from three sites (Kinshasa, Luki, and Mabali) in the western Democratic Republic of Congo were analyzed using regression analysis; rainfall intensities decreased in all three sites. The Congo Basin waters will follow the equation $y = -20894x + 5483.16$; $R^2 = 0.7945$. The model suggests 18%-loss of the Congo Basin water volume and 7%-decrease for fish biomasses by 2025. Financial incomes generated by fishing will decrease by 11% by 2040 compared with 1998 levels. About 51% of women (N= 408,173) from the Lake Tumba Landscape fish; their revenues decreased by 11% between 2005 and 2010. If this trend continues, women's revenues will decrease by 59% by 2040. Decreased waters will severely impact women (e.g. increasing walking distances to clean waters). Increasing populations and decreasing waters will lead to immigrations to this region because water resources will remain available and highly likely ignite social conflicts over aquatic resources.

INTRODUCTION

Climate change is a scientifically established fact now (Walther et al., 2005), even though the magnitude of its multiple, diversified and multidimensional effects remain mostly in the domain of mathematical modeling. Discussions on the mitigation of climate change and adaptation processes (social, cultural and biological) remain at the core of scientific and political debates (Aiken et al. 1992; Fletcher, 1997; Fletcher, 2000; Hobbs & Knausenberger, 2003; Hughes, 1986; Inogwabini et al., 2006) because they are diversified and multidimensional, and will vary throughout the world (IPCC, 2007). The nature and the magnitude of these impacts have, however, yet to be described for different geographic environments

DOI: 10.4018/978-1-5225-1046-8.ch012

and locations. This is even more important for the tropical region of Central Africa where documented evidence is scanty (Halpin, 1998; Inogwabini et al., 2006; Sonwa et al., 2009). Informed guesses and mathematical projections convey the message that climate change's effects in Central Africa may range from drier conditions in areas adjacent to as deserts (Sahara and Kalahari) where water occurs in shortage to high floods in coastal habitats directly adjacent to oceans due to increased water levels as a result of the melting of polar ice (Halpin, 1998).

Unfortunately, however, long-term field data that can document changes which highlight the effects of climate changes over the past years are difficult to find for Central Africa (Edwards & White 2000; Pimm, 2007; Sonwa et al., 2009). Lack of data fosters the use of surrogate data that can predict the effects of climate variation in Central Africa. One of those surrogates has been the linking of phenological data to weather patterns (Tutin & White, 1998), direct measurements of water levels (Colombant, 2005), using direct meteorological data and looking at their trends (Inogwabini et al., 2006). The use of long time series meteorological data as a proper detector of the effects of the climate change is fully justified by the fact that if climate change will affect water level, then drier conditions in the terrestrial Central Africa region are, logically, linked to water cycles, especially rainfall regimes over the continent. However, continuous time series data on rainfalls are difficult to find in Central Africa to document the trends and extents of felt changes (Halpin, 1998). Furthermore, where there are continuous long time series data, the analytical capabilities have made it difficult to determine understandable trends that crude data can convey. Even less available are data on effects of climate change on local economies, health, security and biodiversity (Sonwa et al., 2009; Stern, 2007).

MAIN FOCUS OF THE CHAPTER

This chapter presents three long term data from three locations in the western Democratic Republic of Congo (Figure 1), with the aim of presenting these long time series data in a single analytical framework and discussing patterns that emerge from that analysis in the context of global climate change. The chapter links trends in rainfall and surface area Congo Basin watershed and projects potential effects of these trends to the level of the central Africa region. Combining these projected effects with field data freshwater resources, the paper discusses inferred effects on climate changes on local populations.

STUDY SITES

Rainfall data were collected from (1) Mabali, (2) Kinshasa and (3) Luki. Mabali is located near the equatorial line (Inogwabini et al., 2006). Located in the *Cuvette Central*, weather patterns are believed to be stable (Bultot & Griffiths, 1972). The Cuvette Central is a lowland, flat and heavily flooded equatorial region. It is characterized by *terra firma* forest (forest whose floor remains dry for most of the year), *swampy forest*, and diverse types of savannahs inundated and terra firma (Evrard, 1968; Leonard, 1951 & 1952; White, 1983). The *Cuvette Central* is the lowest point within the Congo Basin and once constituted a water retention point during the late Miocene when many African inland basins were still *endorheic* (Roberts, 1975; Thieme et al., 2005). Kinshasa is a town of nearly 9,000,000 people (Aveling et al., 2003) and located south-west of the Mabali. Adjacent habitats to this mega city are essentially composed of wooded savannas on a sandy soil. Kinshasa is located at Pool Malebo, an open water sur-

Figure 1. Locations from which weather data were collected

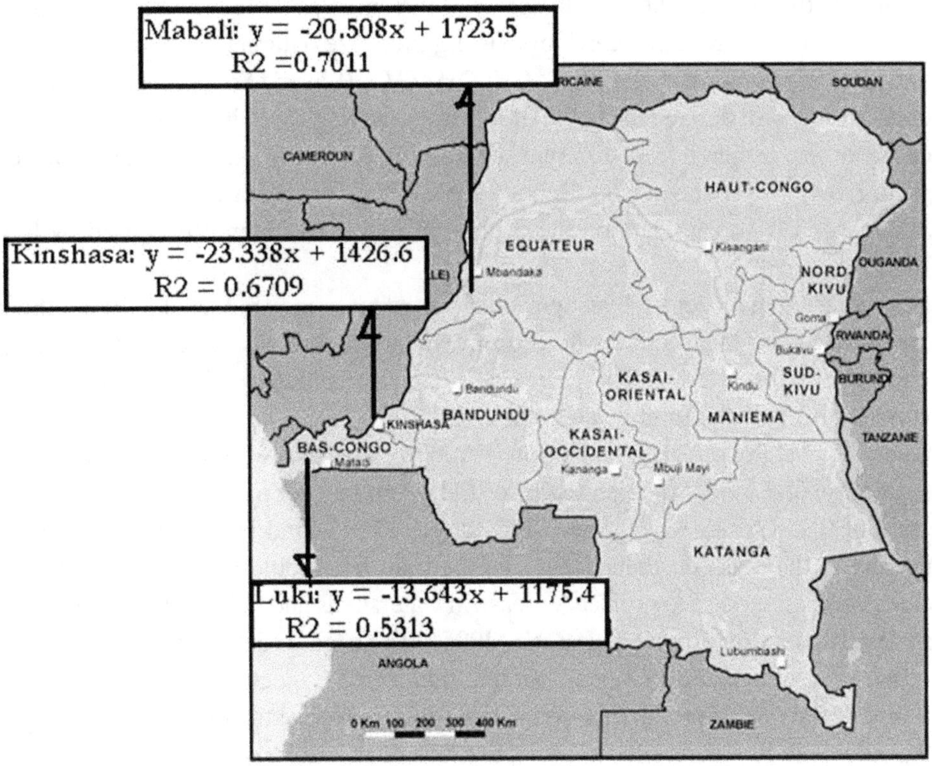

face leading to several multidirectional wind currents. These currents have an impact on rainfall regimes observed in Kinshasa.

Luki are located further south and closer to the Atlantic Ocean. With a population of 150, 4361 people (Iloweka, 2004), Luki is a forest island within a predominantly human landscape and degraded forests. Its position near the Atlantic Ocean implies that rainfall regime and other climatic events occurring in that region are potentially influenced by climatic events affecting the Atlantic Ocean.

The Congo Basin watershed covers 3,730,474 km^2, and is home to ~60 to 78 million people (Sonwa et al., 2009; WRI, 1998). About 62% (37 – 48 million people) of that population is still rural (Sonwa et al., 2009), even though the urban growth rate of the entire region is high and estimated at 5% per year (Atsimadja, 1992; WRI, 1998). This vast zone covers different habitat types, with tropical forest representing ~40% (14,920,190 km^2). The geography of the zone has an impressive biological diversity.

MATERIAL AND METHODS

Rainfall data from Mabali have been collected continuously from 1970 through 2004 (Inogwabini et al., 2006) while those from Luki were collected between 1958 and 2004. The data set from Kinshasa covers the period between 1971 and 2004. Total annual rainfall intensities (mm) of were plotted over time, i.e. 34 years for the Mabali data set, 46 years for Luki and 33 years for Kinshasa (Figure 1). To detect the trends in these data sets, the linear regression was used on annual rainfall intensities (mm). Those data, particularly on the Mabali Scientific Reserve, were published earlier (e.g. Inogwabini et al., 2006).

Other data came mainly from the search of existing literature. Human population data were gathered from Atsimadja (1992), Aveling et al. (2003), CEFDHAC (2001), and Sonwa et al. (2009), WRI (1998), Trends in populations were derived using the model presented by De Saint Moulin (1991 & 2003). Data on fish stocks were also gathered from different sources. Fish biomasses and estimated financial incomes generated from fishing were drawn from Bene (2005), CEFDHAC (2001), Corsi (1984a, b), Colom et al. (2006). Leonard (1987), and Data on the decline in fish stocks were those used by Inogwabini & Lingopa (2006).

These data indicated that over a 34 year period reported by Inogwabini et al., (2006), the water depth of Lake Tumba decreased from 8 m down to 6 m, which is a 25% decrease over 34 years, leading to a rate of decrease of 0.0074%/year. This rate was used in combination with regional elevation data to produce the map depicting the future water situation in the Congo Basin watershed. Also, repeating a survey conducted by the Food Aid Organization (FAO) in 1998, Inogwabini & Lingopa (2006) reported a fish stock decrease of 4.5% for the fish stocks of Lake Tumba over a period of 17 years, which equals a decrease rate of 0.25%/year. As suggested by Huggett (1993) and Hulme et al. (2001), these rates, maintaining each of them as constants and assuming that they would apply to the whole basin, were used in linear regression models to project decrease in watersheds and fish stocks. The financial incomes generated by fish production provided by Bene (2005) were used to determine the per-ton cost of fish. Maintaining the price constant over years, I simply multiplied projected fish stocks per year to obtain trends in incomes. The rate of production was calculated to measure the trend in how many tons of fish will remain available to the increasing rural populations of the region over the next 60 years if current conditions remain constant. This important measure was defined as a simple ratio of projected tons of fish over projected total populations.

Data on effects of shrinking waters in the study region on women were those that came from the socioeconomic study carried out by Colom et al. (2006). These included the gender-sorted fish market data, assessment of felt environmental threat disentangled between women and men, traditional work division, etc. They were collected through a variety of methods, including structured interviews, focus groups, and literature reviews Colom et al. (2006). These data are presented in this chapter as simple means and where projected, they were assumed to progress constantly over years. Data on health came from official reports of the provinces and particular health facilities (MSPPE, 2010; SS, 2007). To confirm that these data reflected the reality on ground, I have checked trends with medical doctors that worked in the health divisions of Bikoro, Ntondo and Mbandaka.

RESULTS

The rainfall intensities decreased in all three sites over time for each sites following the regression equation $y = -20.508x + 1723.5$ ($R^2 = 0.7011$), $y = -13.643x + 1175.4$ ($R^2 = 0.5313$) and $y = -23.338x + 1426.6$ ($R^2 = 0.6709$) respectively for Mabali (Figures 1 and 2), Luki (Figure 1 and 3) and Kinshasa (Figure 1 and 4). The rainfall for Luki and mean rainfall for Kinshasa in 2004 represented only 59% and 61% of its levels 33 years earlier, in 1971 following the drought of 1970 when Luki received only 327.8 mm of rain. The situation has been even more dramatic for Mabali where the maximum rainfall represents > 25% of its levels in 1970 or 34 years earlier.

Figure 2. Rainfall trends in the Mabali Region between 1970 and 2004

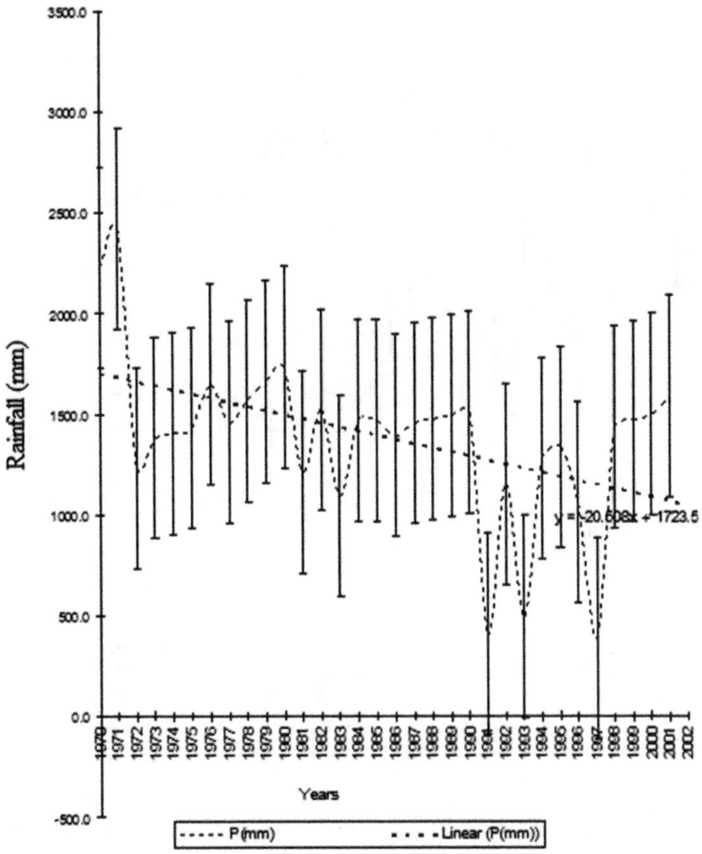

Figure 3. Rainfall trends in the Luki Region between 1958 and 2004

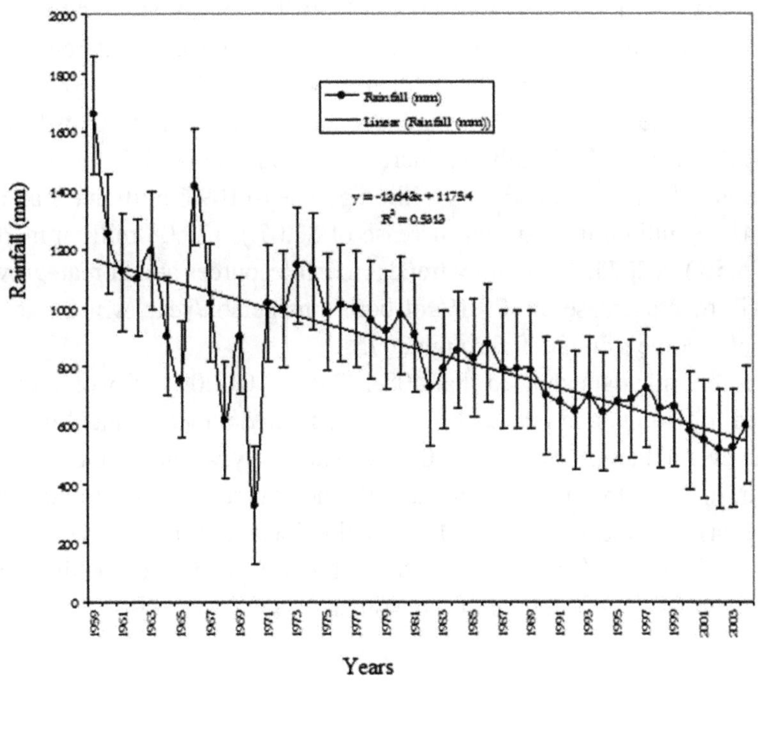

Figure 4. Rainfall trends in Kinshasa between 1971 and 2004

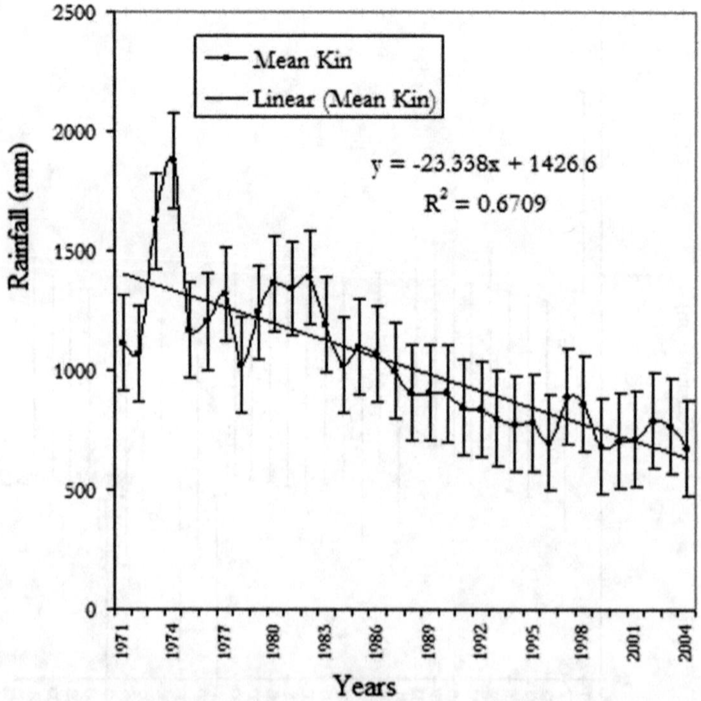

If the rate of water loss in Lake Tumba is applied for the entire region and remains constant, watershed will decrease following the regression equation y = -20894x + 5E+07 ($R^2 = 0.7945$). This model indicates that by 2025, the Congo Basin watershed will have lost 18% of its total water mass as compared with its 1998 levels. The loss will deepen to -26.6% by 2040, -39% by 2065 and then around -43% by 2070. Equally, if the loss in the fish biomasses in Lake Tumba are applied to the whole region and remain constant over years, the fish stock will decrease by 7% between 1998 and 2025. This decrease will deepen down to -11% by 2040, -16% by 2058 and by -20% in 2075. The same proportions will be observed if fish prices in the markets are constant, decreasing from 47.8 million American dollars/year in 2005 (Bene, 2005) down to ~42.5 million American dollars/year (~11%) by 2040. At the same time, the human population will grow from 60 – 78 million people to 109.6 million (or an increase of 40.5%) in 2025, then to 141.18 million in 2040 (an increase of 81.0%), 179.08 million in 2058 (129.6%) and 214.89 million (175.5%) in 2075. Combining trends in fish resources and human growth, these changes imply that there will be a decrease of ~70.1% of both tons/person/year, which parallels the projected magnitude of loss of income from 2009 through 2075.

The 2005 socioeconomic data showed that 408,173 of ~1,012,000 all women (~51%) in the landscape were involved in fishing, and that each woman generated a mean annual income of ~2,100US$. This represented an overall income of 857,164,000 US$ annually; a significant % of the overall national budget for that year. In 2010, the fishing-generated financial income was estimated at ~ 1870 US$ per woman or a net decrease of 11%. If these trends are maintained constant, women generated income will decrease by 25% in 2020 and 59% in 2040 while the number depending on fishing for financial income will be 836,755 women, above the double of the women population in 2005.

Even though the region is a swamped zone, the data showed that more women (85%; N= 360) and girls (90%; N= 200) were involved in the activities of fetching water than men (25%; N = 140) and young boys (50%; N=140). The same socioeconomic data showed that that on average women walked ~ 6 km (round trip) to fetch drinkable water; this was done twice daily. This daily 12-km trip consumed 4 hours on average, representing 40% of a ten hour working day. Logically, three quarters of women reported that fetching water from long distances impeded their ability to conduct more lucrative activities such as commerce or even fishing, which are the most important economic activities in the region. More than three quarters of the interviewed women were afraid of fetching water from long distance sources because they feared being raped by former army soldiers. Interestingly, the same data set also showed reversed trends in girl education patterns; in areas where water sources were far away from the villages, on average only 20% of girls had actually completed their secondary education while in towns where water sources are adjacent to houses, the figure increased to 45%.

Published reports (Secrétariat à la Santé 2007) from the study region indicated a global increase in diarrheas (globally 2.7% for the province of Equateur and 3% for that of Bandundu) and amoebic dysenteries (globally 8% for the province of Equateur and 7.2% for that of Bandundu) at the end of January through the end of February and from end of May through end of July. When looking at reports from Bikoro and Ntondo (near Lake Tumba) these figures are even higher for the same diseases. Diarrheas increase by 25% (Bikoro) and 30% (Ntondo) while amoebic dysenteries increase by 15% (Bikoro) and 24% (Ntondo). Health reports from Bikoro and Ntondo also show an increase in what they call skin itching (40% and 30% respectively for children under the age of 18); and 2007 health report from Ntondo also mentioned an increase in what they shyly called women specific genital ailments and infections by 23.6% during the same periods of the year. Periods from at the end of January through the end of February and from end of May through end of July are dry seasons when clean water is in shortage in the region.

DISCUSSION

General Foresighted Effects

The three data sets on rain intensities (Figures 1, 2 & 3) clearly indicate that there has been 'continuous' drying up of the zone over the last decades. The mean rainfalls for Luki and for Kinshasa in 2004 represented only 59% and 61% of its levels 33 years earlier while a more dramatic change was depicted for Mabali with the maximum rainfall in 2004 representing only $\geq$ 25% of its levels in 1970. These inter-site variations would be logically expected, given the fact that weather patterns are functions of multiple parameters, including micro climates, which are ecosystem-dependent themselves (Salati, 1987). Conditions, ecological and climatic, in those three sites are widely different and therefore account for those differences. In-site changes are rather difficult to explain, even though they may easily be linkable to major human activities in some cases.

Luki is an island of forest within the Bakongo Region, which is the most logged region of the Congo (Iloweka, 2004). With an estimated human density of 237 persons/km^2 (Iloweka, 2004), the population struggles to make a living through agriculture and woodcutting, which are fostered in response to high demands for foodstuffs and the high consumption of wood (for energy and construction) in Kinshasa and the expanding towns of Matadi and Boma. Most of the demand is met by products coming directly from the adjacent Province of Bas-Congo. All these activities increased the deforestation rate and unbalancing of forest ecosystems, which may be feeding the climate changes. Even though there is a need to deepen our knowledge on the correlation between forest cover and rainfall regime in this region, it is fairly logical to infer, based on studies undertaken in other regions of the world (Eltahir & Bras, 1993Gash & Nobre, 1996; Lettau et al.1979,), that changes in the rainfall in this region may be a consequence of deforestation, which is feeding into global climate change and its effects.

The situation appears of Kinshasa seems to be similar to that of Luki. Kinshasa, with a currently estimated population of 9 million people, has seen its immediate environment stripped of all natural forests and bush due to the demand for land needed to accommodate a rapidly increasing population. Forested areas that were left to maintain a healthy urban environment have been either invaded by houses or have seen their trees cut down to provide charcoal for energy. Such is also the case for Luki and even though this question calls for more detailed information to help establish clear causation links, it remains reasonable, based on studies done elsewhere (Eltahir & Bras, 1993; Gash & Nobre, 1996; Lettau et al., 1979), to hypothesize that changes observed in rainfall patterns in Kinshasa may be subsequent to decreasing vegetation cover both within the town itself and its immediately adjacent areas.

The 34-year data set from Mabali intrigues common wisdom and can be fully understood only in parallel with the overall climate change paradigm (Inogwabini et al., 2006). Mabali (Figure 1) is located 75km straight south of the Equator. Within this strip, climatic metrics are constant (Bultot & Griffiths, 1972; Evrard, 1968) there are constant temperatures, stable rainfall regimes and regular seasonal cycles. A first look at the data from Mabali questions the rationality of the data because the site appears to be in an ideal environment for a rainfall pattern to remain constant over years. This remains the question even when you consider the fact that Mabali is also surrounded by active logging activities and many forest clearings due to high human concentration around Lake Tumba because the extent of vegetation clearing are far from being of the same magnitude as in Luki and Kinshasa. Deconstructing the puzzle posed by the Mabali situation, Inogwabini et al. (2006), found that while the most acknowledged consequence of the climate change was global warming (Houghton et al., 1990), the Mabali long-term data indicated a local cooling phenomenon, which they attributed to longer dry seasons. Longer dry seasons mean less precipitation; therefore it is understandable that rainfall had decreased over the period covered by the study. Following Bultot & Griffiths (1972), Inogwabini et al. (2006), explained that in equatorial Africa dry seasons are cooler than normal temperatures due to over-evaporation, leading to persistent clouds. However, while tracking the explanations for longer dry seasons, they concluded that they could be logically explained only by inserting the global climate into the equation. Overall, decreased rainfall in the western Congo (in the three locations of Kinshasa, Luki and Mabali), regardless of reasons that may explain them, do agree with patterns in the other regions of Africa and have been documented previously (IRD, 2002; Mahe et al., 2001), albeit at localized scales.

In conclusion, these trends are clear and should ignite the will to establish causation links if they are to be managed appropriately. This is a call for appropriate research rather than depending on opportunistic data. There is also one precaution in trying to use these data sets and/or extrapolating them: decreased annual rainfall intensities may depict a phenomenon of a long term cyclical event whose frequencies may be difficult to decipher from such a short time-series data. Nevertheless, given the fact that human activities may have pushed natural environmental to the edges of irreversibility, the picture provided by these data series may convey an alarm signal given the fact that we may have gone beyond a certain resilience threshold. This is even more appealing when one considers that these climatic patterns agree with the overall picture of the Congo Basin and Africa, and may be attributable to the global climate change phenomenon (IRD, 2002; Inogwabini et al., 2006).

Previous studies (Hulme et al. 2001; Pearce & Turner, 1990; Sonwa et al., 2009; Stern, 2007) have described effects of climate changes to include economic aspects (decrease in financial incomes), health and sanitation (water quality and decrease in intake of fish proteins) and loss in biodiversity. Inogwabini et al. (2006) argued that the observed decrease in water depth of the Lake Tumba (from maximum depth of 8 m down to 6 m over 34 years) was essentially linked to the dramatic decline in rainfall. They also suggested that, apart from intensive fishing described by Inogwabini et al. (2009), observed decreases in fish stocks may also be linked to climatic conditions prevailing in the lake such as loss of reproduction sites, cooler temperatures and losses in some critical foods due to prolonged dry seasons, as fishes of the region adapted to different conditions and exhibit different requirements to reproduce and grow (Bailey, 1986; Banister, 1986; Chapman, 2001). They also indicated that if the weather change patterns they described remain constant, there would be numerous ecological consequences on forest ecosystems of the western Congo Basin (IRD, 2002; Inogwabini et al., 2006) in the long-term future. Evidence of such effects of climates on biodiversity have been described on several occasions (Fernández & Vrba, 2006; Lemoine & Böning-Gaese, 2003; Lemoine et al., 2007a & Lemoine et al., 2007b) and will likely occur in this region whose biodiversity is one of the most important on earth. To highlight the magnitude of the land affected by drying, a decrease in 18% of the size of the Congo watershed between 1998 and 2025 represents 673,968 km^2. This dried out area represents ~28% of the entire Democratic Republic of Congo and 23 times the size of Equatorial Guinea.

Beyond those consequences on biodiversity, the projected economic effects a drying watershed would also strike human economic activities, not only in terms of fish markets and protein intake but also in terms of agricultural productivity in the whole region. Increasing human populations will need more food and cultivatable land. Shortages in water will push people to find water sources for agriculture where this key resource will still is available, which is in the Central Congo Basin. That pattern would, most likely, lead to conflicts over water resources in the region. Lovett (2006) established that rainfall is one of the main determinants of ecological productivity, which in turn is the basis of agricultural economies. Shortages of rainfall will, therefore, undermine the rich cultural diversity as well by forcing people to migrate to regions that offer better agricultural production and sufficient water sources. The destruction of the resources that support traditional culture is, by any measure, the most dramatic effect of climate change in this region. Climate change would certainly precipitate dramatic, permanent shifts in local cultures. For example, all people in the Lake Tumba region prefer fish as their most important food item. Decreasing fish stocks will most likely push them to adopt new life styles and other techniques to extract natural resources to live on. The introduction of new techniques will affect community views, relationship to the environment and overall stability.

SPECIFIC EFFECTS OF WATER DECREASE ON WOMEN IN THE REGION

Decreasing water levels would likely increase in walking distances to clean water sources, which will ipso facto increase time consumed to fetch water, reduce girls' chances to attend formal education, decrease women's economic powers and expose them to gender-based violence.

While there might be other factors affecting gender-education patterns described in this study, increased walking distances to clean water sources, at least contributed to that picture and will very likely continue to reduce girls' chances to attend formal education unless major cultural and social changes happen. This finding conforms to the results reported from Ethiopia, Ghana, Tanzania (Komba, 1995) and India (UN-DESA, 2005) and other areas across Africa (YPOW, 2006). As reported in many studies (Inter-Agency Commission, 1990; Ilon, 1992; Kinyanjui, 1993), literacy of women is an important key to improving health, nutrition and education in the family and to empowering women to participate in decision-making in society. Naturally, according to the *Childhood Engenderment Theories* (Chafetz, 2006), impeded access to formal education will keep future women under the yoke of inequality and reduce their self-esteem, perpetuating the current imbalanced power toward women (Chafetz & Haggan, 1996).

Socioeconomic data indicate that decreasing water levels will have more severe impact on women commercial activities than compared to the global effect. That the numbers of women will double provides a prelude of heighten conflicts over fishing points resulting from increased competition of fishing points. This, combined with reduced access to education caused by water shortage, will therefore further hinder women's economic powers, limiting the contribution of more than 50% of the population to contribute to economic returns expected by their societies and impact on means of achieving sustainable development and economic growth, which agrees with findings by Hanjra and Qureshi (2010) who projected that climate change effects will hinder the contribution of women to the global development. Reduced economic power means, again, lost self-esteem and will keep the current paradigms of social inequalities between men and women. Ironically, this will happen though women will see their workload increased by doubling walking distances to fetch water. Indeed, the estimated workload for fetching drinking water was a daily 12-km trip, consuming 40% of women working day. This was higher than the average time spent by women across rural Africa (Blackden & Wodon, 2006), which had been estimated at 26% (Water Aid, 2012).

Water Aid (2012) had suggested a cohort of effects of reduced water availability on women in developing countries, principally on their health and that of their families in general. Both structured interviews and focus groups with local communities were not able to document feeling of worsening human health conditions in the region, though 21% of interviewed people talked about health increasing costs. However, official health reports indicated that there was an increase in some diseases during dry seasons, a period when clean water is in shortage in the study region. Diarrheas, amoebic dysenteries, skin and itching in young people are water-borne diseases; and the fact that they increase during the period of clean water shortage indicate that should water decrease, the health conditions of the populations of this region would be affected directly. Diarrheas and amoebic dysenteries and other forms of diseases enumerated above have been reported to be directly linked to quality of water under other circumstances (Fewtrell et al. 2005; Gundry et al., 2004). Hence, the fact both structured interviews and focus groups with local communities were unable to document feeling of worsening human health conditions in the region may, most probably, be due to the sense of plenty of water and the fact that people had many other concerns that were priority to their livelihood than water-borne diseases. Of particular note here is that women specific genital ailments and infections increased by 23.6% during the dry seasons. Though there is no specific name on what these are, it would be sensible to think that decreasing water levels in the region would affect women's health conditions in a specific way. Aveling et al. (2003) and Inogwabini et al. (2009) documented the fact that people in the region of the study had no proper sewage treatment mechanisms and people have been used to dumping garbage into open water, which in turn they fetched for domestic needs. Decreasing water levels will reduce the potential that larger water masses have to dilute effects of dumped sewage and will, therefore, lead to unclean water, which will lead to unhygienic conditions.

It has been shown in the arid African regions (Manyanhaire et al., 2009), decreased water levels will reduce the spread of water-dependent human activities in a limited surface; limited spread zone for human activities alone will likely worsen health conditions. Major human settlements in areas of limited water supplies are always accompanied by worsening health conditions stemming from varied causes such as poor sewage and water treatment infrastructures, people sharing water sources with herds of cattle and other domestic animals. As indicated above, characteristics of this region are that there is no proper sewage treatment mechanisms, people dump garbage into open water, and defecate directly in open water (MSPPE, 2010). Decreasing water levels will reduce the potential that larger water masses have to dilute effects of dumped sewage and will, therefore, lead to unclean water. This will affect, in unforeseen degrees, both human populations and livestock who will be sharing the same water points.

It is worth repeating that in Central Africa, as most cultures, women and girls are primarily responsible for the use and management of water resources, sanitation and health at the household level (Seager et al., 2009; UN-DESA, 2005;); which facts will certainly expose them to gender-based violence. Water-driven conflicts would very likely results in similar impacts, including rape. An important result of this study is that women in this region already torn-apart by armed conflicts of different nature in the recent past have an acute perception of the risks of being raped while going for search of water.

Managing water at times of scarcity will bring women to the forefront of water-ignited conflicts. The recent history, with the Enyele Rebellion Movement on water pounds in the north-western part of the DRC provided proof to this situation (Guerra et al., 2010); decades-old suppressed conflict over fishing rights evolved into ethnic tussle for economic and political power in north-west. Some 200,000 refugees fled the fish-conflict violence since in 2009 (BBC, 2009) of whom 82% were women (IRIN, 2011). Unfortunately, the opportunity provided by the recent Conference of Parties to the Convention United Nations Framework Convention on Climate Change in Paris was not seized to discuss the implications of climate change on freshwater and its social consequences. The word water appears nowhere throughout the text of the Paris Agreement, the importance of water being inferred to only once in the preamble through the use of the word ocean, as a specific ecosystem. Without diminishing the value of other international mechanisms such as the International Water Management Institute and other conferences on the subject, this absence of water in the agreement shows how low the water issues are on the agenda of the global decision-making processes and, hence, lack of global perspectives on how to handle potential conflicts over freshwater sources. Of course, it could easily be argued that water is inferred through article 1 (1) of the Paris Agreement, which refers to the 1992 New York Convention which, in its article 4(e), calls parties to cooperate in preparing for adaptation to the impacts of climate change develop and elaborate appropriate and integrated plans for coastal zone management, water resources and agriculture, and for the protection and rehabilitation of areas, particularly in Africa, affected by drought and desertification, as well as floods. However, this inference is no so obvious and would demand a more refined practitioner of natural resources management to dig it out and make connections. It is suggested here that the best way to account for water and consequences of lack of its proper management is to factor freshwater in the development of the nationally determined contributions (Paris Agreement, article 3). But that will need a much more developed advocacy on behalf of the practitioners of natural resources management toward political decision-makers.

ACKNOWLEDGMENT

I would like to thank Laurent Nsenga for allowing me to use data from Luki. Data from Kinshasa were gathered by students from Institut Bâtiments et Travaux Publics through the supervision of Alpha Egbango. Dr Mwanza Ndunda who co-authored the paper on Mabali shared the data from Mabali scientific Reserve.

REFERENCES

Aiken, J., Moore, G. F., & Holligan, P. M. (1992). Remote sensing of oceanic biology in relation to global climate change. *Journal of Phycology*, *28*(5), 579–590. doi:10.1111/j.0022-3646.1992.00579.x

Atsimadja, F. A. (1992). The changing geography of Central Africa. In G. P. Chapman & K. M. Baker (Eds.), *The changing geography of Africa and Middle East* (pp. 52–79). London: Routledge.

Aveling, C., Bofaya, B., Hall, J.S., Hart, J.A., Hart, T.B., Inogwabini, B.I., Plumptre, A. & Wilkie, D. (2003). *Democratic Republic of Congo – Environmental Analysis*. Wildlife Conservation Society for USAID Kinshasa.

Bailey, R. G. (1986). The Zaïre River system. In Davies & Walker (Eds.), The Ecology of River Systems. Junk Publishers. doi:10.1007/978-94-017-3290-1_6

BBC (British Broadcasting Company). (2013). *Democratic Republic of Congo Profile*. Available at http://www.bbc.co.uk/news/world-africa-13283212

Béné, C. (2005). The good, the bad and the ugly: Discourse, policy controversies, and the role of science in the politics of shrimp farming development. *Development Policy Review*, *23*(5), 585–614. doi:10.1111/j.1467-7679.2005.00304.x

Blackden, C. M., & Wodon, Q. (Eds.). (2006). *Gender, Time Use, and Poverty in Sub-Saharan Africa*. World Bank Working Paper 73.

Bultot, F., & Griffiths, J. P. (1972). The equatorial wet zone. In Griffiths (Ed.), Climates of Africa. Elsevier Publishing Company.

Chafetz, J. S. (2006). The varieties of gender theory in sociology. In J. S. Chafetz (Ed.), *Handbook of the sociology of gender* (pp. 3–23). New York: Springer. doi:10.1007/0-387-36218-5_1

Chafetz, J. S., & Haggan, J. (1996). The gender division of labor and family change in industrial societies: A theoretical accounting. *Journal of Comparative Family Studies*, *27*, 187–217.

Chapman, L. J. (2001). Fishes of African Rain Forests – diverse adaptations to environmental challenges. In Weber, White, Vedder & Naughton-Treves (Eds.), African Rain Forest, Ecology and Conservation: An interdisciplinary perspective. Yale University Press.

Colom, A., Bakanza, A., Mundeka, J., Hamza, T., & Ntumbandzondo, B. (2006). The socio-economic dimensions of the management of biological resources, in the Lac Télé – Lac TumbaLandscape, DRC Segment: A segment-wide baseline Socio-Economic study's Report. Submitted to the World Wide Fund for Nature DRC, Kinshasa Office, Democratic Republic of Congo.

Colombant, N. (2005). *Shrinking Lake Chad sparks tension, ingenuity*. Available at http://www.voanews.com /english/archive/2005-03

Conference on the Central African Moist Forest Ecosystems (CEFDHAC). (2001). *Regional Strategic Action Plan and biodiversity resources on the Congo basin ecosystems*. United Nations Development Program (UNDP), International Union for the Conservation of Nature (IUCN) and the Global Environment Facility (GEF).

Corsi, F. (1984a). *Développement des pêches dans le fleuve Zaïre. Rapport préparé pour le projet d'Assistance à l'Office National des Pêches et Amélioration de la Pêche Artisanale.* Rome: FAO.

Corsi, F. (1984b). *Zaïre: Développement et aménagement des pêches du lac Mai-Ndombe et ses affluents.* Technical report submitted to Food Aid Organization.

De Saint Moulin, L. (1991). Zaïre: Aperçu démographique – Résultats du recensement scientifique de la population en 1984. *Zaire-Afrique, 255*, 227–261.

De Saint Moulin, L. (2003). La perception de la démocratie en République Démocratique du Congo. *Congo-Afrique, 375*, 292–307.

Edwards, A., & White, L. J. T. (2000). Methods for recording the weather. In L. White & A. Edwards (Eds.), *Conservation research in the African rain forests: a technical handbook* (pp. 81–88). New York: Wildlife Conservation Society.

Eltahir, E., & Bras, R. (1993). On the response of the tropical atmosphere to large-scale deforestation. *Quarterly Journal of the Royal Meteorological Society, 119*, 779–783.

Evrard. C. (1968). *Recherches écologiques sur le peuplement forestier des sols hydromorphes de la Cuvette Centrale congolaise.* Série scientifique N°110/Office National de la recherché Scientifique et du Développement – Ministère Belge de l'Education Nationale et de la Culture.

Fernández, M. H., & Vrba, E. S. (2006). Plio-Pleistocene climatic change in the Turkana Basin (East Africa): Evidence from large mammal faunas. *Journal of Human Evolution, 50*(6), 595–626. doi:10.1016/j.jhevol.2005.11.004 PMID:16630645

Fewtrell, L., Kaufmann, R. B., Kay, D., Enanoria, W., Haller, L., & Colford, J. M. Jr. (2005). Water, sanitation, and hygiene interventions to reduce diarrhoea in less developed countries: A systematic review and meta-analysis. *The Lancet Infectious Diseases, 5*(1), 42–52. doi:10.1016/S1473-3099(04)01253-8 PMID:15620560

Fletcher, S. R. (1997). *Global Climate Change Treaty: Negotiations and Related Issues. A CRS Report to Congress.* National Council for Science and the Environment & National Library for the Environment. Available at http://www.ncseonline.org/NLE/CRSreports/Climate

Fletcher, S. R. (2000). *Global Climate Change Treaty: The Kyoto Protocol. A CRS Report to Congress.* National Council for Science and the Environment & National Library for the Environment. Available at http://www.ncseonline.org/NLE/CRSreports/Climate

Gash, J. H. C., & Nobre, C. (Eds.). (1996). *Tropical Deforestation and Climate.* New York: John Wiley.

Guerra, R., Nakwafio, F., Kazadi, S., & Giuliano, M. (2010). *Pooled Fund 2010 Annual Report: Annex 10 - Pooled Fund First special allocation 2010.* Available at http://www.rdc-humanitaire.net/rdchbis/IMG/pdf/annex_10.pdf

Gundry, S., Wright, J., & Conroy, R. (2004). A systematic review of the health outcomes related to household water quality in developing countries. *Journal of Water and Health, 2*, 1–13. PMID:15384725

Halpin, P. N. (1998). Global climate change and natural-area protection: Management responses and research direction. *Ecological Applications*, *7*(3), 828–843. doi:10.1890/1051-0761(1997)007[0828:GCCANA]2.0.CO;2

Hanjra, M. A., & Qureshi, M. E. (2010). Global Water Crisis and Future Food Security in an era of Climate Change. *Food Policy*, *35*(5), 365–377. doi:10.1016/j.foodpol.2010.05.006

Hobbs, M., & Knausenberger, W. (2003). *Burundi Environmental threats and opportunities assessment (ETOA), with emphasis on tropical forestry and biodiversity conservation*. Type-scripted Supplementary report to Integrated Strategic Plan for 2003-2005. USAID.

Houghton, J. T., Jenkins, G. J., & Ephraums, J. J. (1990). *Climate change: the IPPC scientific assessment*. Cambridge, UK: Cambridge University Press.

Huggett, R. J. (1993). *Modelling the human impact on nature – Systems analysis of environmental problems*. Oxford University Press.

Hughes, P. (1986). Ocean color: A key to climate change. *Weatherwise*, *39*(5), 267–270. doi:10.1080/00431672.1986.9929300

Hulme, M., Dohery, R. M., Ngara, T., New, M. G., & Lister, D. (2001). African Climate Change: 1900 – 2100. *Climate Research*, *17*, 145–168. doi:10.3354/cr017145

Ilon, L.Ilon. (1992). Fitting girls schooling into existing economic paradigms: Confronting the complexities. *International Journal of Educational Development*, *12*(2), 147–159. doi:10.1016/0738-0593(92)90036-L

Iloweka, E. M. (2004). The deforestation of rural areas in the Lower Congo Province. *Environmental Monitoring and Assessment*, *99*(1-3), 245–250. doi:10.1007/s10661-004-4028-0 PMID:15641386

Inogwabini, B., & Lingopa, Z. (2006). Les Inventaires des Poissons dans le Lac Tumba, Congo et Ngiri – Paysage Lac Te'le´ – Lac Tumba, Segment RDC. Rapport Produit dans le Cadre du Programme CARPE. Type scripted report to the WWF DRC, Kinshasa (Democratic Republic of Congo).

Inogwabini, B. I., Dianda, M., & Lingopa, Z. (2009). The use breeding sites of *Tilapia congica* (Thys & van Audenaerde 1960) to delineate conservation sites in the Lake Tumba, Democratic Republic of Congo: Toward the conservation of the lake ecosystem. *African Journal of Ecology*, *48*, 800–806.

Inogwabini, B. I., Sandokan, B. M., & Ndunda, M. (2006). A dramatic decline in rainfall regime in the Congo basin: Evidence from a thirty-four data set from the Mabali Scientific Research Center, Democratic Republic of Congo. *International Journal of Climatology*, *31*, 278–286.

Inter-Agency Commission (UNDP, UNESCO, UNICEF, World Bank). (1990). *Final Report of the World Conference on Education for All: Meeting Basic Learning Needs*. United Nations.

IPCC (Intergovernmental Panel on Climate Change). (2007). *Climate Change 2007: Impacts, adaptation and vulnerability. Contribution to of Working Group II to the fourth Assessment Report of the Intergovernmental Panel on Climate Change*. Cambridge University Press.

IRD (Institut de Recherche pour le Développement). (2002). *Congo River Basin: geology and soil types influence drought. European Water Management News*. Available from: http://www.ewaonline.de

IRIN (Integrated Regional Information Networks). (2011). *DR-CONGO: Return delayed for 120,000 refugees*. Retrieved from http://www.irinnews.org/report/92712/drc-congo-return-delayed-for-120-000-refugees

Kinyanjui, K. (1993). Enhancing women's participation in the sciences bases curriculum: The case of Kenya. In J. K. Conway & S. C. Bourque (Eds.), *The politics of women's education. Perspective from Asia, Africa and Latin America*. Ann Arbor, MI: University of Michigan press.

Komba, D. (1995). Declining enrolment and quality of primary education in Mainland Tanzania. In C. Creighton & C. K. Omari (Eds.), *Gender, Family and Household in Tanzania*. Avebury.

Lemoine, N., Bauer, H.-G., Peintinger, M., & Böhning-Gaese, K. (2007). Effects of Climate and Land-Use Change on Species Abundance in a Central European Bird Community. *Conservation Biology, 21*(2), 495–503. doi:10.1111/j.1523-1739.2006.00633.x PMID:17391199

Lemoine, N., & Böning-Gaese, K. (2003). Potential impact of global climate change on species richness of long-distance migrants. *Conservation Biology, 17*(2), 577–586. doi:10.1046/j.1523-1739.2003.01389.x

Lemoine, N., Schaefer, H.-C., & Böhning-Gaese, K. (2007b). Species richness of migratory birds is influenced by global climate change. *Global Ecology and Biogeography, 16*(1), 55–64. doi:10.1111/j.1466-8238.2006.00252.x

Leonard, C. (1987). République du Zaïre - Publié par le Ministère de l'Economie et de l'Industrie. Imprimerie Pirotte, Liège. *Conjoncture Economique, 26*.

Leonard, J. (1951). Contribution a l'étude de la végétation des bains d'éléphants au Congo Belge. Le Rhynchosporeto-Cyperetum longibracteato. *Publication de la Société Royale de Botanique Belge., 84*, 13–27.

Leonard, J. (1952). Cynometreae et Amherstieae. In Flore du Congo Belge et du Ruanda-Burundi. INEAC.

Lettau, H., Lettau, K., & Molion, L. (1979). Amazonias hydrologic cycle and the role of atmospheric recycling in assessing deforestation effects. *Monthly Weather Review, 107*(3), 227–238. doi:10.1175/1520-0493(1979)107<0227:AHCATR>2.0.CO;2

Lovett, J. C. (2006). Climate and Society. *African Journal of Ecology, 44*(4), 421–422. doi:10.1111/j.1365-2028.2006.00708.x

Mahe, G., LHote, Y., Olivry, J. C., & Wolting, G. (2001). Trends and discontinuities in regional rainfall of West and Central Africa: 19511989. *Hydrological Sciences-Journal-des Sciences Hydrologiques, 46*(2), 221–226. doi:10.1080/02626660109492817

Manyanhaire, I. O., Matewa, J. T., Svotwa, E., & Munhuwa, E. (2009). Water shortage Related problems following the closure of Mhangura copper mine in Mashonaland West Province of Zimbabwe. *Journal of Sustainable Development in Africa, 11*, 239–252.

MSPPE (Ministère de la Santé Publique de la Province de l'Equateur). (2010). *Plan Provincial de la santé de l'Equateur*. CEMUBAC.

Pearce, D. W., & Turner, R. K. (1990). *Economics of Natural resources and the environment*. New York: Harvester Wheatsheaf.

Pimm, S. L. (2007). Africa: Still the Dark Continent. *Conservation Biology*, *21*(3), 567–569. doi:10.1111/j.1523-1739.2007.00697.x PMID:17531033

Roberts, T. R. (1975). Geographic distribution of African freshwater fishes. *Zoological Journal of the Linnean Society*, *57*(4), 249–319. doi:10.1111/j.1096-3642.1975.tb01893.x

Salati, E. (1987). The forest and the hydrological cycle. In R. E. Dickinson (Ed.), *The Geophysiology of Amazonia* (pp. 273–296). New York: Wiley.

Seager, J., Robinson, K., Van der Schaaf, C., & Gabizon, S. (2009). *Gender-Disaggregated Data on Water and Sanitation. Department for Economic and Social Affairs (UN-DESA) and UN-Water Decade Programme on Capacity Development*. New York: UNW-DPC.

Sonwa, D., Bele, Y., Somorin, O., Jum, C., & Nkem, J. (2009). Adaptation for forests and communities in the Congo Basin. *European Tropical Forest Research Network News*, *50*, 93–100.

SS (Secrétariat à la Santé). (2007). *République Démocratique du Congo: Annuaire des données sanitaires 2006*. Ministère de la Santé Publique de la République Démocratique du Congo, Kinshasa.

Stern, N. (2007). *Economics of climate change. The Stern Review*. Cambridge University Press. doi:10.1017/CBO9780511817434

Thieme, M. L., Abell, R., Stiassny, M. L. J., Skelton, P., Lehner, B., Teugels, G. G., & Olson, D. et al. (Eds.). (2005). *Freshwater ecoregions of Africa and Madagascar: A conservation assessment*. Island Press.

Tutin, C. E. G., & White, L. J. T. (1998). Primates, phenology and frugivory: Present, past and future patterns in the Lopé Reserve, Gabon. In D. M. Newbery, H. H. T. Prins, & N. Brown (Eds.), *Dynamics of populations and Communities in the Tropics* (pp. 309–338). Oxford, UK: Blackwell Science.

UN-DESA (United Nations Department of Economic and Social Affairs). (2005). *A Gender Perspective on Water Resources and Sanitation*. Background paper NO 2. Commission on Sustainable Development Twelfth Session DESA/DSD/2005/2United Nations, New York.

Walther, G.-R., Hughes, L., Vitousek, P., & Stenseth, N. C. (2005). Consensus on climate change. *Trends in Ecology & Evolution*, *20*(12), 648–649. doi:10.1016/j.tree.2005.10.008 PMID:16701450

Water Aid. (2012). *Women's issues*. Retrieved from http://www.wateraid.org/documents/plugindocuments/women_and_wateraid_issue_sheet.pdf

White, F. (1983). The vegetation of Africa, a descriptive memoir to accompany the UNESCO/AETFAT/UNSO vegetation map of Africa. UNESCO. *Natural Resources Research*, *20*, 1–356.

YPOW (Young People of the World). (2006). *Water rights and wrongs - A Young People's Summary of the United Nations Human development Report 2006. Beyond scarcity: power, poverty and the global water crisis*. Available at http://www.gm.undp.org/Reports/report_youth_hdr.pdf

Chapter 13
Tipaimukh Multipurpose Hydroelectric Project:
A Policy Perspective - Indo-Bangla Priorities, Indigenous Peoples' Rights, and Environmental Concerns

Ali Reja Osmani
Karimganj Law College, India

ABSTRACT

Moving ahead from the freshwater reservoir versus climate change debate, the Indo-Bangla controversy over the Tipaimukh Project exists over the right of riparian states. India needs more energy to propel its economic growth, whereas Bangladesh is worried about downstream impact. The concerns of Bangladesh are based on the experience of severe water shortage and other impacts of Farakka Barrage and Teesta Barrage and also Himalayan Component of the Interlinking of River Project. Over the years some progress was made at bilateral level. But the major problem remains unaddressed i.e. without reconciling the issues of indigenous people a big dam cannot be constructed. This paper highlighted the existing scenario of Bangladesh and the indigenous people of Manipur in India in one hand and ecological, socio-economic concern in other hand i.e. obligation not to cause significant harm. There is no straight way answers available to be choose between a 'Yes' or 'No'; neither depends on the issues of 'might' over 'right' or 'development' over 'destruction', but on the circumstances to come.

INTRODUCTION

... It is not the technical arguments that makes Tipaimukh a threat but the political process that followed it. In fact the dam is tri-partite affair involving the Indian state, the Bangladeshi people and the people and the Manipuris of the North East India. In this case, the Indian government has not received consent from the two protesting people affected by the dam. Over time as protests and resentment have built up, many feel that the dam is really part of a bigger plan of a circle of dams of which Tipaimukh is the first. (Choudhury, 2012)

DOI: 10.4018/978-1-5225-1046-8.ch013

Unlike every energy and water resources management plans, hydropower development projects come with both positive and negative impacts. Issues regarding climate resilience and climate change adaptation have become increasingly important for large infrastructure investments. Every hydropower plant is uniquely designed to fit to the site specific characteristics of a specific geographical location and the surrounding society, environment, ecology, the magnitude of environmental and social impacts as well as the extent of their positive and negative effects is highly site dependent.

India is one of the fastest growing economies in the world and home to almost one third of the world's poor. A country's economic development and energy requirement goes hand in hand. Growth of the economy is very vital for the overall growth of its citizen and poverty elevation and India is committed to its citizens for the same, the idea which is also enshrined in the Millennium Development Goal (MDG) No.1 i.e. Eradicate Extreme Poverty & Hunger (United Nation Development Programme [UNDP], n.d.a). With the concern and commitment for Global Warming and Climate Change under various International Treaties and Convention to which India is a party, opting for less Greenhouse Gas (GHG) emitting technologies to meet its domestic energy needs while fulfilling its international obligations. Hydropower, a well-established and socially desirable renewable energy technology is a sound alternative to meet those challenges. Hydropower is a proven, mature, predictable and typically price-competitive technology. Above all it has among the best conversion efficiencies of all known energy sources: 90 percent efficiency as opposed to up to 50 percent efficiency of e.g. fossil fuel burning (Mendonça et al., 2012).

Whereas Bangladesh's concern is that the Tipaimukh Multipurpose Hydroelectric Project (hereinafter Tipaimukh Project) will have substantial downstream effect. Bangladesh is already suffering due to the Farakka Barrage in western part and due to the Teesta barrage in the northern part of the country, now the proposed Tipaimukh Project will create threat to livelihood in the eastern part as well. The loss of water flow will cause serious ecological problem and is a sustainable development issue which also finds place in the Millennium Development Goal No.7 i.e. Ensure Environmental Sustainability (UNDP, n.d.a).

Ever since the term 'Sustainable Development' has been coined by Brundtland Commission (World Commission on Environment and Development) in its report 'Our Common Future' in 1987, it has been on the forefront of the global agenda. With the recent adoption of the post-2015 Sustainable Development Goals (2030 Agenda for Sustainable Development) at the UN General Assembly in September 2015 to carry further the targets of Millennium Development Goals, which was adopted in 2000 by world leaders. Numerous targets set by Sustainable Development Goals are very much relevant from the perspective of this chapter, especially Goal-6: Ensure access to water and sanitation for all; Goal-7: Ensure access to affordable, reliable, sustainable and modern energy for all; Goal-13: Take urgent action to combat climate change and its impact; Goal-15: Sustainably manage forests, combat desertification, halt and reserve land degradation, halt biodiversity loss; and Goal-16: Promote just, peaceful and inclusive societies (UNDP, n.d.b).

The countries have different priorities for their own country. But in the process of development, countries have to find out that whose priorities are more sustainable, because development without sustainability has no meaning in reality. As the proposed project has led to controversy between India and Bangladesh over water rights as well as controversy within the indigenous people of Manipur, who are to be displaced by the reservoir. Because displacement due to development projects also often restricts indigenous people from availing the benefits of nature's services or ecosystem services as well. Following the signing of the Memorandum of Understanding among Manipur Govt., National Hydroelectric Power Corporation and Satluj Jal Vidyut Nigam Ltd. on April 28th 2010 and signing of the subsequent

Promoter's Agreement with the purpose of setting up a Joint Venture Company on October 22nd 2011 steep opposition started (Mehta, 2012).

The primary objective of this paper is to understand the prevailing situation of the Tipaimukh Project, its genesis, recent developments and initiatives taken by the riparian states and Indo-Bangladesh bilateral relation over the Tipaimukh Project. The paper is important because, since the inception of the Project a deadlock has been created and almost after laps of three decades, both India and Bangladesh came closer to resolve the issue. However, India has to address the concerns of indigenous people, concerns for biodiversity loss, concerns for submergence and displacement and concerns for downstream consequences before finally going ahead with the Tipaimukh Project.

The study gains even more significance after the recently concluded 21st Conference of Parties (COP-21) to the United Nation Framework Convention on Climate Change (UNFCCC) in Paris. The Conference reached a global agreement (Paris Agreement) on action on climate change for post-2020. In the Intended Nationally Determined Contribution (INDC), India has reaffirmed its intention to develop large hydro projects to supplement its energy mix as a source of clean and renewable energy, along with special programmes to promote small and mini hydel projects, new and efficient designs of water mills for electrification of remote villages. With a vast potential of more than 100 GW, a number of policy initiatives and actions are being undertaken to aggressively pursue development of country's vast hydro potential (UNFCCC, 2015). India has a strong policy and highly ambitious programme framework to propel its growth in clean and renewable energy sector, which will certainly help fulfilling India's commitments (Osmani, 2014).

In this paper after the introductory discussion in Section-1, the Section-2 contextualise the origin and evolution of the proposed project, salient features of the project, Joint River Commission meetings and its present status. Whereas the Section-3 will highlight the taxonomy of international legal framework and legal obligation of India towards Bangladesh at the international level and also towards its own people i.e. the indigenous people of Manipur. This section will also try to highlight the concerns of Manipur and Bangladesh, behind opposing the Tipaimukh Project. The Section-4 shall discuss concerns for GHG emission from freshwater reservoir and other environmental aspect of the project, including environmental clearance and Forest Advisory Committee (FAC) meeting's outcome regarding forest clearance, both in Manipur and Mizoram. Finally the Section-5 contains the concluding remarks and is followed by few recommendations.

The methodology adopted is mostly employs policy analysis of the Tipaimukh Project. Primary sources are reviewed to identify the areas of analysis and secondary sources are consulted to construe author's own opinion. The study based on various sources, most of the sources are Government of India sources and the remaining sources are books, journals, news articles etc. In terms of data, those are used in this study; sources belong to Joint Rivers Commission, Ministry of Water Resources, Bangladesh and Ministry of Environment, Forest and Climate Change (MoEFCC), Government of India reports.

EVOLUTION AND CURRENT POSITION OF TIPAIMUKH HEP

The need for a project of such a nature arose due to the fact that the entire Barak Valley had always been hit by recurrent floods causing ravages, human sufferings and miseries. Controlling floods by way of a suitable storage reservoir was in the agenda as early as 1929-30, due to the occurrence of repeated floods in the Valley. Since 1954 the Central Water Commission (CWC)[1] conducted detailed investigations for multipurpose reservoir to find out a long term solution to the problem. For selecting suitable site for integrated development, comprehending hydropower generation, irrigation, navigation etc., the CWC investigated three sites for storage reservoir, which were abandoned subsequently due to poor foundation condition, heavy submergence etc.

The alternative sites and the reasons for their abandonment are as such (MoEFCC, 2006):

1. First site selected at Mainadhar for an earthen dam, but abandoned in 1964 due to unfavourable geological conditions.
2. Second site selected at Naraindhar, but found not suitable due to submergence of large area of cultivable land in Manipur and partial flood control in downstream.
3. Third site selected at Bhuvandhar, but rejected due to soft friable and clayey foundation and submergence of large area of cultivable land and villages in Manipur.

However the investigation continued and in 1974 the CWC identified Tipaimukh as the most favourable site, which fulfils all the techno-socio-economic aspect of such project (Deka, 2012). Initially the project was assigned to the Brahmaputra Board[2], which, subsequently handed over the project to North Eastern Electric Power Corporation (NEEPCO)[3], in 1991 and finally to the National Hydro Power Corporation (NHPC)[4], after a decade (Deka, 2012).

The construction of dam was proposed on the trans-boundary River Barak, flows between India and Bangladesh. The river originates in Manipur, flows across Assam (within India) before entering into Bangladesh through Amalshid in Sylhet. At Amalshid the Barak get divided into two branches namely, the Surma River and the Kushiara River. Both the river again merged together to form Kalni river in Habibganj district (in Bangladesh). The Kalni River further merged with Ghorautra River near Bajidpur in Kishorganj district to become Meghna River. The Meghna River finally joins the Padma Rivern near Chandpur district before falling into the Bay of Bengal ("India and Bangladesh", 2012).

The 1500 MW Tipaimukh Project site is located in Tipaimukh village near Manipur-Mizoram border, 500 meters downstream from the confluence of the Barak and Tuivai River in Churachandpur district of Manipur. The dam will be 390 meters long and 162.8 meters high rock-filled dam with central impervious core, with 6 units of 250 MW each surface powerhouse with Vertical Shaft Francis Turbine (NHPC, n.d.a). With an estimated cost[5] of Rs.8138.79 Crores/ USD1291.71 million (Sept. 08 Price Level) expected to be completed in 7 year and 3 months from the date of sanction by Cabinet Committee on Economic Affairs.

Pursuant to Article 4 of the 'Indo-Bangladeshi Treaty of Friendship, Cooperation and Peace' of 19th March 1972 an Joint Rivers Commission (JRC) has been established to work towards joint studies and joint actions in the field of flood control, river basin development and development of hydro-electric power and irrigation.

India is often burdened with allegations of not disclosing rather communicating appropriate information regarding the Tipaimukh Project. Whereas, both the countries discussed regarding the barrage on the very first meeting of JRC held in New Delhi on 25-26 June 1972 (Bisht, 2012). The Water Resource Minister, Govt. of India revelled that it was Bangladesh that proposed the Tipaimukh Project in 1974 (Zaman, 2012). The minutes of the 6th JRC meeting, held in November 1973 reads as such, "The Commission considered the study of Barak Dam Project in the context of floods in India and Bangladesh and recommended that this project be reviewed by the Superintending Engineers of Bangladesh and India, for flood moderation and protection to Sylhet, Cachar and adjoining areas…". To this the Bangladesh

High Commissioner to India said that the rhetoric was same during 1972-82, but the situation changed in 1982 when it was found the proposed project would be harmful for Bangladesh (Zaman, 2012).

However with security problems in the North Eastern Region, inter-state differences and internal controversies over submergence and displacement in Manipur, the progress was stalled and India's interest shifted to Ganges and Teesta (Verghese, 2009).

Almost after three decades at the JRC in 2005 held in Bangladesh, the Tipaimukh Project became a topic of discussion again, when India proposed the construction of the Tipaimukh Dam at the JRC meeting. Over the decades many things have changed. Global development against dams highlighted the social and environmental concerns. Report released by the 'World Commission on Dams' in 2000 also emphasised upon people's participation in decisions relating to construction of dam. The issues like ecological, hydrological and social costs of large dams gained attention and being increasingly highlighted by various institutes, academia, affected peoples' groups, civil society at large……, Bangladesh expressed categorical concerns on the adverse downstream impact. Since then the Tipimukh Project is surrounded by controversies over implementation-lag and consensus-deficit at various levels. Tipaimukh Project is not an exception to those developments taking place around it during this period (Bisht, 2012).

There has been lots of criticism by the civil society regarding the lack of transparent communication and sharing of knowledge between the countries. However, there are instance which undoubtedly clarify that both countries had shown great interest in solving the issue bilaterally, e.g. in July 2009 an all-party Parliamentary delegation visited India to survey the proposed site (High Commission of India [HCI], Dhaka, 2011). Following the Promoter's Agreement being signed between NHPC, Sutlej Jal Vidyut Nigam (SJVN)[6] and the Government of Manipur on 22nd October 2011, Foreign Affairs Adviser to the Prime Minister of Bangladesh visited India in 3rd December to discuss their concerns ("India Assures Bangladesh", 2011). Subsequently a 10 member group of journalist, lead by External Affairs Publicity Wing, Bangladesh came to visit the site. Despite the fact that they were unable to land on the site due to technical reasons the group was convinced that India is yet to make any construction on the proposed site, after their aerial visit ("Bangladeshi Scribes' Dam", 2012).

There were 13 JRC meeting held from 1972 to 1975, 8 JRC meeting held during 1978 to 1982, 11 JRC meeting held from 1982 to 1991, but there was no JRC meeting held during 1991 to 1996. In between 1996 to 2001 there was 3 JRC meeting held and from 2001 to 2006 only 2 JRC meeting held (Zaman, 2012). The last JRC, i.e. 37th JRC meeting was held in 18-19 March 2010 at New Delhi (HCI, Dhaka, 2010).

At the last JRC meeting held in 2010 the Government of India assured that the "Water Resource Ministry would propose a moratorium on all proposed dams as well as suspension of on-going work on some existing dams till a 'cumulative impact assessment', a 'comprehensive biodiversity impact assessment' and a 'downstream impact assessment' were conducted ("Manmohn to Chair Timaimukh Meeting", 2010). This has given a hope of joint study of the Project by the riparian nations.

In the first Indo-Bangla Joint Sub-Group meeting, India handed over a Five Volume Detailed Project Report (DPR) and after a detailed discussion also finalised the Terms of Reference with a view to facilitating conduct of the joint study of the possible impacts of the proposed project on the downstream areas of Bangladesh. India also assured Bangladesh to extend any additional and supplementary data and information required for the study (Bhattacharya, 2012). Pursuant to this the Government of Bangladesh has also selected two institutions namely the Institute of Water Modelling (IWM) for mathematical modelling in all relevant aspects of the proposed dam including hydrologic and hydrodynamic, morphology, salinity and probable hazards and the Centre for Environment and Geographical Information System

(CEGIS) for impact assessment on hydrology, morphology, agriculture, fisheries, navigation, livelihood, ecosystem etc. The study is expected to be complete in two years (HCI, Dhaka, 2013).

The first Joint Sub-Group meeting on Tipaimukh Project was held on 27-28 August 2012 at New Delhi (HCI, Dhaka, 2012). The second Joint Sub-Group meeting was held on 01-02 February 2013 at Dhaka (HCI, Dhaka, 2013). The Joint Sub-group meeting is being held pursuance of the decisions taken during Foreign office Consultations between both the countries held at New Delhi on 24th July 2012. Earlier in December 2011, Bangladesh Government also demanded a Joint Study on the proposed Tipaimukh Project, to assess the negative and positive impacts of the dam ("B'desh for Joint Study", 2012). The 38th JRC meeting was scheduled and postponed several times earlier, but in the last Foreign Ministry level Joint Consultative Commission Meeting, both the countries resolved to hold the 38th Meeting of the JRC in Dhaka at a mutually convenient date. The two sides also agreed upon the expeditious holding of the third meeting of the Joint Sub-Group meeting on Tipaimukh Project and further exchange of data of ongoing study (Ministry of External Affairs, 2014).

APPLICABLE LEGAL FRAMEWORK

India had unilaterally constructed the Farakka Dam and Barrage on the western border of Bangladesh and the Teesta Dam and Barrage on the northern border of Bangladesh (Alam, 2012). The Farakka Barrage has been taking heavy toll on Bangladesh for almost four decades and the countries entire north-western part is under constant threat of desertification (Askari, 2012).

In 1984 the CWC presented a report on Tipaimukh Project, which was turned down as it failed to have a proper Environment Impact Assessment (EIA) of the submerged areas. After two decades in 2004, India completed the EIA. Popular belief however suggests that India had failed to share the same with Bangladesh.

As per the Article 9 of the Ganges Water Treaty of 1996, unilateral use of Joint River is prohibited. Perhaps privileged India is not giving enough attention to comply with the principle of Equity, Fairness and No Harm. Although India has shared the DPR recently, almost a decade later than it conducted its own EIA, has vandalized all the customary International practices.

The Rio Declaration of 1992, on Environment and Development (UNCED, 1992) talks about the principle of prior notification (Principle 19) in the following words:

Table 1. Catchment area of Ganges, the Brahmaputra and the Meghna River

Rivers	Total Catchment Area (Sq.Km.)	Catchment Area (Sq. Km.)				
		India	Nepal	Bhutan	China	Bangladesh
Brahmaputra	552000	195000	-	47000	270900	39100
Ganges	1087300	860000	147480	-	33520	46300
Meghna	82000	47000	-	-	-	35000
	1721300	1102000	147480	47000	304420	120400
	(100%)	**(64.02%)**	**(8.57%)**	**(2.73%)**	**(17.69%)**	**(7%)**

Source: (Joint River Commission, n.d.a)

State shall provide prior and timely notification and relevant information to potentially affected States on activities that may have a significant adverse transboundary environmental effect and shall consult with those States at an early stage and in good faith.

The Article 12 of the Convention on The Law of the Non-Navigational Uses of International Watercourses, also known as UN Watercourse Convention (UNWC) says as follows (Rieu-Clark et al., 2012, p. 139):

Before a watercourse state implements or permits the implementation of planned measures which may have a significant adverse effect upon other watercourse state, it shall provide those states with timely notification thereof. Such notification shall be accompanied by available technical data and information, including the results of any environmental impact assessment, in order to enable the notified states to evaluate the possible effects of the planned measures.

The duty to provide notification under the convention arises at the event when the planning state have reasons to believe that the planned measures in question may have a "significant adverse effect" upon other states, the obligation is triggered and not when the state planning measures or issuing any permit regarding the planned measures, believes that those measures may result in "significant harm" (McCaffery, 2007, p. 473). The lower threshold provided by the term "significant adverse effect" purportedly avoids legitimising the presumption that the planned measures fall under Article 7 of the UNWC i.e. Obligation not to cause significant harm (Tanzi & Arcari, 2001, p. 201), which has been derived from the theory of Limited Territorial Sovereignty (Rahaman, 2009a).

Therefore Article 12 obliges the planning state to give notice of planned measure which may have a "significant adverse effect" upon the other riparian state 'before' the implementation of the work, most importantly in a timely manner. By the term 'timely' it is intended to require notification sufficiently early in the planning stage to permit meaningful consultation and negotiations under the provisions of UNWC, if any necessity arises (Tanzi & Arcari, 2001, p. 202).

The Articles 12 to Article 19 of the UNWC concern measures which may have a "significant adverse effect", the provisions in this regard refers to all possible effects of planned measures comprising both negative and positive effects (Tanzi & Arcari, 2001, p. 201).

The Article 3 of the Complementary Rules of International Law Association (ILA) applicable to the international resources (adopted at the 62nd conference held in Seoul in 1986) states that (International Law Association, 1986, p. 26):

When a basin State proposes to undertake, or to permits the undertaking of, a project that may substantially affect the interest of any co-basin State, it shall give such State or States notice of the project. The notice shall include information, data and specifications adequate for assessment of the effects of the project.

In the Pulp Mill case the court observed that the notification must take place before the state concerned decides on the environmental viability of the plan, taking due account of the EIA submitted to it........ The obligation to notify is there for an essential part of the process leading the parties to consult in order to access the risks of the plan and to negotiate possible changes which may eliminate those risks or minimize their effect...... The fact that the parties have complied with their substantive obligation does not mean that they are deemed to have ipso facto with their procedural obligation, or are excused from doing so (International Court of Justice, 2010).

Despite the fact that between India and Bangladesh there is no specific Agreement exists on River Barak, the general rule of prior notification has reached the status of a customary international law legal obligation being applicable regardless of the fact whether a special agreement between the state initiating and the potentially affected state exist or not (Rieu-Clark, 2012: p.135). However, it is noteworthy to mention here that, subject to compliance with the procedural obligations of the UNWC, states do not necessarily have a veto power over the development of an international water course (Vinogradov, Wouters & Jones, 2003: 19).

Therefore vast literature and available evidence, including state claims, the manner in which disputes between states have been resolved, state treaty practice, instrument adopted in intergovernmental forum, the work done by expert bodies, and the writings of commentators, suggests that the prior notification is required by customary international law (McCaffery, 2007, p. 473).

The foundation of the principles of international water law in relation to transboundary water resources management has been evolved from different theories and doctrines, which are as follows (Rahaman, 2009a):

1. **Theory of Absolute Territorial Sovereignty:** As per this theory every nation could utilise the water of an international watercourse flowing on its territory as it wishes, regardless of the consequences or needs of the riparian state and also without any obligation to consult. However, most of the modern experts do not agree to this doctrine.
2. **Theory of Absolute Territorial Integrity:** As per this theory downstream riparian of an international watercourse has the right to a full flow of water of natural quality and any interference with the natural flow by the upstream riparian requires the consent of the downstream riparian. Unlike the previous doctrine this doctrine has also limited acceptability among the experts.
3. **Theory of Limited Territorial Sovereignty:** As per this theory every state is free to use its shared rivers flowing on its territory as long as such use dose not affects the rights and interest of other riparian state. The riparian state sharing waters of common watercourse have reciprocal rights and duties in the utilization of its international watercourse and also entitle to an equitable share of its benefits and reasonable use of its watercourse. This theory is widely accepted by the experts and forms the basis of modern international water law.

There are certain principles which govern Transboundary Water Resource Management (Rahaman, 2009b), such as:

1. Principles of Equitable and Reasonable Utilization
2. Obligation not to cause Significant Harm
3. Principles of Notification, Consultation and Negotiation
4. Principles of Cooperation and Information Exchange
5. Peaceful Settlement of Disputes

The aforesaid customary and general principles of international law applicable to transboundary water resources management that are accepted globally and incorporated in modern international conventions, agreements and treaties. These principles form the basis of the 1966 Helsinki Rules on the Uses of the

Waters of International Rivers (Helsinki Rules, 1966) (International Law Association, 1967) and the 1997 UN Convention on Non- Navigational Uses of International Watercourses (UN Watercourses Convention, 1997) (United Nations, 1997), 2002 Revised Southern African Development Community Protocol on Shared Watercourse Systems (SADC Protocol, 2000) (Southern African Development Community, 2000), 2002 Sava River Basin Agreement (International Sava River Basin Commission, 2002), 1995 Mekong Agreement (Mekong River Commission, 1995), 1960 Indus Waters Treaty (Ministry of Water Resources, River Development & Ganga Rejuvenation, 1960) etc. For ready reference of the readers major provisions relating to transboundary water management principles could be considered at Table 4 at APPENDIX.

Indo-Bangladesh Framework Agreement on Cooperation for Development

The Prime Ministers of India and Bangladesh signed Framework Agreement on Cooperation for Development at Dhaka on 6th September 2011 (Ministry of External Affairs, 2011). Both India and Bangladesh expressed their desire in promoting trans-border cooperation in the management of shared water resources, hydropower potentials and eco-systems and accordingly agreed as per Article 2 "To enhance cooperation in sharing of the waters of common rivers, both Parties will explore the possibilities of common basin management of common rivers for mutual benefit. The Parties will cooperate in flood forecasting and control. They will cooperate and provide necessary assistance to each other to enhance navigability and accessibility of river routes and ports." and Article 7 "To harness the advantages of sub-regional cooperation in the power sector, water resources management, physical connectivity, environment and sustainable development for mutual advantage, including jointly developing and financing projects." The Framework Agreement seems to be a fruitful in the years to come in resolving the differences between the riparian nations.

Position in Bangladesh

Bangladesh is criss-crossed by nearly 405 rivers out of which 57 are of transboundary nature, 54 Indo-Bangla and 3 Bangla-Mayanmar shared watercourses (Joint Rivers Commission, n.d.b). Each of the 54 international water courses that exist between India and Bangladesh are dealt individually and not on the basis of a uniform international water sharing mechanism. Regarding Teesta River an ad-hoc water sharing agreement was made between India and Bangladesh in 1983, where by the countries are allocated 39 percent and 36 percent of water flow respectively. The only comprehensive river pact signed by India and Bangladesh is the Ganges Water Sharing Treaty signed on 12th December 1996, establishes a 30 year water sharing agreement between the riparian nations (Ray, 2012).

The concerns of Bangladesh are based on the experience of severe water shortage and other impacts of Farakka Barrage and Teesta Barrage. The projects are under criticism and often face strong opposition from environmentalists, social activists and scholars due their non-observance the norms and state practices for water resources management of an international watercourse. There are concerns regarding staggering environmental degradation, economic crisis (livelihood issues) and hydrological drought.

The Farakka Barrage was built by India in 1974 nearly 10 km away from Bangladesh border, to control the flow of the Ganges and diverting some of the water to Hooghly River through feeder canal to increase the water depth at Kolkata port, which was threatened by siltation. In November 1977 both countries proposed a five year agreement on water sharing, which was expired by 1982 and since then

Figure 1. Indo-Bangla Transboundary River Map
Source: (Joint Rivers Commission. n.d.b)

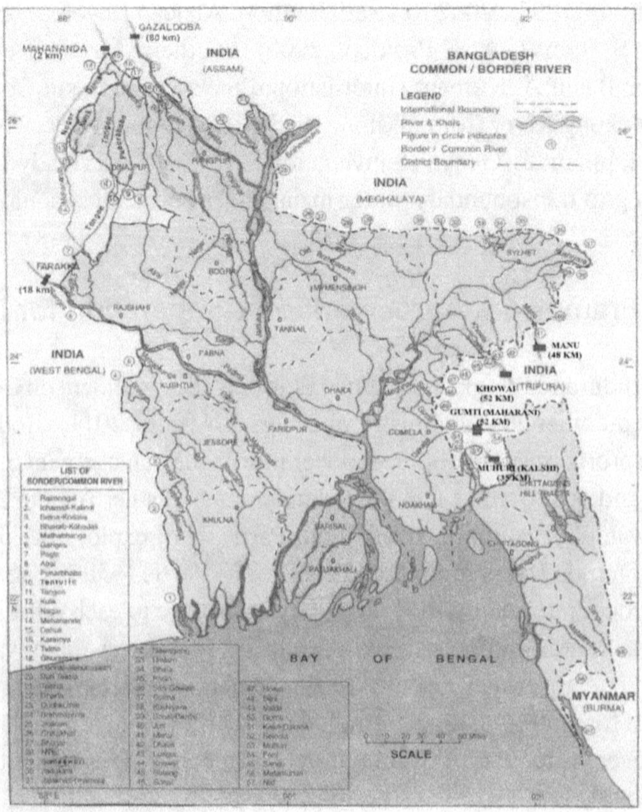

the issues remained the same until signing of the Ganges Treaty in 1996 (Dutta, 2010, p. 4). The trans-boundary river Teesta is the fourth major river in Bangladesh after Ganges, Brahmaputra and Meghna and major source of water for agriculture in North Bengal (India) and North-West Bangladesh. Bangladesh constructed a barrage on the Teesta River in 1990 to provide irrigation water for crop etc. (Islam, et al., n.d.). India also subsequently constructed upstream dams on the Teesta river Basin, such as Rangti-III Dam and Teesta-V Dam commissioned in 2000 and 2008 respectively (NHPC, n.d.b). The upstream barrages in India reduced the free-flow of water during the lean season and there are plans for construction of many more such dams in the basin. However, for the interest of the present paper, I shall be limiting any detailed discussion on Farakka Barrage and Teesta Barrage.

Damming the river Barak may have serious impact over free flowing Surma and Kushyara rivers and could disrupt agriculture, irrigation, drinking water supply, navigation etc. It could also reduce recharge of ground water during the lean season, affecting all dug wells and shallow tube wells. Bangladesh receives nearly 7 percent to 8 percent of its waters from river Barak through mazy tributaries and distributaries of Surma and Kushyara, which are supporting agriculture, irrigation, navigation, drinking water supply, fisheries, wildlife etc. in the entire Sylhet division and also in the peripheral areas of Dhaka division. In addition to this there are millions of people, who are engaged in fishing, agriculture etc. and many industries that are also dependent on the river Barak, directly or indirectly (Yumnam & Koijam, 2014).

Figure 2. People in Bangladesh protesting against the Tipaimukh Project
Source: Jishnu Dutta, June 22, 2012[7]

In the other front, India's high priority plan of linking water surplus Himalayan Rivers with water scarce parts of western and peninsular India through the mega project of Inter-Linking of Rivers (ILR) is also a major concern for downstream Bangladesh. The mission of the project is to ensure greater equity in the distribution of water by enhancing the availability of water in drought prone and rain-fed area. The implementation of ILR Project would enhance benefits of 35 million hectares of irrigation, raising the ultimate irrigation potential from 140 million hectare to 175 million hectare, which will ensure sustained food production and also generation of 34000 MW of power, apart from the incidental benefits of flood control, navigation, water supply, fisheries, salinity and pollution control etc (Ministry of Water Resources, River Development & Ganga Rejuvenation n.d.).

The ILR Project intend to build 30 links to connect 37 Himalayan and Peninsular rivers to form a gigantic South Asian water grid, which are divided into 14 links under Himalayan River Component and 16 links under Peninsular River Component respectively to envisages the transfer of water from 'water surplus' to 'water deficit' regions. The Himalayan Component of the ILR Project is the major cause of Bangladesh worry, especially the Manas-Sankosh-Teesta-Ganga link (MSTG), which envisages diversion of 43 billion cubic metres (BCM) of surplus water of Manas, Sankosh and intermediate rivers, for augmenting the flow of Ganges and provide 14 BCM of water in Mahanadi basin for further diversion to South through Peninsular link system and potentially catastrophic for Bangladesh. The water diversion will severely diminish the water flow into Bangladesh's major rivers like the Teesta, Jamuna and the Padma. The Pre-feasibility Report of MSTG link was completed in 1996 and Feasibility Report of the MSTG link for Alternate alignment (original alignment could not be completed as Manas-Sankosh-Teesta link is passing through Manas Tiger Reserve, Buxa Tiger Reserve and other forests) is under final stage of preparation. The Himalayan component would consist of a series of dams built along the Ganga and Brahmaputra, Nepal and Bhutan for the purposes of storage. So, from Bangladesh perspective the Himalayan Component of ILR Project (diversion of water from the Brahmaputra and the Ganges) will significantly reduce the downstream river flow, which provide 85 percent of the country's fresh water flow in the dry season, would result into an ecological disaster (Mehta & Mehta, 2013). However, in the Foreign Ministry level Joint Consultative Commission meeting held in September 2014, India has reiterated its earlier commitment that it would not take any unilateral decision on the Himalayan Component of the proposed IRL project which may adversely affect Bangladesh (Ministry of External Affairs, 2014).

Position in Manipur

"The attempt to secure energy in this fringe geography has already made indigenous people insecure who were put at the altar without any democratic consultations. This has severed the Tipaimukh villagers whose ancestral land and resource were being targeted as urban fodder." (Buhril, 2009)

In Manipur the concern over the construction of the proposed dam are basically based on three aspects,

1. The direct physical aspect of displacement, loss of biodiversity, loss of economic activities of the indigenous people, social and environmental impact etc.,
2. Procedural lapses i.e. absence of holistic impact assessment, constrains of developmental and environmental regulations, weak enforcement mechanisms and lack of people oriented accountability norms, and
3. Ambiguous benefits of the project towards the indigenous people and to the people of Manipur at large (Yumnam & Koijam, 2014).

Apart from Tipaimukh Project there are some other multipurpose hydroelectric projects which are under criticism and facing strong opposition from indigenous people, forest dwelling communities, environmentalists, social activists and scholars as well, such as the Loktak Multipurpose Hydroelectric Project, the Singda Multipurpose Hydroelectric Project and the Thoubal Multipurpose Hydroelectric Projects. But for the interest of the present paper, the author will very briefly highlight those concerns without going much into the detail.

The Loktak Multipurpose Hydroelectric Project was commissioned in 1984. The project is responsible for devastation of the ecosystem of the largest freshwater lake in the Northeast and the only floating lake in the world. The project had disturbed the Loktak wetland's biotic ecosystem leading to extinction of several endemic plants, animal and aquatic species. The project remains irresponsible and unaccountable submergence of vast tract of agricultural land and also failed to resettle and rehabilitate communities affected by project even after three decades (Yumnam & Koijam, 2014). The Singda Multipurpose Hydroelectric Project started in 1975 and commissioned in 1995. The Singda Dam Project is a multipurpose Project intended to provide drinking water, irrigation and also for the generation of 750 Kw of electricity. The project is also widely criticized for its failure of irrigation project and underperformance of drinking water supply component of the project (Failed Dams in Manipur, 2014). The Thoubal Multipurpose Hydroelectric Projects (Mapithel Dam) was approved by the Planning Commission in 1980 and construction of the dam began in 1989. But following a RTI[8] revelation it is found that the project started without obtaining the required forest clearance Under the Forests Rights Act, 2006 (Scheduled Tribes, 2007), which is required for identification and settlement of rights of forestland for non-forest purposes. On the move of the project affect peoples, National Green Tribunal stopped the construction of the Mapithel Dam on 20th November 2013 (Shrivastava, 2013).

The India Government's plan to construct more than 169 dams in the North Eastern Region and name the region as *'India's Power House'* has been met with stiff opposition from the region (Yumnam & Koijam, 2014). There are numerous social organization in Manipur, Mizoram and Assam are protesting together against the proposed dam such as Action Committee Against Tipaimukh Dam Project (ACTIP), Citizens Concern for Dams and Development (CCDD), North East Dialogue Forum (NEDF), Sinlung Indigenous People Human Rights Organisation (SIPHRO) and Committee on Land and Natural Resources (COLNER) etc.

ENVIRONMENTAL AND FOREST CLEARANCES OF THE PROJECT

The controversy of GHG emission from freshwater reservoirs and their contribution in increasing GHG concentration leading Climate Change is still a matter of debate. The major GHGs responsible for climate change, such as carbon dioxide (CO_2), methane (CH_4) and nitrous oxide (N_2O) are emitted from both natural aquatic (lakes, rivers, estuaries, wetlands) and terrestrial ecosystems (forest, soils) as well as from anthropogenic sources. The concern regarding the GHG emissions by hydroelectric reservoirs has steadily increased since the early 90's, even though it remains unclear what their actual emission is (Mendonca, 2012). Usually reservoirs are collection points for material coming from the whole upstream drainage basin area. As a natural course, organic matter is flushed into these collection points from the surrounding terrestrial ecosystems. In addition to this, domestic sewage, industrial waste and agricultural pollution may also enter these systems and produce GHG emissions.

In the context of hydropower reservoir Carbon dioxide (CO_2) is formed by the decomposition of organic carbon in the reservoir, Methane (CH_4) (a GHG that is 25 times more potent than CO_2) is formed by bacteria that decompose organic matter in oxygen-poor water and reservoir-bottom sediments; and

Figure 3. Indigenous people in Manipur protesting against the Tipaimukh Project
Source: Jishnu Dutta, June 22, 2012

Nitrous oxide (N_2O) is a powerful greenhouse gas formed by the bacterial breakdown of nitrogen. However, studies indicate that hydroelectric power reservoirs can emit substantial amounts of methane under certain conditions (International Rivers, 2008). Hydroelectric power projects do produce some GHG from decay and from cement and steel manufacture; fossil fuel fired equivalents typically, but not generally generate much. Study suggests that GHG resulting from the manufacture of the dams cement and steel, plus the energy used in the construction amount to less than 10 percent of the annual CO_2 emissions of the fossil fuel equivalent (UNEP, 2000). Hydropower projects create no direct atmospheric pollutants or waste during operation, and GHG emissions associated with most lifecycle stages are minor.

There are studies based on impact of reservoirs, research and field surveys on GHG balances of freshwater systems to understand the relationship between dam and climate change. Variability in estimates stems from differences in study context (e.g., climate, carbon stock of flooded area), technological performance (e.g., turbine efficiency, lifetime, residence time of water) and methods (e.g., Lifecycle Assessment system boundaries). For instance, the assumed operating lifetime of a dam can significantly influence the estimate of lifecycle GHG emissions as it amortizes the construction and dismantling related emissions over a shorter or longer period. Lifecycle Assessment studies is needed to increase the number of estimates and the breadth of their coverage especially in terms of climatic zones, technology types, dam sizes etc. to understand the impact (Kumar et al., 2011). Nevertheless, the important role of hydroelectric reservoirs in the global GHG dynamics is unquestionable even after decades of scientific research (is complex and subject to numerous uncertainties) into this issue, the belief that hydropower is inherently believed to be a climate friendly technology and hence is still common among policy makers. The International Hydropower Association's Hydropower Sustainability Assessment Protocol could be a guiding base for future hydropower developments.

Any project of this nature and dimension, such as, Tipaimukh Project, requires two kinds of clearances such as, Environmental Clearance and Forest Clearance. The proposal for environmental clearance of Tipaimukh Project has been accorded on 24th October 2008 by Ministry of Environment, Forest and Climate Change (MoEFCC) in accordance with the provisions of EIA, subject to strict compliance of certain terms and conditions, which shall remain valid for a period of 10 years from the date of issue of the clearance letter for commencement of construction work. The environmental clearance was accorded, after the proposal being considered by the Expert Appraisal Committee for River Valley and Hydroelectric Projects (MoEFCC, 2008). The MoEFCC received proposals for prior approval of the Central Government under Section 2 of the Forest (Conservation) Act 1980 (FC Act) (The Forest (Conservation) Act, 2014), for forest clearance from the State Governments of Manipur and Mizoram for 22,777,50 hectors (ha) and 1551.60 ha of forest land respectively for non-forest purpose, all together 24329.10 ha. Both proposals were being dealt separately by MoEFCC.

Manipur

The State Government of Manipur initially submitted a proposal for diversion of 20,464 ha of forest land for construction of Tipaimukh Project. Subsequently they submitted a revised proposal of 25,822.14 ha on 4th April 2007. However, on the perusal note and recommendation of the North East Regional Office of the MoEFCC the proposed forest area for diversion for construction of the project has been further reduced to 22,777.50 ha from 25,822.14 ha by the user agency i.e. NHPC (Ministry of Environment, Forest and Climate Change (MoEFCC, 2013c).

Table 2. Item wise break-up details of the forest land proposed for diversion (in hectors)

Sr. No.	Project Component		Forest Land	
			Manipur	Mizoram
1.	Reservoir Submergence Area		21952.64	1489.00
2.	Project Area (Dam including colony and approach)		759.46	50.00
3.	Land diversion for National Highways (N.H.)	N.H. 53	15.60	–
		N.H. 150	49.80	12.60
	Total		**22777.50**	**1551.60**

The Forest Advisory Committee (FAC) after examining the proposal in its meeting held on 11-12 January 2012 noted as such (MoEFCC, 2012):

1. The forest land proposed for diversion in Manipur alone contains more than 78 lakh trees and around 27 thousand bamboo culms. The proposed forest land harbours several endangered species of flora and fauna.
2. The Principal Chief Conservator of Forests, Manipur has observed in his report that ''no compensatory measure would help in mitigating the adverse impact caused by loss of such large forest tracts on the habitat, flora, fauna, biodiversity, micro-climate and environment unless additional Non-forest areas in affected districts or adjoining districts are taken up for compensatory afforestation.
3. The area of forest land proposed for diversion is perhaps one of the largest ever for a single project. The per-megawatt requirement of forest land (16 hectares of forest land per megawatt) for the above project of 1500 MW installed capacity is much higher than the average per-megawatt forest land for the 497 projects (*approx.* 1 hectare per MW of installed capacity after excluding the forest land diverted for Sardar Sarovar Project) for which approval under the FC Act has been accorded so far.

Keeping in view al, the above issues the FAC recommended as, that a sub-committee of FAC along with domain experts may visit the project site to make an onsite assessment of the impacts, the project is likely to have on flora, fauna and socio-economic conditions of local residents and the sub-committee may also suggest appropriate measures, including reduction in dam height, to minimize requirement of forest land in execution of the project.

Following the recommendation of the FAC, MoEF constituted a five member committee on 6th February 2013 under the Chairmanship of Dr. Mohammad Firoz Ahmed, non-official member of the FAC (MoEFCC, 2013c) and accordingly a visit of the sub-committee was scheduled for 2nd to 5th June 2013.

Table 3. Division wise forest areas for diversion (in hectors)

Legal Status	Western Divn.	Jiribam Div.	Southern Div.	Total
Reserved Forest	37.37	1,612.00	35.53	1,684.90
Unclassed Forest	7,200.13	6,181.60	7,710.87	21,092.60
Total	**7,237.50**	**7,793.60**	**7,746.40**	**22,777.50**

(MoEFCC, 2013c)

However, due some unavoidable circumstances the sub-committee could not make onsite inspection of the proposed land for diversion.

Considering the fact of repeated requests from the Ministry of Power and NHPC to expedite the decision on the proposal, MoEFCC placed the proposal before FAC in its next meeting scheduled for 11-12 July 2013 for re-examination. The following facts were kept before the FAC in addition to the earlier notes of FAC meeting of 11-12 January 2014 (MoEFCC, 2013c):

1. The project requires 24,329 ha of forest land which is more than one-fifth of the total 1, 18,184 ha of forest land diverted for execution of 497 hydroelectric projects in the entire country since the FC Act came into force.
2. The forest land required for the project is more than 100 times the average rate of diversion forest land of (238 ha per project) for 497 hydroelectric project for which approval under the FC Act has been accorded by the MoEFCC.
3. Requirement of 24,329 ha of forest land for the project is the second largest for a single hydroelectric project for which approval under the FC Act has been sought, so far.
4. The forest land required for the project is almost two-third of the average annual rate of diversion of forest land for non-forest purpose (35,890 ha per annum) during the 32.50 years of the existence of the FC Act.
5. As per the information provided by the Ministry of External Affairs the project is a very sensitive issue in Bangladesh.

The FAC after re-examining the proposal of the State Government and through examination and interaction with the representatives of NHPC, who made presentation at the meeting, concluded and decided that requirement of forest land for the project is large and is disproportionate to its power generation capacity and the very high ecological, environmental and social impact/cost of the diversion of the vast tract of forest land will outweigh the benefits likely to accrue from the project (MoEFCC, 2013b). Among many other things the FAC noted as follows:

1. As per the proposal submitted the fauna of the proposed submerged area are reported to include Jungle fowl, Barking deer, Wild Boar, Assemese macaque (Macaca assamensis) Animals such as Leopard (Panthera pardus), Clouded Leopard (Neofelis nebulosa), Slow Lorries (Nycticibus coucang), Golden cat (Felis tenmincki), Hoolock Gibbon (Hylobates hoolock), Capped Langoor (Presbytis pileatus), Pangolin (Manis Crassicacaudata) Hog Badger (Arctonyx caloris), Himalyan Black Bear (Selenarctos thibetanus), etc, and birds like Great Indian Hornbill, Bamboo Patridge, etc. are also reported to have been sighted by the local people.
2. The Chief Conservator of Forests (Central), North Eastern Regional Office, Shillong in the site inspection report of 02nd May 2008 commented that in his opinion such a diversion proposal involving a huge stretch of critically important forests & wildlife habitat in a state where compensatory measures may not prove to be effective due to problems in effective monitoring and enforcement, such diversion may not be advisable.
3. The project involves displacement of 557 families having a population of around 2027 Scheduled Tribes spreading over 12 villages.
4. The project is likely to generate employment opportunities for 826 individuals and therefore it appears that employment opportunities likely to be created by the project is not commensurate with

the loss of land and natural resources, which are otherwise the main source of livelihood for the tribal population of the state.

Based on all the above facts and notes the FAC strongly recommended not to accord the diversion of the propose forest land for the project and also recommended that in case the user agency i.e. NHPC desires, they may explore feasibility to construct smaller dams, involving diversion of smaller forest area commensurate with their power generation capacity.

Mizoram

On the other hand the State Government of Mizoram submitted a proposal for diversion of 1551.60 ha forest land for the Tipaimukh Project in Mizoram on 7th October 2011 (MoEFCC, 2013d). The site inspection was done on 8th January 2013 (MoEFCC, 2013e). The 1551.60 ha of forest land falls under the Tuivai Riverine Reserve Forest, Tuivawl Riverine Reserve Forest, Tuirang Riverine Reserve Forest and Inner Line Reserve Forests. The FAC after examining the proposal in its meeting held on 13-14 August 2013 and concluded that since the proposal for clearance of 22,777.50 ha of forest land in Manipur part of the project has not been granted, which constitutes the major part of the project, therefore the proposal seeking diversion of 1,551.60 ha of forest land may not be recommended for approval (MoEFCC, 2013a). As per the enumeration list submitted by the Forest Department, numbers of trees to be felled of girth bellow 60 cm accounts for 11,306 and girth above 60 cm accounts for 3,328. The user agency information shows that, there are 14 villages and 340 people, which will be affected by the project and therefore a detailed examination has to be conducted by the state government (MoEFCC, 2013d).

The total submergence area for the Tipaimukh Project is around 29,150 ha, i.e. 27,550 ha in Manipur (forest land 21,952.64 ha and non-forest land 5597.36 ha) and 1,600 ha in Mizoram (forest land 1,489 ha and non-forest land 111 ha). As per the user agency information the dam construction will simultaneously start from both the river bank and commissioned within 7 years and 3 months from the date of Government sanction. The project also provides power sharing to Manipur and Mizoram respectively. For Manipur staggered free power 5 percent for 5 years, 10 percent for following 10 years and 15 percent for following 20 years and for Mizoram 01 percent free power throughout the life of the project (MoEFCC, 2013d).

The environmental clearance required the user agency to conduct a study on the downstream impacts of the project in the state of Assam and a comprehensive study of the possible GHG emissions from the project should be carried out, as reservoirs in tropical countries are a major source of GHG emissions and the project involves a large submergence (MoEFCC, 2008). However, these studies are yet to be conducted.

RECOMMENDATIONS

There could be the following way out to resolve the differences between both the countries sharing common international watercourse by upstream India to gain the confidence of downstream Bangladesh:

1. Provision for Power Sharing: India as an upstream nation could assure Bangladesh a reasonable share of power generated from the disputed Hydroelectric Power Project based on principle of equity.
2. Assurance for Food security: At the event of any drought or any unforeseen event threatening the food security in Bangladesh, upstream India could assure for food security.
3. Knowledge sharing and R&D for agriculture and fisheries: Upstream India could extend R&D facilities and sharing of knowledge regarding developing high yielding crops, fisheries etc.
4. Livelihood security support: Upstream India as a developed nation and being a super power in waiting could assist and facilitate less privileged Bangladesh through providing vocational training, skilled development training for alternative livelihood support.
5. Infrastructure projects: Upstream India could take up development of infrastructural projects in less privileged downstream Bangladesh.
6. Transboundary watercourse management agreement: A comprehensive transboundary watercourse treaty/agreement between riparian states, based on individual river basin, could resolve existing as well as any future differences of interest likely to arise.

Apart from observing the norms and standards set by international treaties, agreements and state practices, upstream India could adopt these measures to solve the disputes mutually.

CONCLUSION

Despite the controversy over hydropower reservoir's contribution to GHG responsible for Global Warming, even after decades of studies no clear Conesus on scientific conclusion has been achieved. It is true that there are evidences of GHGs emissions at freshwater reservoirs, but is debateable and dependent on site specific factors. Even accepting for a while that there are strong evidences establishing linkage between freshwater reservoirs and contribution to GHGs; methane as well as carbon dioxide emissions tend to decline during the lifetime of a dam and even including methane emissions total GHG per KWh generated from hydropower is still at least half that from the least polluting thermal alternatives available. Therefore from the perspective of global warming mitigation, dams are the most attractive alternative to fossil fuel based energy sources and certainly going to be one of the major source of renewable energy in future as well.

Moving ahead from the freshwater reservoir versus climate change, from the through study of the paper it could be easily understood that the Tipaimukh Multipurpose Hydroelectric Project is the brain child of both the countries. The idea behind the conception of the Tipaimukh Project was to moderate floods and to protect Sylhet, Cachar and adjoining areas. But later on the differences created regarding the Project and riparian India and Bangladesh took stand for and against. Both the countries stands are justifiable, as Indo-Bangla concerns find place in the Millennium Development Goal and in the Sustainable Development Goals. Although in the initial period, there was a deadlock in sharing the information with downstream Bangladesh which remained static over decades. However, the status has been changed significantly and the recent developments are a great achievement for both the riparian nations. It seems that both nations will find some mutually acceptable way out to resolve the differences.

There is another angle in to the debate, that the issue should not be viewed only from the perspective of Bangladesh. India is welfare state by it conception and it must respect the opinion of the indigenous people of North East Region. That in the rush of development, which is not all about deriving benefits only, the country has to face many challenges and these challenges are not always external. To fulfil the growing energy need of the country the indigenous people will be looser. Therefore India must try to settle the issues with the local people first and then go at the international level. Otherwise an unrest society will be greater threat to nation's internal security and overall development. The greatest hurdle in front of the government will be ensuring the rights of forest dependent/dwelling communities entrusted under the Forest Rights Act and the forest itself under the Forest (Conservation) Act. The government has to ensure livelihood security to those communities through smooth functioning cross-sectorial programmes and ensuring less damage to the forest. Because, if the Government fails to consider these concerns in the race of becoming an economic super power the financial loss will be lesser than the loss of our heritage, culture and biodiversity, which is very well quoted by W. H. Auden as "A culture is no better than its woods".

The careful observation of the politico-economic scenario suggests that with all possibility India will hopefully secure the required consent of downstream Bangladesh for the Tipaimukh Project. Now the question arises as to, weather the government is desperate to go ahead with the project even at the cost of snatching the rights of indigenous people. There is no straight way answers available to be choose between a 'Yes' or 'No'. It neither depends on the issues of 'might' over 'right' or 'development' over 'destruction', but on the circumstances to come. Although criticisms are being raised from all the corners, it will be too early to come to any conclusion. Despite the fact, India has reaffirmed development of hydropower projects through INDC for meeting post-2020 Climate Change obligations; country has to give a careful thought from the perspective of Indo-Bangla priorities, indigenous people's right and most importantly environmental concern.

REFERENCES

Alam, M. S. (2012, January 11). *Tipaimukh Dam in India: A Death Trap for Banglades.* Retrieved from http://www.parisvisionnews.com/articles/1959-tipaimukh-dam-in-india-a-death-trap-for-bangladesh-dr-muhammad-shah-alam.html

Askari, R. (2012, January). Tipaimukh Dam and Indian Hydro-politics. *Forum, 6*(1). Retrieved from http://archive.thedailystar.net/forum/2012/January/tipaimukh.htm

B'desh for Joint Study On Cross-Border Dam in India. (2012, December 9). *Press Trust of India.* Retrieved from http://www.hindustantimes.com/world/b-desh-for-joint-study-on-cross-border-dam-in-india/story-UZvr0kxlRcPWkMTBPFiEgM.html

Bangladeshi Scribes' Dam Mission Fails to Take off. (2012, June 8). *Times of India- Guwahati,* Retrieved from http://timesofindia.indiatimes.com/city/guwahati/Bangladeshi-scribes-dam-mission-fails-to-take-off/articleshow/13910971.cms?referral=PM

Bhattacharya, P. (2012, August 29). Tipaimukh Dam: Dhaka, Delhi Finalise Joint Study Terms. *The Daily Star.* Retrieved from http://www.thedailystar.net/newDesign/news-details.php?nid=247472

Bisht, M. (2012, April 25). *Tipaimukh Dam: Some Myths, Some Facts. Institute of Defense Studies and Analyses.* Retrieved July 10, 2016, from http://www.idsa.in/idsacomments/TipaimukhDam_mbisht_250412

Buhril, D. (2009, February 2). Tipaimukh Dam: Development or Destruction – I. *E-PAO*. Retrieved from http://www.e-pao.net/epSubPageExtractor.asp?src=leisure.EI.David_Buhril.Tipaimukh_Dam_Development_or_Destruction_1

Choudhury, A. (2012, January 9). Is Bangladesh doing enough about Tipaimukh? *bdnews24.com*. Retrieved from http://opinion.bdnews24.com/2012/01/09/is-bangladesh-doing-enough-about-tipaimukh/

Deka, M. (2012). Mega Dams, Environment and History in North-East India. *Dialogue*, *13*(4), 2012. Retrieved from http://www.asthabharati.org/Dia_Apr%20012/me.htm

Dutta, P. (2010). India-Bangladesh Relations: Issues, Problems and Recent Developments. *Institute of Peace and Conflict Studies*. Retrieved July 10, 2016, from http://www.ipcs.org/pdf_file/issue/SR97.pdf

Failed Dams in Manipur? Any Lessons Learnt. (2014, September 3). *The Sangai Express*. Retrieved from http://www.thesangaiexpress.com/failed-dams-in-manipur-any-lessons-learnt/

India and Bangladesh: Headed Towards International Legal "War" Over The Tipaimukh Dam?. (2012, January 13). *Water Politics*. Retrieved from http://www.waterpolitics.com/2012/01/13/india-and-bangladesh-headed-towards-international-legal-war-over-the-tipaimukh-dam/

India Assures Bangladesh on Tipaimukh. (2011, December 3). *Times of India*. Retrieved from http://articles.timesofindia.indiatimes.com/2011-12-03/india/30471587_1_tipaimukh-dam-tipaimukh-issue-bangladesh-government

International Law Association. (1986). Complementary Rules Applicable to International Water Resources. Adopted by the International Law Association at its 62nd Conference held at Seoul in 1986. In *The 1966 Helsinki Rules and Complementary and Supplementary Rules (1972-1996)*. Retrieved from http://water.cedare.int/files15%5CFile2917.pdf

International Rivers. (2008). *Dirty Hydro: Dams and Greenhouse Gas Emissions*. Retrieved from http://www.internationalrivers.org/files/attached-files/dirtyhydro_factsheet_lorez.pdf

International Sava River Basin Commission. (2002). *Framework Agreement The Sava River Basin.* Retrieved from http://www.savacommission.org/dms/docs/dokumenti/documents_publications/basic_documents/fasrb.pdf

Islam, N., Azam, A., & Islam, Q. R. (n.d.). *Teesta River Water Sharing: A Case Study In Teesta Barrage Project*. Retrieved from http://www.watertech.cn/english/islam.pdf

Joint Rivers Commission. (n.d.a). *Basin Map. Ministry of Water Resources*. Govt. of Bangladesh. Retrieved from http://jrcb.gov.bd/new/index.php/9-link-page/12-basin-map

Joint Rivers Commission. (n.d.b). *Ministry of Water Resources.* Govt. of Bangladesh. Retrieved from http://jrcb.gov.bd/new/

Kumar, A., Schei, T., Ahenkorah, A., Caceres Rodriguez, R., Devernay, J.-M., Freitas, M., & Liu, Z. et al. (2011). Hydropower. In O. Edenhofer, R. Pichs-Madruga, Y. Sokona, K. Seyboth, P. Matschoss, S. Kadner, & C. von Stechow et al. (Eds.), *IPCC Special Report on Renewable Energy Sources and Climate Change Mitigation*. Cambridge, UK: Cambridge University Press. doi:10.1017/CBO9781139151153.009

Manmohn to Chair Timaimukh Meeting. (2010, October 13). *Bangladesh News*. Retrieved from http://www.independent-bangladesh.com/2010101312037/country/manmohan-to-chair-tipaimukh-meeting.html

McCaffery, S. C. (2007). *The Law of The International Watercourses* (2nd ed.). New York: Oxford University Press.

Mehta, D., & Mehta, N. K. (2013). Interlinking of Rivers in India: Issues & Challenges. *Geo-Eco-Marina*. Retrieved July 10, 2016, from http://www.geoecomar.ro/website/publicatii/Nr.19-2013/12_mehta_web_2013.pdf

Mehta, N. (2012, July 23). Dialectics of Tipaimukh Dam: Issues and Concerns. *Vivekananda International Foundation*. Retrieved July 10, 2016, from http://www.vifindia.org/article/2012/july/23/dialectics-of-tipaimukh-dam-issues-and-concerns

Mekong River Commission. (1995). *Agreement on The Cooperation For The Sustainable Development of The Mekong River Basin*. Retrieved from http://www.mrcmekong.org/assets/Publications/policies/agreement-Apr95.pdf

Mendonca, R., Barros, N., Vidal, L. O., Pacheco, F., Kosten, S., & Roland, F. (2012). Greenhouse Gas Emissions from Hydroelectric Reservoirs: What Knowledge Do We Have and What is Lacking? In *Greenhouse Gases - Emission, Measurement and Management*. InTech. Retrieved from http://cdn.intechopen.com/pdfs-wm/32342.pdf

Ministry of External Affairs. (2011). *Framework Agreement on Cooperation for Development between India and Bangladesh*. Retrieved from http://mea.gov.in/bilateral-documents.htm?dtl/5218/Framework+Agreement+on+Cooperation+for+Development+between+India+and+Bangladesh

Ministry of External Affairs. (2014). *Joint Statement on the Third Meeting of the India-Bangladesh Joint Consultative Commission*. Retrieved from http://mea.gov.in/bilateral-documents.htm?dtl/24024/Joint+Statement+on+the+Third+Meeting+of+the+IndiaBangladesh+Joint+Consultative+Commission

Ministry of Water Resources, River Development & Ganga Rejuvenation. (1960). *The Indus Water Treaty*. Adopted on 9th September 1960. Retrieved from http://wrmin.nic.in/forms/list.aspx?lid=346

Ministry of Water Resources, River Development & Ganga Rejuvenation. (n.d.). *Inter Linking of Rivers*. Retrieved from http://wrmin.nic.in/forms/list.aspx?lid=1279

MoEFCC. (2006). *Form-A: Project Details - Justification for locating the project in the forest area*. Retrieved from http://forestsclearance.nic.in/search.aspx?cat_id=FP/MN/HYD/1589/2012

MoEFCC. (2008). *Tipaimukh Multipurpose HEP Project (1500 MW) in Manipur & Mizoram States – Environmental Clearance*. No.J-12011/63/2006-IA.I. Retrieved from http://www.indiaenvironmentportal.org.in/files/html/Environment_Clearance/Tipaimukh%20Multipurpose%20HEP%20Project.html

MoEFCC. (2012). Agenda No.2: Diversion of 25822.14 hectares Forest land for construction of Tipaimukh Hydroelectric Project in Manipur (File No. 8-63/2005-FC). *Proceeding of the Forest Advisory Committee Meeting*. Retrieved from http://forestsclearance.nic.in/writereaddata/FAC_Minutes/jan2012.pdf

MoEFCC. (2013a). *Agenda Item No.1: Diversion of forest land for 1551.60 ha of forest land under the forest (Conservation) Act, 1980 for construction of Tipaimukh Hydro Electric Project in Mizoram* (File No. 8-02/2012-FC). Minutes of the Forest Advisory Committee Meeting held on 13-14 August 2013. Retrieved from http://www.ercindia.org/files/fac/FAC%20Minutes%2013-14%20Aug_2013_9113124312181minutesAugust.pdf

MoEFCC. (2013b). Agenda No.3: Diversion of 22777.50 hectares Forest land for construction of Tipaimukh Hydroelectric Project in Manipur (File No. 8-63/2005-FC). *Proceeding of the Forest Advisory Committee Meeting*. Retrieved from http://www.internationalrivers.org/files/attached-files/forestadvisorycommittee_july_2013_minutes.pdf

MoEFCC. (2013c). *Fact Sheet: Diversion of 25822.14 hectares Forest land for construction of Tipaimukh Hydroelectric Project in Manipur* (File No. 8-63/2005-FC). Retrieved from http://forestsclearance.nic.in/search.aspx?cat_id=FP/MZ/HYD/2187/2012

MoEFCC. (2013d). *Fact Sheet: Diversion of forest land for 1551.60 ha of forest land under the forest (Conservation) Act, 1980 for construction of Tipaimukh Hydro Electric Project in Mizoram* (File No. 8-02/2012-FC). Retrieved from http://forestsclearance.nic.in/search.aspx?cat_id=FP/MZ/HYD/2187/2012

MoEFCC. (2013e). *Site Inspection Report: Diversion of forest land for 1551.60 ha of forest land under the forest (Conservation) Act, 1980 for construction of Tipaimukh Hydro Electric Project in Mizoram* (File No. 8-02/2012-FC). Retrieved from http://forestsclearance.nic.in/search.aspx?cat_id=FP/MZ/HYD/2187/2012

National Hydroelectric Power Corporation. (n.d.a). *Tipaimukh*, Retrieved from http://www.nhpcindia.com/projectdetail.htm?CatId=6&ProjectId=53

National Hydroelectric Power Corporation. (n.d.b). *Power Stations*. Retrieved from http://www.nhpcindia.com/project.htm

Osmani, A. R. (2014). Conventional Energy to Renewable Energy: Perspective for India. *The NEHU Journal, 12*(2). Retrieved July 10, 2016, from http://www.nehu.ac.in/Journals/Journal_Jul_Dec14_Art3.pdf

Rahaman, M. M. (2009a). Principles of International Water Law: Creating Effective Transboundary Water Resources Management. *Int. J. Sustainable Society., 1*(3), 209. doi:10.1504/IJSSOC.2009.027620

Rahaman, M. M. (2009b). Principles of Transboundary Water Resources Management and Ganges Treaties: An Analysis. *Water Resource Development, 25*(1), 159–173. doi:10.1080/07900620802517574

Ray, A. (2012, June 8). India-Bangladesh: Water Dispute and River Diplomacy. *International Security Relations and Security Network ETH Zurich*. Retrieved from http://isnblog.ethz.ch/global-voices/india-bangladesh-water-disputes-and-river-diplomacy

Rieu-Clark, A., Moynihan, R., & Magsi, B. (2012). *UN Watercourse Convention: User's Guide*. IHP-HELP Centre for Water Law, Policy and Science.

Rules, H. (1966). *The Helsinki rules on the uses of the waters of international rivers*. International Law Association.

Shrivastava, A. (2013, December 10). Green Tribunal Halts Mapithel Dam in Manipur. *Down to Earth*. Retrieved July 10, 2016, from http://www.downtoearth.org.in/content/green-tribunal-halts-mapithel-dam-manipur

Southern African Development Community. (2000). *Revised Protocol on Shared Watercourse*. Retrieved from http://www.sadc.int/files/3413/6698/6218/Revised_Protocol_on_Shared_Watercourses_-_2000_-_English.pdf

Tanzi, A., & Arcari, M. (2001). *The United Nations Convention on The Law of International Watercourse: A Framework for Sharing*. London: Kluwer Law International.

The Forest (Conservation) Act 1980. (2014). No.69 of 1980, 27th December 1980, in Environmental and Pollution Laws. New Delhi: Universal Law Publishing, 2014.

The International Court of Justice. (2010). *Pulp Mills on the River Uruguay (Argentina v. Uruguay). Summaries of Judgments 20 April 2010*. Retrieved from http://www.icj-cij.org/docket/files/135/15873.pdf

The Scheduled Tribes and Other Traditional Forest Dwellers (Recognition of Forest Rights) Act. (2007). *No2 of 2007, 29th December 2006, The Gazette of India, Extraordinary Part-II Section-1*. New Delhi: Ministry of Law and Justice. Retrieved from http://www.egazette.nic.in/WriteReadData/2007/E_2_2010_050.pdf

UNDP. (n.d.a). *The Millennium Development Goal*. Retrieved from http://www.undp.org/content/undp/en/home/sdgoverview/mdg_goals.html

UNDP. (n.d.b). *Sustainable Development Goal*. Retrieved from http://www.undp.org/content/undp/en/home/sdgoverview/

UNEP. (2000). *Climate Change and Dams: An Analysis of the linkages between the UNFCCC Legal Regime and Dams*. Prepared for Thematic Review II.2: Dams and global change. Retrieved from http://re.indiaenvironmentportal.org.in/files/DAMs%20and%20climate%20change.pdf

UNFCCC. (2015). *India's Intended Nationally Determined Contribution: Working Towards Climate Justice*. Retrieved from http://www4.unfccc.int/submissions/indc/Submission%20Pages/submissions.aspx

United Nations. (1997). *Convention on the Law of the Non-navigational Uses of International Watercourses*. Retrieved from http://legal.un.org/ilc/texts/instruments/english/conventions/8_3_1997.pdf

United Nations Conference on Environment and Development. (1992). *Rio Declaration on Environment and Development*. UN Doc. A/CONF.151/5/Rev. 1, 31 ILM 874.

Verghese, B. G. (2009, August 6). Straight Talk on Barak. *The Indian Express*. Retrieved on from http://archive.indianexpress.com/news/straight-talk-on-barak/498574/

Vinogradov, S., Wouters, P., & Jones, P. (2003). *Transforming Potential Conflicts into Cooperation Potential: The Role of International Water Law*. UNESCO. Retrieved from http://unesdoc.unesco.org/images/0013/001332/133258e.pdf

Yumnam, J., & Koijam, P. (2014). *Tipaimukh Dam Plan and Uncertainties in Manipur.* Centre for Research and Advocacy, Manipur and United Voluntary Youth Council. Retrieved September from http://static1.squarespace.com/static/50ec44f5e4b050fcaabb6e6f/t/54946e5de4b0547f4620fe3a/1419013725530/Uncertainties+of+Tipaimukh+dam+plan+in+Manipur.pdf

Zaman, S. S. (2012, June 4). Tipaimukh Dam 'Only if Dhaka Agrees'. *bdnews24.com*. Retrieved from http://bdnews24.com/bangladesh/2012/06/04/tipaimukh-dam-only-if-dhaka-agrees

ENDNOTES

1 The CWC is a premier Technical Organization of India in the field of Water Resources under the Ministry of Water Resources, came into existence in 1945. The Commission is entrusted with responsibilities of initiating, coordinating and furthering in consultation with the concerned State Govt., schemes for control, conservation and utilization of water resources throughout the country for the purpose of Flood Control, Irrigation, Navigation, Drinking Water Supply and Water Power Development. Retrieved from http://www.cwc.gov.in/

2 The Brahmaputra Board, an autonomous statutory body under the Ministry of Water Resources, Govt. of India set up in 1980. The jurisdiction of the Board includes both Brahmaputra and Barak Valley and covers all the states of North Eastern Region, Sikkim and part of West Bengal, which falls under the Brahmaputra basin. Retrieved from http://www.brahmaputraboard.gov.in/NER/Organisation/organisation.html

3 The NEEPCO is a Govt. of India enterprise under the Ministry of Power, came into existence on 2nd April 1976 to play a significant role in the integration and development of hydroelectric and thermal power through planning, investigating, designing, constructing, generating, operating and maintaining power stations in the North Eastern Region. Retrieved from http://neepco.co.in/neepco/#

4 The NHPC (formerly known as National Hydroelectric Power Corporation Ltd.), a Govt. of India enterprise set up in 1975 with a objective to plan, promote and organize an integrated and efficient development of hydroelectric power in all aspect. Subsequently NHPC expanded its object to include development of power in all aspects through conventional and non-conventional sources in India and abroad. Retrieved from http://www.nhpcindia.com/about-overview.htm

5 Updated estimated cost excluding cost on a/c of Flood Moderation, External Security and Diversion of National Highway, but including internal security.

6 The SJVN is a Joint Venture of Govt. of India and Govt. of Himachal Pradesh was set up on 24th May 1988 with a motive to plan, investigate, organize, execute, operate and maintain Hydro-electric power projects. Retrieved from http://sjvn.nic.in/index.htm

7 Social Activist, Lakhipur, Dist.-Cachar (Assam); Information gathered through exchange of e-mails.

8 The Right to Information Act, 2005: An Act to provide for setting out the practical regime of right to information for citizens to secure access to information under the control of public authorities, in order to promote transparency and accountability in the working of every public authority, the constitution of a Central Information Commission and State Information Commissions and for matters connected therewith or incidental thereto.

APPENDIX

Table 4. Provisions relating to the Ttansboundary Water Management Principles

Principles	Helsinki Rules, 1966	UN Watercourse Convention, 1997	International Treaties/Agreements
Reasonable & Equitable Utilization	Articles IV, V, VII, X, XXIXX	Articles 5, 6, 7, 15, 16, 17, 19	2002 Revised SADC Protocol on Shared Watercourse Systems [Article. 3(7)(a), 3(7)(b) & 3(8)], 2002 Sava River Basin Agreement (Articles 7–9), 1995 Mekong Agreement (Articles 4–6, 26)
Not to cause Significant Harm	Article V, X, XI, XXIX (2)	Articles 7, 10, 12, 15, 16, 17, 20, 21(2), 22, 26(2), 27, 28(1), 28(3)	2002 Revised SADC Protocol on Shared Watercourse Systems [Article 3(10)], 2002 Sava River Basin Agreement (Articles 2, 9), 1995 Mekong Agreement (Articles 3, 7, 8)
Cooperation & information exchange	Articles XXIX (1), XXIX (2), XXXI	Articles 5(2), 8, 9, 11, 12, 24(1), 25(1), 27, 28(3), 30	1960 Indus Waters Treaty (Articles VI-VIII), 2002 Revised SADC Protocol on Shared Watercourse Systems [Article 3(5), 3(6) & 4(1)(a)], 2002 Sava River Basin Agreement (Articles 3–4, Articles 14–21), 1995 Mekong Agreement (Preamble, Articles 1, 2, 6, 9, 11, 15, 18, 24, 30)
Notification, consultation & negotiation	Articles XXIX (2), XXIX (3), XXIX (4), XXX, XXXI	Articles 3(5), 6(2), 11–19, 24(1), 26(2), 28, 30	1960 Indus Waters Treaty (Articles VII [2], VIII), 2002 Revised SADC Protocol on Shared Watercourse Systems [Article 4(1)(b) & 4(1)(g)], 2002 Sava River Basin Agreement (Part Three and Four, Article 22), 1995 Mekong Agreement (Articles 5,10, 11, 24)
Peaceful settlement of disputes	Articles XXVI–XXXVII	Article 33	1960 Indus Waters Treaty (Article IX, Annexure F, G), 2002 Revised SADC Protocol on Shared Watercourse Systems [Article 7], 2002 Sava River Basin Agreement (Articles 1, 22-24, Annex-II), 1995 Mekong Agreement (Articles 18.C, 24.F, 34, 35)

Source: Rahaman, M. M. (2009). Principles of Transboundary Water Resources Management and Ganges Treaties: An Analysis. *Water Resource Development.* Vol.25 No.1 March. 159-173.

Chapter 14
An Approach to Sustainable Watershed Management:
Case Studies on Enhancing Sustainability with Challenges of Water in Western Maharashtra

Sneha Kumari
Symbiosis International University, India

Yogesh Patil
Symbiosis International University, India

Prakash Rao
Symbiosis International University, India

ABSTRACT

Western Maharashtra consists of drought prone villages, which around seventy years ago were facing drastic challenges regarding water requirements. People living in such areas started to migrate from these villages in need of food and water. The chapter focuses on few case studies of drought prone areas of Western Maharashtra which made a move towards sustainability and has created an excellent model of watershed projects. Several practices like ban on cutting trees, terrace farming, contour bunding, drip irrigation, sprinkle irrigation, introduction of cooking stoves, rotational grazing, efficient water distribution and change in cultivation of agricultural crops according to the climate were followed which led the village to become a model for Aadarsh Gaon (An Ideal Village). The author has used secondary data from previous research work and critically analyses the case studies on water management in drought prone areas of Western Maharashtra which managed to become a model for sustainable use of water.

DOI: 10.4018/978-1-5225-1046-8.ch014

INTRODUCTION

The State of Maharashtra is located in the Western parts of India and is classified into five regions namely, Konkan, Desh (Pune region), Vidharba, Khandesh and Marathwada. Along the Arabian Sea the state has a coastline of 720 km. Maharashtra is home to renowned historical monuments, caves, temples, festivals and a bustling lifestyle. Moreover, it is housed by large number of educational institutions, industries and service sector organizations and 45% of total population in the state lives in urban area. As the second most populous state, Maharashtra is one of the most developed states of India and plays a key role in the overall economic development of country as major educational and financial institutions are located in this state. The current population of Maharashtra is 121 million people accounting for over 9.29% of the total population of India.

With time large population has been shifted towards urban areas, while rural areas have undergone limited shifts in their boundaries. Cities like Thane, Mumbai, Raigarh, Ratnagiri and Sindhudurg are separated by the Western Ghats which receive heavy rains (>1000 mm) whereas the rain-shadow region towards the east of the Western Ghats suffers from low rainfall. Furthermore, far Eastern regions of Maharashtra again receive a good rainfall. Western region comprises of Nashik, Satara, Sangli, Kolhapur, Solapur, Ahmednagar, Aurangabad and Pune which has comparatively low rainfall than other district of Pune since they lie in the rain shadow region of the Western Ghats. The rain shadow regions receive an average rainfall of 500-600 mm. The state has a pleasant climate throughout its season with rainfall exceeding to 6000 mm in Western Ghats to 500 mm in the rain-shadow regions. The state has been surrounded by Krishna, Godavari, Tapi, Narmada and Konkan River providing water for the need. About 50% of the Maharashtra region lies in the drought prone area of Deccan plateau (nidm.gov.in/pdf/dp/Maharashtra.pdf). In the past few years it has been noticed that the Western region of the state is facing severe water scarcity problems. Most of the water supply in this region is happening through water tankers. Water being the life is essential at every step of day-to-day activity and therefore must be sustained for a better living.

This chapter focuses on the key approach towards watershed management, a topic of great concern in present era of climate change. Watershed management is an approach towards life of the earth since excess water flow leads to soil erosion, floods and other natural calamities which ultimately destroy the resources of living and scarcity of water again concludes to drought (Dixon & Easter, 1991). Western Maharashtra consists of drought prone villages which around seventy years ago was facing drastic challenges regarding water requirements. People living in such areas started to migrate from these villages in need of food and water (Foster et al., 2009). Then a commerce student Mr. Popatrao Pawar took an initiative to become Sarpanch (Head) of the Hiware Bazar village and tried to change the history of the village. Several practices like ban on cutting trees, terrace farming, contour bunding, drip irrigation, sprinkle irrigation, introduction of cooking stoves, rotational grazing, efficient water distribution and change in cultivation of agricultural crops according to the climate were followed which led the village to become a model for Aadarsh Gaon (Phand & Arya, 2007). Mr. Anna Hazare thirty years ago also took an initiative to achieve watershed resources by implementation of Programs through public participation in Ralegan Siddhi village of Parner (Sharma, 2006).

The authors in the present chapter have used secondary data from previous research work and critically analyses the case studies on water management in drought prone areas of Western Maharashtra which managed to become a model for sustainable use of water. The objective of the chapter is to focus on case studies in the drought prone areas which managed to survive efficiently with a successful water management and to discuss measures for a sustainable watershed management with reference to Maharashtra.

GROUNDWATER MANAGEMENT IN WESTERN MAHARASHTRA

Groundwater is defined as the water available in the aquifer below the ground at the zone of saturation which can be dug through machines and water extracted as bore well, tube well or by other means. The groundwater table is abundant in Eastern part of the country than in the Western part. The Western region has limited water leading to depletion of water resources.

NABARD has been utilizing the finance for developing wells in the rural areas. The increasing tariffs on water can make a sustainable use of water for irrigation by the use of diesel and electric wells (Moench & Kumar, 1995). This electricity tariff will lead to increase the efficient utilization of groundwater and use of electricity (Kumar & Singh, 2001). With time due to variation in the climate, the groundwater level has gone down due to less rainfall and the unhealthy use of water.

L.A Bill No. XLII 2009 has been introduced to ensure the sustainable water management in Western Maharashtra and established state groundwater authority and district groundwater authority to protect drinking water resources and managing the groundwater exploitation within the districts of Maharashtra. This has led to forward a law for managing the demand and supply of water for various purposes at different sectors. This act is commonly called as the Maharashtra Groundwater Development and Management enacted in the year 2009. This act focuses on valuable terms required for sustainable management of water as described in Table 1.

It is the responsibility of the state authority to check the polluters in the ground water both in the rural as well as in the urban areas and make it safe for drinking purpose for the population. The State Authority and the State Government ensures the implementation of integrated watershed development and management. It is the responsibility of the State Authority to charge water for the highly water in-

Table 1. Valuable terms used in sustainable management of water

Rainwater harvesting	This is the essential part for water management which cannot be ignored. The rain water must be stored or collected instead of being wasted. This water can be conserved for future purposes.
Aquifer	This refers to a structure for storing saturated water underground.
Area of influence	Refers to the area beneath the ground which is modified for pumping purpose.
Artificial water recharge to groundwater	refers to the water being available due to the natural climatic conditions
Deep well	This is the groundwater dug deep in the soil covering twenty- five strata of the soil and includes bore well, tube well and even sixty meter deep.
District watershed management committee	Enrolled under section18 for the better management of groundwater in the districts.
Drinking water purpose	This was included in the bill for assessing the demand and supply of drinking water for the human population of the state.
Groundwater surveys and development agencies	Under the water supply and sanitation measures the survey is made underground in Maharashtra and the developing agencies manages the water supply in the state.
Integrated watershed development and management	is the key for the sustainable management of water in the state

tensive crops grown by the farmers. This practice has been made to ensure the sustainable management of water. Training Programs have been held by the state authority to ensure awareness in the rural areas of Maharashtra regarding integrated water management. The Program discusses on the best practices for the implementation of watershed management. Watershed Water Resource Committee supplies ground water for the highly water intensive crops with terms and conditions. The District Authority shall take the responsibility to preserve the drinking water from the groundwater. At times of water scarcity in the areas, the District Authority regulates the extraction of water from any well and manages to supply water within the area of one kilometer of the public drinking water source. The inspection of wells is carried out and the water is tested in terms of quality and any contamination.

The State Authority has a watershed water review committee who is responsible for water management in the villages of Maharashtra affected by water scarcity. The Watershed Water Resource Committee comprises of a chairman of the Panchayat Samiti, one representative from the water supply and sanitation measure, one from the water use association, elected members of the Panchayat Samiti, block development officer and member of village and water committee. It is mandatory that one-third of the members of the water resource committee must be women. The Watershed Water Resource Committee allocates the water budget of every area and on the basis of availability of water all the year around, the cropping pattern of that particular area is decided. The groundwater and developing agency carry out the hydro-geological study of a particular area for the management of water.

The Maharashtra Groundwater Act 1993 has been enacted to protect the drinking water for the rural population and also to regulate the exploitation of groundwater. The Government of Maharashtra has enacted a new law under Groundwater Act to ensure the sustainable supply of groundwater, protecting drinking water, conservation of groundwater resources and provide beneficiary schemes in rural as well as urban areas to support the ground water protection. The bill also provides the following proposals as mentioned in Table 2.

CASE STUDIES ON ENHANCING SUSTAINABILITY OF WATER IN WESTERN MAHARASHTRA

Maharashtra occupies a total area of 3,07,713 square kilometers with an average annual rainfall of 1433 mm. Majority of the area of the state lies in the Deccan belts. The rocks of different areas are confined with different characteristics which divides the area into unconsolidated, consolidated and semi-consolidated.

Wadgaon

This is a village which is suffering from water scarcity. Roof top rainwater harvesting was found as the best measure for conserving water in this village. This has been implemented in most of the areas of the village and the measure was also supported by the Panchayat of the village by constructing several water storing structures in the local nullah.

Watershed Warud Taluka in Amravati

Due to water scarcity the groundwater has declined to a very lower level in this village. This led to the construction of three percolation tanks at Manikpur, Benoda and Bhindi and ten cement plugs or nullah

Table 2. Clauses of the bill proposed in 2009 for enhancing sustainable water management

Clause	Sub-clause	Objectives
1		The state Government has been assigned with the power to bring out the act of watershed resource development.
2		Such authorities must be specified by the state Government which can provide water from the public drinking water resources.
3		The state groundwater authority must exercise the power on discharge of water to different sectors. The state authority must be assisted by the groundwater surveys and development agencies, one expert from the field of groundwater and one women representing the user of groundwater.
4	1	An official gazette must be notified in the area
5		The area where the groundwater has improved must be denotified by the state authority.
6	3,5	The state Government is assigned with the power to restore the groundwater quality with the minimum water standards and it's preservation for drinking purpose.
7		Registration of the owners of the wells in the notified as well as non-notified areas of the state must be ensured by the state authority.
8	1, 3, 5, 4	Permission should be granted to any person by the state authority for digging well in the area for drinking purpose. The use of existing deep wells is ensured and permissions should be obtained by the district authority for the users of groundwater.
9	3,6,7, 8,10	District watershed management committee should look after the efficient implementation of integrated watershed management plan. The state Government shall issue the guidelines for disincentives the use of water for water intensive crops. The State Government must prescribe the efficient manner for rain water harvesting by the top roof method in urban areas. Incentives shall be provided to the water resource committee by the state Government.
10	1, 3	The state Government must give directions regarding the cropping plan in a particular area based on its water requirement. In case for the cultivation of water intensive crops, the farmers may approach to the water resource committee for discharge of groundwater.
12		Compulsory registration must be done by the rig owners and other operators in consultation with the groundwater survey and developing agency.
14		The state authority can direct special order to water resource committee or the village groundwater surveys and developing agency for sustainable management of water
16		The state Government can assign responsibility of the state watershed management council to the state water council
18	1, 2, 3	The state Government can constitute a district committee called as district watershed management committee as per need. The members of the district watershed management committee can be appointed by the state Government. The decisions carried out by the district watershed committee shall be then implemented by the district authority.
19		For artificial recharge of water Integrated watershed management plan should be prepared by the district watershed committee.
20		Power has been assigned to the district authority for public drinking water source.
21		The district authority shall see the influence of the area under public drinking water source.
22		The excess discharge of water can be prohibited by the district authority from any well for a particular period of time. The state Government also holds the power to prohibit the discharge of water from a well for a particular period of time.
23		The extraction of groundwater can be regulated by the district authority in the village
25		The district authority shall declare any area as water scarcity area for a specific period of time
26		The district authority can place an order to extract water from any well in case of water scarcity.
27		The state Government looks after the method of payment charged by the district authority.
28	1	The state Government is assigned with the power to look after the manner in which a person sinks a well and the preservation of the soil

continued on next page

Table 2. Continued

Clause	Sub-clause	Objectives
29	1, 5, 6	The state Government shall look after the work of the water resource committee as per required. The terms of the official members of the watershed water resource committee are assigned by the state Government
30	1, 4, 5	The state Government has the power to look after the manner in which the watershed water resource committee prepares the groundwater plan, which is notified by the district authority. The cropping pattern will be then decided on the basis of the water budget by the watershed water resource committee with consultation to the state Government.
32	1, 4	The state Government has the power to give permission for the artificial recharge of water.
33		The watershed water resource committee shall promote the participation of people for sustainable and equitable management of water.
34		The watershed water resource committee implements plans in coordination with Panchayat.
35		The watershed water resource committee can regulate or prohibit the mining of sand under the recommendation of district authority.
36	2	The service charge and fees shall be collected by the watershed water resource committee for the use of groundwater
37		It is mandatory for the watershed water resource committee to submit its report to the Gram Sabha
39		The state Government prescribes the time for the water budget to be prepared for the watershed water resource committee.
40		Rules are made for the account statement submitted annually.
43		The groundwater surveys and developing agencies shall assist the watershed water resource committee for watershed preparation.
45	1	The state Government shall prescribe the rules for the hydrogeological studies carried out by the groundwater surveys and developing agencies.
46	1	The state Government has the power to look after the manner in which the Panchayat and other local bodies looks after the maintenance of the water bodies and the water budget. The artificial water recharge structure is also taken care of.
47		The state Government shall look after each and every rule and the manner in which it is followed for a sustainable water management.
56	1, 2	An appeal can be made to the state Government with fees for going against the orders of the district authority.
58		Difficulties must be removed in following such rules for the ground watershed management

Source:http://www.slideshare.net/ruralindia/maharashtra-groundwater-development-and-management-act-2009.

bunds at Alora, Benoda, Mamdapur, Malkhed, Temburkhed, Loni, Sendurjanaghat. Five underground water storage tanks were also constructed for conserving water.

Watershed at Yuvraj Taluka in Jalgaon

The Yuvraj taluka has a declining groundwater by one meter every year due to cultivation of banana. For sustaining water in the region a total of three percolation tanks were constructed at Ichkeda, Haripura and Dongaon. Injection well, recharges shift, dug wells and nullah's were constructed in the village to fulfil the water requirement of the village.

Watershed at Yaval Tehsil of Jalgaon

Percolation tanks and recharge shifts have been constructed at Borkheda, Wadri, Sangvi and Dongarda to allocate water in twenty-eight villages of Jalgaon district. The storage capacities of these tanks were found to be seventy-one to ninety-eight thousand cubic meters and a catchment area of 0.425 to 4.273 sq.km each and a submergence area of 12.35 to 42 thousand sq.m.

Warud at Amravati

Roof top rain water harvesting was initiated as a project in the village by the Panchayat samiti. Roof top area of 500 sq.m was used for rain water harvesting. This water has been used for storage of water in the ground well. The ground water well level showed a substantial increase in its level due to roof top rain water harvesting. The water stored from the rain water harvesting was an essential measure for storing water at times of need and a major portion of the water stored was used to increase the level of groundwater.

RALEGAN SIDDHI

Ralegan Siddhi is a small village of Parner taluka of Ahmednagar District, Maharashtra. The village was once under the hands of poverty followed by the scarcity of water and limited agricultural growth. The village was very badly shattered due to climate change and unsustainable management of water. Agricultural crops were badly ruined and farmers faced few bad days.

Figure 1. Scenario of Ralegan Siddhi village in 1975

THE NEW BEGINNING FOR CHANGE IN RALEGAN SIDDHI

It was Mr. Anna Hazare who made the first step to bring out a change in the village some thirty years ago. With consistent and committed efforts of Mr. Anna Hazare, the village came out to be named as the first ideal village of Maharashtra. Mr. Anna Hazare made a successful implementation of sustainable practices in the village.

Prohibition on grazing of agricultural crops, cutting of trees, consumption of alcohol, shramdan and regeneration of water resources were some of the practices for the development of the village. In the present scenario availability of water resources in the village for irrigation in agriculture land has been found to be the luxurious asset for the village. Phand & Arya (2007) made a detailed study on the developing trends of Ralegan Siddhi before and after sustaining practices and came to a conclusion that the village has twelve hundred acres of land out of seventeen hundred which are under irrigation. This has led to increase in the agricultural yield and ultimately this has led to increase the number of livestock animals. In 2007, it was found that dairy sector in the village contributes to Rs. one crore per annum (USD 1,50,212/- pa). Practices of this nature have made the village full-fledged with drinking water, school facilities, roads, medical hospitals and sanitation thereby making it sustainable. Further the model of Ralegan Siddhi village was taken as the Aadarsh Gaon Yojana for other villages. This model focused on practices for making water safe for drinking and was taken as Aadarsh Gram Yojana in about three hundred villages across Maharashtra.

The implementation became famous and was based on five principles namely shramdan, ban on cutting of trees, ban on grazing, ban on consumption of alcohol and introduction of family planning as one of the sustainable practice. With shramdan, forty thousand contour trenches were built for conserving water and increasing the groundwater table. Along with contour trenches several trees were planted around. The trees like neem (*Azadirachta indica*), babul (*Acacia nilotica*), kashid (*Cassia siamea*) and shisam (*Dalbergia sissoo*) were given preference.

Figure 2. Hind Swaraj Trust run by Mr. Anna Hazare

Figure 3. Ralegan Siddhi as an example of Model Rural Village

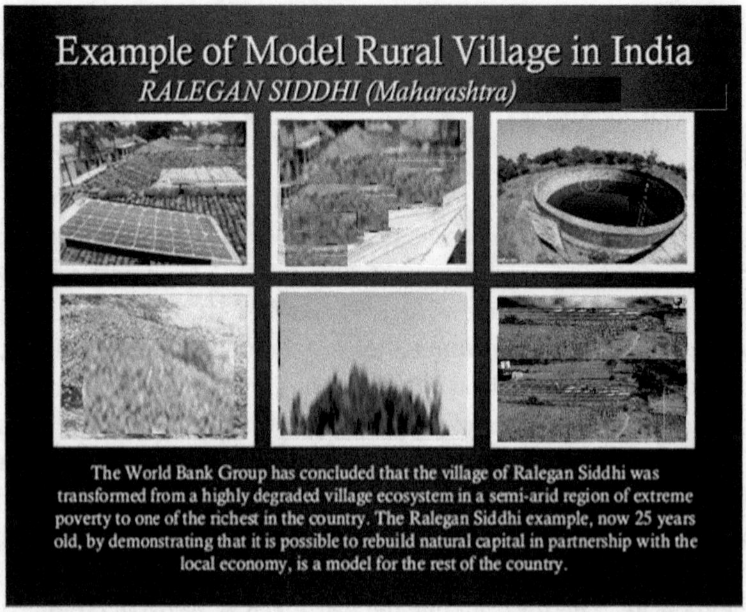

HIWARE BAZAAR

Hiware Bazaar is one of the village located in Ahmednagar District which has been suffering from drought conditions. During British rule, Hiware Bazaar village was full of joy and peace since it was surrounded by the hills therefore it had a pleasant weather during rainy seasons which resulted into wells and groundwater table to be filled with water all the year around.

This promoted agricultural crops with good yield and the farmers of the village chose agriculture as their main occupation for living and trading their crops in Mumbai. With time the things did not remained

Figure 4. Drought conditions in Hiware Bazaar

the same and with the growing population trees were cut down for wood and agricultural land were shifted towards construction of houses. With the decrease in the number of trees the soil left barren with excess of erosion and maximum water being wasted into the floods. This resulted into wash down of the fertile soils into rain and the village drowned down to adverse conditions. In 1972, things did not remain the same and the wells were left dry with the groundwater table went significantly down. Shortage of potable water led the people to lead a miserable life. People began shifting away from the village and the farmers' crop lands converted into deserted ones. Losing the means of their life farmers got addicted to alcohol and few committed suicide. By 1973, things got even worse with 80% people being addicted to alcohol and agricultural land turned into desert. With no means of survival Government officers were scared of being posted in the village, therefore only blacklisted candidates were posted there as a punishment. By the year 1988, the village was ruined in terms of economic, social and environmental aspects. The teachers in the school rushed away from the village and the schools were locked in waiting for the good teachers to return. This resulted into adverse education system in the village. Therefore it was found that not only elders but children also suffered from the scarcity of water in the village.

NEW BEGINNING IN THE WESTERN REGION

Hiware Bazaar Village

As discussed in the background it is clear that the village wanted a change for the need of water as their survival. The young generations saw some ray of hope for sustaining the village and requested Mr. Popatrao Pawar for becoming the sarpanch (Head) of the village in the election. By 1990 Popatrao Pawar was elected as the sarpanch on a condition that villagers must not quarrel among themselves. Pawar called out Gram Sabha for hearing the needs of the villagers and found out that the basic needs of the farmers in the village was water for drinking and agricultural purpose which had resulted an adverse impact on their health and environment.

The need for water in the village was given prime importance with other needs. To begin with Mr Pawar planted tamarind saplings which were destroyed by the villagers because it was not considered relevant for them. This promoted the sarpanch to educate people as his first step for the betterment of the village and also for sustaining water. The schools were painted with playground and education was started as an initial step in the village. The primary school led to high schools where teaching was made by the youngsters without charging any fee from them. The next step was to generate availability of water in the village. The villagers requested the forest department for initiating afforestation in their village but the overgrazing practice made by the villagers had made the farm barren. Therefore, their request was not taken by the forest department. In 1993, the villagers went to Ralegaon Siddhi and Aadgaon in Kolhapur to understand the water management practices implemented there. This visit taught them measures to fight with scarcity of water in the drought regions. In 1993, with the support of forest department around seventy hectare of village forests was regenerated and sustainable practices were made to conserve water. Contour trenches around the hills were found to be the most important measure for saving water from wastage and also for maintaining the fertility of the soil. Contour trenches in Government land led to plantation of about 45000 trees. Plantation of trees was carried out on hills which helped to conserve water. Pits were dug to store water and grazing was prohibited. The practice of cooking food in chulhas was also banned and a shift towards LPG (Liquified Petroleium Gas) was made in order to

Figure 5. Popatrao Pawar transformed the village from drought to a sustainable one

save the trees and avoid cutting down of wood. In 1994, a NGO called 'Yashwant Krishi Ram Va Panlot Vikas Sanstha' was established consisting of eleven members who planted local trees and limited the plantation of Eucalyptus and Subabul which consume a lot of water from the soil. In order to review the initial steps taken and continue for further development, the villagers felt that it was necessary to have a social audit body at place in the village. This led to audit in the village and by 1994 the transparency and accountability were regarded as the key for the village development. In 1994, twelve agencies were implemented under employment guarantee scheme for addressing water scarcity problems. Hence then Hiware Bazaar was selected as the Aadarsh Gaon Yojana by the Government of Maharashtra (Sakhuja, 2008). The Aadarsh Gaon Yojana was based on prohibition of grazing, cutting down trees consumption of alcohol and promotion of family planning and labors (Sangameswaran, 2006). Following were the main five principles which made Hiware Bazaar as the Aadarsh (Ideal) village.

1. **Ban on Tree Cutting:** Cutting down of trees for wood as fuel or fodder was strictly prohibited in the village. This led to the promotion of long lasting trees and thereby replacing the conventional chulhas with LPG.
2. **Restrictions on Grazing:** Grazing was also restricted as it resulted into tremendous soil erosion and wastage of water. As a remedy for the farmers having animals rotational grazing was thought to be a better means.
3. **Family Planning:** According to Malthusian Theory of Population it was found to control the growing population to meet the need for food. This practice had a limitation that family planning was not made a part of the health concern.
4. **Ban on Consumption of Alcohol:** The basic purpose of this step was to save the people from the evil effects of alcohol and also to minimize the gap between their debt and consumption. People addicted to alcohol were provided bank loans for purchasing animals in order to run their life.

5. **Shramdam:** This was a common practice which led to promotion of agricultural practices and water conservation measures for the village.

Several water harvesting structures and wells have been constructed in the village for storing water in times of need. The production of grass also increased from 100 to 6000 tons by the year 2004, which resulted to increase the livestock animals with the farmers and this resulted in increase in the milk supply (Singh, 2010).

Implication of Water Rules in the Village

Water rules were made for the distribution and use of water. The rules for distribution of water were bifurcated into two, one being for people who get access to water and another for the distribution of water. Water is supplied to lands as per the ownership and groundwater is accessed through private wells.

Impact of Watershed Management in the Village

The main effect of watershed management is an increase in the groundwater table and the percentage people being benefitted from the water also increased tremendously. This also led to increase in employment since the practices involved a great number of labours. Drinking water was supplied to every households and this resulted into good health of people. During drought situation water was available both for domestic as well as agriculture purposes.

Phand & Arya (2007) conducted a research study on Hiware Bazaar village of Ahmednagar District to find out the impact of Aadarsh Gaon Yojana in the villages and also the status of the village before and after the implementation of Aadarsh Gaon. The Aadarsh Gaon Yojana was implemented on 15 August, 1994. From the research conducted it had been found that the total village area remained the same before and after. The total area for plantation of trees had increased with good storage structures, nullah bunding, increasing number of wells and contour trenches. The cropping pattern grown in the agricultural land was also found to change as per the availability of water in the area throughout the season. With increase in the number of animals the total production of milk also increased. The areas having no schools were found to be led by two or more schools.

As regards to adaptation of sustainable measures in Hiware Bazaar for sustaining water, the contour bunding was practiced in the area of 414 hectare, cement nullah bunding in an area of 4 hectare and nullah bunding in an area of 51 hectare. The two measures helped in assuring conservation of soil and water for future needs. Afforestation was also considered as an important aspect for enhancing sustainability in the village in an area of 404 hectares. Tree plantation was made along road side covering a length of six kilometers. Storage as well as earthen structures was constructed in an area of 9 sq.km for conservation of water

SUSTAINABLE WATERSHED MANAGEMENT

A sustainable water use is defined as the use of water beneficial for all the users without reducing the need for other users being human beings, animals and also natural resources (Gleick, 1998). The demand for water is not just limited for fulfilling the need of water but also the wants of people for water in future.

Variation in the stock and flow of water that changes the availability of water at proper time and space and variations in the demand for water use disturbs the sustainability of watershed management. The disturbance in sustainability of water is mainly caused due to increasing human population and changing technologies. The first and foremost task to be done for sustaining water is to control the above two problems. Since the increasing human population requires enough water for meeting their needs and the technologies which is changing every now and then decreases the amount of water required for the social benefit. The sustainability of water can be managed by controlling the disturbances created by the four parameters namely, water requirement for human being, water requirement for environment, maintenance of water quality standards and the renewable water resources maintenance (Gleick, 1998). Maharashtra is the first state in India to take an initiative for sustaining water and prepared a state disaster management plan giving importance to water requirement in the state. It is also the first to regulate water resource authority for regulating water requirements and its availability.

Balawadi and Tandulwadi: These are the drought prone villages which faces problems for water availability for the agro based industries and plantation of sugarcane crops which requires a lot of water. Baliraja Memorial dam was constructed in South-Western Maharashtra which allocated water to the villages for meeting the farmer's irrigation needs as well as the household domestic needs of the villagers. The dam changed the living style of the village and addressed the water issues in the village.

AN APPROACH TO SUSTAINABLE WATERSHED MANAGEMENT

Das (2006) reported in his research different approaches and Programs for sustainable management of water. In the present era with the upcoming challenges of water, the availability of water for the fast growing population of the country is a matter of deep concern. In many parts of the rural areas availability of safe water is a major problem while few areas are facing scarcity of water. With this thought it has become very essential to take some initiatives for sustaining water for different uses. Various centrally sponsored schemes have been working for accelerating water in the rural areas by the implementation of minimum water tax for conserving water. During 1981-91, United Nations International Drinking Water Supply took an initiative step for sustaining water and concluded that drinking water must be paid or charged by the user in order to give it a value. Government also took an initiative step for making water available to the rural areas and came out with Sector Reform Program. The Sector Reform Program Swajaldhara comprised of four basic elements namely, demand for water, accountability, contribution of people and community participation (Das, 2006). The Constitution 73rd and 74th amendments act in the year 1992 focused on drinking water supply, water management and sanitation measures in rural areas from the Panchayats and Nagarpalikas (Corporation). The drinking water supply in rural areas are managed by the Ministry of Rural Development; Ministry of Water Resources at the central level; Public Health Engineering Department at the state level; Zilla Parishad, executive engineer from public health engineering department and district rural development agencies at the district level; and taluka Panchayat, gram Panchayat and village water and sanitation centre at the block level. The five year plan of Indian Govt. has brought different policies and Programs to come over from the issues of water scarcity. The problems of rural water are managed by the respective bodies as mentioned above in the state, central, district and block level. In 1950, constitution of India was set up and first national water supply was started in 1954. Then in the year 1956-1972, public health engineering department was set up at the state level to look after the rural water supply in towns and villages. Rural water schemes were sanctioned by financial terms in 1968 followed by the accelerated rural water supply Program in 1972 for handling the problems of the village related to rural water supply. In 1974, the Programs related to minimum need of water supply were introduced in states. The accelerated rural water supply Program identified the problems of the village for supply of drinking water throughout the year with a sustainable management of water in 1977. National Drinking Water Mission was initiated in the year 1986 for the application of conservation measures for sustaining supply of water. Ultimately the National Water Policy evolved in the year 1987 to look after the planning of different water resources and supply of drinking water in India. In 1991, the National Drinking Water Mission was renamed as Rajiv Gandhi National Drinking Water Mission. This Mission put forward standards for the supply of drinking water to the rural population. The year 1996-1999 began with audit of water resources with World Bank and other agencies. All the developments led to Swajaldhara Program in the year 2002 and the national water policy was revised by the National Development Council.

Several Government agencies, World Bank, German Development Bank and other non-Government organizations are indulged in the implementation of an efficient water supply in Maharashtra. The effort started with the dug well era from1960 to 1970, which provided several dug wells with power pumps. The year 1971-1985 led to the bore well era where bore well technologies came into existence leading to several hand pumps and power pumps. Rural piped water supply era arrived from the year 1985 to 1997 which led to increase in the number of piped water supply scheme in the rural areas. Several water projects were handled in this era. The year 1997 to 2000 began with the Master plan era which led to expenses for different water supply schemes. Since the year 2000 the policy reform era was evolved for delivering water in the rural areas and providing several sanitation measures in the rural area (Das, 2006). This made Maharashtra to be the first state known for launching Program for allocating water resources and sanitation services. The Government of Maharashtra has implemented several Programs for conserving groundwater. Shivkalin pani sathawan yojana is one such Program started by the Government of Maharashtra for storing rain water, rainwater harvesting, construction of tanks for water storage and other measures which can sustain water. A large number of Programs and schemes have been developed by Government of Maharashtra for conserving water.

The Government has also implemented two programs namely the Sant Gadge Baba Gram Swachatha Abhiyan and the Total Sanitation Campaign (TSC) based on the sanitation measures. The Government has focused the hygiene and sustainable practices for clean rural area.

CHALLENGES OF WATER IN WESTERN MAHARASHTRA

In rural areas of Maharashtra, water has been suffering with several challenges. The major issue of concern is about the water industry to work efficiently meeting the demand of water. It has been found that the groundwater has been over wasted for the private productive use and must be controlled. The first challenge faced by the Maharashtra sector for the availability of water is the technologies which interfere with the sustainable water management (Das, 2006). In 2002 the state Government of Maharashtra took an initiative to address the water resources in rural areas of thirty three districts of Maharashtra. After testing of water it was found that the water in such districts was contaminated. The districts like Jalgaon, Buldana, Akola, Washim, Wardha, Nagpur, Bhandara, Chandrapur, Yavatmal, Parbhani, Jalna, Nashik, Bid, Latur, Ratnagiri, Sangli, Sindhudurg and Solapur are facing problems regarding groundwater quantity as well as quality. The water table has been found to decline (increase in vadose zone) in the districts of Akola, Beed, Bhandra, Buldana, Chandrapur, Gadchiroli, Gondia, Hingoli, Jalgaon, Jalna, Kolhapur, Latur, Nagpur, Nashik, Nandurbur, Thane, Sindhudurg, Sangli, Solhapur, Satara, Ratnagiri, Parbani, Osmanabad, Yavatmal, Wardha and Washim with problems in taluka. The TDS, fluoride and nitrate content in water is high causing severe problems (Scott et al., 2005).

Eastern districts of Maharashtra are facing problem of high presence of iron in water while the ground water in Akola, Amravati and Buldhana are facing problems of salinity. One third of the area under Maharashtra lies in the rain shadow region mainly from the Western part of Maharashtra and Marathwada. The problem arises when such regions lying in the rain shadow regions cultivate high water intensity crops. Most of the rain shadow regions of Maharashtra consist of porous rocks which drains water quickly (Das, 2006). Government has led to Programs for increasing water supply for rural areas. Many wells have been dug to overcome water scarcity problems in Maharashtra. The limitation to such a program is that majority of the wells dug by the Government scheme are left dry in the summers with few of them aquifer polluted due to deposition of debris. Most of the problem arises due to lack of appropriate information about the status of water in the area or there is delay in the implementation of rural water scheme. In most of the rural areas people are resistant to take up the Programs or policies implemented by the Government easily (Das, 2006).

Farmers are very sensitive to water prices. The lowest water use efficiency is that point when farmers continue to apply irrigation till the net marginal productivity becomes equal to zero. The price shifts in the market results in the shifts in the selling price of water. The farmers generally do not go for water markets because the demand and supply do not match among themselves and the returns obtained from the crops are higher than the selling price of water. The farmers have to pay for the water on hourly basis and pay a percentage of their crop yield as a charge for water. The well owners charge water on the basis of the opportunity cost for excess water supply for irrigation. If water is charged on volumetric basis then the farmers would generally shift towards the cropping pattern which gives higher yield return from per unit of water consumed. This results into a limitation that supply of water for agriculture has been found to be limited so water is charged at a higher price for agriculture. The table given below reflects the status of availability of water in different districts of Maharashtra:

INITIATIVES TAKEN BY THE STATE MAHARASHTRA REGARDING MANAGEMENT OF SUSTAINABLE WATER REQUIREMENT

Maharashtra state is the first to carry out an initiative for sustaining water requirement in the state and is the first to prepare a state disaster management plan. The state has undertaken several managing plans regarding a sustainable use of water which are given in the table below (Table 3).

CONCLUSION

The Maharashtra state has several challenges regarding availability of water due to the structural location of the regions since the area has been classified into Eastern and Western Maharashtra. It has been found that there are few regions in the state which are drought prone while few of them are affected by floods, cyclones and earthquakes. Some regions receive adequate rainfall where as in few region rainfall is very limited. Diverse climatic conditions are seen in Maharashtra due to its location thereby

Table 3. Status of availability of water in different districts of Maharashtra

Sr.no.	Districts	Status of Water Availability
1	Nashik	Declared as drought by Government of Maharashtra in 2003. Affected by flood in the monsoon season.
2	Yavatmal	Flood prone areas
3	Parbhani	-
4	Raigad	-
5	Pune	Declared as drought by Government of Maharashtra in 2003
6	Ratnagiri	Can be hit by cyclones
7	Satara	Declared as drought by Government of Maharashtra in 2003
8	Sangli	Declared as drought by Government of Maharashtra in 2003
9	Solapur	Declared as drought by Government of Maharashtra in 2003
10	Sindhudurg	-
11	Beed	Declared as drought by Government of Maharashtra in 2003
12	Chandrapur	-
13	Gadchiroli	-
14	Kolhapur	Flood prone areas
15	Nagpur	Affected by earthquake
16	Bhandara	Affected by earthquake
17	Dhule	-
18	Jalna	-
19	Mumbai	Affected by floods, cyclones and earthquake.
20	Nanded	Affected by flood in the monsoon season
21	Ahmednagar	Declared as drought by Government of Maharashtra in 2003
22	Amravati	Flood prone area
23	Akola	Largest flood prone area
24	Thane	Flood in monsoon and drought in summers.
25	Wardha	Flood prone areas
26	Aurangabad	Declared as drought by Government of Maharashtra in 2003
27	Dharashiv	Declared as drought by Government of Maharashtra in 2003
28	Latur	Declared as drought by Government of Maharashtra in 2003

affecting the availability of water in few regions. Inspite of the above challenges the state has been able to overcome the challenges of water scarcity by adapting sustainable practices in the state for enhancing availability of water for future needs. In rural areas of Maharashtra, water has been suffering with several challenges. The major issue of concern is about the water industry to work efficiently meeting the demand of water. It has been found that the groundwater has been over wasted for the private productive use and must be controlled. The first challenge faced by the Maharashtra sector for the availability of water is the technologies which interfere with the Sustainable Water Management (Das, 2006). In 2002, the Government of Maharashtra took an initiative to address the water resources in rural areas of

Table 4. Managing plans undertaken by the State of Maharashtra for sustaining water

Sr. No.	Managing Plans	Characteristics
1	State Disaster Managing plan	Maharashtra is the first to prepare a state disaster management plan. The plan addresses the risks due to floods, cyclones, accidents and other disasters in the state. Such issues are undertaken by the state to solve the problem. The water disaster is also undertaken in the plan for the proper management.
2	District Disaster Managing Plan	The districts facing problems of disasters are undertaken in the plan by the Government of Maharashtra. These problems include earthquakes, floods, cyclones, industrial disasters and accidents which must not be ignored.
3	Disaster Management plan	The center for disaster management was set up in the month of august in 1996 with an objective to develop plans for preparing the state from any disaster like earthquakes, floods, cyclones for a sustainable water management.
4	District disaster management plan	The district disaster management plan has been developed to analyze the occurrence of disasters in the district and improve the preparation for overcoming the challenges of the disaster.
5	District Disaster Management committee	The district disaster management committee analyses the occurrence of disasters reviewing the threats in order to improve the fulfilment of the demand for water in the state.
6	Maharashtra state disaster management authority	Under section 14 of the disaster management act 2005 Maharashtra state disaster management authority came into existence in 24th May, 2006 involving chief minister, deputy chief minister, ministers, official members and the chief secretary.
7	State emergency operation center	The emergency operation center tackles the hazardous events which can lead to emergency situation.
8	Maharashtra water resource regulatory authority	It has been evolved with an objective to regulate water resources within the state and analyze the issues regarding use of water.
9	State executive committee	This committee was developed with chief secretary, additional chief secretary, member secretary and other secretary.
10	Hydrological projects	It was developed in the year 1996 in Nashik with an objective to improve hydrological data and assess the availability of water.
11	Dam safety organization	It evolved in 1980 in Nashik for the inspection of large dams.
12	Government department soil conservation	It was organized by Marathwada Sheti Sahayak Mandal in Adgaon at Aurangabad
13	Government department soil conservation and soil forestry	It was organized by local organization under Mr. Anna Hazare at Ralegan Siddhi in Ahmednagar
14	Aadarsh gaon yojna	It was organized by Yashwant agriculture and watershed development at Hiware bazar in Ahmednagar
15	Indo-German watershed Program	This Program was organized at Vijju Bhulgaon in Ahmednagar
16	Indo-German watershed Program	It was organized by SEWA at ambewadi in Beed
17	N.G.O	Organized by Manavlok at Bhavthan in Beed
18	N.G.O	Organized by Afarm at Damoli in Beed
19	Drought prone area Program	Organized by Sarwodaya welfare at nandgaon khadeshwar in Amravati
20	Drought prone area Program	Organized by Sarwodaya welfare at Nandgaon Khadeshwar in Amravati
21	Drought prone area Program	Organized by Chandrapuri maharaj shikshan prasarak at Pimpalgaon in Amravati
22	Drought prone area Program	Organized by Prabhodankar Thakre at Dharani Chikaldhara in Amravati
23	Drought prone area Program	Organized by Mahila Utkashtha at Hirangimangulpur in Wasim
24	Drought prone area Program	Organized by Shriram Sukhshainik at Sanglud in Akola
25	Drought prone area Program	Organized by Gomukh trust at Sangamner in Ahmednagar

continued on next page

Table 4. Continued

Sr. No.	Managing Plans	Characteristics
26	National watershed development Program in rainfed areas	Organized by Prabodhani at Bariewadi in Ahmednagar
27	National watershed development Program in rainfed areas	Organized by agriculture department at Chale in Pune
28	Agakhan foundation	Organized by Dilasa at Yevat in Yavatmal
29	Shivalin Pani Sathawan Yojana	The system of rain water recharge was launched. Drinking water was stored by different rain water harvesting structures.
30	Artificial recharge of water	This scheme was initiated by the central ground water board at viii plan at Warud
31	Recharge of aquifers	This scheme was initiated by the central ground water board at viii plan at Yuval
32	Maharashtra Artificial Recharge to groundwater	This scheme was initiated by the central ground water board at ix plan at Jalgaon district
33	Maharashtra roof top rain water harvesting	This scheme was initiated by the central ground water board at ix plan at Warud in Amravati district
34	Maharashtra roof top rain water harvesting	This scheme was initiated by the central ground water board at ix plan in KITS campus at Ramtek
35	Artificial Recharge structures	This scheme was initiated by the central ground water board xi plan in Rajbhawan in Nagpur

Source: http://nidm.gov.in/pdf/dp/Maharashtra.pdf

http://www.ctara.iitb.ac.in/water/waterplanning/maharashtra.pdf_

Watershed Development in Maharashtra: Present Scenario and Issues for Restructuring the Program. ö Technical Report. Pune: SOPPECOM.

http://cgwb.gov.in/documents/case-studies-rwh-ar-may2011.pdf_

thirty three districts of Maharashtra. After testing of water it was found that the water in such districts is contaminated. The districts like Jalgaon, Buldana, Akola, Washim, Wardha, Nagpur, Bhandara, Chandrapur, Yavatmal, Parbhani, Jalna, Nashik, Bid, Latur, Ratnagiri, Sangli, Sindhudurg and Solapur are facing problems regarding groundwater quantity as well as quality. Western part of Maharashtra has been affected by several measures of water scarcity. There have been several sustainable measures being adopted by the people of the villages of the Western Maharashtra which has made their living easier by allocating water for irrigation as well as household domestic needs. Not only the people of the villages even the Panchayat samiti and central Government has supported several schemes for sustaining water management in the Western part of the state. There has been a great participation of the central groundwater board regarding the sustainability of water in the eighth, ninth, tenth and eleventh five year plan. With instant application of the sustainable practices it has been found that several villages located in the Western part of the state are doing great work for enhancing water availability through storage of water in earthen structures as well as through rain water harvesting. Several other measures like afforestation, terrace farming, contour bunding, tree plantation along road side and limitations to grazing of animals have led to conservation of water as well as the soil. Villages like Ralegan Sidhdhi, Hiware Bazaar, Wadur, Yaval, Yuvraj and Wadgaon have come up with sustainable practices for allocation of water for future needs. Villages like Ralegan Siddhi and Hiware Bazaar which once had severe water problems are now called as the Aadarsh Gaon (Ideal Village) and are the true symbols for sustainable water management in a drought prone area. The Maharashtra Government has also come up with several schemes which have supported the sustainability of water. Several initiatives have been taken by the Maharashtra Government for overcoming the challenges in availability of water in the Western regions of the state. Supportive mechanisms has been followed to encourage agriculture in the Western areas and to control floods by the construction of bunds and dams which otherwise may have destroyed the villages and also will lead to soil erosion making the soil infertile for agriculture. Thus, in the current era of climate change, sustainable water management practices at the local level is the need of the hour to resolve the problems at state and national level.

REFERENCES

Das, K. (2006). *Drinking Water and Sanitation in Rural Maharashtra: A Review of Policy Initiatives*. Gujarat Institute of Development Research.

Dixon, J. A., & Easter, K. W. (1991). Integrated watershed management: an approach to resource management. *Watershed resource management: Studies from Asia and the Pacific*, 3-15.

Foster, S., Limaye, S., Mandavkar, Y., & Msangi, S. (2009). *A Hydrogeologic and Socioeconomic Evaluation of Community-based Groundwater Resource Management–The Case of Hivre Bazaar in Maharashtra-India*. World Bank GW-MATE Case Profile Collection, 22.

Gleick, P. H. (1998). Water in crisis: Paths to sustainable water use. *Ecological Applications*, *8*(3), 571–579. doi:10.1890/1051-0761(1998)008[0571:WICPTS]2.0.CO;2

Government of India. (2011). *Central Ground Water Board (CGWB)*. Ministry of Water Resources, River Development and Ganga Rejuvenation. Retrieved January 25, 2016, From http://cgwb.gov.in/documents/case-studies-rwh-ar-may2011.pdf

Government of India, Ministry of Agriculture. (n.d.). *Water resources of Maharashtra State*. Retrieved January 25, 2016, From http://www.ctara.iitb.ac.in/water/waterplanning/maharashtra.pdf

Kumar, M. D., & Singh, O. P. (2001). Market instruments for demand management in the face of scarcity and overuse of water in Gujarat. *Water Policy*, *3*(5), 387–403. doi:10.1016/S1366-7017(01)00081-2

Moench, M., & Kumar, D. (1995). Distinctions between "Efficiency" and "Sustainability": The role of energy prices in groundwater management. *The Challenge of the Balance: Environmental Economics in India*, 305-309.

Most Urbanised State, & the The Times of India. (2011). *NIDM Maharashtra National Disaster Risk Reduction Portal*. Retrieved January 25, 2016, From http://nidm.gov.in/pdf/dp/Maharashtra.pdf

Phand, S., & Arya, H. P. S. (2007). Aadarsh Gaon Yojana-Way Towards Gandhi's Dream Village: Self-Sufficient, Prosperous and Peaceful. *Journal of Rural Development*, *26*(2), 207.

Sakhuja, N. (2008). Hiware Bazaar-A Village with 54 Millionaires. Down. *Earth (Waukesha, Wis.)*, 16.

Sangameswaran, P. (2006). Equity in watershed development: A case study in Western Maharashtra. *Economic and Political Weekly*, 2157–2165.

Scott, Wilson, and JPS. (2005). *Rural Water Supply and Sanitation Project.* Maharashtra with World Bank Assistance Sector Status Study.

Sharma, M. (2006). The making of moral authority: Anna Hazare and watershed management Program in Ralegan Siddhi. *Economic and Political Weekly*, 1981–1988.

Singh, S. (2010). Hiware Bazaar - A village of millionaires. Maharashtra.

Western Ghats as world heritage site. (2012, July 2). *The Times of India.*

Chapter 15
Participation Framework to Sustainability:
The Undercurrents in Bottled-Water Production and Consumption

Taksina Chai-Ittipornwong
Muban Chombueng Rajabhat University, Thailand

ABSTRACT

The topic of bottled water is particularly timely since water crisis and waste plastics are shifting from one of problem-framing to what is much more concerned to climate agenda. As plastics are everywhere to make life possible for a faster and convenient pace of society, bottled water has become one of the most disposable products to be consumed and disposed of, on a regular basis, both at and away from home. These eventually translate to consumer wastes causing a profound impact to the environment. The more bottles mean more greenhouse gases from production, transportation, waste bottles management, including fresh water supply. These adverse impacts are compounding a framework to justify as to how producer and consumer apply, individually and within the whole product life cycle, to sustainable consumption and production practice (SCP).

INTRODUCTION

The world has been pushing the planet away from sustainability. Humanities are shaped their lives on materially intensive economies and consumerist lifestyle regardless of living in a finite world. Many disposable products are developed to make life possible for a faster and convenient pace of society. Bottled water has become one of the most common plastic products to be consumed and disposed of, on a regular basis, both at and away from home. These bottles eventually translate to consumer wastes (Amano & Ness, 2004; Chamornmarn, 2008; Huber, 2010). Water and plastics are two major materials of bottled water. More bottles mean more greenhouse gases from production, transportation and waste management, including continued risk of ecological vulnerability and water conflict with commercializing

DOI: 10.4018/978-1-5225-1046-8.ch015

a public resource. Without greater understanding of how to produce and how to consume bottled water in sustainable ways, it may even worsen the most disagreeable impacts to the planet and future livelihoods.

Thailand was ranked near the top in bottled-water consumption in Southeast Asia and is among the top 9 consumers globally (Rodwan, 2010) In global context, bottled water has been the top end-used application of Polyethylene terephthalate (PET) plastic since its introduction into beverage packaging (Willett, 2003) According to the survey during 2008-2010, bottled-water business in Thailand maintains annual growth by 15 percent in total volume terms, with the consumption of 40,000 tons of PET plastic and 2.4 billion liters of water (Chamornmarn, 2008; Euromonitor International, 2010; Ratthanin, 2010).

The bottles can constitute various pollutions to air, water and soil quality as carbon substance is the key element of PET resin formulation. Increasingly, the ways to capture surface water supplies and underground sources are compounding the problem of water inadequacy into environmental degradation. Precisely, this chapter deals with an evidence-based case of sustainable consumption and production (SCP) implementation for PET-bottled water. Consumers' environmental awareness is evaluated whereas an input-output approach to bottling system is included to address the role of bottlers in an SCP. The undercurrents of bottled-water production and consumption are therefore closely examined for filling the gaps between socio-economic development and environmental protection.

The presumption is that if enough bottlers implement sustainable production whereas more and more consumers are aware of sustainable consumption, they can drive change towards a sustainable consumption and production of PET-bottled water. The guiding research questions are as to

1. Whether the bottled-water businesses and the consumers perform the SCP for PET-bottled water?
2. How the practice is? and
3. What the differences are in terms of resource efficiency and impact reduction?
4. How is an LCM approach contributed to the SCP success?
5. What is the most practice for bottling drinking water done in Thailand?
6. What are the environmental and health consequences occurred from PET bottles?
7. Is there any different impact compared between virgin and recycled bottles?

BACKGROUND

Sustainable Consumption and Production

In 2002, all the nations facing the climate crisis agreed not only to develop a 10-year framework of the SCP programs (the Marrakech Process)[1], but also had the SCP addressed as a prerequisite for sustainable development. The vision of sustainability was first referred to in *Our Common Future* in Bruntland Report defining sustainable development as a pathway to meet the present needs without compromising the ability of future generations to meet their own needs (Elliott, 2006).When environment lost capacity to continuously support the overexploitation placed upon it, a stronger and sensitive linkage between sustainability and socio-economic growth is truly required (Akenji & Bengtsson, 2009). Hence, sustainable practices in resource efficiency from less material consumption and less pollution have been universally addressed to handle the consequences of economic globalization and bring about sustainable development (Jones, Hillier, & Comfort, 2014).

Definitions of SCP have been internationally discussed in many respects. SCP is linked to the adoption of sustainable lifestyle to meet the increasing demand for goods and services of both present and next generations, while improving the level of emissions and energy used or released over a product lifecycle (Giljum, Burger, Hinterberger, Lutter, & Bruckner, 2011). SCP implies the ways that humanities who live in a finite world protect the planet away from sustainability (United Nations Department of Economic and Social Affairs [UNDESA], 2010). SCP becomes an effective framework for improving resource efficiency and impact reduction occurred from the production and consumption of products and services (Cellura, Longo, & Mistretta, 2011). As such, SCP is not only essential for decoupling socio-economic growth from environmental degradation, but is also behind the success of sustainable development (Economic and Social Commission for Asia and the Pacific [ESCAP], 2011; United Nations Environment Programme [UNEP], 2015).

Many empirical studies exhibit the outcomes of SCP approach. The interaction analysis of consumption and production in China revealed that consumer's demand for energy-intensive product was the driving force of pollutant emissions (Liu, Wang, Yang, & Shi, 2010). It was substantial to apply the SCP framework for identifying energy and environmental impacts from household consumption (Cellura, Longo, & Mistretta, 2011; Zhao, 2009). An SCP offered better solution for impact management and sustainability objectives (Tukker, 2008). Resource efficiency and pollutant emission reduction were the basis of energy sustainability under the SCP domain (Sampattakul, 2010).

In completing the SCP of products and services, the roles are shared among four parties: producer (business), consumer, government and other stakeholders like retail and distributer (ESCAP, 2011). Businesses and consumers can apply individually and within the entire product life cycle for a reach of SCP (Hoff et al., 2014; Parent, Cucuzzella, & Reveret, 2013). The contributions involve responsible industrial development from supply side as well as an individual change to sustainable lifestyle in demand side (Brodhag, 2010; Raskin, Electris, & Rosen, 2010). State entity is a regulator to deliver a top-down policy to the concerns, facilitate the treatments and enforce the law and regulation. The other three elements those from supply and demand sides stand for bottom-up participation. Businesses can respond to the SCP of PET-bottled water at the early stage of product design, during bottling processes, product distribution, or even extend to the end of the useful life of product (Akenji & Bengtsson, 2009). Consumers can adopt the SCP through the patterns of use and waste handling.

A trade-off between resource efficiency and impact reduction targets is referred to in every step of SCP implementation because the valuation of resource efficiency is actually linked to a reduction of environmental impacts. Thus, the approach to impact prevention and reduction is often discussed as a component of resource consumption practice.

1. **Resource Efficiency:** On an everyday basis, natural resources are extracted and transformed to provide goods and services for supporting all social and economic activities. When people use fewer resources and generate fewer emissions, they minimize natural resource consumption as well as maximize its all benefits (Niccolucci et al., 2011; Prakash, Yogesh, Viraja, & Shilpa, 2015). Resource efficiency could be expressed in accordance with the domestic environment and import materials and goods. Consumption efficiency is based on a reduction of resource intensity, material and product flexibility, utilization of all resource benefits and applying saving technology (ESCAP, 2011). The adoption also deals with possibilities of reuse, recycling and recovery approaches, reduced risk to resource scarcity, lower environmental cost on less wastes and pollution generation and lower prices for materials and products (Liu et al., 2010; Pogutz & Micale, 2011).

Eventually, resource efficiency can deliver the rebound effects not only to increase a demand for the goods and a growth in resource consumption, but also contradict efforts to save resources for greater efficiency.

2. **Impact Reduction:** Treatment on impact reduction implies the ways of utilizing resources and processes to eliminate pollutants and wastes generation both at the source and within the use (Vasami, 2010). Environmental and human health impacts are caused by both natural events and human activities those involving with procedures in producing and consuming goods and services. Reducing environmental impact profile includes the attempts to conserve finite resources and prevent natural resource depletion together with decreasing possibilities of greenhouse gases emission, pollutants and wastes generation (Brodhag, 2010; Veshagh & Obagun, 2007). Additional is energy used or released over the life cycle of goods and services. Potential impacts could decrease if businesses prioritize clean technologies in producing goods and services, whereas societies are aware of the precedent consumerist-lifestyle (Raskin et al., 2010; Termpittayapaisith, 2011).

PET-Bottled Water

1. **Life Cycle Management:** On a linear basis, a product is developed, produced, distributed, consumed and disposed. A product life cycle management (LCM) is very close to a reduction of carbon and ecological footprints by minimizing socio-economic burdens as well as maximizing environmental value and sustainability (Power, 2009). When LCM is reconciled with products and services, a set of activities and processes is applied to the whole product life cycle for improving impact reduction, environmental protection and sustainability (Ramani et al., 2010). LCM helps in improving the impact profile generated from every interaction with environment, directly or indirectly, both at present and in future (Harris, 2011; Hertwich, 2005; Kuczenski & Geyer, 2010).

   LCM can be diversified to three perspectives: a cradle to gate, a cradle to grave, and a cradle to cradle (DuPont, 2011; Gironi & Piemonte, 2010) A cradle-to-gate LCM involves with the stages of material acquisition and manufacture as belonging to inputs and outputs from production operation in factory only. A cradle-to-grave LCM, which is an extension of the cradle-to-gate boundary, includes contributions to the retail, the use and the end of the useful life of the product. When businesses possess a partnership with customers and suppliers, the LCM is implemented from a cradle of product to a grave by thinking about sustainable processes to access sources of raw materials, to deliver materials and products and to foster environmentally-friendly practices for the end-of-useful life management. Towards a cradle-to-cradle LCM, the involvement with all stages in a product life cycle is required so that all wastes are sustainably discharged and recovered for starting a new product life cycle.

   In identifying potential impacts from PET-bottled water, the LCM begins with an initial stage of material extraction (plastics and water) to material processing (productions of PET bottles and water disinfection), product manufacture (bottling system), distribution (retail of bottled water), use and reuse (bottled water consumption) and end-of-useful life management (waste bottles management) (Nestlé Waters North America, 2010). A profile of impact inventory relates to certain amount of finite resource, a series of transformations and releasing emissions of diverse natures from landfill, burning, incineration, recycling and recovery used for handling the wastes (Hauschild, Jeswiet, & Alting, 2005; Huber, 2010; Song & Hyun, 1999; Takahashi, 2010).

In view of many researchers, toxic substances, such as antimony, aldehydes and carcinogenic chemicals, are reasoned to health consequences caused by improper use and multiuse (Schmid, Kohler, Meierhofer, Luzi, & Wegelin, 2008; Shotyk & Krachler, 2007; Westerhoff, Prapaipong, Shock, & Hillaireau, 2008). The cause relates to direct contact between mouth and bottle after leaving it in sunlight exposure at longer time because time is dominant factor to the organic substances leaching out from PET bottle (Mutsuga et al., 2006). Logistics and transportation in distribution management are also the inclusive factors to reach efficient use of fuel energy and carbon dioxide emission reduction in the LCM of PET-bottled water (Amano & Ness, 2004; Coelho, Castro, & Gobbo, 2010; Foolmuan & Ramjeeawan, 2012; Huber, 2010; Takahashi, 2010)

a. Life cycle inventory of PET bottles. Basically, PET resin is a thermoplastic condensed between ethylene glycol with terephthalic acid forming long chains of repeating units of carbon, oxygen and hydrogen. When producing PET bottles, the resins are melted and polymerized till feasible for a pre-form injection in latter conjunction with a blow molding process. The processes consume both materials and energy while emitting much of chemical substance, including dust and debris of pollutions and acetaldehyde from the polymer degradation (Foolmuan & Ramjeeawan, 2012). Much of the fossil fuel used in the PET bottle derives from feedstock energy, which is truly bound within the product (Zhang & Kimura, 2007).

   Referring to ISO 14040[2] and 14044[3], life cycle inventory (LCI) deals with the interpretation of impact categories. It means LCI implies all input-output unit processes quantified along with a product system so that environmental impact profile can appear in classification and characterization (Gironi & Piemonte, 2010). Hence, LCI refers to a compilation and quantification of input-output data against the environmental profile of a product. For examples, the impact category of global warming represents the inventory item of carbon dioxide (CO_2) while cadmium (Cd) and lead (Pb) inventories stand for the impact category of eco-toxicity and human toxicity when exposed to toxic substances.

   As with the data in many researches, the LCI of PET bottles increased to the highest level when 100 percent of used bottles went to landfill (Takahashi, 2010) The waste bottles approximated 14 percent of post-used bottles were recycled (Huber, 2010) The balances ended up in landfills, lakes and streams leading to many impact consequences. PET bottles could be recycled in multiple times and into a variety of end products (Hurd, 2000) The production processes and transportation of PET polymerization and PET bottles generated higher impact on global warming than did the categories of nonrenewable energy, acidification and eutrophication (Gironi & Piemonte, 2010).

b. Recycled PET. Among all plastics types, PET resin is defined as the most recyclable plastic corresponding to the symbol of number 1 in the loop of chasing arrows. In early 1978, PET bottle was first recycled into a fiber product (Willett, 2003). It is evident that recycled-PET plastic is possibly used as 100 percent virgin material for new applications as the scraps from PET bottles could recapture resin properties for a second life either in bottles or new applications (Coelho et al., 2010; Shen, Nieuwlaar, Worrell, & Patel,2011; Vakili & Fard, 2010). In case of bottle to bottle application, the outcomes are based on the efficient use and cost effectiveness, except environmental profile (Amano & Ness, 2004; Vasami, 2010).

   LCI of recycled PET can be changed towards many conditioning factors those including recycling rates, recycling system, sorting process, modes and distances for transport of the wastes (Vakili & Fard, 2010). Recycled PET could be optimized within three recycling trips.

The multiple-recycling trips (the useful number of lives of the material) also influenced on the LCIs of both fuel energy and electricity used for processing operation and greenhouse gases emissions (Gleick & Cooley, 2009; Song & Hyun, 1999). However, there was no difference of CO_2 released from producing 10,000 units of 12-ounce PET bottle compared between virgin PET and a mix of recycled resin (Franklin Associates Ltd [FAL], 2010a, 2010b). It was revealed that the recycled resin alternated better consequences than the reuse, in terms of cheaper recycled cost and easier recycling process (Gironi & Piemonte, 2010)

c. Recycling system. Recycled PET is actually processed by two methods: mechanical recycling and chemical recycling (Hurd, 2000; Ida, 2006). Mechanical recycling is defined for material recycling, whereas chemical recycling serves feedstock recycling. Both methods have a typical processing sequence beginning with a collection and transporting, sorting and separation, and reprocessing.

In mechanical system, PET plastics are crushed and grinded into small flake for further contaminants separation. The flakes are washed and dried before feeding into an extruder where they are heated to melting state and converting into recycled PET manufacturing.

Feedstock recycling, which waste PET is chemically recycled, refers to a de-polymerization process, in which plastic polymer is broken down into the constituents-monomers. There are three main methods in PET chemical recycling: glycol for glycolysis, methanol for methanolysis, and water for hydrolysis (Vakili & Fard, 2010). In identifying the best device, the results of desired products yield and environmental impact profile are first accounted. Particularly, the category of global warming is the major impact from chemical recycling because the inventory of CO_2 derives from fuel-related emissions, not process emissions (Hurd, 2000; Vasami, 2010). Mechanical recycling is best to recover the waste bottles for textile and yarn applications whereas the bottle application is limited to chemical recycling. A standardized procedure also influences to have waste PET recycled effectively. Development of recycling technology for waste bottles is still continued in demand to be reconciled with the growing market of PET-bottled water.

2. **Bottling System:** Upon a basis of bottling processes, the inputs of plastics and water resources are required, including utilities of electricity and fuel energy, while delivering the outcomes of bottled water and the impacts. The bottling system featured in Figure 1 displays the used resources and the outcomes against the input-output approach to PET-bottled water. Bottling system of drinking water in PET container is technically classified as the open-loop and the closed-loop processes. The difference is that the close loop begins with producing PET stretch-blow-molding bottles from PET pre-forms while the open loop sets the production of PET bottle apart.

Towards the closed-loop process, production lines of bottles manufacture and filling stage are automatically connected together. In filler room, the closed loop begins with using PET pre-forms appearing in tube-formed objects as intermediate materials to produce the formed-bottles. Then, the bottles fed by electronically-automatic system are moved forward to dust detection and elimination with an infrared light scanner and a high speed blower before filling. The infrared light is also equipped to limit the volume of filled water in every bottle prior capping and labeling stages. The closures and labels are generally sourced from plastic molders, instead of the bottler's own production. Label application is an intermediate step following capping stage in the closed-loop process. The open loop is a semi-manually-conventional process, in which returnable bottles are possibly refilled. The loop initially proceeds with a feeding of the formed-bottles, which are usually hand fed onto a conveyor to be filled with drinking water passing through orifice at a continual rate of flow. As would be true, the open loop produces the lost-filled water approximates 30 percent compared to the close loop. On the other, the bottlers investing in the open-loop process can both exclude PET bottles production from bottled-water business, or separate filling and capping stages (filler room) from labeling and bundle packaging. In case that closure wrapper is specially required, the open loop enables a separate operation of cap-sealing. It means the open loop allows the bottle to be labeled either before or after filling step. The water is also shared for cleaning the inside of the bottles before filling.

ENVIRONMENTAL AWARENESS

Environmental impacts and human health risk would decrease when societies are aware of the precedent consumerist-lifestyle for daily pleasure. Humanities generally behave to satisfy the opportunity to consume and the sense of fulfillment and satisfaction upon the experience of social life. Consumer behavior is therefore mediated through the consumption and production practices those certainly covering emissions, energy and water use, and waste produced.

When unsustainable consumption and production pattern stands out in public realm, environmental awareness becomes a stronger and sensitive device to sustain the planet. A consumption and production pattern exhibits the degree to which the market economy has shaped the surroundings and eventually the way of living (Shobeiri, 2005). In contemporary society, humanities are induced in adjusting their lifestyle towards low-carbon pattern by consuming reduced material and less energy of functional utility. This is due to the fact that consumption activities are demanding nature's resources and producing greenhouse gases at a faster rate than nature regenerated and reabsorbed (Liu et al., 2010; Sze, 2010)

Figure 1. Bottling system of PET-bottled water on the input-output approach
Source: Author, doi:10.4028/www.scientific.net/AMR.807-809.2897

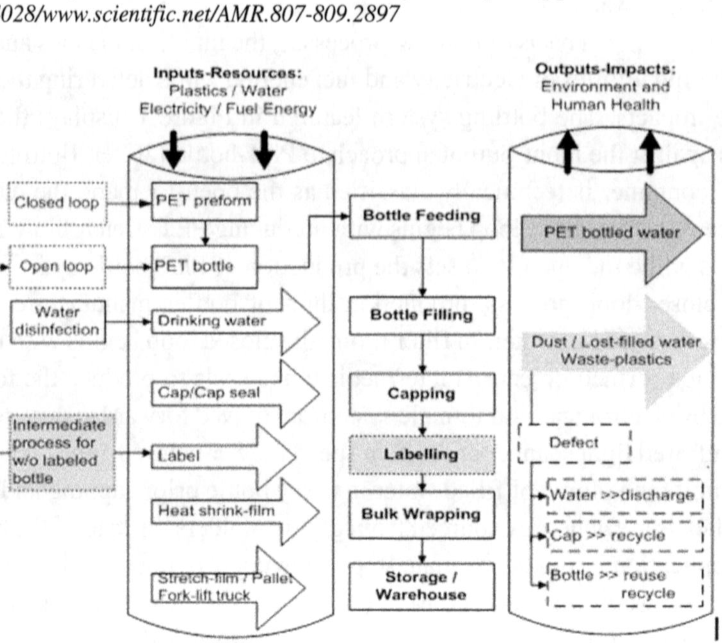

Consumer behavior is not simply about the act of purchase but is rather a thoroughly cultural phenomenon that serves to legitimate capitalism on an everyday basis (Poonuchaphai, 2010) . The social experiences eventually constrain a wider recognition of sustainable lifestyle. When asked to rank what was most important to Thai people aged 13-18, playing LINE application in smart phone came first. Saving environment employed the lowest interest (Suwajittanont & Chai-ittipornwong, 2011). A sense of responsibility to care, protect and enhance environmental quality can be concerted to people if more and more environmental issues are documented through mass media, with the eco-journalism basis those relevant to verifying, informing, educating and giving a warning to news receivers. (Kaweeya, 2006).

Humanities are actually shaped their lives on social action with various factors, including the external influences of norms and values as everyone wants to be inside the mainstream behavior (Berger & Corbin, 1992) Level of environmental awareness can interpret as to how people view themselves on social agenda. The more people have education and information, the more they could account for environmental protection and a change in behavioral perception (Larijani, 2010). Environmental awareness can be expressed in the manners of environmental prevention and protection on waste and pollutant emissions, coupled with the increasing efficiency in resource consumption (Harris, 2011; Peattie, 2010). A new kind of consumers' relationship with the public realm could be constructed if environmental protection and management are truly converged with economic and humanistic values those within bio-geophysics, way of thinking and social interaction in a social system (Flor, 2004).

In psychological view, the stages of covert and overt activities and experiences, which individuals engage in their behavioral patterns, are conceptualized for an interplayed set of thought and action (Peacocke, 2006). A conceptual form of human behavior is actually linked between perception and action. Person's actions are expressed with the intention to environmental protection can represent an individually-environmental behavior linked between areas of action and the readiness to act (Brodhag, 2010). Thus, the gap between bodily action and mental action belongs to the interplay of awareness and action against consumption behavior.

In business scale, the institutional attributes to environmental issues deal with a corporate social responsibility expressed in the formats of policy, plan, project and activity. The responsibilities determine the level of environmental awareness consistent with reducing unnecessary resource depletion and sustaining resource benefits together with creating social value along a product LCM (Power, 2009; Schultz, W., Nolan, J.M., Cialdini, R.B., Goldstein, N.J., & Griskevicius, V., 2007). When businesses are conscious to reducing environmental disruption and integrating environmental care into a corporate vision, a product value chain is simultaneously reconciled with business sustainability (Po gutz & Micale, 2011; Raskin et al., 2010).

MAIN FOCUS OF THE CHAPTER

Potential Impacts from PET Bottles

Within characteristics of food container, PET bottle is considered a safe, healthy and convenient packaged food product as much as a more sustainable container, in terms of CO_2 emissions and energy consumption, compared to aluminum can and glass bottle (Feron, Jetten, De Kruijf, & Van Den Berg, 1994; Lehmann, Vilaplana, Strömberg, Suliman, & Rodriguez, 2005). In this section, the author set an assessment against a content ratio of resin quality for greater understanding of the impact variation from PET bottles. A tested scenario was simply arranged to three parameters: virgin PET resin, half-by-half between virgin and recycled resin, and 100 percent recycled PET resin, with functional unit of one kilogram of 500-milliliter bottle. As it was an additional part in this chapter, the parameters of virgin resin and recycled PET were based on the literature LCIs from the SimaPro 7.1 software supported by the National Metal and Materials Technology Center in Thailand. These data were computed to support the half-by-half parameter (a mix ratio of virgin-PET resin and recycled PET by 50 percent for each). Three potential impacts were focused on the categories of global warming potential (GWP), eco-toxicity and human toxicity. GWP was expressed as kg CO_2 eq. inventory (kilogram of CO_2 equivalent), while contamination of para-dichlorobenzene (1,4-DB) substance in fresh water aquatic, marine aquatic and terrestrial ecologies were translated to inventory data of both eco-toxicity and human toxicity.

As drawn in Table 1, producing PET bottles with the virgin PET delivered better result in impact reduction of every category whereas the recycled PET valued for finite resource efficiency. The highest amount of GWP (2.8108 kg CO_2 eq.) came from the recycled PET, while the mix ratio was highly sensible for the impact categories of eco-toxicity (675.5046 kg 1,4-DB eq.) and human-toxicity (0.9852 kg 1,4-DB eq.). Towards the eco-toxicity result, the marine aquatic resource was much more vulnerable than did the fresh water aquatic and terrestrial ecologies, in all parameters. It was obvious with regard to the test that PET bottles from 100 percent of virgin resin exhibited better environmental profile. Nevertheless, the discrepancy in the parameters is possibly due to quantities and types of input resources; material and utilities used upon the recycling methods. Either are selection of assigned functional unit and the coverage on PET recycling method. Sources of CO_2 emission those including process energy, transportation and non-process emissions during material management can also differ the impact profile of recycled PET. The chemical recycling requires a lot of energy in spite that there is no energy possibly recovered by landfill. The impact category of GWP between the energy use in landfill operation and the recycling system are different. That's why waste-PET bottles are mostly valuable for recycling to new applications of textiles, sheeting for vacuum packaging, office stationery and detergent bottles, instead of bottle to bottle in drinkable water application. However, LCI data varies within five percent is not considered significantly different (FAL, 2010a).

Role of Bottlers in SCP

As could be expected, bottled-water manufacturing is one of the fastest-growing businesses in Thailand. The business value approximates USD 300 million has the growth rate maintained by 15 percent since 2007. Among 6,000 bottlers throughout nation, 65 percent of the market is captured by four leading brands consisting of two local entities; Crystal and Singha, and two international brands; Namthip and Purelife. The remaining 35 percent are shared by the bottlers regarded as small and medium enterprises (SMEs) that partly own their brands and also provide bottling service as an original equipment manufacturer (OEM).

PET-bottled water has widely dominated Thai consumers, and caused the decline in returnable glass bottles and the opaque bottles made of high density polyethylene polymer (HDPE bottles). The SME bottlers are most likely to produce a 500-milliliter-water weighted bottle. Either are the individual cups by 150 milliliter, or smaller bottle sizes of 330-350 milliliter. The larger pack sizes of 1000-1500 milliliter serve the highest sale volume of leading brands.

It is interesting that varieties of sustainability framework are particularly addressed for bottled-water businesses in Thailand. The measures to say exclude the FDA registration and good manufacturing prac-

Table 1. A test scenario of potential impacts from PET bottles based on content ratio of resin quality set for three parameters: virgin PET resin, half-by-half between virgin and recycled resin, and recycled PET

Parameters (% of Quality Resin)	Impact Sensitivity (Functional Unit:1 kg of 500 ml-PET Bottles)					
	GWP ($KgCO_2$ eq.)	Eco-Toxicity (Kg 1,4-DB eq.)				Human-Toxicity (Kg 1,4-DB eq.)
		Fresh Water Aquatic	Marine Aquatic	Terrestrial	Total	
Virgin PET	0.0494	0.0014	11.5351	0.0003	11.5368	0.0088
Virgin + Recycled 50:50 (%)	2.7364	0.0962	675.3979	0.0105	675.5046	0.9852
Recycled PET	2.8108	0.0033	544.8582	0.0022	544.8637	0.9210

tice (GMP) certification as they are mandatory bottled-water manufacture. Global quality certifications, such as ISO (International Standards Organization) 9001 for quality management system, ISO 14000 for environmental management, ISO 22000 for food safety and ISO 26000 for social responsibility, are accounted by all leading brands and some SME brands.

1. Resource efficiency:
    a. Plastics. In PET-bottled water business, efficiency in plastics consumption mostly begins with a lighter-weight bottle. A bottle with thinner wall, shorter neck (from 18 to 12 millimeter) and narrow cap (from 32 to 28 millimeter) is primary designed for reducing PET plastic consumption. PET-bottled water can employ a sustainable design or design for environment from the initial stage of development of bottle to bottling process and logistic operation. In drinking water business, the design in bottle shape and bottle body together with bottling process really compromise with the efficiency in reducing waste-in-process plastics, which can be caused by several threats, such as bottle-mold quality, uneven thickness, deviations in the spiral neck, bottle clarity, leakage and off-centered bottoms.

       As results, the efficient use of plastics material is initiated from bottle-mold preparation. The design in bottle neck is also advantageous for common cap of all bottle shapes and size. Included is increasing toughness of bottle bottom for better stiffness, which is beneficiary for multi-layer stacking and storage of bundle packaging normally holding twelve bottles. The design for multi-layer stacking can helps in decreasing trip of distribution that affects reduced fuel energy consumption and distribution cost.

       Among the SMEs, a device of bottle-shape harmonization across brands is another efficiency to enhance sustainability in plastic resource consumption. The leading brands do the contrary. The varieties of bottle shape and sizes, which are not valued by the SMEs, are used among the leading brands to manage consumer perception to brand identity and satisfy the requirement at all aspects. Neither leading nor SME brands own the accessory production of closures, films, labels and bulk-wrapper though some SMEs experience in plastic molding. Selection of bottling process also affects the success of SCP implementation for PET-bottled water. The open loop is generally worthwhile for the brands familiar with plastic injection and blow-molding production. They can utilize their knowledge and skills to establish a cross-sector business regarding plastics injection and blow-molding and bottled-water manufacturing service.

Additionally, the most lighter-weight bottle approximates 12-13 grams for reduced plastics consumption by 30 percent and easier squeezing after use for reduced space of landfill volume. This green bottle is neither attractive nor feasible for the bottlers, except a leading brand. Due to the thinnest wall and the slightest bottle bottom, these special bottles require additional investment in nitrogen feeding connected to the bottling system. The process deals with an installation of gas feeding system and nitrogen gas reserve in a tank those requiring a more safety operation.

Recovering waste-PET bottles to new applications is less taken. The interest in advancing the recycling technology is quite low, too. A huge demand for PET-bottle scraps can distort the bottlers' interest in renewable PET technology. Thus, waste PET is less frequently recycled among Thai bottlers, with remarking on the relevant laws and regulations prohibiting use of recycled plastics for food packaging. As waste plastics are a major concern in the bottling system, amounting to 2.0 percent of all materials, the limitation of waste-in-process bottles is arranged at 0.2-0.3 percent. These wastes are sorted and sold to recycling agents since these flakes are mostly required for export market in China and India. Indeed, there is no difference between the leading brands and the SME bottlers in handling waste plastics from bottling system. The wastes are directly turned to money, instead of renewal approach.

b. Water a rapid growth of bottled water business is truly related to depletion of surface-water supplies and groundwater sources. The leading brands in Thai market mainly use underground water to produce bottled water while the SMEs bottle tap water. Efficiency in water resource is not the major concern among Thai bottlers because of the low cost of water acquisition. Nevertheless, tap water is the option providing insurance against shortages, as does a reserve of underground water that should supply the companies for approximately five years.

The open loop processes 65 -70 percent drinkable water ends up in the bottle. The rest is wasted. The lost-filled water can be reused for chilling the process before drainage, however, it cannot be re-circulated to clean empty bottles or bottled and sold. The bottlers using the open loop should be aware of the substantial waste of drinkable water, which can reach 30-35 percent, which can be reduced through the close loop.

The other sensible point of water efficiency in bottling system is that the water is both a final product and a cleaning product. The open loop consumes drinking water for cleaning the inside of the bottles while the close loop processes with a high-speed blower. The discharged water is treated as clean is not appropriate for filling bottles under food-contact regulations. The closed-loop process, which seems newly for Thai-based bottling manufacturing, is equipped for all the leading brands. The SMEs, including every branch of leading brands, are common to the open loop. However, it would be inconceivable for thousands of the SMEs to switch to a closed-loop system.

2. Impact Reduction:

a. Plastics. The amount of PET plastics used in bottled-water businesses in Thailand is quite interesting. The SME bottlers consume PET plastic amounting to 120 tons per month, while the leading brands approximate 400 tons, excluding the peak period in the hot season (when consumption reaches 500 tons per month). It is evident that reduced use of plastics and reduced defects are major concerns to the bottlers. Except for a multi-plant policy taken by all leading brands, a truck-loaded management is accounted by the manufacturers who prefer the efficiency in reduced energy consumption and unit cost reduction to impact reduction from uses of electricity and fuel energies.

Most bottlers assert that the bottling system generates no CO_2 emissions to the air, water and soil. In case of PET pre-form blow-molding in the filler room, the closed-loop process is equipped with a cleaner technology because it enables use of electrometers and infrared device for debris and dust elimination, instead of emitting to the air. For the open loop, the bottles are supplied by PET-bottle manufacturing. The outsource supply becomes an excuse by the bottlers that no emission is generated in the open-loop bottling system.

Based on LCM perspective, the attempts to impact reduction direct to a cradle-to-grave module. The cap-seal wrapper is modified to every bottle to ensure a primary use in health concern. Additional is the communicated information via the bottle label about sustainable use for environmental protection from the waste bottles. However, recycling the post-used bottles for new purposes is slightly accounted, either encouraging the consumers or by the bottlers themselves. A legal constraint on using recycled material for food packaging, enforced by the Food and Drug Administration (Thailand) is addressed for a low literacy in recycling technology in bottled-water businesses. Precisely, reduced use of plastics and reduced defects in bottling system that are the major concerns to the bottlers can be converged with the corporate contribution to environmental prevention and protection.

b. Water. Bottled-water business requires a shift to the SCP. In Thailand, most bottling plants are located in the provinces along the Chao Phraya River and its tributaries that can cause the profound impact on ecological system. The water is mostly extracted from underground water or artesian wells, with a fee paid per cubic meter to the office of municipal affairs or to the Ministry of Natural Resources and Environment at USD 0.40 per cubic meter for the right to capture the water. The water of any quality discharged from factories in the provinces near the Chao Phraya River and its tributaries must legally undergo filtration before drainage. Increasingly, PET-bottled-water manufacturing is simply established for the SMEs because they bottle tap water, with a normal filtration.

c. According to unsustainable water resource management spread over the country for many years, the surface water supplies are currently limited as well as groundwater sources are highly vulnerable to depletion. The belief that the discharged drinking water is neither hazardous nor wastewater is not considered justified by those aware of the water crisis.

Consumer Behavior of SCP

Consumer behavior towards the SCP of PET-bottled water is evaluated by the strength of agreement to avoid the manners, which are known to be damaging to the environment. The analysis contains comparative statements about the contributions to improving greater efficiency in consuming PET-bottled water and decreasing the impact from the post-used bottles.

In most cases, Thai consumers moderately perform the SCP, with more emphasis on the resource efficiency rather than the impact reduction. In their opinion, PET-bottled water is purified, clean and safe for drinking. Increasing use of tap water is totally declined due to negative perception of the impurity. Meanwhile, the refrigerated water from coin-machine and freezer services, which is tap water, is considered appropriate for drinking due to a filtration in chilling step. The convenience in use of PET-bottled water, either in or away from home, becomes the normal practice in working days and during shopping and travelling.

The idea of zoning gas station and shopping mall is most credited for improving waste bottles management. A forbidden use in the offices is totally declined, so does the enactment of the Polluter Pays Principle (PPP) when asked to treat them as the SCP behavior. It is interesting that business sector is expected by nearly all consumers to take the leading role in the SCP development by incorporating environmental-awareness rising to product campaigns. Two histograms as shown in Figure 2 and 3 present consumers' performances in between resource efficiency and impact reduction for bottled water, respectively.

1. **Resource Efficiency:** Drinking up a whole bottle of water is considered best effective to resource efficiency target; otherwise, consumers tend to choose a smaller pack size of 330-350 milliliter that fits to demand, so does taking the leftover for further use. Consumers also like to optimize all product benefits and minimize resources used with the reuse. Additional are preparing personal flask when away from home and avoiding the use of bottled water for coffee, tea and cooking. Selling post-used bottles to junk shop is the preferable way for handling the wastes bottles, rather than recycling approach. According to consumers' feedback, a change to tap water is neither performed nor considered valuable for increasing efficiency in bottled-water consumption.
2. **Impact Reduction:** Thai consumers are willing to accompany the post-used bottle until finding a litterbin but hesitate to recycle the waste bottles for decreasing impact generation. In their opinion, changing to ordering glass bottles helps in decreasing the impacts from PET-bottled water as little as choosing refillable jug when eating out. Consumers prefer using personal flask for sake of decreasing the volume of waste bottles

Figure 2. Consumer behavior of resource efficiency in consuming PET-bottled water and disposing of the post-used bottles to fulfill the SCP

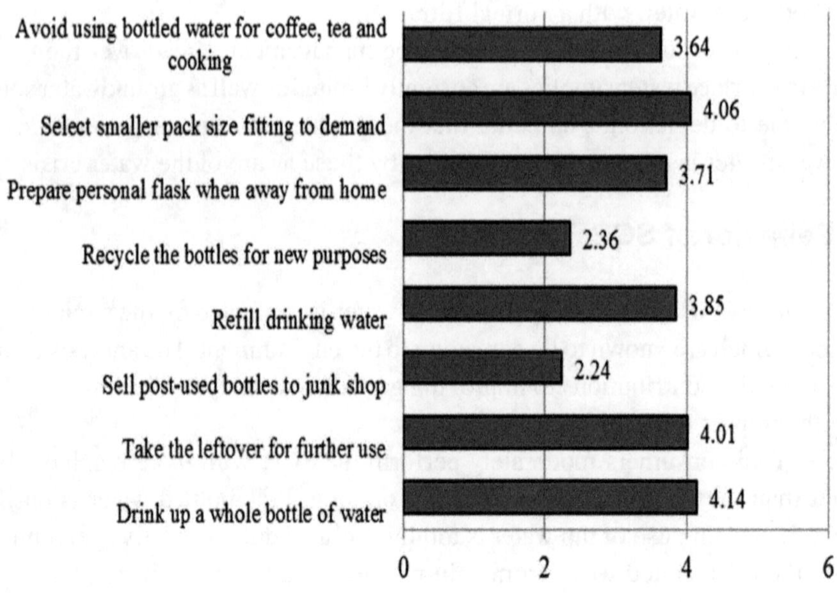

Figure 3. Consumer behavior of impact reduction in consuming PET-bottled water and disposing of the post-used bottles to fulfill the SCP

In most cases, the post-used bottle are sorted for recycle bin rather than disposed in the environment inappropriately. Additional addressed is that the inadequacy of recycle bin for plastics waste, compared to consumers' need, becomes a significant burden to sort out the waste bottles. Among the concerns to waste-PET bottles, a large number of door-to-door collectors represent the key role of junk dealer in the logistic flow of recycling waste bottles.

In view of health concerns, most consumers do not avoid using the bottle that has been opened and exposed to direct sunlight contact. Nor is the reuse without proper cleaning that can cause the health risk. On the other, performing a mouth-to-bottle, without a straw or pouring into a glass is least considered related to health risk. Notwithstanding, PET refillable bottles could be safely reused, if the bottles derive from the good manufacturing procedures and the appropriate cleaning prior refilling. In brief, the consumption pattern of PET-bottled drinking water is somewhat away from impact reduction implication due to the speeding up of lifestyles together with increasingly disposable products for consumerism.

SOLUTIONS AND RECOMMENDATIONS

The conflict agenda of water resource; agriculture and fishery sectors against industries, habitants against businesses, and within the community, is heavily increasing, nationwide. Towards the undercurrents in the LCM of bottled water, a paradigm shift to the holistic view and sustainable actions is truly required so as to bring the implementable solutions to and from all concerns. As a production and consumption pattern is actually related to sustainability objectives, the increasing demand for bottled water shows what social gains for better living is what the world has lost. We must accept that without environmental awareness, the SCP will not advance the larger development in industry practice and consumer behavior. This is important to the SCP implementation, which several variables are associated, which public participation and regulatory policies have not been fully developed.

Under the leadership of the government, its role remains an option to foster a consumption and production pattern away from or toward sustainability. The tasks shall be responded by many governmental entities like the Department of Industrial Works, the Department of Industrial Promotion, the Department of Environmental Quality Promotion, the Department of Health, the Department of Mineral Resources and the Metropolitan, or Provincial, Waterworks Authority. Business would go beyond a product value chain to sustainability objectives in order to reach a viable solution that helps in reducing environmental disruption and compromising economic and social value at the same time. These concerns can be fixed by a removal of unsustainable goods and service from the market. The eco-tax or a refund fee shall be rewarded to those who produce and consume sustainable products and services or eco-products. If putting harm to environmental quality, the polluter could be charged a fee towards funding environmental rehabilitation, and finally edited out of the market – Polluter Pay Principle (PPP). Any related governmental entities could put a driving force to the SCP by either shaping the business practice to influence consumer behavior, or formulating solution taken by each stakeholder. In such case, they should separately set minimum recycling standards and indicate material combinations with recycled PET for the new packaging those covering nonfood-contact products.

Environmental themes in advertising and marketing processes should be employed for raising awareness of sustainable practices. Both business and consumers should be alert to the duties and responsibilities when there is no legislation, nor strict enforcement of disposing of waste bottles. Technological knowledge about recovery of waste-to-energy device should be widely disseminated to communities. Coupled with marketing-based instruments, public policies that support cleaner technologies for a value chain of bottled water could be established. The waterworks authority shall also be due on the establishment of tap-water supply throughout nation. There has to be a thematic agenda inspiring people for the quality of tap water that is appropriate for drinking. The consideration shall also direct to the role of recycling agents and retailers in responsibility for the logistics of waste bottles. These should be well organized in order to prevent the reverse channel from inappropriate approach at landfills and dumping areas.

Finally, environmental education has to be driven so that a positive change in production and consumption pattern of products and services like PET-bottled water can translate to larger sustainability gains. This is concurrent to the agreements across the globe adopted from the 2015 Paris Climate Conference (COP21) that sustainable lifestyle in the SCP approach is an important role in addressing climate change (United Nations Framework Convention on Climate Change [UNFCCC], 2015).

FUTURE RESEARCH DIRECTIONS

The content in this chapter is by no means exhaustive, but enables a potential for the methods and datasets of sustainability framework. More extension could be observed in new areas of study and possible pathways that are branched in various directions. The approaches to soft drinks, energy drinks and health drinks could be developed to identify the differences of causes, effects and solutions on the SCP target themselves. There should be opportunities for future studies about the SCP approach that provide greater interdisciplinary expectations, especially externality cost analysis and a power of eco-journalism to public participation in SCP development. In addition, the improvement of assessment on a linkage between globalization and sustainability from a multi-disciplinary perspective could be set for reframing current public awareness of mitigation and adaptation manners towards climate agenda.

CONCLUSION

PET bottles are potential to the impact categories of global warming, eco-toxicity and human toxicity, especially waste-bottles management that causes a profound impact to the planet. The SCP of PET-bottled water is accounted by the businesses and consumers, with emphasis on efficiency in water and plastics resource use. Contributions to impact prevention and reduction are the contrary. The advance in plastics recycling technology is required, so does consumer behavior in recycling waste bottles. Due to the most practice in bottling the water by open-loop system, the LCM of PET-bottled water can be direct and indirect burden to sustainability target in water resource management, both at present and in future. In general, particular information about demand and supply of fresh water resource varies geographically and culturally. The undercurrents of bottled water are in evidence when the production and consumption pattern are radically addressed through the bottling system and consumer behavior towards the level of environmental awareness. A rapid growth of bottled-water business in Thailand that are captured by many thousands bottlers becomes the interplay of social, economic and environmental development. The concerns are in need to reconsider the impact on water supply, use and management. Businesses and consumer are challenging for greater efficiency in water resource consumption and the improvement of plastics recycling. Shortly, this chapter is particularly timely as water resource and waste plastics are shifting from one of environmental issues to what those are much more concerned to sustainability success.

REFERENCES

Akenji, L., & Bengtsson, M. (2009). *Is the customer really king? Stakeholder analysis for sustainable consumption and production using the example of the packaging value chain.* Retrieved from www.iges.com

Amano, M., & Ness, B. (2004). *PET bottle system in Sweden and Japan: an integrated analysis from a life-cycle perspective*. Lund University.

Berger, I. E., & Corbin, R. M. (1992). Perceived consumer effectiveness and faith in others as moderators of environmentally responsible behaviors. *Public Policy and Advertising*, *11*(2), 79–89.

Brodhag, C. (2010). A differentiated approach for sustainable consumption and production policies. *Natural Resources Forum*, *34*(1), 63–70. doi:10.1111/j.1477-8947.2010.01276.x

Cellura, M., Longo, S., & Mistretta, M. (2011). The energy and environmental impacts of Italian household consumptions: An input-output approach. *Renewable & Sustainable Energy Reviews*, *15*(8), 3897–3908. doi:10.1016/j.rser.2011.07.025

Chamornmarn, W. (2008, June). *Bottled water: External environmental analysis.* Retrieved from www.eiolca.net/LCAexperts/index.html

Coelho, T. M., Castro, R., & Gobbo, J. A. Jr. (2010). PET containers in Brazil: Opportunities and challenges of a logistics model for post-consumer waste recycling. *Resources, Conservation and Recycling*, *55*(3), 291–299. doi:10.1016/j.resconrec.2010.10.010

DuPont. (2011). *Sustainability best practice series: Life cycle thinking*. DuPont Packaging Graphics.

Economic and Social Commission for Asia and the Pacific (ESCAP). (2011). *Tools for low carbon development in Thailand.* Paper presented at the First National Seminar on Green Growth Policy.

Elliott, J. A. (2006). *An introduction to sustainable development*. New York, NY: Routledge.

Euromonitor International. (2010). *Bottled water-Thailand. Country sector briefing*. Retrieved from http://www.euromonitor.com/bottled-water-in-thailand/report

Feron, V. J., Jetten, J., De Kruijf, N., & Van Den Berg, F. (1994). Polyethylene terephthalate bottles (PRBs): A health and safety assessement. *Food Additives and Contaminants*, *11*(5), 571–594. doi:10.1080/02652039409374258 PMID:7835471

Flor, A. G. (2004). *Environmental communication*. The Philippines: University of the Philippines.

Foolmuan, R. K., & Ramjeeawan, T. (2012). Disposal of post-consumer polyethylene terephthalate (PET) bottles: Comparison of five disposal alternatives in the small island state of Mauritius using a life cycle assessment tool. *Environmental Technology*, *33*(17), 2007–2018. doi:10.1080/09593330.2012.655321 PMID:23240194

Franklin Associates Ltd (FAL). (2010a). *Life cycle inventory of 100% postconsumer HDPE and PET recycled resin from postconsumer containers and packing*. Prairie Village, KS: Franklin Associates, A Division of ERG.

Franklin Associates Ltd (FAL). (2010b). *Cradle-to-gate life cycle inventory of nine plastic resins and two polyurethane precursors*. Prairie Village, KS: Franklin Associates, A Division of ERG.

Giljum, S., Burger, E., Hinterberger, F., Lutter, S., & Bruckner, M. (2011). A comprehensive set of resource use indicators from the micro to the macro level. *Resources, Conservation and Recycling*, *55*(3), 300–308. doi:10.1016/j.resconrec.2010.09.009

Gironi, F., & Piemonte, V. (2010). Life cycle assessment of polylactic acid and polyethylene terephthalate bottles for drinking water. *Environmental Progress & Sustainable Energy*, *30*(3), 459–468. doi:10.1002/ep.10490

Gleick, P. H., & Cooley, H. S. (2009). Energy implications of bottled water. *Environmental Research Letters*, *4*(1), 1–6. doi:10.1088/1748-9326/4/1/014009

Harris, P. G. (2011). *Ethics and global environmental policy: cosmopolitan conceptions of climate change*. Hong Kong: Edward Elgar Publishing. doi:10.4337/9780857931610

Hauschild, M., Jeswiet, J., & Alting, L. (2005). From life cycle assessment to sustainable production: Status and perspectives. *CIRP Annals-Manufacturing Technology*, *54*(2), 1–21. doi:10.1016/S0007-8506(07)60017-1

Hertwich, E. G. (2005). Life cycle approaches to sustainable consumption. *Critical Reviews in Environmental Science and Technology*, *39*(13), 4673–4684. doi:10.1021/es0497375 PMID:16053063

Hoff, H., Doll, P., Fader, M., Gerten, D., Hauser, S., & Siebert, S. (2014). Water footprint of cities-indicators for sustainable consumption and production. *Hydrology and Earth System Sciences*, *18*(1), 213–226. doi:10.5194/hess-18-213-2014

Huber, M. K. (2010). *Bottled water: The risks of our health, our environment, and our wallets*. Indiana University.

Hurd, D. J. (2000). *Best practices and industry standards in PET plastic recycling*. Retrieved from www.napcor.com

Ida, H. (2006). *Current status of plastics recycling in Japan*. Paper presented at The IEA/CEFIC Workshop: Feedstock Substitutes, Energy Efficient Technology and CO2 Reduction for Petrochemical Products, Paris, France.

Jones, P., Hillier, D., & Comfort, D. (2014). Sustainable consumption and the UKs leading retailers. *Social Responsibility Journal*, *10*(4), 702–715. doi:10.1108/SRJ-04-2013-0046

Kaweeya, S. (2006). *EARTH Journalism*. Bangkok: Ratanakosintr House.

Kuczenski, B., & Geyer, R. (2010). Material flow analysis of polyethylene terephthalate in the US, 19962007. *Resources, Conservation and Recycling*, *54*(12), 1161–1169. doi:10.1016/j.resconrec.2010.03.013

Larijani, M. (2010). Assessment of environmental awareness among higher primary school teachers. *Human Ecology*, *31*(2), 121–124.

Lehmann, B., Vilaplana, F., Strömberg, E., Suliman, W., & Rodriguez, C. L. (2005). Comparative LCA on plastic packaging. Prairie Village, KS: Franklin Associates, A Division of ERG.

Liu, J., Wang, R., Yang, J., & Shi, Y. (2010). The relationship between consumption and production system and its implications for sustainable development of China. *Ecological Complexity*, *7*(2), 212–216. doi:10.1016/j.ecocom.2010.02.003

Mutsuga, M., Kawamura, Y., Sugita-Konishi, Y., Hara-Kudo, Y., Takatori, K., & Tanamoto, K. (2006). Migration of formaldehyde and acetaldehyde into mineral water in polyethylene terephthalate (PET) bottles. *Food Additives and Contaminants*, *23*(2), 212–218. doi:10.1080/02652030500398361 PMID:16449065

Nestlé Waters North America. (2010). *Environmental life cycle assessment of drinking water alternatives & consumer beverage consumption in north America*. Retrieved from http://beveragelcafootprint.com/Nestlé

Niccolucci, V., Botto, S., Rugani, B., Nicolardi, V., Bastianoni, S., & Gaggi, C. (2011). The real water consumption behind drinking water: The case of Italy. *Environmental Management*, *92*(10), 2611–2618. PMID:21741755

Parent, J., Cucuzzella, C., & Reveret, J. P. (2013). Revisiting the role of LCA and SLCA in the transition towards sustainable production and consumption. *Life Cycle Assessment, 18*(9).

Peacocke, C. (2006). *Mental action and self-awareness*. Blackwell.

Peattie, K. (2010). Green consumption: Behavior and norms. *Annual Review of Environment and Resources, 35*(1), 195–228. doi:10.1146/annurev-environ-032609-094328

Pogutz, S., & Micale, V. (2011). Sustainable consumption and production. *Society and Economy, 33*(1), 29–50. doi:10.1556/SocEc.33.2011.1.5

Poonuchaphai, W. (2010). *Country experience: Sustainable consumption and production in Thailand.* Paper presented at a Path of Good Governance Seminar, Seoul, South Korea.

Power, W. (2009). *Life cycle management.* Retrieved from http://www.unep.fr/shared/publications/pdf/DTIx1208xPA-LifeCycleApproach-Howbusinessusesit.pdf

Prakash, R., Yogesh, P., Viraja, B., & Shilpa, K. (2015). Environment and social sustainability approaches in policy and practice: A case study of symbiosis international university, Pune, India. In W. L. Filho (Ed.), *Transformative Approaches to Sustainable Development at Universities, Working Across Disciplines* (pp. 471–488). Springer International Publishing.

Ramani, K., Ramanujan, D., Bernstein, W.Z., Zhao, F., Sutherland, J., Handwerker, C., … Thurston, D. (2010). *Integrated sustainable life cycle design.* doi: 10.1115/1.4002308

Raskin, P. D., Electris, C., & Rosen, R. A. (2010). *The century ahead: Searching for sustainability.* doi:10.3390/su2082626

Ratthanin, S. (2010, October). Bottled water. *Marketeer, 11*(128), 114–117.

Rodwan, J. G. (2010). *Bottled water 2009: Challenge circumstances persist, future growth anticipated, US and international developments and statistics.* Retrieved from www.bottledwater,org/files/2009BWstats.pdf

Sampattakul, S. (2010). *Climate Thailand.* Paper presented at the Climate Thailand Conference, Bangkok, Thailand

Schmid, P., Kohler, M., Meierhofer, R., Luzi, S., & Wegelin, M. (2008). Does the reuse of PET bottles during solar water disinfection pose a health risk due to the migration of plasticisers and other chemicals into the water? *Water Research, 42*(20), 5054–5060. doi:10.1016/j.watres.2008.09.025 PMID:18929387

Schultz, P. W., Nolan, J. M., Cialdini, R. B., Goldstein, N. J., & Griskevicius, V. (2007). The constructive, destructive and reconstructive power of social norms. *Psychological Science, 18*(5), 429–434. doi:10.1111/j.1467-9280.2007.01917.x PMID:17576283

Shen, L., Nieuwlaar, E., Worrell, E., & Patel, M. K. (2011). Life cycle energy and GHG emissions of PET recycling: Change-oriented effects. *Int J Life Cycle Assess, 16*(6), 522–536. doi:10.1007/s11367-011-0296-4

Shobeiri, S. M. (2005). *A comparative study of environmental awareness and attitude of teachers and students of secondary schools in India and Iran.* University of Mysore.

Shotyk, W., & Krachler, M. (2007). Contamination of bottled water with antimony leaching from polyethylene terephthalate (PET) increase upon storage. *Environmental Science & Technology, 41*(5), 1560–1563. doi:10.1021/es061511+ PMID:17396641

Song, H. S., & Hyun, J. C. (1999). A study of the comparison of the various waste management scenarios for PET bottles using the life-cycle assessment (LCA) methodology. *Resources, Conservation and Recycling*, *27*(3), 267–284. doi:10.1016/S0921-3449(99)00022-1

Suwajittanont, W., & Chai-ittipornwong, T. (2011). *Role of mass media for enhancing the sufficiency economy*. Bangkok, Thailand: Public Relations Department.

Sze, T. L. (2010). *Spaces for consumption*. Retrieved fromhttp://www.satepub.com/upm-data/35384_01_Miles_CH_01.pdf

Takahashi, Y. (2010). *Sustainable PET bottle recycling system in Japan for sound material cycle*. Japan: The University of Tokyo, Graduate School of Frontier Sciences.

Termpittayapaisith, A. (2011, February). *Low carbon development in Thailand.* Paper presented at the First National Seminar on Green Growth Policy Tools, Bangkok, Thailand

Tukker, A., Emmert, S., Charter, M., Vezzoli, C., Sto, E., Munch Andersen, M., & Lahlou, S. et al. (2008). Fostering change to sustainable consumption and production: An evidence based view. *Journal of Cleaner Production*, *16*(11), 1218–1225. doi:10.1016/j.jclepro.2007.08.015

United Nations Department of Economic and Social Affairs (UNDESA). (2010). *Trends in sustainable development towards sustainable consumption and production*. Retrieved from https://www.un.org/development/desa/en/

United Nations Environment Programme (UNEP). (2012). *Global outlook on sustainable consumption and production policies: taking action together.* Retrieved from http://www.unep.org/publications/contents/pub_details_search.asp?ID=6251

United Nations Environment Programme (UNEP). (2015). *Sustainable information forum 2015*. Retrieved from http://www.cop21paris.org/

United Nations Framework Convention on Climate Change (UNFCCC). (2015). *Adoption of the Paris agreement: draft decision -/CP.21*. Paper presented at the Conference of the Parties Twenty-First Session, Paris, France. Retrieved from https://unfccc.int/resource/docs/2015/cop21/eng/l09r01.pdf

Vakili, M. H., & Fard, H. M. (2010). Chemical recycling of poly ethylene terephthalate wastes. *World Applied Sciences*, *8*(7), 839–846.

Vasami, R. (2010). *New study gives "green" light to PET bottles over glass or aluminum.* Retrieved from http://www.petresin.org/news_GreenLighttoPETBottles.asp

Veshagh, A., & Obagun, A. (2007). *Survey of sustainable life cycle design and management.* Paper presented at the 14th CIRP-Conference on Life Cycle Engineering, London, UK. doi:10.1007/978-1-84628-935-4_41

Westerhoff, P., Prapaipong, P., Shock, E., & Hillaireau, A. (2008). Antimony leaching from polyethylene terephthalate (PET) plastic used for bottled drinking water. *Water Research*, *42*(3), 551–556. doi:10.1016/j.watres.2007.07.048 PMID:17707454

Willett, C. (2003, March). *PET: is this as good as it gets?* Paper presented at the World Petrochemical Conference: XVIII Annual, Houston, TX.

Zhang, B., & Kimura, F. (2007). *Framework research on the greenness evaluation of polymer materials.* London: Springer-Verlag. doi:10.1007/978-1-84628-935-4_51

Zhao, M. (2009). *Consumption behavior in China 2009.* Shanghai: CBC Marketing Research.

KEY TERMS AND DEFINITIONS

Bottled Water: A disposable product that serves one-way usage and appears in the form of PET bottle with the filled water for drinking.

Bottling System: An input-output approach for filling drinking water in the bottle. The system consumes the input of several resources, such as plastic bottles, purified and disinfected water together with electricity and fuel energy in order to deliver the output of bottled water and environmental impact profile.

Bottom-up Participation: Role of business and consumer in environmental protection and prevention from the interactive processes between production and consumption.

PET Bottles: A plastic container appears in bottle body which is made of polyethylene terephthalate polymer.

Sustainability in Bottled Water Production and Consumption: An integration of sustainable consumption and production approach into the whole product life cycle of bottled water.

Sustainable Consumption and Production Pattern: The ways of producing and consuming products and services with environmental awareness of improving efficiency in resource consumption and reducing potential impact generation.

Undercurrents of Bottled Water: Environmental impact profile occurs from PET-bottled water directly and indirectly, both at present and in future, in many categories, such as global warming, ecotoxicity and human-toxicity.

ENDNOTES

1. The Marrakech Process is a global process driven by the United Nations Environmental Programme (UNEP, 2015) and the United Nations Department of Economic and Social Affairs (UN DESA) to support the elaboration of a 10-Year Framework of Programs (10YFP) on SCP according to Johannesburg Plan of Action in the World Summit on Sustainable Development (WSSD).
2. ISO 14040 (2006) is the leading standard to describe the principles and framework for life cycle assessment (LCA) that can be followed by the steps of environmental management - life cycle assessment - principles and framework.
3. ISO 14044 (2006) is the international standard to specify requirements and provide guidelines for life cycle assessment (LCA), with the following steps; environmental management - life cycle assessment - requirements and guidelines.
4. SimaPro 7.1 is the software developed by PRé Consultants as the tool for a product life cycle assessment through the phase of LCI. The software provides several literature-based inventory

data covering the compilation and quantification of inputs and outputs for a given product system throughout its life cycle, plus various impact assessment methods, e.g. Eco-indicator, CML (Center of Environmental Science of Leiden University). The software and the literature LCIs used in this chapter are supported by the National Metal and Materials Technology Center (Thailand) for academic purpose only.

5 GWP is the abbreviation of global warming potentials. Due to the International Panel on Climate Change (IPCC), CO_2 emission is normalized as an equivalent weight of carbon dioxide for the relationship between greenhouse gases (GHGs) emission and the 100-year time horizon Global Warming Potentials (GWP).

Chapter 16
Industrial Wastewater Management in the Context of Climate Change Adaptation in Selected Cities of India:
A Business Approach

Rahul Hiremath
SCMHRD, Symbiosis International University, India

Gurudas Nulkar
SCMHRD, Symbiosis International University, India & Trustee Ecological Society, India

Bimlesh Kumar
Indian Institute of Technology Guwahati, India

Sharmila S. Patil
Walchand Institute of Technology, India

Sheelratan S. Bansode
Solapur University, India

J. Murali
Environmental Solutions and Consultancy, UAE

ABSTRACT

Climate change and wastewater control are one of the foremost demanding situations for Indian cities. Urbanization and unparalleled growth of cities across India continue to create immense pressure on land and water resources. This uncontrolled growth continues to produce growing volumes of wastewater. Climate change, impacts inclusive of, intense storm events in summer time or extended moist periods in wintry weather are quite visible in India. In urban and peri-urban areas, wastewater use for agriculture is an emerging precedence. Due to susceptible enforcement of regulatory, most of the wastewater generated is permitted off untreated/ partially treated. While many previous studies have checked out the global modifications and associated impacts of climatic variations on water resources, few have targeted at the evaluation of the particular effects and adaptation priorities for water systems in towns. Proper reuse of wastewater for irrigation will significantly lessen the shortage, offer a sustainable water source, improving farming productiveness, lessen pollution, generate livelihood potential for

DOI: 10.4018/978-1-5225-1046-8.ch016

low earnings city households along with contributing to their each day food needs. There are tradeoffs which need to understand which includes problems to individual's health, and surroundings. Through suitable treatment methods, water users' cooperatives, policy shift and the introduction of market based approaches, treated wastewater use in agriculture can be enhanced and all associated risks can also be curtailed. This chapter focuses on use of treated urban wastewater and its management for agriculture in selected Indian cities.

INTRODUCTION

Urban wastewater management is a challenge for many developing countries including India. Rapid urbanization and unparalleled growth of cities across India continue to create immense pressure on land and water resources. This uncontrolled growth continues to produce growing volumes of wastewater. Due to weak enforcement of regulatory measures, most of the wastewater generated let off untreated/ partially treated. While many preceding research have checked out the global modifications and climate related variability on water sources, few have targeted at the assessment of the specific consequences and version priorities for water systems in cities. Proper reuse of wastewater in irrigation can significantly lessen water shortage, offer a sustainable source of water, improving agricultural productivity, lessen pollution, create livelihood potential for low earnings city households, put in to their each day food needs. Hence a needed focus to understand and review the wastewater supply and treatment for Indian cities is required. The variety of challenges associated with climate, water and cities are very pressing, which depends on its demography, economics and administrative potential. The urban wastewater use for irrigation is a growing global phenomenon. Such irrigation method can build livelihood opportunities and strengthen food security. But the major hurdle is to come up with the sensible, low-cost safeguards that do not hamper farmer's livelihoods and thus providing a useful resource in meeting food needs and supplying cheap produce to expanding towns. Through suitable treatment methods, water users' cooperatives, policy shift and the introduction of market based approaches, treated wastewater use in agriculture can be enhanced and all associated risks can also be curtailed. This chapter focuses on use of treated urban wastewater and its management for agriculture in selected Indian cities.

CHANGE IN CLIMATE AND ITS EFFECT ON WATER RESOURCES AND AGRICULTURE

Climate change as per the United Nations Framework Convention on Climate Change (UNFCCC) has been defined as, "a change of climate which is attributed directly or indirectly to human activity that alters the composition of the global atmosphere and which is in addition to natural climate variability observed over comparable time periods". Increased emission from industrial sectors along with changed agricultural pattern combined with widespread deforestation has effected a change in the gaseous composition of earth's atmosphere (Table 1).

The global yearly mean temperature toward the end of the twentieth century, as an after effect of GHG aggregation in the climate, has expanded by 0.4–0.7 °C over that recorded toward the end of the nineteenth century. The previous 50 years have demonstrated an expanding pattern in temperature @

Table 1. Greenhouse gases in surroundings

Elements	CO_2	CFCs	CH_4	N_2O
Avg. concentration 100 years ago (ppbV)	290x 10^3	0	900	270
Present concentration (ppbV) (2007)	380 x 10^3	3 to 5	1774	319
Proposed concentration in the year 2030 (ppbV)	400 x 10^3 500 x 10^3	3 to 6	2,800 to 3000	400 to 500
Atmospheric life (years)	5 to 200	75	9 to 15	114
Global warming ability years relative to CO_2	1	4750 to 10900	25	298

Source: IPCC (2007)

0.13 °C/decade, while the ascent in temperature amid the previous one and half decades has been much higher (Figure 1). The Inter-Governmental Panel on Climate Change (IPCC) has anticipated the temperature increment to be between 1.1 °C and 6.4 °C before the end of the 21st Century (IPCC, 2007). The worldwide temperature alteration is relied upon to prompt other local and worldwide changes in the atmosphere related parameters, for example, precipitation, soil dampness, and ocean level. Snow spread is additionally answered to be bit by bit diminishing.

Along these lines, coordinated endeavors are required for moderation and adjustment to decrease the helplessness of farming to the unfavorable effects of environmental change and making it stronger. With more than sixty percent of its populace subject to atmosphere touchy exercises, for example, farming, the effects of environmental change on horticulture accept huge significance for India. Environmental change projections made up to 2100 for India, demonstrate a general increment in temperature by 2-4°C combined with expansion in precipitation, particularly amid the rainstorm time frame. Mall et al. (2006) give a great audit of environmental change sway examines on Indian farming for the most part from physical effects point of view. The accessible proof shows critical drop in yields of essential grain crops

Figure 1. Trends in global temperature over the years
(IPCC, 2007)

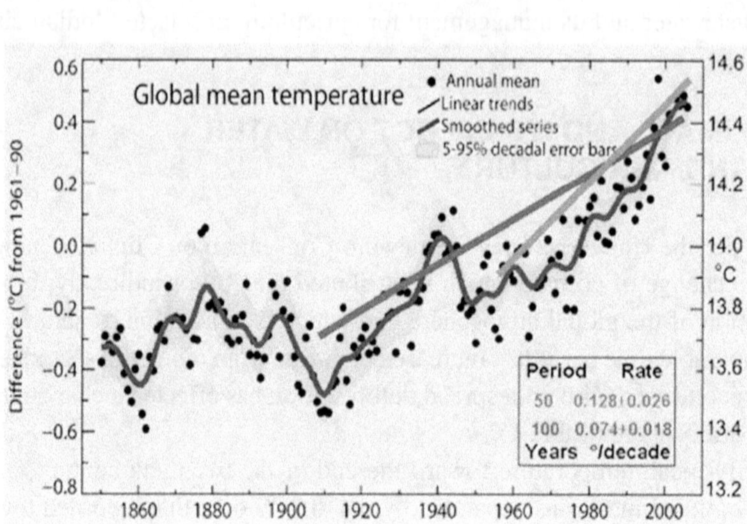

like rice and wheat under environmental change conditions. Biomass and yield tend to decay with expanding temperature, as higher temperatures abbreviate crop length, upgrade breath and diminish time for radiation interference (Rawson, Gifford, & Condon, 1995). As yields in probably the most profitable areas of the world are drawing nearer a level or notwithstanding declining (Pathak et al., 2003), the possible impact of environmental change on harvest generation adds to the officially complex issue. It is a noteworthy test to assess the effect of rising temperature on product yield. Indian farming is profoundly reliant on the onset, retreat and extent of rainstorm precipitation, especially in the downpour sustained territories of east, north-east and south India. Atmosphere modelers and IPCC records have anticipated conceivable outcomes of expanding variability in Asian Monsoon flow in a hotter world. Regardless of extension of range under watering system, dry spells, created by insufficient and uneven dispersion of precipitation, keep on being the most imperative climatic distortions, which impact the horticultural generation in India. The seriousness of a dry spell will be heightened in a hotter world. Extreme and incessant flooding because of environmental change would be a noteworthy issue in the Indian subcontinent. Soil temperature influences the rates of natural matter disintegration and arrival of supplements. At high temperatures, however supplement accessibility will increment in the short-term, over the long haul natural matter substance will reduce, bringing about a decrease in soil ripeness.

INDUSTRIAL AND MUNICIPAL WASTEWATER IN MAJOR CITIES

There is an extensive literature in context to industrial and municipal (sewage) wastewaters, its sources, volumes generated, wastewater characterization, impacts on life forms, legal norms and diverse treatment methods/strategies; however, most studies have not considered explicitly the implications of climate change for the delivery of water management objectives (Arnell et al., 2015; Gebrezgabher et al., 2015; Grit et al., 2015; McCarl et. al., 2015; Neumann et. al., 2015; Peirson et al., 2015; Zouboulis & Tolkou, 2015; Lewis, 2016; Valta et al., 2016). In the recent times, there have been research papers that are tackling the issues of industrial wastewaters treatment and management by novel methods, technology adoption, etc. taking climate change aspects into account (Itankar & Patil, 2015; Patil & Rao, 2015; Rao & Patil, 2015). For betterment of the class of life and improvement of productive competency of the people city water supply and sanitation are important basic needs. Also in many arid and semi-arid countries water is becoming an increasingly limited resource and planners are forced to consider any sources of water which might be used economically and effectively to promote further development. In urban areas, water is tapped for domestic use (78% of population has access to safe drinking water and 38% to sanitation as per CPCB 2009) and industrial uses from streams, rivers, lakes and wells. Municipal sewage is defined as "waste originating from a community; may be composed of domestic wastewaters and/or discharge from industry" and is major source of water pollution in India, particularly in and around large urban centers (Central Pollution Control Board, 2009). Almost 80% of the water supplied for domestic purposes, comes out as wastewater. In most of the cases wastewater is let out in the same form and it either gets soaked into the ground as a possible contaminant of ground water or is discharged into the natural sewage system ensuing in the contamination in downstream areas. According to the Central Pollution Control Board (CPCB, 2001) the Infrastructure Development Finance Corporation estimated that 73% of urban wastewater in India is untreated; with the increase in the drinking water supply to urban areas, the wastewater increases. Tables 2 shows the quantity of water requirement in future for various sectors

Table 2. Sector wise water demand

Sector	Water Demand in km³ (BCM)					
	2010		2025		2050	
	A*	B#	A*	B#	A*	B#
Drinking water	43	56	62	73	111	102
Irrigation	557	688	611	910	807	1072
Energy	19	5	33	15	70	130
Industry	37	12	67	23	81	63
Others	54	52	70	72	111	80
Total	**710**	**813**	**843**	**1093**	**1180**	**1447**

Source: A* National Council for Integrated Water Resource and Development (NCIWRD), B#Website of Ministry of Water Resources, Govt. of India, National.

whereas the Table 3 gives the sewage water generation from class I cities and Class II towns which also includes the upcoming smart cities as recently announced by the Government of India

If wastewater is not treated or disposed, it will create problem (Misra, 2014). It also poses great threat to the hydrology as well as water quality of natural water bodies. It is linked with a number of hazards, counting community's physical condition, agronomic, and ecological. Microbial health hazards are harsh

Table 3. Class I Cities and Class II Towns Sewage Generation

Name of the City	Sewage Generation (in MLD*) *Million Liters Per Day
Hyderabad	426.1
Vishakhapatnam	135
Vijayawada	128.4
Patna	279.1
Delhi	3799.9
Ahmadabad	472.02
Surat	432.1
Rajkot	109
Vadodara	1179.85
Bangalore	772
Indore	204.02
Bhopal	334.8
Jabalpur	143
Mumbai	2671.02
Pune	474.01
Nagpur	379
Solapur	149.5

continued on next page

Table 3. Continued

Name of the City	Sewage Generation (in MLD*) *Million Liters Per Day
Nasik	227.8
Ludhiana	235
Amritsar	191.5
Jaipur	451.7
Chennai	158.01
Kanpur	417.4
Lucknow	364
Agra	260.4
Kolkata	705.86
Jamshedpur	199
Faridabad	164.01
Asansol	147.1
Meerut	177.1
Madurai	98
Coimbatore	120.01
Varanasi	230.2
Kochi	188
Allahabad	176
Dhanbad	192
Total	15794

Source: *Status of sewage in India(CUPS/61/2005-06)- Central Pollution Control Board* (http://www.sulabhenvis.nic.in/Database/STST_wastewater_2090.aspx)

in low and middle-income nations, which frequently involve the straight or indirect use of raw wastewater and/or impure waters from rivers to water edible-crops.

The wastewater treatment requires an investment of US$65 billion which is ten times bigger than what is proposed by government (Kumar, 2003). India ranked as the 3rd largest user of untreated wastewater for agriculture in the world (Scott et al. 2010). In the year 2012, Israel and India used wastewater for agriculture to the tune of 64% and 24%, respectively (Grover, 2012). India will soon cross the 50:50 urban-rural population threshold and reach 750 million urban Indians by 2050 (Scott et al. 2010). A total of 38,254 million liters per day (MLD) of wastewater discharge in the year 2009 is projected to increase 120,000 MLD in 2051 (Planning Commission, 2012). There are various issues like treatment technology, different sizes of urban settlements and the land availability for setting-up treatment facilities. Mara (2009) suggests that the waste stabilization ponds (oxidation ponds, maturation ponds and duckweed ponds) can be appropriate for small towns having land availability. Afifi (2011) points for large urban settlements with land scarcity for the establishment of sewage treatment plants (STPs) and less demand for treated sewage for farm application, mechanical treatment systems viz. activated sludge process, trickling filter, up-flow anaerobic sludge blanket (UASB) and aerated pond are suitable to create good results.

OPPORTUNITIES AND CHALLENGES OF WASTEWATER RECLAMATION AND REUSE

Total municipal wastewater generated by Indian industries is 13,468 MLD out of which 60% is treated and the small and medium industries being major contributor to untreated wastewater. As it is known, that the industrial waste effluent is typically more poisonous thus the efforts for its use in agricultural needs to be focused mainly on municipal sewage for the predictable prospect. This difference is also seen in the research writing, which is filled with the studies on sewage-fed agriculture but barely the reuse of industrial wastewater for agriculture (although much investigation is being undertaken on water use competency in industries and on-site use by recycling).

The exception is industries producing comparatively less unhealthful effluent such as those within the food/beverage sector, for example breweries, sugar mills and fruit processing units. Data on brewery manufacturers shows that two million KL (Kilo Liters) of beer was consumed in India in 2012. According to the CPCB, about ten KL of effluent is generated per KL of beer created in India on an average which results in 20 million KL of effluent per year or about 55 MLD. It may seem that this is a small number but as per statistics the beer industry in India is experiencing growth in the coming future which invites a large amount of waste generation. In dry season or in the water scarce regions a small amount of reuse of wastewater can make difference. Thus these industries are more interested in wastewater reuse which has been seen in EU sponsored W4C project.

Most of the industries are practicing the reuse of wastewater in agriculture. Treated wastewater is being distributed by the SABMiller brewery in Andhra Pradesh for free of charge to local farmers. The treated effluent from the sugar mill is being sold by the Ugar Sugar Works in Karnataka at a cost of Rs. 3,000/ha to farmers. These are also generating the media coverage in positive way but the actual situation is more complicated. The main problematic issues are that there is no monitoring and of these practices. Most of the industries are situated in non urban areas.

State Pollution Control Boards (SPCBs) in India do not have the assets to check them continuously. Due to lack of any monitoring, there is no way to assure a certain quality of treated wastewater to farmers. Without technical studies it is not possible to state whether such wastewater reuse methods are having a negative effect on soils, crops or farmers' physical condition. In addition, farmers' needs frequently differ by season; for example, they would desire wastewater in the dry season stopping wastewater supply to farmers and putting more stress on reuse on-site and in-process. This may sometimes result into the evaporation of waste matter which can be energy serious and creates solid waste that needs removal.

An appropriate monitoring method can solve this crisis and help in agricultural reuse of wastewater from these industries. If an self-governing body can be set up containing of local agricultural extension officers, scientists from local universities, representatives from local farmers' and involvement of cooperative and industries, this committee can draw up official preparations and be able to check class and amount of wastewater delivered. Ideally, financial support for this should be shared justifiably among the various stakeholders; while the involved industry should cover a comparatively big share of financial support; if it is only industry sponsored there may be perceptions of disagreement of interest. Some primary steps are being taken in this way through the W4C project: ICRISAT is researching on SABMiller's wastewater reuse in Andhra Pradesh and the University of Agricultural Sciences, Dharwad is investigating on Ugar Sugar Works' wastewater reuse in Karnataka. The results of this beginning research will optimistically guide to testing with establishing a feasible model for checking agricultural reuse of water from the industries.

Use of wastewater in agriculture is a budding priority in India under water shortage. The influential issues are accountable for this are - growing water stress in a lot of areas of tier-I cities as well as tier-II cities; rising urbanization increasing wastewater flows connected with the growing water delivery and sewage; also more city households engaged in agricultural events, chiefly as foodstuff prices carry on to enlarge. The unforeseen consumption of raw wastewater helps to increasing risks to society, flora, and the surroundings. In these conditions, state administration is more and more worried to make wastewater reuse for agriculture an optimistic power for growth as component of their water resources management schemes, is about to be an major key region for development. The succeeding foremost messages look for to direct governments, practitioners, and stakeholders, water professionals, willing to construct and/ or make stronger national as well as local capacities for protected wastewater irrigation.

A study of main issues concerned to wastewater irrigation can be stated by segregating cities into four heads by state of financial development, characterizes the variation of issues across cities in diverse steps of growth, depicts probable ways and prospect importance for developing wastewater handling. Correctly designed, managed and implemented, wastewater irrigation policies can give numerous paybacks that accumulate in the agricultural, as well as environmental sectors. Facing such problems and benefits, cities looking better wastewater use for agriculture should follow the under mentioned objectives:

I. reduce danger for community health;
II. reduce surroundings risk;
III. advance livelihoods for city agriculturists; and
IV. put together wastewater as part of water resources management context

One or more combinations of these objectives can be followed by the city depending upon the economic growth (Table 4). As an example the cities with low income can put more concern on reducing microbial hazards to physical condition while developing the livelihoods of urban farmers, whereas on the other hand cities with medium and high income can give importance to reduce chemical and environmental hazards, mostly when they are stressed out with water, a complete collaboration of wastewater and the water resources management schemes.

Table 4. Wastewater irrigation objectives in Indian Cities

Economic Development	Minimize Public Health Risk		Objective 2: Minimize Environmental Risk	Urban Agriculture and Improvement in Livelihood	Integration of Water Use in Wastewater Irrigation
	Microbial	Chemical			
Cities with low income	Vital	Short	Short	Vital	Short
Lower-middle income cites	Elevated	Emerging	Emerging	Elevated	Incipient
Upper-middle income cities	Elevated	Vital	vital	Elevated	Evolving
High-income non-OECD cities	Elevated	Elevated	Elevated	Short	Advanced
High-income OECD cities	Short	Elevated	Elevated anthropogenic compounds is focused	Nil	Advanced

Checking of Microbial Risk and Establishing Pathogen Decreasing Targets

The World Health Organization's "Guidelines on Wastewater Use in Agriculture" of 2006 focuses on a risk evaluation and organization approach which follows the Stockholm Framework. Following are the advancements:

1. Definition of bearable highest extra load of sickness;
2. To get bearable risks of infection;
3. To lay down health-based goals to reduce pathogen;
4. To decide how the necessary pathogen reductions can be achieved; and
5. To put in a scheme for confirmation monitoring.

The 2006 strategy provide nations at all level of growth with the way to sensibly take embattled steps to decrease health risks. Even when complete wastewater treatment is the solution, it presents a multi-barrier move toward shielding the health of farmers and consumers.

Gradually Apply Industrial Pre-Treatment and Control

To reduce the chemical risk and the desired function of the treatment set and wastewater irrigation policies the effluent generated from the industries discharging considerable amount of wastewater needs to be controlled. To confirm that the heavy metals, salts, organic toxins and other pollutant produced by industrial processes do not attain levels which may destroy pipes, slow down biological treatment processes, contain in the wastewater in large amount than the allowable for irrigation use or discharge in the surroundings or get collected in the sludge, thus limiting its reuse for this standards and discharging restrictions must be stated for the industries. The rising industrialization of city regions requires that chemical hazards of wastewater for irrigation be accounted, mainly through the preface, gradual execution, enforcement of industrial wastewater pre-treatment and manage programs.

Approach for Wastewater Use in Agriculture

Summers of 2016 in India, have brought the water scarcity issue at the forefront, thus making treated wastewater as the best commodity for trade for meeting industrial and irrigation needs of the cities. To make wastewater a viable resource governments investment and policies would inspire and encourage private sector to get involved through pedicle business models, (like the 2006 WHO tips). An approach to move from unstructured to structured wastewater use for irrigation and recognizing that dreams can't be met in a day or in single phase, the strategic plan have to be multi-phased and within an agreed and practical time-frame.

Resourceful Wastewater

The wastewater reuse system, establishment of equivalent authoritarian framework, and the execution of structured wastewater irrigation requires embedded use of agricultural wastewater inside the broader water property manage background. A quantity of integrated water property control thoughts should be accomplished, in particular: accept a multi- sectored method to water management in affiliation with river basin manage; inspire stakeholder participation and devolution of duty; promote non-public region involvement; and rent financial gadgets.

Wastewater Reuse Case Studies

Beneficial use of wastewater has been practiced in California since 1890s, when raw sewage was implemented on 'sewer farms'. Over a period of time there was a steady increase in the sewage utilization and by the year 1987, roughly 900 m3/d of municipal wastewater (7-8% of the production) was getting used for the various applications (California State Water Resources Control Board, 1990). Historically, agricultural use has dominated, and maintains to do so, but during the last decade reclaimed wastewater has been more and more used for landscape irrigation in urban regions and for groundwater recharge. Maximum of the reclaimed water (78%) is used inside the principal Valley and South Coastal regions of California. This has saved around 0.759 Mm3/d of clean water. In the period 1981 to 1987, the Ministry of Agriculture and Public Health, with assistance from the United Nations Development Program (UNDP), carried out studies designed to assess the effects of using treated wastewater and dried, digested sewage sludge on crop productivity and on the hygienic quality of crops and soil. Treated wastewaters and dried, digested sludge from the La Cherguia (Tunis) and Nabeul (SE4) activated sludge plants were used in the studies and irrigation with groundwater was used as a control. At La Soukra, tests were conducted on sorghum (*Sorghum vulgare*) and pepper (*Capsicum annuum*) using flood irrigation and furrow irrigation, respectively. Clementine and orange trees were irrigated at OuedSouhil (Nabeul). Use of treated wastewater and sewage sludge at the L. A. Soukra and Oued Souhil experimental stations, wherein the soils are alluvial and sandy-clayey to sandy, had no longer adversely affected the physical or bacterial best of the soils. The use of wastewater led to annual and perennial crop yields higher than yields produced via groundwater irrigation. Sewage sludge application increased the production of sorghum and pepper and resulted in the plants containing better concentrations of N, P and K and some minor factors (Fe, Zn and Cu). Untreated sewage has been used for decades to irrigate forestry tasks a long way from the inhabited areas of Kuwait. Effluent from the Giwan secondary sewage treatment plant became used to irrigate plantations on an experimental farm from 1956 (Agriculture Affairs and Fish Aassets Authority, Kuwait 1988). Use of raw sewage for irrigation within the Mezquital Valley of the Tula River Basin began in 1886 (Sanchez Duron 1988). But, it become now not until 1945 that the Ministry of Agriculture and Water assets installed the number 03 Mezquital Irrigation District to manage the distribution of wastewater from Mexico town for irrigation functions. Irrigation is important in this Irrigation District due to the fact rainfall is constrained and poorly allotted over the year, most falling between July and September. Sewage from Mexico town blended with variable proportions of surface water gathered in reservoirs in the basin has enabled farmers within the Mezquital Valley to provide agricultural produce for the capital city. The East Calcutta sewage fisheries are the biggest unmarried wastewater use machine regarding aqua-way of life inside the global. An historical account of the development of this gadget has been given by using Edwards (1985 and 1990). Ghosh (1984) offered the facts on the range of size and numbers of sewage fisheries in Calcutta as shown in desk 45. In 1945, the location of sewage-fed fish ponds become approximately 4628 ha, in a wetlands region of about 8000 ha, however the fish pond location have been reduced to about 3000 ha through 1987 because of city reclamation and conversion of fish ponds to rice paddies. Possession of the ponds is in the control of around 160 urban dwellers, who appoint almost 4000 households as fishermen, and there are several fishermens' cooperatives (Strauss & Blumenthal, 1989).

Consequences of Use of Wastewater

There are many winning wastewater use policies throughout the globe where nutrient reuse is a main benefit to the project (Pescod & Arar, 1988; FAO, 1992). Rarely, however, is a system laid out or intended on the base of nutrient recycling the main restraint to any wastewater use project is community health.

Depending on its composition, the treatment it has undergone, the extent to which it is irrigated and the regulations and principle guidelines under which it is being utilized, wastewater use in agriculture can be viewed as both a benefit, providing water and nutrients for the cultivation of crops and ensuring food supply to the cities, as well as a source of pollution, a threat affecting the health of users, consumers and the environment (Javier et al., 2013). Wastewater, especially domestic wastewater, contains pathogens which can cause disease spread when not managed properly. The survival of pathogens and how they infect a new host needs to be understood in developing a program to eliminate or minimize health risks. The primary objective of any wastewater use project must therefore be to minimize or eliminate potential health risks. As stated earlier, wastewater contains pathogenic microorganisms inclusive of bacteria, viruses, and parasites, which have the ability to reason disorder mainly, human parasites such as protozoa and helminth eggs are of unique importance on this regard as they show to be most hard to eliminate by using treatment procedures and had been implicated in some of infectious gastrointestinal illnesses in both developed and growing nations (Hussain, Raschid, Hanjra, Marikar, & Hoek. 2002). However, in evaluating health impacts it must be remembered that it is the actual risk that make people fall ill that must be quantified and not the presence of pathogens in water. Whilst the potential risk may be quite high, the actual risk depends on many other factors. The use of untreated wastewater for irrigation, no doubt, poses a high risk to human health in all age groups. However, the degree of risk may vary among the various age groups. Untreated wastewater irrigation leads to relatively higher prevalence of Ascariasis infections with children (Cifuentes et al. 2000; Habbari et al., 2000), hookworm (Feenstra, Hussain, & Hoek, 2000). A lot of cities are facing key challenges in organizing the hazards linked with unrestrained wastewater irrigation, mainly water-shortage cities and low-income cities. Many have also begun to concentrate on those problems and progressed toward some degree of intended wastewater use in agriculture some with budding ways and others with definite plans. Each nation is sole, and experiences of other countries and cities can give some instruction as to how to get better wastewater irrigation practices. Cities can become accustomed to the techniques and schemes developed somewhere else to go with their own situation.

Development Partners

The support from the different organizations was sought by the cities for implementing their own ideas, skill development institutions, NGO's, etc. Many of the organizations have generated awareness among the people to avoid the usage of untreated wastewater. Most of them helped to maintain the cities with the secured wastewater irrigation and also selling of the same. Many volunteers are being motivated to offer more attention on this kind of initiative and add an asset to his/her nation. The different organizations that helped in the above context are - World Health Organization, The United Nations Development Program, International Development and Research Centre, International Water Management Institute, Resource Centre for Urban Agriculture and Forestry (RUAF), Water Partnership Program (WPP) to name a few.

BUSINESS MODEL FOR WASTEWATER REUSE IN AGRICULTURE

Fifty percent of the world's inhabitant's lives in urban areas and this is likely to reach two thirds by 2050. In many developing countries, urban populations are fraught with inefficient waste and used water management systems and these spill over into the environment, often with dire consequences. A walk through any Indian city would reveal open sewages and trash-filled drains. Moreover, rivers in many cities like Pune have turned in to veritable sewers. These cities have an immediate and urgent need for improved waste management and sanitation; however, the local authorities are neither financially nor managerially ready to meet the burgeoning challenge. Untreated wastewater inadvertently enters agricultural fields in India and is a prime reason for the spread of pathogens. However, for the entrepreneur urban waste presents a business opportunity and for municipal bodies this could be a potential market approach to solve an everyday urban problem. Due to India's monsoon based climate and consequent seasonality of its rivers, most parts receive their natural supply of fresh water only about four months in the year. In such a condition urban wastewater generated daily, turns out to be a resource which can bridge the supply gap. Wastewater, can be, in many cases a source of rich materials which can be recovered after some treatment. The nutrient value of domestic wastewater in agriculture is well studied but carries significant health risks. Involving the private sector in wastewater treatment and developing a business model can potentially turn this activity operationally efficient and distributionally effective.

The two major applications of urban wastewater, or used water, are in agriculture and in industry. Significant developments are seen in both cases, especially in arid regions of the world like Australia, Middle East and United States. Industry can safely use treated wastewater in a variety of applications like cooling system make-up water, boiler feed water, process water, wash down water and uses like site irrigation, fire protection, road cleaning. A successful example is seen in the mining industry, where an increasing number of mining companies are using reused water from residential or industrial areas (Workshop on industrial water reuse in the mining sector, 2013). In India, an example of urban wastewater use is seen in the city of Bangalore where it is used to recharge ground water sources. Urban wastewater is used for refilling exhausted irrigation tanks in rural neighborhoods and this in turn replenishes ground water levels. Such a depleted tank situated in Hoskotte in rural Bangalore receives excess water from an urban tank and serves as a storm water and wastewater reservoir throughout the year (Drechsel, Qadir, & Wichelns, 2015). Another ongoing project of the International Water Management Institute is being run with the objective of analyzing risk and opportunities of fecal sludge reuse in Indian agriculture. The project will examine options for optimizing private sector engagement in the valorization of fecal sludge and urine as agricultural inputs. The project focuses on the areas around the cities of Bangalore and Hubli-Dharwad in the State of Karnataka, but will also explore the scale of fecal sludge reuse across at least five other states within India. Of particular interest is private sector participation and search for scalable business models.

Similarly, for rural waste management, which is perhaps a problem of lesser proportion but similar consequences, the Decentralized Waste Management Systems model recommended by ICRISAT (ICRISAT, 2015) can be implemented through the green business model (Figure.2).

In African countries small-scale entrepreneurs are using human waste to make a profit. In Ghana, for instance, urban and peri-urban farmers use nutrient-rich sludge and wastewater to grow crops (International Water Management Institute, 2015). While this requires careful attention to health concerns, they can be overcome by regulation and third party audits. Planners must ascertain that there are enough end-users for a waste product.

For assessing the viability of any urban or rural wastewater use projects, planners must evaluate two aspects of the proposed project. Economic justification – are the total benefits arising out of the project higher than the total costs? There is a high probability of economic justification for most Indian urban wastewater projects for a reason – the untreated wastewater finds its way to farms and ground water sources and poses a massive health hazard. This cost, while difficult to quantify, is nevertheless a real cost and one which must be factored in this evaluation.

Financial viability – who pays and how? This is not an easy challenge. The urban wastewater requires energy intensive technology for treatment while the output is valuable to agriculture and industry. Does the municipal authority pay for the treatment with the urban tax payer's money or must the beneficiaries of the treated water pay for it. Successful examples from several places in the world can be looked to find an effective and equitable way to manage the costs.

Arguments Favoring Private Sector Participation in Used Water Treatment

Studies and examples from other countries provide these compelling arguments, which are summarized below. Profit motive can lead to efficient operations, which is not possible when operations are undertaken by the local governing body. The introduction of a profit motive can lead to increasing the efficiency as compared to public management of the water system.

Private enterprises can get in managerial skills and capabilities which may not be available with local governing bodies. The private sector has improved managerial capabilities and sources of financing leading to better planning and control systems. Moreover, efficient operating and accounting procedures and effective distribution must be incorporated for securing profits.

Assurance of customer satisfaction is a strong influencer to efficient operations. This is a key objective of any business. Businesses run on customer satisfaction, while public authorities woefully lack this objective.

Figure 2. Decentralized Waste Management Systems model recommended by ICRISAT
Source ICRISAT

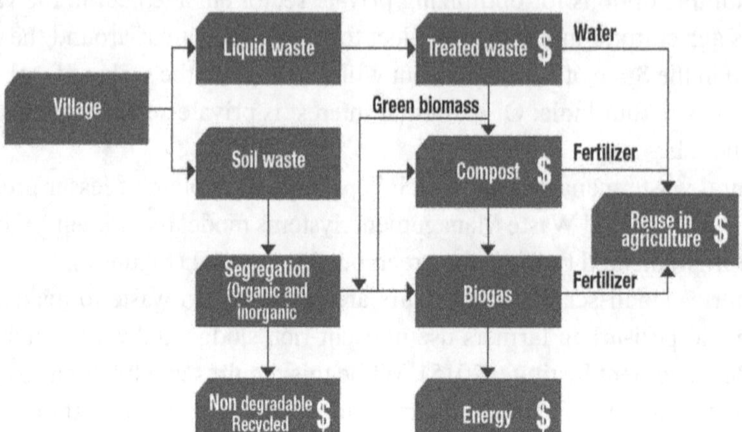

Lessons from Other Countries

Various countries have successfully used private partnerships to create sustainable business models in urban wastewater management. United Nations has worked with several African countries and developed economies in the European Union and has implemented models which seek business profits and social good. In these cases local bodies provide the services using a mixture of financing sources which suit local conditions. The public-private liasons (PPP) models with contractors can include management contracts, service contracts, leases, divestures concessions (build-operate-transfer), and public-private organizations (Hutton & Wood, 2013). In many nations, wastewater management service are not supported, low valued and rules (where they exist) may not be meticulously applied, thus cost revival may be difficult. Cost upturn options include creating business opportunities and economic instruments. Murray et al. (2011) have recommended that public-private liasons based on cost-recovery from the recycle of human waste could help to co-finance the sanitation/wastewater sector and to incentivize also, at the same time, promoting small and medium scale entrepreneurs. They put ahead four waste- based business models (involving aquaculture, biogas recovery, compost production and the use of fecal waste as an industrial fuel) and they projected a number of competence indicators in order to permit a comparison between different reuse options and hence allow a monetary assessment to be made of different reuse business concepts.

Critical Success Factors for Successful Public-Private-Partnerships in Wastewater Treatment

A good number of works by experts and several winning examples demonstrate the usefulness of partnering with the private sector. But, these tie-ups are complex to design, to practice and function. Therefore, PPPs must be carefully checked in the background of the project, the social benefit and the concerned gains to be attended under various ways. Some of the critical factors for a successful PPP are discussed below.

1. **Availability of Market:** Private participation cannot be attractive where there is no prevailing demand. While it may seem that treated wastewater may be readily in demand, there have been cases where farmers have declined to use such water. This could be due to low awareness and may require government campaigns to promote its use.
2. **Multiple Stakeholder Benefit Objectives:** Profit is certainly the motive for private participation, however, it must be met with social benefits, which may not always be quantifiable. The stakeholder analysis often requires data and strong managerial capabilities to lay out the multiple benefits that the PPP must address.
3. **Competitive Procurement Process:** While this is a fundamental requirement of social projects, there have been instances where private companies, on the basis of technological uniqueness, have driven specifications to make the bidding process favorable to them.
4. **Thorough Risk Assessment:** Many studies have shown the local governing bodies to be ineffective to quantify project risks. On the other hand, some successful projects have employed the services of third party risk assessors. This could be an option when the local body lacks the required skills.
5. **Shared Authority and Responsibility:** Holding the private partner responsible for the outcome is effective when they have the required authorities in place. Where such authority has been lacking, the projects have encountered undue delays in the bureaucratic decision making processes.

6. **Social Support:** This is an essential condition to any public-private project. The society which stands to benefit from the project must have a buy-in of the project. In urban wastewater treatment projects, this includes the polluters and the beneficiaries of the treated water.

Lower-average-income nations (like Iran and Jordan), higher-center-profit nations (like Chile, Argentina, Mexico), and high-earnings nations (like Kuwait and Qatar) give best examples of the successful use of wastewater. It can be the examples that a major role in treatment of wastewater can be played by the private sector by providing private investments, design, build and operate the treatment plants so as to attain the standards for irrigation. Many water scarce nations practice the involvement of private sector for the same and use in irrigation has been practice in many water scarce countries.

A study by Ernst & Young indicates that the Indian water sector can require venture of around $130 billion between 2011 and 2030. Wastewater management, in particular, is emerging as a key thrust area. Currently, only 60% of industrial and 26% of domestic wastewater is treated in India. Metros and large cities are treating only about 30% while smaller cities treat a minuscule 3.7% of their wastewater. Wastewater treatment has been a largely neglected area because municipal bodies cannot charge water users for the service. Thus, while municipal bodies often do get funding for wastewater treatment projects, operating and maintaining treatment plants is still a challenge. Recycling and reuse of wastewater has been encouraged by the Water Policy 2012. Industries in coming years may have to rely mostly on reuse of water for which recycled water can fetch a good price. A Chennai based company has bagged a project in Tuticorin from the Tamil Nadu government to set up a sewage treatment and solid waste treatment plant. The contract includes building the plant free, taking ownership and operating it. The treated water will be sold to industries which will bring in the revenues.

CONCLUSION

Climate change and wastewater control are one of the foremost demanding situations for Indian cities. Urbanization and unparalleled growth of cities across India continue to create immense pressure on land and water resources. This uncontrolled growth continues to produce growing volumes of wastewater. Climate change impacts inclusive of, intense storm events in summer time or extended moist periods in wintry weather are quite visible in India. In urban and peri-urban areas, wastewater use for agriculture is an emerging precedence. Due to susceptible enforcement of regulatory, most of the wastewater generated is permitted off untreated/ partially treated. While many previous studies have checked out the global modifications and associated impacts of climatic variations on water resources, few have targeted at the evaluation of the particular effects and adaptation priorities for water systems in towns. Proper reuse of wastewater for irrigation could significantly lessen water shortage and offer a sustainable source of water, improving agricultural productiveness, reduce pollution, and create livelihood possibilities for low earnings city households and contribute to their each day food needs. However, there are tradeoffs that need to be understood which includes risks to human health, and environment

The success stories from Israel and Chile indicates that for achieving secure and successful wastewater irrigation requires slow, steady and determined efforts on all fronts for several years. Learning from these examples Indian wastewater projects can create big opportunities for countering urbanization and helping economic development.

Possible solution to achieve this is to help municipal authorities to transport large volumes of water from long distances (150 km) which are a part of the rural agricultural system. During this process couple of hurdles gets created like over excessive fees of lifting water, river pollution and groundwater contamination. In every panorama, water augmentation must be considered at the side of special wastewater remedy strategies for multiple makes use of, and ought to be backed up with public policy and social incentives. This will surely guard the downstream customers and also offer financial benefits and help the surroundings offerings that constitute an imperative part of all varieties of lifestyles. A state and city wise approach for wastewater use in agriculture ought to seize the variety seen within the Indian context; this bottom up approach will identify nodal agencies for systematic data series and collection. Data collected can be effectively used for wastewater control plans thus preserving and reducing continued dispute between states for freshwater use .The reuse of wastewater generated will conserve groundwater, ecosystems, and reduce downstream affects thus helping in restoring livelihoods. However, wastewater treatment is crucial stages and hence should be designed to meet the requirements of end users, but these needs to be elaborated based on the wastewater supply at that location. Such sector based study can become a part of municipal improvement plan, making best use of wastewater generated in the towns. If wastewater is taken into consideration and included as a vital factor into national water plans and water cycles catchments, doubtlessly this will protect the downstream users and provide economic opportunities through alternative use of wastewater within towns.

The crude approach for India estimated that over 1.1 Mha of land can be irrigated using the available wastewater resource from different cities. Water can be collected through dedicated channels and infrastructure. Wastewater needs to be diluted with fresh water for its better usability,

Because dilution changes the water class, and estimations may require a exceptional modeling approach, technology driven systems like Remote Sensing and Geographical Information Systems (GIS) and unique mapping of drainage networks can offer better basic consequences which could help to investigate the nutrient loads leaving the metropolis. GIS-based mapping make land-use as part of their utility, to increase baselines, upon which future research may be modeled. Wastewater irrigation may be a dynamic system in the urban areas, and land-use patterns can exchange with development and socioeconomic exchange; consequently, to study techniques to capture this dynamism, spatially and temporally. Case studies confirmed that because of wastewater use farmers spent plenty less on inputs, and wherein the nutrient resources can be balanced the final results modified into more excessive exceptional (Delhi, Kanpur and Kolkata) in phrases of price monetary financial savings and economic returns. This become most effective primarily based on agricultural production, and a further holistic economic analysis desires to be completed to seize the personal advantages to the households and social benefits to the organizations. Wastewater agriculture is but no longer without horrible externalities, and fitness affects on farmers and clients are of huge state of affairs as mentioned above. From an Indian context more research are required in the regions of wastewater irrigated agriculture, health and food protection, and health economics, particularly at the farm and purchaser stages, to capture the diverse settings wherein the problems exist. Threat evaluation device like Quantitative Microbial chance evaluation (QMRA) and Quantitative Chemical chance evaluation (QCRA) may be used to evaluate the capability risk, which want to then be addressed via a couple of barrier approaches with fitness-based totally dreams for risk discount (WHO, 2006). In evaluation in India, greater emphasis wishes to be located on wastewater treatment processes that take away heavy metals, which seem to have a great deal better tiers than in maximum components of Africa (Raschid-Sally & Jayakody, 2008). Wastewater management and remedy cannot be planned in isolation. They should be a center a part of the strategic plans for water deliver and sanitation, irrigation and drainage, strength, and environmental offerings (World Bank, 2004). Moreover, it will become very essential to recall the ones additives in light of water availability for cities, and to highlight the need for continuous inter-sectoral communicate and action plans to address the ever-developing water demands (Global Economic Organization, 2010). Integration of water assets improvement with water services can provide greater help for agricultural water control. India being in recent times greater urban and peri-city than rural, it's time for safe wastewater.

REFERENCES

Afifi, A. A., Abd El-Rheem, & Youssef. (2011). Influence of Sewage Water Reuse Application on Soil and the Distribution of Heavy Metals. *Nature and Science*, *9*(4), 82–88.

Arnell, N. W., Halliday, S. J., Battarbee, R. W., Skeffington, R. A., & Wade, A. J. (2015). The implications of climate change for the water environment in England. *Progress in Physical Geography*, *39*(1), 93–120. doi:10.1177/0309133314560369

California State Water Resources Control Board. (1990). *California Municipal Wastewater Reclamation in 1987*. Office of Water Recycling.

Central Pollution Control Board, (2005). *A report on Status of Water Supply, Wastewater Generation and Treatment in Class-I Cities & Class-Ii Towns of India*. Author.

Cifuentes, E., Gomez, M., Blumenthal, U., Tellez-Rojo, M. M., Romieu, I., Ruiz-Palacios, G., & Ruiz-Velazco, S. (2000). Risk factors for Giardia intestinalis infection in agricultural villages practicing wastewater irrigation in Mexico. *The American Journal of Tropical Medicine and Hygiene*, *62*(3), 388–392. PMID:11037783

Drechsel, P., Qadir, M., & Wichelns, D. (2015). *Wastewater: Economic Asset in an Urbanizing World*. Springer. doi:10.1007/978-94-017-9545-6

Environmental Information System (ENVIS) Centre on Hygiene, Sanitation, Sewage Treatment Systems and Technology. (n.d.). Retrieved from http://www.sulabhenvis.nic.in/Database/STST_wastewater_2090.aspx

Feenstra, S., Hussain, R., & van der Hoek, W. (2000). Prevalence of intestinal parasites in the Southern Punjab, Pakistan. IWMI Pakistan Report 101.

Gebrezgabher, S., Rao, K., Hanjra, M. A., & Hernández-Sancho, F. (2015). Business Models and Economic Approaches for Recovering Energy from Wastewater and Fecal Sludge. In Wastewater (pp. 217-245). Springer Netherlands. doi:10.1007/978-94-017-9545-6_12

Grit, R., Jörg, L., Steffen, D., & Gerel, O. (2015). Integrated urban water management: Development of an adapted management approach. *Environmental Earth Sciences*, *73*(2), 709–718. doi:10.1007/s12665-014-3701-z

Grover, M. (2012). *Degremont Water Treatment Technologies and A Case study of 635 MLD WTP at Sonia Vihar–Delhi*. Presented to working group of Planning Commission on Working Group on Urban and Industrial Water for Twelfth Plan of Government of India.

Habbari, K., Tifnouti, A., Bitton, B., & Mandil, A. (2000). Geohelminthic infections associated with raw wastewater reuse for agricultural purposes in Beni-Mellal, Morocco. *Parasitology International, 48*(3), 249–254. doi:10.1016/S1383-5769(99)00026-4 PMID:11227765

Hussain, I., Raschid, L., Hanjra, M. A., Marikar, F., & van der Hoek, W. (2002). *Wastewater use in agriculture: Review of impacts and methodological issues in valuing impacts. (With an extended list of bibliographical references)*. Working Paper 37. Colombo, Sri Lanka: International Water Management Institute.

Hutton, G., & Wood, A. (2013). *Development financing for tangible results: A paradigm shift to impact investing and outcome*. Academic Press.

International Crops Research Institute for the Semi-Arid Tropics. (2015). *Waste management recycling and reuse.* Retrieved from ICRISAT: https://www.google.co.in/url?sa=t&rct=j&q=&esrc=s&source=web&cd=4&cad=rja&uact=8&ved=0ahUKEwi4v5-9uL3LAhWUUI4KHbdFCw4QFggzMAM&url=http%3A%2F%2Fwww.icrisat.org%2Fidc-flyers%2F3_Waste%2520management_CSR%2520Flyers.pdf&usg=AFQjCNFBTLfK7Mu_ng0uljn_TzJO0eAyMQ&

International Water Management Institute. (2015, March 10). *Wastewater Business models.* Retrieved from: http://www.iwmi.cgiar.org/issues/wastewater/business-models/

Itankar, N., & Patil, Y. (2015). Biosorptive Removal of Hexavalent Chromium from Aqueous Industrial Solutions by A Novel Wood Apple Shell Powder Biomass. *Advanced Materials Research, 1130*, 689–692. doi:10.4028/www.scientific.net/AMR.1130.689

Lewis, N. D. (2016). Sustainable development through a gendered lens: Climate change adaptation and disaster risk reduction. *Reviews on Environmental Health, 31*(1), 97–102. doi:10.1515/reveh-2015-0077 PMID:26943600

Mara, D. (2009). Design Manual for Waste Stabilization Ponds in India. University of Leeds. Available at http://www.leeds.ac.uk/civil/ceri/water/tphe/publicat/wspwarm/wspwarm.html

McCarl, B. A., Musumba, M., Smith, J. B., Kirshen, P., Jones, R., El-Ganzori, A., & Bayoumi, M. et al. (2015). Climate change vulnerability and adaptation strategies in Egypts agricultural sector. *Mitigation and Adaptation Strategies for Global Change, 20*(7), 1097–1109. doi:10.1007/s11027-013-9520-9

Mekala, G. D., Davidson, B., Samad, M., & Boland, A. M. (2008). *Wastewater reuse and recycling systems: a perspective into India and Australia.* Colombo, Sri Lanka: International Water Management Institute (IWMI). Retrieved from www.iwmi.cgiar.org/Publications/Working_Papers/working/WOR128.pdf

Misra, A. K. (2014). Climate change and challenges of water and food security. *International Journal of Sustainable Built Environment, 3*(1), 153–165. doi:10.1016/j.ijsbe.2014.04.006

Mohan Kumar, R. (2003) Financing of Wastewater Treatment Projects. Infrastructure Development Finance Corporation and Confederation of Indian Industries, Water Summit 2003, Hyderabad, India.

Murray, A., Cofie, O., & Dreschsel, P. (2011). Efficiency indicators for waste-based business models: Fostering private-sector participation in wastewater and faecal-sludge management. *Water International, 36*(4), 505–521. doi:10.1080/02508060.2011.594983

Neumann, M. B., Rieckermann, J., Hug, T., & Gujer, W. (2015). Adaptation in hindsight: Dynamics and drivers shaping urban wastewater systems. *Journal of Environmental Management*, *151*, 404–415. doi:10.1016/j.jenvman.2014.12.047 PMID:25594746

Pathak, H., Ladha, J. K., Aggarwal, P. K., Peng, S., Das, S., Singh, Y., & Gupta, R. K. et al. (2003). Climatic potential and on-farm yield trends of rice and wheat in the Indo-Gangetic Plains. *Field Crops Research*, *80*(3), 223–234. doi:10.1016/S0378-4290(02)00194-6

Patil, Y., & Rao, P. (2015). Industrial waste management in the era of climate change - A smart sustainable model based on utilization of active and passive biomass. In W. L. Filho (Ed.), *Handbook on Climate Change Adaptation* (pp. 2079–2092). Springer-Verlag Berlin Heidelberg. doi:10.1007/978-3-642-38670-1_49

Peirson, W., Davey, E., Jones, A., Hadwen, W., Bishop, K., Beger, M., & Gray, L. et al. (2015). Opportunistic management of estuaries under climate change: A new adaptive decision-making framework and its practical application. *Journal of Environmental Management*, *163*, 214–223. doi:10.1016/j.jenvman.2015.08.021 PMID:26321531

M. B. Pescod, & A. Arar (Eds.). 1988. Treatment and use of sewage effluent for irrigation.*Proceedings of the FAO Regional Seminar on the Treatment and Use of Sewage Effluent for Irrigation.*

Rao, P., & Patil, Y. (2015). Climate Resilience in Natural Ecosystems in India: Technology Adoption and the Use of Local Knowledge Processes and Systems. In Handbook of Climate Change Adaptation. Springer Berlin Heidelberg.

Raschid-Sally, L., & Jayakody, P. (2008). *Drivers and characteristics of wastewater agriculture in developing countries: Results from a global assessment. IWMI Research Report 127*. Colombo, Sri Lanka: IWMI.

Rawson, H. M., Gifford, R. M., & Condon, B. N. (1995). Temperature Gradient Chambers for Research on Global Environment Change. I. Portable Chambers for Research on Short-Stature Vegetation. *Plant, Cell & Environment*, *18*(9), 1048–1054. doi:10.1111/j.1365-3040.1995.tb00616.x

Sanchez Duron, N. (1988). *Mexican experience in using sewage effluent for large scale irrigation. Treatment and Use of Sewage Effluent for Irrigation* (M. B. Pescod & A. Arar, Eds.). Sevenoaks, UK: Butterworths.

Scott, C., Drechsel, P., Raschid-Sally, L., Bahri, A., Mara, D., Redwood, M., & Jiménez, B. (2010). Wastewater irrigation and health: challenges and outlook for mitigating risks inlow-income countries. In P. Drechsel, C. Scott, L. Rashid-Sally, M. Redwood, & A. Bahri (Eds.), *Wastewater Irrigation and Health: Assessing and Mitigating Risk in Low-Income Countries* (pp. 381–394). London: Earthscan.

Strauss, M., & Blumenthal, U. J. (1989). *Human waste use in agriculture and aquaculture: utilization practices and health perspectives*. IRCWD Report No. 08/89. International Reference Centre for Waste Disposal.

Valta, K., Moustakas, K., Sotiropoulos, A., Malamis, D., & Haralambous, K. J. (2016). Adaptation measures for the food and beverage industry to the impact of climate change on water availability. *Desalination and Water Treatment*, *57*(5), 2336–2343. doi:10.1080/19443994.2015.1049407

Workshop on industrial water reuse in the mining sector. (2013). *Report on Workshop on industrial water reuse in the mining sector.* Windhoek: Water Research Commission. Retrieved from www.wbcsd.org/web/water/2013Report-IWA-WBCSD-WRC-Mining-Water-Reuse-Workshop.pdf

World Bank. (2010). *Improving Wastewater Use in Agriculture: An Emerging Priority, Energy Transport and Water Department Water Anchor*. ETWWA.

Zouboulis, A., & Tolkou, A. (2015). Effect of climate change in wastewater treatment plants: Reviewing the problems and solutions. In *Managing Water Resources under Climate Uncertainty* (pp. 197–220). Springer International Publishing. doi:10.1007/978-3-319-10467-6_10

Chapter 17
Adaptive Coevolution:
Realigning the Water Governance Regime to the Changing Climate

Dauglas Wafula Juma
Tongji University, China

Hongtao Wang
Tongji University, China

Makomere Reuben
Tongji University, China

Fengting Li
Tongji University, China

ABSTRACT

Climate change is having a significant impact on water resources globally; more so in developing countries due to lower resilience and weakened coping mechanisms. Water resource and supply management systems need to be adaptive to the uncertainty and unpredictability posed by this challenge. The current water crisis is mainly that of governance within the water sector. On this premise, this chapter seeks to explore the adaptiveness of water management institutions and systems to climate change impacts, with a focus on Kenya's case. Kenya continues to face numerous challenges relating to water supply, which are further exacerbated by climate change. Prior to the development of key legislative and institutional frameworks, through adoption of the Water Act 2002, the water sector grappled with gross mismanagement, manifested via inefficiency in water service provision and resource degradation. Deploying the conceptual model of adaptive (co)-management institutional prescription of polycentric governance, experimentation, public participation, and management of bioregional scale, this chapter stands to contribute novel insights into designing sustainable, adaptive water governance regimes.

WATER SECURITY AND CLIMATE CHANGE

Water is a vital component of an ecosystem and is crucial for survival, yet scarce with only 0.1% earth's water being both accessible and in usable form. Despite this critical resource being a prerequisite to society's development, water security has become a global challenge in 21st century (Gunawanasa & Bhullar, 2013; Hanjra & Qureshi, 2010) Water scarcity is being experienced across the globe, with the same being exacerbated by the adverse impacts of climate change on water resources. In the recently

DOI: 10.4018/978-1-5225-1046-8.ch017

concluded climate change negotiations (COP21) held in Paris in the year 2015, where parties to the United Nations Framework Convention on Climate Change (UNFCCC) ushered in the Paris Agreement, the issue of especially the most vulnerable countries' resilience and adaptation to the adverse effects of climate change (UNFCCC,2015).

The climactic nature of water security was apparent with a dedicated side event at the Paris talks, organized to specifically discuss the issue, in addition to mitigation and adaptation to climate change. One of the critical outcomes of the conference was the Paris Pact on Water and Climate change Adaptation, which aims towards making the water system a core foundation of sustainable human development, more resilient to climate change. This was essentially a water resilience focus event; vide the Lima to Paris act on climate change, which involved about 290 water basins globally. This underscores the importance of water in the socio-ecological ecosystems since water is the medium through which climate change affects the human and non-human lives (UNFCCC, 2015; Waterblog, 2015). Studies show that climate change is expected to account for about 20% of global increase in water scarcity (UN-Water, 2007)

Globally, the impacts of climate change are being felt disproportionally and with differing magnitudes. Developing countries continue to bear the brunt of climate change impacts as compared to developing countries due to their low resilience and adaptive capabilities. In developing nations, research on vulnerability to climate change shows that the marginalized groups of people are most vulnerable as often they are excluded from the planning and decision making process (Tompkin & Adger, 2004). Poor communities in the developing world thus face serious challenges from climate change impacts on water, as this water insecurity has direct impacts on poverty alleviation. Consequently, their pathways to sustainable development become jeopardized. The developing world thus suffers from increased water insecurity manifested vide unpredictability and high variance in precipitation as a result of climate change (Goswami et al., 2006).

Global South nations continuously face an uphill task with regard to realizing water availability for their populace; a fundamental human right (Gunawansa & Bhullar, 2013). Studies indicate that about two billion people globally currently live in either water scarce or water stress areas (JMP, 2012). Countries that are water-scarce or water-stressed require concerted efforts in managing the few available resources, more so in the wake of climate change impacts. Climatic variability is affecting the ecosystem through water as a medium, with manifestations ranging from prolonged droughts in La Nina to excessive flooding during El Nino periods (Moser & Boykoff, 2013) This consequently translates into water insecurity across the globe, with local effects like prolonged drought in Horn of Africa (Eriksen & Lind, 2009; Huho et al., 2011) and destructive weather events like the typhoon Haiyan in Philippines (Chiu, 2013).

Despite Millennium Development Goals (MDGs) target to halve the number of people without access to clean water, having been met by 2010, approximately 780 million people do not have access to it as at 2012 (JMP, 2012) Notwithstanding this achievement, the same is disproportionate with some regions continuing to lag behind with. Most of the improvement in water access has occurred in China and India, while Sub-Saharan African countries (including Kenya) continue to lag behind with only 60 percent of the population being able to access the same (Gunawansa & Bhullar, 2013). Most of the improvement in supply targets has come from Asian countries coincides with the exponential economic growth seen in China and India. This highlights the importance of water security in the socio-ecological development of society as a whole. Water security is highly regarded in international development to an extent that out of the 17 newly agreed upon Sustainable development Goals (SDGs), three of them are about water; SDGs number 6, 14, and 15 which provide;

SDG 6: Ensure availability and sustainable management of water and sanitation for all
SDG 14: Conserve and sustainably use the oceans, seas and marine resources for sustainable development
SDG 15: Protect, restore and promote sustainable use of terrestrial ecosystems, sustainably manage forests, combat desertification, and halt and reverse land degradation and halt biodiversity loss

To achieve these goals and water security in general, it is imperative to address climate change impacts in relation to water needs. Adaptation of water governance systems to climate change would be the key. It is on this premise, that this chapter seeks to explore the adaptiveness of water management institutions and systems to climate change impacts, with a case of Kenya through the lens of adaptive (co)-management prescriptions. The study deployed various approaches and methods to achieve this. To gain more understanding of the historical water governance issues in Kenya, the author carried out archival review of documents, and then literature review in the line of inquiry was carried out. To gain insights into the field, the author did participant observations with unstructured interviews where there was a need for further understanding. Analysis and synthesis of the information and data was also done.

WATER GOVERNANCE IN THE GLOBAL SOUTH

The current water crisis is a crisis of governance of water resources rather than water resource endowment (Gulyani et al. 2006). Thus, water management systems in place heavily influence water access, which in turn would affect the available water resources. Management problems range from failure in superintending water resources sustainably to poor management of water supply systems. In developing countries, poor water resources management is existent across multiple avenues ranging from poor land management, including deforestation to unregulated abstraction, water resource pollution, and unsustainable agriculture resulting in resource pollution and degradation (Juma et al., 2014; Patil & Rao, 2015).

The available water resources are further abridged with impacts of climate change. On one side, increased population and consumerism are depleting the water resources; while on other hand climate change is causing increased variability in water resource availability. This increases susceptibility of the systems that are in place due to increased uncertainty posed by climate change. This subsequently calls for adaptive and community-based resources management which builds resilience both to people and ecological systems, increasing the effectiveness of coping with environmental changes characterized by future uncertainties and unforeseen risks (Rao & Patil,2015; Tompkins & Adger, 2004).

ADAPTIVE (CO-) MANAGEMENT

Adaptive co-management is conceptualized as a governance-based approach for managing complex adaptive systems. It's the synthesis of two governance approaches of adaptive management and collaborative or co-management. The approach is expected to achieve this by marrying the strengths of adaptive and collaborative or co-management through the focus on adaptive learning and linkages between actors and organizations, working at different multiple levels (Armitage, 2005; Olsson et al. 2007).

Adaptive management is a management system that emphasizes learning and uses structured experimentation in combination with flexibility as ways to achieve this. Tompkins and Adger (2004) argue that adaptive management processes are informed by iterative learning about the ecosystem and earlier management successes and failures in addition to increases in present-day resilience, which can in turn increase the ability to respond to the threats of long term climate change. This could help in pursuit of greater ecological stability and more flexible institutions that would be essential for resources management (Lee, 1999; Tompkins & Adger, 2004) with the prevailing unpredictability of water resources and high variability of precipitation as a result of climate change; calls for an adaptive management system that is flexible are valid.

Co-management on the other hand could be defined as a management system that emphasizes the sharing of rights, responsibilities, and power between different levels and sectors of government and civil society (Huitema et al. 2009; Pomeroy & Berkes, 1997) defined co-management as sharing responsibility and authority between the government or state and the community to manage the resource. In their view, co-management regimes have emerged as a partnership arrangement that uses the capacities and interests of the local users and the community, which is complemented by the ability of the government to provide enabling legislation, enforcement and conflict resolution and other assistance (Pomeroy & Berkes, 1997). State and community stakeholders complement each other in the co-management system of governance. Singleton (1998) rightly puts in his definition of co-management as; the term given to governance system that combine state control with local, decentralized decision making and accountability and which, Ideally, combine the strengths and mitigate the weakness of each (Singleton, 1998).

Co-management as a governance theory falls under the common property theory, in which common property resources intrinsically exhibit non-exclusivity properties. These properties of non-exclusivity manifest themselves through difficulties in exclusion and sub tractability, where each user subtracts from welfare of others (Pomeroy & Berkes, 1997) Water resource management is a good example of common property resource which requires collaborative management (co-management) between government and other community stakeholders working together. Here participation is essential with shared responsibilities in water management between state and communities that are using the resource (Huitema et al. 2009).

In this era of climate change impacts where there's increased degree of uncertainty in water availability, adoption of both the adaptive and co-management concept in water resource management would be prudent. Thus, adaptive co-management is a novel combination of the learning dimension of adaptive management and the linkage dimension of co-management (Huitema et al. 2009).

The co-management governance paradigm would enhance the adaptive capacity of a population and ecosystem in two ways: co-management helps in building networks that are important for coping during extreme events, and helps in retaining resilience of the underpinning resources and ecological systems (Tompkin & Adger, 2004) On the other hand, adaptive capacity is akin to a capital asset but can only be put into play through appropriate institutions. These institutions need legitimacy in addition to being in harmony with the wider social goals, if the adaptation is to be sustainable (Tompkin & Adger, 2004) Co-management helps in legitimizing the institutions and creating harmony through state engagement with community stakeholders in participatory governance. It's worth noting that much of adaptation to climate change occurs through collective action to mediate collective risks as Adger (2003) rightly puts.

The Adaptive (co-)management system has several institutional prescriptions: polycentric governance, collaboration and participatory, experimentation, and bioregional resource management (K. Lee, 1993; K. Lee, 1999) Polycentric governance is a management system that has multiple centres of power (polycentric) rather than one centre of control (monocentric-that usually have tendencies of giving "one blanket solutions" that fit all). It can also be defined as governance systems in which "political authority is dispersed to separately constituted bodies with overlapping jurisdictions that do not stand in hierarchical relationship to each other" (Skelcher, 2005). The concept stems from an understanding that local communities face their problems, and their skills and local knowledge, place them in a better position to address the problems they face as compared to someone from outside the community (Ostrom et al., 1961). Originally this was a concept of self-governance system in which the community takes lead in resource management but current scholarships view polycentric governance as having institutional diversity and it thus offers considerable advantage when addressing complex and uncertain problems resource (Huitema et al. 2009).

Polycentric governance regimes are thought to be more resilient and better able to cope with change and uncertainties like those posed by climate change. It offers this in three perspectives:

1. Issues with different geographical scopes could be managed at different scales hence reducing risks involved with handling the whole region from centralized point;
2. polycentric systems has a high degree of overlap and redundancy, this makes them less vulnerable (Ganovetter, 1983, Perrow,2011) and
3. large number of units make it possible to experiment with new approaches so that the units can have opportunity to learn from one another (Ostrom, 2009).

In water resource management, this polycentric governance is practiced when every water basin or ecosystem has a body (which is constituted from communities using the resource and groups with interests) that manages the resource at local level. To avoid the complexities of several units of polycentric system there is need to have improved coordination mechanism in place.

Collaboration and participatory, comes from the fact that different government entities have to work together to manage issues that are cross jurisdictional boundaries and fall in different policy frameworks or sectors. Collaboration would also entail the State working together with non-government stakeholders like resource users and community at large. Participation is essential in adaptive co-management in that it engages the community in the management which increases the ownership of the process of resource management hence increasing sustainability of the resource. It improves the quality of decision-making by opening up the decision-making process and making better use of creativity that is available in the society resource (Huitema et al. 2009).

In so doing it increases transparency in the system, and legitimizes the whole governance system more so when the community is involved in the whole process of governance. This improves trust between institutions of governance which in turn would help improve social capital. Social capital influences the ability of groups to work collectively during a crisis (Cundill & Fabricius, 2010) including extreme weather events like prolonged droughts or flash floods. This thus affects the adaptive capacity of the society more so in times of uncertainty as espoused by climate change.

Experimentation here is used to imply the probing of the system to be managed, monitoring its response, and adjusting interventions on the basis of the findings resource (Huitema et al. 2009). This could be approached from two dimensions: as either a research methodology or as an approach to management (Huitema et al. 2009). As a research methodology, experimentation is a means to test hypothesis on ecosystem response to different interventions and in order to come up with scientific basis for management (K. N. Lee, 1999). The second approach is where management itself is taken as a form of experimentation. This is where the governance system learns from its successes and its failures as it improves iteratively (Walteres, 1997). On this, the unexpected outcomes are not seen as failures but as an opportunity for learning. Thus it encourages an iterative process for learning from the past and the present consequently helping anticipate the future. This is crucial element in adaptive co-management more so with the high uncertainty in resources like water due to climate change. Pahl Wostl (2007) rightly defines the change towards adaptive management as "learning to manage by managing to learn" (Pahl-Wostl. 2007).

Adaptive co-management calls for a focus on a bioregion resource management, more so when such a bioregion crosses administrative boundaries like a river system. In water management the bioregional perspective translates into managing the resource at a basin level, like river-basin management. Rivers do not follow administrative boundaries and thus the essence of managing them at a bioregion scale would give complete picture of the issues faced in a particular basin. This management prescription would be helpful in trans-boundary water resource management, that traverses several states boundary (Dinar et al., 2005) Sustainable managing of water resources requires perceiving the entire water basin as a whole rather than categorizing it from a political boundary spectrum. For effective implementation of this at basin scale, unitary river-basin organizations can be founded, or collaborations between existing organizations at the river-basin level instituted. Although the foundation of unitary river-basin organizations ("strong" arrangements) is often recommended, recent scholarship has started emphasizing collaboration between existing organizations (Huitema et al., 2009).

A bioregional resource management approach requires seeing the ecosystem as a whole before managing it, as clearly articulated by Lee (1993) who posited; "seeing the ecosystems as a whole must precede efforts to manage it." He implied that focus should be placed on the bioregion, more so when such a region crosses multiple administrative borders a case of trans-boundary water bodies as earlier noted.

Adaptive (co-) management system would be key to managing water resources in this era of climate change-induced uncertainties and water insecure countries like Kenya. Having an institution that cultivates these institutional prescriptions would help in achieving adaptive co-management in water sector, essential for sustainable resource management.

CASE OF WATER GOVERNANCE IN KENYA

Kenya, with a land area of approximately 582,000 km^2, it has 11,000 km^2 surface areas consisting of water bodies (Onjala, 2006). Kenya is currently classified as a chronically water scarce country, with 20,500 million m^3 of renewable surface water which translates to approximately 647 m^3 per capita as of the 2006. This figure is expected to drop to 340 m^3 per capita in 2025 when the population is predicted to grow to about 59 million (Onjala, 2006). The country is divided into 5 water basins, Lake Victoria, Rift valley, Athi River, Tana River and Ewaso Ng'iro basins; out of which only two basins Victoria and Tana River basins are deemed surplus while the rest have to rely on inter-basin transfers.

Kenya is highly vulnerable to the effects of climate variability and the steady degradation of the nation's water resources. Studies reveal that the country experienced the El Nino associated floods during the 1997/98 period and La Nina associated drought in the years 1983/84, 1991/92, 1995/96, 1999/2000, 2004/5, 2009/2010 which have had devastating impacts on socio-ecological systems (T21, 2012). The most recent El Nino event in 1997/98 and La Nina event of 1999/2000 were the most severe in 50 years. This was a rare phenomenon since the El Nino year was followed by a La Nina year which made adaptation very difficult, further illustrating variability and uncertainty associated with a changing climate (Harding & Devisscher, 2009; T21, 2012)

Generally, precipitation has become rapid and sporadic, exacerbating water insecurity both spatially and temporally. Studies reveal that rainfall intensity and variability is increasing; this is in line with Numerical models General Circulation Models, GCMs, of climate projections of the east African region, which anticipates an increase in precipitations thus occasioning increased flooding incidences. Kenya suffers from two twin problems of cyclic droughts and floods from time to time. However the frequency of occurrence and the intensity has increased in the last three decades. Several districts in Kenya have been greatly affected by floods like Garissa, Mandela, Wajir, Kilifi, Tana and Busia with vast resources lost and sometimes lives lost (Harding & Devisscher, 2009) Water resources are constantly vulnerable in Kenya, majorly due to mismanagement, overexploitation, forest clearance (Harding & Devisscher, 2009) climate impacts on the resource, and poor management of water resources. (Gunawansa & Bhaullar, 2013).

Water governance in Kenya has evolved through many processes, from a centralized governance system prior to the Water Act 2002, through to the promulgation of the Constitution of Kenya 2010 that recognized water as a basic fundamental human right (GoK, 2002, 2010). The aim was improving water resource management and water supply service provision. Prior to the Water Act 2002, water management in Kenya was primarily administered by the government and municipal or city councils. This was a system in which decisions were centralized and had tendencies of having a "one size fits all" solution to all problems across the scale. This period was marred with inefficiency and mismanagement of water resources and water supply systems (K' Akumu, 2004).

The adoption of the Water Act 2002 revolutionized the management by putting in place the legal, policy and institutional framework for water governance in Kenya. In 2010, Kenya adopted a new Constitution 2010, which enshrined access to water as a basic fundamental human right (GoK, 2010). The Constitution ushered in devolved governance comprising of the central government and 47 county governments headed by governors. With this came also devolution of state organs and ministries of governance to county level. This created polycentric governance units at County levels. Water resources management system changed with adoption of the water Act 2002, from a centralized system to a decentralized system of lower centres of governance at county level; thus being more polycentric.

Polycentric governance of water resources and supply in the country decentralized water management units and set up different institutions from national to local community levels as shown in Figure 1. This revitalized the management of water resources and water supply services country wide. The

Figure 1. Institutional set-ups under Water Act, 2002
(GoK, 2002)

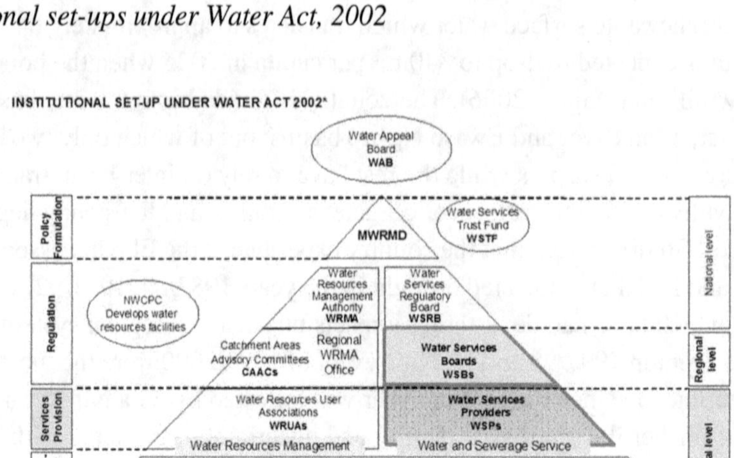

Act's reforms thus revolved around the following themes: the separation of the management of water resources from the provision of water services; the separation of policy making (the State) from day to day administration and regulation; decentralization of functions to lower level state organs; and the involvement of non-government entities in the management of water resources and in the provision of water services. The new water laws thus became more polycentric as regards governance of water resources, an essential characteristic of adaptive co-management.

To engage the community in participation in day-to-day management of the water resources the Water Act established formation of Water Resource Users Associations (WRUAs) that comprises of the community water resource users and other stakeholders. The WRUAs usually signs a Memorandum of Understanding (MoU) with the Water Resources Management Authority, WRMA (an authority mandated to ensure good management of the country's water resources), to run and manage the water resources in their locality on behave of the state. This depicts a form of co-management governance in application. They usually have a constitution and leaders are chosen amongst themselves. The WRMA not only comes in to assist the WRUAs with technical advice where it is needed but also on policy advice.

WRUAs activities are usually co-financed by the community (at least 30 percent of project value) and by the government through the Water service Trust Fund, WSTF (an authority responsible for funding government water projects aimed at improving access to water and sanitation in areas without adequate services), and WARMA. As at June 2010, there were about 236 WRUAs registered by the Attorney General and about 139 WRUAs that were registered under the District social services office who were in operation (WARMA, 2011). Since the communities are the ones that form the WRUAs, they are the ones that manage particular the water resources within their mandate. This has increased the accountability on water resource use among the community members and has built a self-governance model in which the community members themselves have self-monitoring and policing exercise (E. Ostrom, 2005). This enhances leads to sustainable water resource management exemplifying collaboration between the State and the community, in addition to participation of the community in water resource management.

Every water basin has its own WRUA formed from a community of water users in the water basin and interest groups. This has given rise to bioregions management model in which every water basin ecosystem is managed as a whole and not divided into administrative units. The Act established the Catchment Advisory Committee to give advice on management of the water catchment resources within their catchment. The committee includes (among others) members of the community in the catchment, who helps by proffering advice about the problems they face in the catchment. All the catchment areas are based on the water body catchment and not political boundaries and they are all managed at the catchment level depicting a form of management at bioregional scale. This has improved the management system as the catchment problems are tackled at a catchment level and with every catchment tackling their own problems that they clearly understand.

There are several polycentric units of governance in water resource governance in Kenya. To encourage learning from one another, exchange visits are usually organized between catchment advisory committees (CACs) and WRUA members from one region and those from another. This has fostered unity and increased learning from one another on sustainable water resource management. The whole process is taken as learning by managing the resource, where the WRUAs and CACs learn from the past, present and from each other to improve the management of their water resource. This is an iterative and continuous process of learning and improving the management system. Although it requires the willingness of the community to readjust to new strategies to increase resilience to help in building the social capital that is essential to increasing community adaptiveness and resilience.

The Kenyan case represents a form of adaptive co-management in water resource management. The country transitioned from a State centralized management of water resources to an adaptive co-management system through the adoption of the Water Act 2002. To the knowledge of the authors, there has been significant improvement of Water and Sewage service delivery throughout the country. According to JMP (2012) access to safe drinking water supplies in Kenya as at 2012 stood at 59% (83% in urban and 52% in rural) and improved sanitation at 31% (27% in urban and 32% in rural) as at 2008. According to the JMP estimates, access to improved water sources in urban areas decreased from 91% in 1990 to 83% in 2008. This could be due to increasing number of sprawls of slums in city and high density of population overwhelming the WSS infrastructures and service systems in cities. However, in rural areas where adaptive co-management is prevalent access increased from 32% to 52% during the same period (JMP, 2012).

With Kenya being a water scarce country, her ever increasing population and impacts of climate change on resources and systems, there is a significant threat to water security of Kenyans for now and for future. With the adoption of the Water Act of 2002 and by extension the adaptive co-management prescriptions, there has been positive community engagement in water resource management and increased awareness of the water resource as a scarce commodity and need to manage it sustainably. There seems to be a willingness by the community to take care of the water resource since there is increased ownership of the resource within the community. This is as a result of the State letting the community to manage the water resource with a guidance of the of WARMA experts. Through various agencies, the Government of Kenya for the last decade has invested massively in water infrastructure (with help from donors) and service provision training all aimed at improving service delivery.

CONCLUSION

Climate change is impacting society through water resources, ranging from heavy floods to severe prolonged droughts. It is causing unprecedented susceptibility in water security to countries both developed and developing. The developing world is being affected much more because of its low resilience and less adaptiveness capabilities. With this uncertainty in water availability and erraticism in precipitation, adaptive (co-) management system of water resource governance greatly improves the management of scarce water resources. Kenya's water resource management case highlights the empirical application of the adaptive co-management concept. The system provides insights to the adaptation and mitigation strategies in water management in the era of climate change. Through the adoption of the Water Act 2002, and by extension adoption of elements of adaptive governance, there has been a significant improvement in Water and Sewerage service provision in the country. The deployment of the institutional prescriptions of polycentric governance, collaboration and participatory, experimentation, and bioregional resource management helps the system learn and improve at the same time, which is significant in era of climate change-induced uncertainty in water security.

REFERENCES

T21. (2012). *Kenya Threshold 21 (T21) Model.*

Adger, W. N. (2003). Social aspects of adaptive capacity. *Climate change, adaptive capacity and development*, 29-49.

Armitage, D. (2005). Adaptive capacity and community-based natural resource management. *Environmental Management, 35*(6), 703–715. doi:10.1007/s00267-004-0076-z PMID:15940398

Chiu, Y.-T. (2013). Typhoon Haiyan: Philippines faces long road to recovery. *Lancet, 382*(9906), 1691–1692. doi:10.1016/S0140-6736(13)62380-6 PMID:24278979

Cundill, G., & Fabricius, C. (2010). Monitoring the governance dimension of natural resource co-management. *Ecology and Society, 15*(1), 15.

Dinar, A., Kemper, K., Blomquist, W. A., Diez, M., Sine, G., & Fru, W. (2005). *Decentralization of river basin management: a global analysis.* World Bank Policy Research Working Paper (3637).

Eriksen, S., & Lind, J. (2009). Adaptation as a political process: Adjusting to drought and conflict in Kenyas drylands. *Environmental Management, 43*(5), 817–835. doi:10.1007/s00267-008-9189-0 PMID:18726051

GoK. (2010). The Constitution of Kenya 2010. Author.

Goswami, B. N., Venugopal, V., Sengupta, D., Madhusoodanan, M., & Xavier, P. K. (2006). Increasing trend of extreme rain events over India in a warming environment. *Science, 314*(5804), 1442–1445. doi:10.1126/science.1132027 PMID:17138899

Granovetter, M. (1983). The strength of weak ties: A network theory revisited. *Sociological Theory, 1*(1), 201–233. doi:10.2307/202051

Gulyani, S., Talukdar, D., & Potter, C. (2006). *Inside informality: poverty, jobs, housing and services in Nairobi's informal settlements: Report*. Academic Press.

Gunawansa, A., & Bhullar, L. (2013). *Water governance: an evaluation of alternative architectures.* Edward Elgar Publishing. doi:10.4337/9781781006429

Hanjra, M. A., & Qureshi, M. E. (2010). Global water crisis and future food security in an era of climate change. *Food Policy, 35*(5), 365–377. doi:10.1016/j.foodpol.2010.05.006

Harding, B., & Devisscher, T. (2009). Review of the Economic Impacts of Climate Change in Kenya. Academic Press.

Huho, J. M., Ngaira, J. K., & Ogindo, H. O. (2011). Living with drought: The case of the Maasai pastoralists of northern Kenya. *Educational Research, 2*(1), 779–789.

Huitema, D., Mostert, E., Egas, W., Moellenkamp, S., Pahl-Wostl, C., & Yalcin, R. (2009). Adaptive water governance: Assessing the institutional prescriptions of adaptive (co-) management from a governance perspective and defining a research agenda. *Ecology and Society, 14*(1), 26.

JMP. (2012). *Progress on D inking Water 2012 UPDATE*. Retrieved from www.unicef.org/media/files/JMPreport2012.pdf

Juma, D. W., Wang, H., & Li, F. (2014). Impacts of population growth and economic development on water quality of a lake: Case study of Lake Victoria Kenya water. *Environmental Science and Pollution Research International*, *21*(8), 5737–5746. doi:10.1007/s11356-014-2524-5 PMID:24442964

Kakumu, O. (2004). Privatization of the urban water supply in Kenya: Policy options for the poor. *Environment and Urbanization*, *16*(2), 213–222. doi:10.1177/095624780401600212

Lee, K. (1993). *Compass and gyroscope: integrating politics and science for the environment*. Washington, DC: Island Press.

Lee, K. N. (1999). Appraising adaptive management. *Conservation Ecology*, *3*(2), 3.

Moser, S. C., & Boykoff, M. T. (2013). *Successful adaptation to climate change: Linking science and policy in a rapidly changing world*. Routledge.

Olsson, P., Folke, C., Galaz, V., Hahn, T., & Schultz, L. (2007). Enhancing the fit through adaptive co-management: Creating and maintaining bridging functions for matching scales in the Kristianstads Vattenrike Biosphere Reserve Sweden. *Ecology and Society*, *12*(1), 28.

Onjala, J. (2006). *The 5th Human Development Report for Kenya 2006: Human Security and Human Development*. Academic Press.

Ostrom, E. (2005). *Self-governance and forest resources. In Terracotta reader: a market approach to the environment* (pp. 131–155). New Delhi: Academic Foundation.

Ostrom, E. (2009). *Understanding institutional diversity*. Princeton University Press.

Ostrom, V., Tiebout, C. M., & Warren, R. (1961). The organization of government in metropolitan areas: A theoretical inquiry. *The American Political Science Review*, *55*(04), 831–842. doi:10.1017/S0003055400125973

Pahl-Wostl, C. (2007). Transitions towards adaptive management of water facing climate and global change. *Water Resources Management*, *21*(1), 49–62. doi:10.1007/s11269-006-9040-4

Patil, Y., & Rao, P. (2015). Industrial waste management in the era of climate change - A smart sustainable model based on utilization of active and passive biomass. In W. L. Filho (Ed.), *Handbook on Climate Change Adaptation* (pp. 2079–2092). Springer-Verlag Berlin Heidelberg. doi:10.1007/978-3-642-38670-1_49

Perrow, C. (2011). *Normal accidents: Living with high risk technologies*. Princeton University Press.

Pomeroy, R. S., & Berkes, F. (1997). Two to tango: The role of government in fisheries co-management. *Marine Policy*, *21*(5), 465–480. doi:10.1016/S0308-597X(97)00017-1

Rao, P., & Patil, Y. (2015). Climate Resilience in Natural Ecosystems in India: Technology Adoption and the Use of Local Knowledge Processes and Systems. In Handbook of Climate Change Adaptation. Springer Berlin Heidelberg.

Singleton, S. G. (1998). *Constructing cooperation: the evolution of institutions of comanagement*. University of Michigan Press.

Skelcher, C. (2005). Jurisdictional integrity, polycentrism, and the design of democratic governance. *Governance: An International Journal of Policy, Administration and Institutions*, *18*(1), 89–110. doi:10.1111/j.1468-0491.2004.00267.x

The Water Act, 2002 No. 8 of 2002, (2002).

Tompkins, E. L., & Adger, W. (2004). Does adaptive management of natural resources enhance resilience to climate change? *Ecology and Society*, *9*(2), 10.

UN-Water. (2007). *Coping with water scarcity challenge of the twenty-first century.* Retrieved from http://www.unwater.org/downloads/escarcity.pdf

UNFCCC (Producer). (2015). *2015 Paris Climate Conference: Paris Agreement*. Retrieved from http://unfccc.int/meetings/paris_nov_2015/items/9445.php

Walters, C. (1997). Challenges in adaptive management of riparian and coastal ecosystems. *Conservation Ecology, 1*(2), 1.

WARMA. (2011). *Performance Report 2 A report to the public from the Water resource management Authority*. Retrieved from http://freshwaterblog.net/2015/12/04/water-and-climate-change-at-cop21-in-paris/

Chapter 18
Climate Change and Agriculture:
Time for a Responsive and Responsible System of Water Management

Eshwar Anand Ventrapragada
Symbiosis International University, India

Neela Rayavarapu
Symbiosis International University, India

ABSTRACT

This chapter is an attempt to study the impact of climate change on water and agricultural production in India and abroad. While analyzing best practices in climate change adaptation and water management, the chapter examines regional issues and challenges. Policy interventions, success stories and new initiatives to tackle drought, boost rained agriculture as well as increase the irrigation potential have been studied together with the need for necessary course corrections. Leveraging technology for crop forecasting, inter-State river water disputes and measures needed to resolve them in the light of international experience are other areas of focus. In fine, the chapter calls for a comprehensive water policy that will not only recognize water as a national resource but also help bridge all differences for making world a worthy place to live in. The research methodology adopted in this chapter is primarily historical-analytical. Research papers, journal articles, official reports and newspaper clippings have all been consulted for analysis and interpretation.

INTRODUCTION

Day after day, day after day, we stuck, nor breathe nor motion; As idle as a painted ship Upon a painted ocean. Water, water, everywhere, and all the boards did shrink; Water, water, everywhere, not any drop to drink (Samuel Taylor Coleridge in the Rime of the Ancient Mariner)

DOI: 10.4018/978-1-5225-1046-8.ch018

This poem, which signaled a significant shift to modern poetry and the beginning of British Romantic literature as far back as 1798, will no more remain a poem but become a stark reality if various studies in the recent past on the impact of climate change on water are to be believed. Indeed, these studies need to be considered as warning signals. In the Tenth Global Risk Report, the World Economic Forum has examined the acute water crisis in the world and maintained that the problem needs to be tackled with a sense of urgency (World Economic Forum, 2015).

India has been a victim of climate change impact on changing weather patterns in South Asia. Over the past five decades, rising temperatures have led to a 10% reduction in rainfall so crucial to Indian agriculture (Antholis, 2014). The melting of the Himalayan glaciers threatens India's water resources. The rising sea levels are a cause for concern for those living in low-lying areas in coastal cities of Kolkata (former Calcutta) and Chennai (former Madras) as demonstrated during the 2004 tsunami.

Water scarcity, tropical cyclones like Phailin in Odisha in 2013, Hudhud in Andhra Pradesh in October 2014 and Super Cyclone in Odisha in 1999, contamination of drinking water and groundwater, increasing number of deaths due to both heat wave and cold wave in North India and a rise in vector-borne diseases are major fallouts of climate change in India. Other related aspects of climate change can pose a major threat towards India's achievement of Millennium Development Goals and economic development. These are, water resources availability and river water disputes, change of monsoon pattern, reduced output of agricultural commodities such as rice, wheat and maize, distribution of water as per requirement and demand, extreme weather events such as floods and drought, the reduced green line in high altitude and latitude areas, and the reduced green line in low altitude and rain security areas. This, in turn, could trigger problems of governance, political discontent and internal security threats.

Climate change effects are visible in neighboring countries of India as well. Pakistan is one of the world's most water-stressed countries. It has emergency water reserves for only 30 days as against the recommended 1000-day supply for countries with similar climates. (Asian Development Bank, 2013). Owing to low snowmelt as a result of climate change, the Indus River, Pakistan's main source of fresh

Figure 1. Impact of climate change

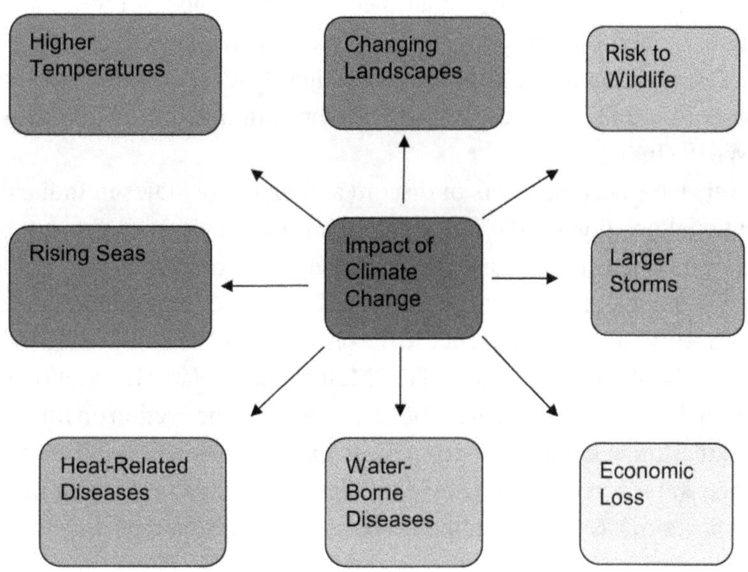

water, has seen reduced water flow in recent times. As for Bangladesh, according to the Maple croft Report (Maplecroft is a British risk consultancy) the economic impact of climate change will be "most keenly felt" by it. Disturbingly, of the 50 cities ranked in this report based on the Climate Change Vulnerability Index (CCVI), the top five "extreme risk" cities include Dhaka along with Mumbai (former Bombay) and Kolkata (Maplecroft, 2013). Climate change impacts such as these affect the natural distribution of water with scarcity and drought in regions that may, in turn, lead to conflicts over water.

Clearly, the mitigation of climate change requires major efforts at the global level under the United Nations Framework Convention on Climate Change and its derivative, the Kyoto Protocol. Unfortunately, as the Lima Conference on Climate Change in December 2014 reveals, the action taken on the ground so far by countries collectively is weak. Yet, India has no choice but to adapt to climate change owing to increasing water scarcity. A more responsive and responsible system of water management has become imperative. Indian agriculture will also require development of new species and strains that are drought-resistant and salt-tolerant to ensure that agricultural yields do not go down.

The research methodology of this chapter, being historical-analytical, is primarily based on content analysis. Articles, editorials, special stories, features and news analyses in leading newspapers, magazines and periodicals, book chapters, reputed journals, important reports of the Government of India and various state governments under study such as Maharashtra, Andhra Pradesh, Karnataka, Odisha and Tamil Nadu, live situations such as speeches of Chief Ministers and media interviews of experts have all been consulted for a comprehensive assessment of the subject.

The researchers have not only examined the impact of climate change on agriculture because of unseasonal rains and increasing desertification but also studied initiatives by the Government of India through the Centrally-sponsored program such as the Drought Prone Area Program and the Command Area Development Program to meet acute water shortage. While referring to a few notable experiments such as the Coimbatore Story, the Shirpur Model, the National Rainfed Area Authority and Cloud Seeding in Karnataka, the researchers have particularly examined success stories in Ralegan Siddhi and Hiware Bazar in Maharashtra and the Adarsha Watershed in Kothapally, Telanagana. In addition, new initiatives such as the Jalyukta Shivar Abhiyan in Maharashtra, the Neeru-Chettu Project in Andhra Pradesh and the smart card project, based on the information communication technology (ICT) solution in Andhra Pradesh have also been examined. Having underlined the importance of leveraging technology with a view to maximizing agricultural output and efficient water management, the chapter makes a special mention of FASAL (Forecasting Agricultural output using Space, Agro-meteorology and Land-based observations) project of the Indian Space Research Organization and the Water Resource Information Systems of India, WRIS (India).

As inter-State rivers have become rivers of discord among many States in India, the chapter seeks to explain the constitutional position on the inter-State river waters vis-à-vis the authority of the Central Government and the role of Parliament and the Supreme Court of India. Case studies of international treaties such as the Colorado Agreement, the La Plata Pact (together with new research in this area), Agreements on Shortages with special reference to the one between the Southern Nevada Water Authority with the Arizona Water Banking Authority and the Metropolitan Water District of Southern California and the Murray-Darling Basin in Australia have been discussed in the section on International Experience with the expectation that India could learn from these experiments and replicate them, wherever possible, for resolving the water problem and providing this vital source to teeming millions. The question of interlinking of rivers, too, has been examined. The chapter ends on a positive note in the light of the resolve of the 2015 Paris Conference on Climate Change (COP21) to tackle global warming and water crisis. It is hoped that the chapter will contribute to ongoing research in this critical area and discover new frontiers of knowledge.

IMPACT ON AGRICULTURE

The increasing threat to agriculture in India as a result of climate change is disturbing because agriculture in India primarily depends on the quantum and timing of rainfall. Rising temperatures in India will impact farm yields (for instance, the output of tea, Assam's highly weather-sensitive crop and one of India's largest exports, has fallen despite an increase of the area under cultivation), an increase in the aridity of land (through evapotranspiration), reduction in soil infertility, increase in the frequency of extreme weather conditions like drought, heat waves and floods, and a lowering of the underground water table. The problem is acute because more than 60% of India's groundwater is rainfed.

Interestingly, India is home to extraordinary variety of climatic regions, ranging from tropical in the south to temperate and alpine in the Himalayan north while elevated regions receive sustained winter snowfall. The Himalayas and the Thar Desert influence India's climate. Studies reveal that rising levels of greenhouse gases (GHGs) are likely to increase the global average surface temperature by 1.5-4.5°C over the next 100 years. The difference of average surface temperature between the ice age and the present climate is 6°C. Consequently, this is expected to raise sea levels, shift climate zones pole ward, decrease soil moisture and storms (Senapathi *et al*, 2013).

There is the threat of El Nino every year, the unusual weather pattern that often weakens rains in India, which, in turn, adversely affects agricultural production. The Climate Prediction Centre of the United States' National Oceanic and Atmospheric Administration (NOAA), in its update of April 9, 2015, has forecast an "approximately 70% chance" of an El Nino through this summer and a "greater than 60% chance" of it lasting through autumn during September-November (Damodaran, 2015). Yet, the saving grace is NOAA's pointing to a "weak" El Nino as against the "moderate to strong" one in 2009. This comes close on the heels of its earlier forecast of "50:50 chances" this summer (Bhattacharya, 2015). In March 2015, the NOAA and the Australian Meteorological Bureau announced that an El Nino had formed over the Pacific Ocean in February 2015 and that it may continue in late summer when the south west monsoon hits India.

Reports and surveys by international institutions of repute on the consequences of global warming and climate change on agricultural production are revealing in that they underscore the gravity of the situation (Jayaraman, 2011). Senapathi *et al* (2013) found a 10-15% increase in monsoon precipitation in many regions, a simultaneous precipitation decline of 5-25% in drought-prone Central India and a sharp decline in winter rainfall in northern India implied changes in output of winter wheat and mustard crops in northwestern India. Quoting Pune's Indian Institute of Tropical Meteorology, Ministry of Earth Sciences, Government of India, they observed that changes in temperature and precipitation could have a significant impact on over 350 million people who are largely dependent on rain-fed agriculture.

Rice, an important staple of most Indians, will be the worst victim of climate change. The International Rice Research Institute (IRRI) and Stress Tolerant Rice for Africa and South Asia (STRASA) have forecast 20% reduction in rice yields per degree Celsius of temperature rise (Kleemans & Sadoulet, 2012). They found drought, which occurs in every weak monsoon year in India's Odisha, Chhattisgarh and Jharkhand, can result in 40% loss in total rice production. It is said that rice becomes sterile if exposed to temperatures above 35 degrees for more than one hour during flowering and, consequently, produces no grain at all.

UNSEASONAL RAINS

Unseasonal rains and hailstorms in many States of India during March 2015, apparently because of global warming and climate change, have hit the farmers hard. According to the Agriculture Ministry, Government of India, quoting Government of India's Chief Economic Adviser Arvind Subramanian, rain and hail damaged Rabi (winter-sown) crops in about 113 lakh hectares across the country (*The Economic Times*, 2015). The entire Northern India has recorded loss of wheat, mustard, potato, pea, barley and several other crops. Rains have caused extensive damage to standing crops in States like Punjab, Haryana, Rajasthan, Himachal Pradesh, Madhya Pradesh, Maharashtra, Uttar Pradesh and Uttarakhand (Niyogi, 2015).

Of particular concern is the havoc caused by heavy rains in the Bundelkhand region of Uttar Pradesh and Madhya Pradesh, the Vidarbha region of Maharashtra and the Saurashtra region of Gujarat. States like Punjab and Haryana, which produce the largest amount of food grains in the country, faced the fury of unseasonal heavy rains inasmuch as about 20% of the crop has been damaged (Choube, 2015). In Telangana, strong thunderstorms and incessant rain have destroyed various crops planned in 11,628 hectares. Crops to suffer worst damage are paddy, bajra, jowar and green gram. The mango crop in Andhra Pradesh is badly hit with strong gales causing severe fruit dropping (*The Times of India*, 2015).

INCREASING DESERTIFICATION

Alarmingly, climate change is leading to desertification. Official figures suggest that 25% of India's total land is undergoing desertification while 32% is facing degradation. This has affected productivity, critically impinging on the livelihood and food security of millions of people across the country. According to the Government of India's Fifth National Report on Desertification, Land Degradation and Drought, the government has admitted that land degradation is a major environmental concern for the country with consequent implications for sustainable development. According to the report, which has been submitted to the Secretariat of the United Nations Convention to Combat Desertification, to which India is a party, 105.19 million hectares (mha) of India's total geographical area of 328.73 mha is undergoing desertification. Figure 2 highlights the major causes of land degradation. According to the report, Rajasthan accounts for the most desertified land (23 mha), followed by Gujarat, Maharashtra and Jammu and Kashmir (13 mha each) and Odisha and Andhra Pradesh (5 mha each). The report mentions that 68% of India is prone to drought which will be further heightened because of the change, particularly in dry land (Government of India, 2014).

Figure 2. Land degradation: A major environmental concern
Source: Government of India, Ministry of Environment, Forests and Climate Change (2014)

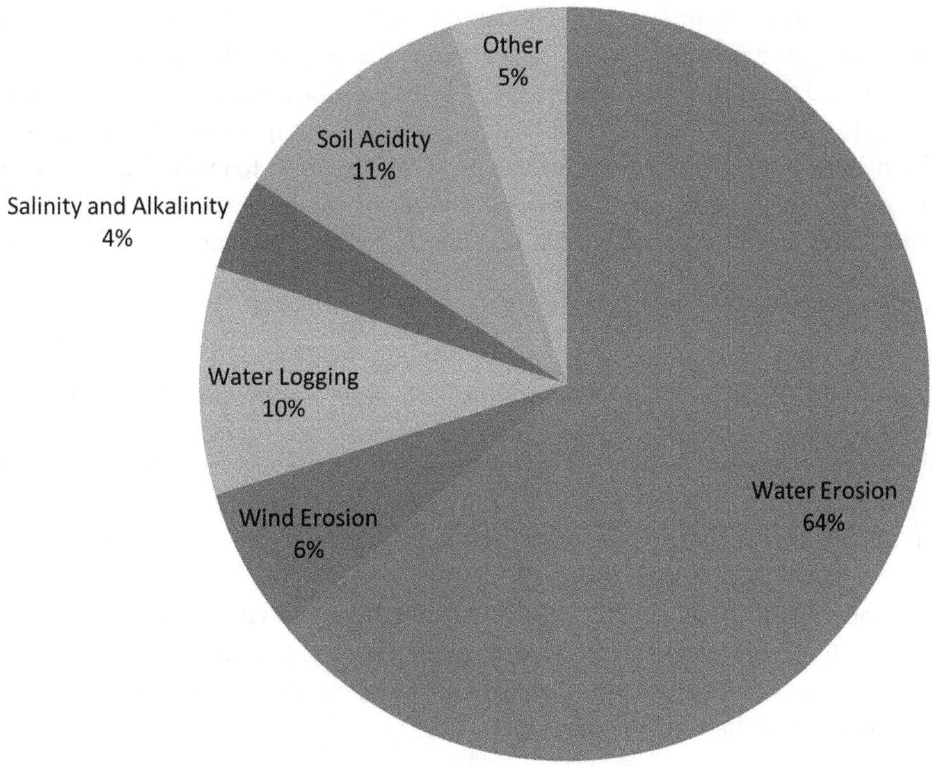

With increasing pressures of growing human and cattle population, deforestation, soil erosion and indiscriminate use of water, chemical fertilizers and pesticides, there is a serious threat to the ecological system. This has resulted in over-extraction of groundwater with little concern for commensurate improvements in harvesting and use of the increasingly precious water resources available. The impact of nature on agriculture can be felt in rainfed areas as rainfall is variable and it occurs in torrential downpours. In the semi-arid tropics of India, most of the rainfall is lost as runoff causing severe soil erosion. Rainfall exceeds evaporation only for about five months in a year (June to October) at Patancheru near Hyderabad, India. Rainfall use efficiency is as low as 30-45%, for crop production. This results in spells of excessive moisture and drought during the crop-growing period (Government of India, 2014).

ADAPTING TO WATER SHORTAGE

A few of Government of India's Centrally-sponsored programs have received increasing attention over the years to tackle drought and raise the irrigation potential in most States of India. They are the Drought Prone Area Program (DPAP) and the Command Area Development Program (CADP). These programs need to be examined in detail for a realistic appraisal of the problems at hand and the imperative need for course corrections.

The Drought Prone Area Program

The Drought Prone Area Program (DPAP), introduced by the Centre in 1971, marks a significant break in conventional developmental strategy in that it incorporates the very real physical constraints in an area. It attempts at tailoring developmental strategy to tangible results rather than trying to fit economic realities into conventional textbook prescriptions. The basic theoretical prescription behind the DPAP is that some areas are chronically drought-prone as against areas which are prone to cyclical and sporadic drought. Thus, an inherently drought prone area requires a package of solutions which may be at variance with developmental policy in other regions (Lal Bahadur Shastri National Academy of Administration, 1979).

DPAP Objectives

1. Preparation of an inventory of resource endowments in the drought prone districts and the uses to which the resources were put.
2. Development of resources and patterns of their optimal use for conservation and production improvement.
3. Preparation of watershed plans for implementation within the practical farm level, institutional, technological and managerial constraints.
4. Undertaking research to solve the problems experienced at individual farm and watershed levels.
5. Developing stronger local institutions (credit institutions, livestock cooperatives, panchayats, etc.).
6. Synchronizing and dovetailing the allocations for planned and relief expenditure for solving the problems of drought, unemployment.

Functional Constraints

1. The DPAP lacked an adequate data base on resource endowments. Problems of integration seemed to be inherent in the program. For example, afforestation was not linked with plantations in catchment areas of irrigation tanks or with the upper reaches of watersheds.
2. As for the shelf of projects, there was confusion between routine drought relief works which were seasonal or temporary in nature and the DPAP works which aimed at permanent reversion of arid regions.
3. There was duplication of effort and repetition in the variety of works undertaken year after year. Unmetalled (*kachcha*) roads were easily built and the soft option was often to re-sanction road building almost on a perennial basis. The priorities were not well defined.
4. On account of the indifferent attitude of the bank staff and their insistence on security and independent sureties, there was often considerable delay in setting up of projects. As a result, though projects were often technically feasible and the margin money and subsidy were deposited with the banks, the beneficiaries were denied the benefits.
5. Inadequate attention was given to the extension of infrastructure. Beneficiaries had to rely heavily upon institutional or independent sources in a manner that the system was complete. A case in point was the inadequate veterinary services in animal husbandry programs.

6. The various programs under the DPAP should have been taken up after a careful survey of resource endowments in compact areas. This is to ensure the necessary impact and proper tie-up with related and supporting activities such as extension, input provision, marketing support, etc.

The Command Area Development Program

The Government of India launched the Command Area Development Program (CADP) in 1974-75 as a Centrally-sponsored program. It envisaged, among other things, execution of on-farm development works like field channels, land leveling, field drains and conjunctive use of ground water and surface water, the introduction of *Warabandi* or the rotational system of water distribution to ensure equitable and timely supply of water to each command area. At the initial stage of implementation, the focus was on infrastructural development required to deliver the water to the farmers' fields. But the progress in implementing the full package of on-farm development works was found to be limited (Government of India, 1992).

Identification of a host of constraints such as absence of up-to-date land records, farmers' resistance to land consolidation, inadequate flow of institutional credit and organizational weaknesses may not altogether be considered beyond the scope of this paper. However, suffice to mention, despite the official line of thinking, that the planning and design criteria for the system adopted at the time of the formulation of the program was out of sync and did not keep pace with the changing times, the modernization program and other innovative practices did not meet the expected results. It was only the Periyar-Vagai scheme in Tamil Nadu, completed with the World Bank assistance, which held out promise.

In another initiative, the Government of India decided that the States needed to be encouraged and assisted to set up organizations to undertake a proper study of existing systems to identify deficiencies and formulate project-specific packages. Today, while all the large irrigation systems are by and large managed by the States themselves, issues like the deployment of personnel, collection of water rates and allocations for maintenance of each system are centralized in each system. The users' participation in the whole process was found to be very little. Soon official and non-official agencies started experiments in organizing farmers' cooperatives for regulating water collections, maintenance and revenue collection. Since 1978, all irrigation schemes having up to 2000 hectares of cultivable command area have been classified as "minor irrigation". All ground water structures are also classified as minor irrigation. Ground water development is under the private sector to the extent of 90% and is financed by loans from the land development and commercial banks with re-finance facility from the National Bank for Agriculture and Rural Development (NABARD).

The Planning Commission, Government of India, recommended that efforts needed to be stepped up for renewing and improving the traditional local systems. This could be fitted into the employment guarantee and other schemes of land and water improvement as part of local area planning by panchayats with technical help from the respective Irrigation Departments. It also observed that surface lift irrigation schemes were "less expensive" compared to major, medium or minor surface schemes with storage systems. When water is available in canals, drains, streams, rivers and when power is also available, these schemes should be preferred and be taken up on an individual or cooperative basis (National Portal Content Management Team, 2011).

Functional Drawbacks

1. The CADP was launched with an avowed mission to improve production in irrigated areas and those substantially different from rain-fed agriculture. It was specifically launched to integrate the various engineering, irrigation and agricultural activities in an area development approach and not in isolation. However, it has failed to achieve the desired results for various reasons.
2. In Karnataka, except the Upper Krishna Project, the program in other areas has been marred by poor maintenance of field channels, ayacut roads, and violation of irrigation rules and cropping pattern and lack of functional coordination among the personnel of various departments in the command areas.
3. A joint study by the Government of India's Union Ministry of Water Resources and Ahmedabad's Indian Institute of Management (1987) revealed several functional drawbacks in six selected CADP projects: Jayakwadi (Maharashtra), Indira Gandhi Nahar (Rajasthan), Tawa (Madhya Pradesh), Kosi (Bihar), Mahanadi (Odisha) and Lower Bhawani (Tamil Nadu). (Government of India & IIM-A, 1987).
4. According to the joint study, the drawbacks are: inadequate organizational support, under-utilisation of funds earmarked for various activities; area covered by survey in far excess of the volume of field channels; carrying out land levelling/ shaping works on much larger scale than planned; and very little work on field drains. Karnataka Legislature's Public Accounts Committee has also pointed out similar drawbacks in the State's projects (Anand, 1988).
5. The Command Area Development Authority's organizational set-up itself has too many disadvantages. It is top-heavy and bottom-thin with weak linkages. It was originally expected to ensure sound technical linkage from the field to the headquarters. But its administrative and managerial linkages are not too strong with no multi-disciplinary staffing pattern at all levels.
6. The very pattern of location of authority and decision-making is so complicated that it is difficult to know where the boundary limits of CADA start for coordination. For a long time, Karnataka staff was not under the direct control of the Administrator and functioned through district and State level officers. Karnataka reversed this practice only in early Nineties.
7. Experience in Gujarat and Maharashtra was no better as CADAs were under the Agriculture Department with separate secretariats. Lack of political and bureaucratic will for synergism has largely been responsible for CADA's poor performance in the States (Anand, 1992).

Notwithstanding these drawbacks, the CADP should not be scrapped as the command areas are vitally essential to provide the necessary infrastructural support to the farmers. Since irrigated agriculture is yet to take off properly, a different activity mix will have to be adopted for the command areas without disturbing the organizational set-up. This calls for a permanent and efficient cadre of CADA personnel and not those on deputation from other departments and multi-disciplinary staffing patterns at the headquarters, divisional and sub-divisional level. Proper coordination between the line and staff units will help strengthen the administrative and managerial linkages at various levels of the CADP administration. The CADA officers lack powers to check violation of the irrigation rules.

As irrigation is a State subject, the States will have to design and develop appropriate distribution network from reservoir to the field level by determining the planning areas and mapping the resources. This ultimately calls for closer cooperation between the CADAs and Zilla Panchayats in the formulation and execution of all activities relating to agriculture, horticulture, fisheries, sericulture, marketing, etc.

The Coimbatore Story

In Coimbatore district of Tamil Nadu in India, the DPAP was implemented to mitigate the adverse effects of drought, employment generation and development of the human and economic resources of watershed and encourage restoration of ecological balance in the watershed area through sustained community action. The District Rural Development Agency (DRDA) implemented simultaneously the Integrated Wasteland Development Program (IWDP) to promote the economic development of the people/village community (Palanisami *et al.,* 2002).

The reason why Coimbatore has been selected for the DPAP project is that it has low rainfall areas. It has just 20-30 rainy days in a year. Its annual rainfall is 600-650 mm. Groundwater is extracted through tube wells most of which dry up during summer. Dry land crops such as sorghum, maize and pulses like black gram, green gram or cowpea are grown during north-east monsoon (October to December) period.

The data on the impact of water harvesting structures on water level in the wells, change in cropping pattern, productivity of major crops, incomes of the farm household have been encouraging. As 170-mha of land area is classified as degraded, half of this falls in undulated (wave-like movement or appearance; a number of wave-like curves or slopes) semi-arid areas where rainfed farming is practiced. This proves that there is a vast potential to improve the land to increase crop production.

As climate change and global warming are affecting rainfall every year, there is an imperative need for rainwater conservation, especially in drought prone areas. Rainwater conservation plays an important role in dry land agriculture or farming for which watershed management has been accepted as the most rational approach in preventing the deterioration of the ecosystem, restoration of degraded lands and improving the overall productivity for sustainable use (Venkateswarlu, 2000).

Some of the significant pointers of the Coimbatore success story are:

1. Higher water level – a maximum of 7.6 ft in May (peak summer month).
2. Increased storage of rainwater during the north-east monsoon season helped raise the water level in the wells up to 38.5 ft during November.
3. Efficient harvest of runoff water resulted in maximum rise in the water level during October-December.
4. Following change in the cropping pattern of the watershed area, farmers raised new crops like sugarcane, turmeric, beetroot, jasmine, tobacco and bitter gourd. Cropping intensity increased.
5. There was higher yield and good income for vegetables such as brinjal (eggplant), chillies and bitter gourd and cash crops like sugarcane and tobacco and for high water-requiring crops like coconut, tomato, onion and banana.

Studies undertaken by institutions like Mussoorie's Lal Bahadur Shastri National Academy of Administration, Government of India, suggest that these programs, well-intended though, have turned out to be considerably narrower than originally conceptualized. The progress has so far been marginal and so has the effort and research in evolving and propagating various patterns and practices for optimal use of water under conditions of drought and those prevailing in each irrigation command. (Palanisami *et al*, 2002).

The Shirpur Model

Though the Shirpur Watershed Model, implemented in the Tapi basin in Shirpur Tehsil of Dhule District of Maharashtra, enjoyed the State Government's support and coverage in the local Marathi Press, the Mukund Ghare Expert Committee was not impressed with the claims made by its sponsor, Suresh Khanapurkar, a retired officer of Pune's Groundwater Survey and Development Agency. The Committee refuted his claim that the measures he had taken had raised the water level in the bore wells on either bank of the stream (up to a distance of 2 km) by 150 feet. In its report to the State Government, it maintained that his claims of recharge and storage impacts were "exaggerated". There were no detailed plan and cost estimates for various interventions which were "scientifically and technically wrong". Subsequently, it rejected his suggestion for the replication of the model elsewhere (Joy, 2013).

The National Rainfed Area Authority

The first major step by the Government of India in addressing the problems of rainfed areas in the country is the establishment of the National Rainfed Area Authority (NRAA) on November 3, 2006. It does not confine to water management alone but comprehensive and holistic development of rainfed areas. (Government of India, 2006).

The rationale of NRAA lies in the fact that India's 86% pulses, 77% oilseeds and 50% cereals are contributed by rained agriculture. Export of castor oil, guar gram, seed spices and soybean cake as non-GM product are other important commodities. Conceptually and organizationally, NRAA is a step ahead of the DPAP. It has a Governing Board, the Chairman and Co-Chairman of which are Union Ministers of Agriculture and Rural Development. In addition to Union Ministers of Water Resources and Environment and Forests, Secretaries to the Government of India, Agriculture, Cooperation, Agriculture Education and Research, Rural Development, Water Resources, Environment and Forests, Panchayati Raj are all members of the Governing Board. Similarly, the Executive Committee of NRAA is also a high power body with experts on water management, agriculture, forestry and watershed management, including

Figure 3. Increasing demand for Rainfed agriculture
Source: National Rainfed Area Authority, Government of India, 2006

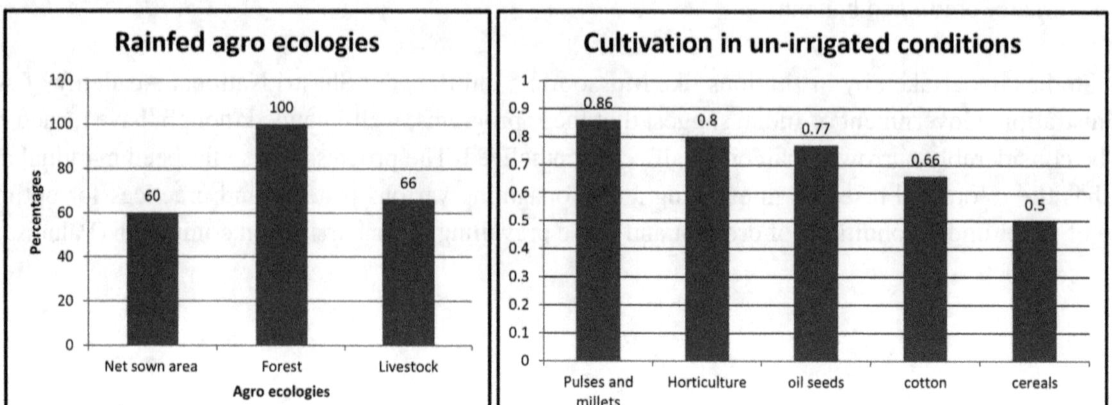

Directors of Jodhpur's Central Arid Zone Research Institute (CAZRI) and Hyderabad's Central Research Institute for Dry land Agriculture (CRIDA) as members.

Its main objectives are, to identify rainfed areas in different States which need priority attention and prepare watershed development programs for integrated natural resource management in consultation with States; to evolve common guidelines for all schemes of different Ministries including Externally Aided Projects for development of rainfed/ dry land farming systems; to guide the implementing agencies on priority setting and monitor the specific interventions required; to establish institutional linkages with prioritized watersheds; and to develop plans for capacity building of Centre/State Government functionaries in rainfed areas (Venkateswarulu, 2000).

The NRAA is an improvement over both the DPAP and CADP as it has taken care of inter-departmental coordination and institutional linkages which are vital for achieving the desired results. It has published common guidelines for watershed development projects with a fresh framework for the next generation watershed programs. These guidelines provide a framework for the planning, design, management and implementation of the country's watershed development projects. Watershed projects from April 1, 2008, are being implemented in accordance with these guidelines. The NRAA has also prepared A Vision Document for Harnessing Opportunities in Rainfed Areas. It has prepared a detailed report on Mitigation Strategy for the Bundelkhand Regions of Uttar Pradesh and Madhya Pradesh (Government of India, 2010).

Cloud Seeding Project

Cloud seeding is a technique of stimulating or enhancing precipitation by distributing dry ice crystals or silver iodide particles over storm clouds in a specified region. According to Prof. T. Shivaji Rao, (former Director, Centre for Environment Studies, Andhra University, Visakhapatnam, and Expert Member, Cloud Seeding Project, Government of Andhra Pradesh) if a warm cloud does not contain sufficient number of giant size water drops or hygroscopic particles, the cloud cannot give 10 to 20% of its moisture as rainfall. In cold clouds, whose tops attain freezing level in the sky; insufficient number of ice-nuclei prevents the clouds from giving more than 20% of the water content as in the form of rainfall or snowfall. Dr Rao maintains that if warm clouds have to give more rain, we have to inject into them chemicals like hygroscopic common salt or calcium powder into such clouds. Silver iodide particles will have to be injected into cold clouds which extend into the freezing zone for about 15 km into the sky. Dr Rao says that the injection of seeding chemicals into the clouds causes them to produce additional rainfall up to 25% (Rao, 2005).

In the absence of adequate research in India, it would be difficult for one to say whether the cloud-seeding experiments in Maharashtra and Karnataka were a success. Cloud seeding, its limited use in India notwithstanding, is not new. In countries such as the US, South Africa, China, Russia, Australia, Indonesia, Latin American and European States, cloud seeding is an integral part of their respective water resource management strategies. Government departments, insurance companies and the corporate sector do advance planning in these countries. Cloud seeding is done for various purposes: increasing annual rainfall for drinking and agriculture; dispersing fog in airports and metropolitan city roads; boosting hydro-power generation at a cheaper cost; suppressing hailstorms to reduce damage to life and property and crops; tackling recurring droughts, global warming and rising summer temperatures; and increasing annual rainfall for improving the forests, wildlife and environment.

However, in India, only a few states have used it as a last resort. Interestingly, in July, August and September, 2015, the Maharashtra Government had executed the cloud seeding project to bring artificial rain in the drought-hit Marathwada region. Though Maharashtra's Agriculture and Rehabilitation Minister Eknath Khadse admitted before the media that the project did not yield significant results, he said that the experiment would be continued in anticipation of positive results. Mr. Khadse claimed that Latur and Beed Districts had seen rain, varying between 5 mm and 30 mm, following cloud seeding. However, according to Mr. Khadse, the project was a disappointment in Osmanabad District which did not have rains. Subsequently, cloud seeding was undertaken in Aurangabad, Jalna, Parbhani and Solapur. For the execution of the project in Marathwada region, the Maharashtra Government allotted Rs 27 crore (USD 4 million), excluding the cost of hiring a pilot and a plane that cost Rs 3 lakh (USD 4400) a day (Khapre, 2015).

The Maharashtra cloud seeding project, issued through a tender, was awarded to Bangalore-based Kyathi Climate Modification Consultants, in collaboration with a similar firm in the USA, Weather Modification Inc. (WMI). Incidentally, the WMI is said to be the world's largest private aerial cloud seeding company, based in Fargo, North Dakota, USA (Little, 2015). Luckily, rains during the third week of September eased the water crisis and provided relief from drought (*Indian Express*, 2015).

Just as the Maharashtra experiment in 2015, the Karnataka experiment, too, had created lot of enthusiasm among the policymakers and general public. Indeed, Karnataka executed the project as far back as 1999. The S.M. Krishna Government implemented it between 1999 and 2003 to meet deficient monsoon. It had first conducted seeding experiments in 1999 for nine consecutive days. North Karnataka experienced rain. Interestingly, while the experiment was conducted in Dharwad, precipitation occurred 100 km away, i.e. in Belgaum. It rained for an hour after the experiment (Aiyappa, 2012). The policymakers and the farmers were disappointed after the limited success. Yet, the government continued it though in varying degrees. After Mr. Krishna demitted office, his successors carried out two other experiments but the results were no different.

In 2012, the Sadananda Gowda Government revived the project. It faced a piquant situation – the directives by the Supreme Court and the Cauvery River Authority to the State Government to release water to Tamil Nadu even though water inflow into the Krishnaraja Sagar Dam near Mysore was 10,000 cusecs and outflow was around 9,000 cusecs. Water Resources Minister Basavaraj Bommai said though Karnataka needed water to the tune of 150 TMC for drinking and irrigation purposes, the storage level at the Krishnaraja Sagar Dam was estimated at 68 TMC. Subsequently, the Karnataka Government planned cloud seeding in Cauvery catchment areas – Kodagu and Hassan (*The Hindu Business Line*, 2012). Having set aside Rs. 5 crore (USD 735000) for the project, it sought technical assistance from Pune's Indian Institute of Tropical Meteorology (IITM) and New Delhi's Water Resources Development Organization (WRDO). However, the Government abandoned the project following heavy rains in the catchment area.

Interestingly, the Karnataka Government mooted the ground-based cloud seeding technology, popularly known as "Rain Rockets", an easier and cheaper version of the project in which small rockets carrying about 200-250 gram of sodium chloride (common salt) are launched from the ground to induce silver iodide and dry ice for precipitation of clouds. These Rain Rockets are usually launched from skyscrapers or temporary towers close to catchment areas or reservoirs. Each rocket costs Rs 5,000 (USD 78) and about 50 rockets would be sufficient for a single cloud-seeding operation (Aiyappa, 2012). Airplanes, helicopters and radars have the advantage of pinpointing clouds, but this operation is expensive. For the success of the mini-rocket method, the weather condition should be favorable.

States like Uttar Pradesh, Tamil Nadu, Gujarat and the Brihan Mumbai Municipal Corporation have been toying with the idea of cloud seeding at one point or the other. And there is no dearth of companies desirous of undertaking the cloud-seeding project. The Hyderabad-based Agni Aero Sports Adventure Academy, Myavani, Bangalore's Kyathi Climate Modification Consultants, as has been referred to in the case of the Maharashtra experiment in 2015, the Bangalore-based Agni Aero Academy, the Vedanta Resources, the Hindustan Zinc have all made a beeline to pursue the project to its logical conclusion (Sally, 2014). While Kyathi has tied up with WMI, Myavani has teamed up with the United States National Centre for Atmospheric Research for the Mumbai project. Pune's Indian Institute of Tropical Meteorology has been conducting cloud-seeding experiments as part of the Cloud Aerosol Interaction and Precipitation Enhancement Experiment (CAIPEEX) program (Sally, 2014).

SUCCESS STORIES

The Ralegan Siddhi Experiment

"Earlier, the water used to run everywhere. But now, we have made it walk, crawl and even stand at various places." This is the common saying in Ralegan Siddhi, the village which has successfully defied drought, year after year, in Maharashtra's Marathwada-Vidarbha region. Tucked away in Parner Tehsil of Ahmednagar District, near Pune, Ralegan Siddhi's residents are lucky in summer months. Though the entire region reels under drought every year, thanks to the initiative and enterprise of eminent social worker, Gandhian and crusader against corruption in public life, Anna Hazare, the people of this village, have plenty of water for use. In summer, there is no sign of distress or panic. A rare display of collective wisdom has more than offset the customary vagaries of the monsoon (Mehta & Satpathy, 2011).

Today the village's percolation tank, which caters to the needs of both people and cattle, is a symbolic representation of a well-planned oasis. While the villagers turn on their taps for refreshing drink of water to slake their thirst, the cattle, too, are well fed as there is no dearth of fodder. Ralegan Siddhi was once a barren land. It had hardly 30 wells in the late 1970s. In the absence of impounding facilities, rain water ran off into an east-west stream out of the village. Consequently, while only 50 to 60 acres of the total 2,200 acres of cultivable land were assured of seasonal irrigation, most villagers had to willy nilly take recourse to water tankers and the Employment Guarantee Scheme (EGS) during summer. Here are a few pointers on the changing face of Ralegan Siddhi:

1. The once bald hilltops have been greeted with ample vegetation. The cultivation of grass, intermittent shrubs and the construction of contour bunds along the hill slopes have arrested the rain water runoff and doubly enhanced the soil moisture.
2. The process has been supplemented by nullah bunds, underground check dams, tree plantation and permanent bunds at a series of strategic locations throughout the village and its surroundings.
3. Percolation tank, nullah bunds, about 400 acres of grass land and over three lakh well-maintained trees have considerably increased groundwater levels.
4. The resultant groundwater surcharge keeps the wells and bore wells viable all through the year.
5. This has brought a whopping 1,200 acres of farm land under assured irrigation.
6. Nearly 75% of the total funds spent on these soil and water conservation works have come from various government schemes and bank loans.

7. Voluntary agencies like the Dorabjee Trust have also chipped in with liberal assistance for the wellbeing of the villagers.
8. With extraordinary discipline, commitment and dedication to duty, the villagers took turns to work on the respective sites without remuneration. As the share of the villagers' voluntary work (called *Shramdan* in local parlance) on these schemes came to 12% of the total investment, this brought down the total project cost.
9. As the villagers were truly committed, their personal interest ensured quality work.
10. The work got completed in a shorter period of time, thus saving the money, energy and cost overruns.
11. The *Shramdan* offered by villagers created a strong emotional bond between the villagers and the various structures they shaped.
12. This prompted the Government of Maharashtra to accord Ralegan Siddhi top priority when it came to introducing new developmental plan programs in the region.

A chief contributory factor for effective water management in Ralegan Siddhi is the notable role played by the Gram Sabha (Village Assembly). All decisions are taken unanimously and there is no dissent. Differences, if any, are sorted out through debate, dialogue and discussion in the regular sessions of the Gram Sabha.

Here are a few pointers on the role played by the Gram Sabha in effective water management in the village:

1. The Gram Sabha periodically met to discuss issues relating to village welfare and took decisions of far-reaching significance.
2. It prohibited sugarcane cultivation by drawing water either from the private wells or the community wells in the village. This helped in equitable use of available water for irrigation.
3. The Gram Sabha banned tree felling and cattle grazing on fodder lands.
4. It introduced water-saving techniques like drip irrigation and sprinklers, wherever possible.
5. It also laid down guidelines for the cropping pattern depending upon water availability every year.

Hiware Bazar in Ahmednagar District of Maharashtra is yet another example of change, as a result of people's access to water for agriculture. As in Ralegan Siddhi, the lives of the people of Hiware Bazar underwent spectacular transformation following their active participation in the task of their village development. Indeed, in both Ralegan Siddhi and Hiware Bazar, the people implemented in letter and spirit the time-tested principles of human development – equity, administrative efficiency, sustainability, inclusive approach and effective people's participation. Just as Anna Hazare played a notable role in the socio-economic transformation of the people of Ralegan Siddhi; Popatrao Pawar, twice elected as the Sarpanch, significantly helped in the all-round development of Hiware Bazar (Sangameswaran, 2008).

The Adarsha Watershed

The Adarsha Watershed in Kothapally, Ranga Reddy District, Telangana, is another success story. In a comprehensive study, agricultural scientists found it as the country's most innovative farmer participatory integrated watershed management model (Wani, *et al* 2002). Adarsha is considered a model watershed for the following reasons:

1. Private contractors were not involved in the watershed activities.
2. Community participation was ensured in all activities through facilitation which, in turn, had accrued tangible economic benefits.
3. Farmers conducted on-farm trials with technological support from the International Crop Research Institute for Semi-Arid Tropics (ICRISAT) at Patencheru near Hyderabad and other collaborative research institutes such as the Asian Development Bank (ADB) at Manila in The Philippines and the Central Research Institute for Dry land Agriculture (CRIDA) at Hyderabad.
4. Farmers were empowered through training programs and workshops.
5. While the non-government organization (NGO) was only a social mobilizing agency, the watershed commission and agency took charge of the project implementation.
6. Monetary disbursement was by the watershed commissions under the supervision of the DPAP functionaries.
7. It was the villagers' responsibility to do social audit.
8. There was a convergence of various activities in the watershed. Scientific tools were used for watershed development and management.
9. Farmers adopted improved cropping systems including legumes, shifting from traditional crops such as rice and cotton.
10. Farmers practiced self-learning and have become trainers themselves.

The Adarsha Watershed has also created a record in monitoring the progress of works. Here are a few examples:

1. Changes in cropping pattern and systems in farmers' fields were monitored.
2. An automatic weather station installed to collect data on rainfall, maximum and minimum temperatures and solar radiation.
3. As many as 64 open wells in the watershed were geo-referenced and regular monitoring of water levels was done. Water quality was monitored in all the wells.
4. Nutrient budgeting studies are also undertaken.
5. Runoff and soil loss were monitored by using automatic water level recorders and sediment samples.
6. Satellite monitoring was done. Pest monitoring, too, was carried out.

The impact of all this is manifold: Effective management of natural resources and improvement in the livelihood of the rural people; increased crop productivity (the maize crop yield recorded two-to-three fold increase) and rise in farmers' incomes; improved greenery and increase in vegetation cover; improved groundwater levels; and reduced runoff and soil loss (the runoff was 12% of the rainfall in the undeveloped watershed while it was only 6% in the developed watershed where soil and water conservation measures were undertaken).

NEW INITIATIVES

The Jalyukta Shivar Abhiyan

The Jalyukta Shivar Abhiyan (Campaign to provide villages with permanent source of water) is Maharashtra Chief Minister Devendra Fadnavis' new strategy to tackle drought in the State. Amid increasing reports of farmers allegedly committing suicide because of chronic drought condition in the State, Mr. Fadnavis has launched the program with a view to providing moisture security for farmers together with developing sustainable agricultural practices over a period of five years, i.e. during 2015-19. While 82% of Maharashtra's cultivable area is under rainfed farming, over 52% has been reeling under recurrent drought situation. Maharashtra, situated in the central part of India, is a semi-arid zone. Climate change has hit the State hard. It receives uncertain and erratic rainfall which, in its turn, has crippled agriculture and drinking water supply. In 2014, the State received 20% deficit rain as against the normal rain (Government of Maharashtra, 2014).

Currently, while over 24,000 out of 43,700 villages in the State are facing the brunt of drought, the Marathwada region is worst affected. In Yavatmal district, 90% villages are drought hit. According to official figures, the final *paisewari* (a survey system to measure yield per rupee) for the 2014 kharif crop showed all 8,139 kharif villages in Marathwada as having a yield below 50 paise to the rupee, the margin to be officially declared by the State Government as drought-hit. The corresponding number for the 2012 kharif season was 3,299 drought-hit villages. A disturbing signal is that this time even once successful agriculturists, too, are facing hardship. Of the 20,77,429 hectares in Marathwada with a Rabi crop, sowing has taken place on only 61.25% or 12,72,396 hectares. Parbhani district has seen only 38.47% Rabi sowing. Hingoli and Aurangabad districts are marginally better at 58.67% and 49.31% respectively. As a total of 2,839 villages in Marathwada sow rabi crop, families who subsist on agriculture, are passing through a harrowing time (Iyer, 2015).

Having realized the crux of the problem, Mr. Fadnavis feels that the issue needs a long-term solution and not a short-term one. Consequently, through the Jalyukta Shivar Abhiyan, the Government seeks to focus on water conservation to make the State drought-free within a specific timeframe. A step in this direction is the plan to integrate various State departments connected with water resources which have been working at cross purposes for decades. For instance, there has been no functional coordination between departments like Agriculture, Water Conservation, Irrigation, Soil Conservation, Forest, Social Forestry, etc. As a result, one department did not know what the other was doing. Through effective coordination between these departments, the government intends to push for sustainable farming. The Jalyukta Shivar Abhiyan will help unite various smaller schemes in order to deal with water problems. At the district level, the District Collector will function as the man on the spot. He will coordinate water conservation departments and will submit a report to the Chief Minister periodically. At the apex level, the Chief Minister's Transformation Office (CTMO) will monitor the scheme while the ground-level delivery change foundation will supervise the scheme and post aerial photographs of the area in the public domain before and after the completion of the project. This will enable people to see whether the work has been completed properly or remains on paper.

The State Government has also taken an important decision: factories that use sugarcane from farmers will have to provide drip irrigation facilities for farmers producing the crop. This is logical and rationale: farmers grow crops that are sent to sugarcane factories. Consequently, it is the responsibility of the sugar factories to put in place an effective drip irrigation system for the farmers concerned.

The Chief Minister's call for providing security to the farmers has not come a day too soon. Clearly, it would take a very long time for a government to go in for big projects like construction of big dams. Even otherwise, such projects are not only time-consuming but also require huge funds. It is in this context that the significance and importance of a program like the Jalyukta Shivar Abhiyan needs to be appreciated. Mr. Fadnavis wants the officials to construct small dams, barrages and wells as an alternative to dams. Worthy of mention in this context is the Madhya Pradesh Government's strategy in constructing three lakh wells across the State which enabled them to improve the irrigation capacity to 24% in a span of five years.

Significantly, the State Government, through a resolution, has listed out the following salient features of the Jalyukta Shivar Abhiyan:

1. The State Government has targeted to make 5,000 critical villages drought free by the end of March, 2016.
2. Village-specific micro planning will focus on water budgeting.
3. Approval of village work plan by the Gram Sabha.
4. Preparation of integrated village plan by the active involvement of various line departments of the government.
5. Effective convergence of various governmental schemes for availability of funds.
6. Institutional arrangement from the State to the village level for effective planning, execution and monitoring of the mission.
7. The Chief Minister's Transformation Cell for achieving results.
8. For effective implementation, the Delivery Management Officer is assigned in the office of the line department concerned at the State level (Government of Maharashtra, 2015).

The Neeru-Chettu Project

Gummasamudram village in Tambapalle taluka of B. Kothakota Mandal in Chittoor District of Andhra Pradesh is most drought-prone. Successive governments have shown lackadaisical attitude towards this neglected region. As a result, the problem recurs every summer.

Andhra Pradesh Chief Minister N. Chandrababu Naidu, who hails from Chittoor District, has selected the Gummasamudram village tank for his ambitious Neeru-Chettu (water tree) project. Among other things, this project aims at encouraging plantations, water conservation and improving the ground water table to make the State drought-proof within five years. The administrative and political machinery will also be geared up to transform the flagship program in to a "people's movement" and take up repairs and maintenance of canals, reservoirs, lift irrigation schemes, check dams and de-silting of tanks in a campaign mode across the State. The Chief Minister hopes that all this would help increase their capacity to store rain and surface water. The program not only seeks to help farmers take up desilting works but also guides them in the equitable utilization of the mud as the fertiliser for their farms (*The Hindu*, 2015).

Just as Maharashtra's Jalyukta Shivar Abhiyan, Mr. Naidu does not want to treat the Neeru-Chettu project as a standalone program. Mr. Naidu has tagged on a host of other programs to it – Handri Neeva, Pattiseema Etthipothalu, Galeru Nagari, Vijayanagaram, Thotapalli and Velugonda. This will help the government to utilize about 3,000 TMC of water for both agriculture and drinking water purposes. As the State will have to wait for a few more years for the completion of the Polavaram project, the government intends to draw water from the Godavari river through the Pattiseema Etthipothalu project (*The New Indian Express*, 2015).

An important feature of this project is its focused approach to various issues relating to water conservation. Equally important is the government's emphasis on effective coordination of various departments for achieving better results – Agriculture, Rural Development, Forest, Water Resources, Ground Water and Urban Development. The Chief Minister held a meeting on the project with all the heads of these departments after inaugurating it on February 19, 2015. The District Collectors are accountable for the smooth implementation of the project. Mr. Naidu keeps interacting with them through video conference. Significantly, the government is keen on involving all the stakeholders in the speedy execution of the project. What lends credence to the Chief Minister's assertion to make it a "people's movement" is the Government's efforts to involve the Janmabhoomi committees from village, mandal and district level as also the water level associations (*The New Indian Express*, 2015).

Though Andhra Pradesh had 36% deficit rainfall during 2013-14, owing to effective supervision and constant monitoring by officers under the Polam Pilusthondi (the land is calling) program, the agriculture sector in the State has registered a growth of 11.32%. Through water conservation programs like the Neeru-Chettu project, the government aims at achieving a growth rate of 20 to 25%. Of 395 lakh acres in the State, only 200 lakh acres are cultivable or fit for agriculture. Even otherwise, only 50% of it is irrigated area. The government aims at bringing the remaining 100 lakh acres too under irrigation. The Chief Minister laments that of 1,557 TMC of dependable yield, only 1,299 TMC of water has been utilised. Mr. Naidu expressed his concern on the fact that though 3,525 TMC of flood water goes into the sea, "hardly anything" has been utilized for agriculture and drinking water purposes (*The New Indian Express*, 2015).

ICT INITIATIVES

Integrated Watershed Management Program in Andhra Pradesh

The Andhra Pradesh Government has given a digital push to the implementation of the Government of India's Integrated Watershed Management Program (IWMP). This program, launched in 2009-10, is aimed at restoring the ecological balance by harnessing, conserving and developing degraded natural resources such as water, soil and vegetative cover and create sustainable livelihoods for the poor. The IWMP was launched in the combined state of Andhra Pradesh (before it was bifurcated in June 2014

Table 1. Some facts of the Integrated Watershed Management Program (IWMP) in India

Number of Projects	432
Total Project Area	17,89,775 hectares
Sanctioned Amount (in lakhs)	2,26,290
Expenditure incurred (in lakhs)	58,509
Number of GO WCCs	104
Number of NGO WCCs	58

Source: www.iwmp.ap.gov.in

paving the way for the creation of the new Telangana State as India's 29th State). Significantly, there is no burden on the States as the Government of India gives 90% financial assistance to them for the implementation of the IWMP (Pilla, 2015).

What led to the digitization of the IWMP is that within a year of its launch, the Andhra Pradesh Government realized that due to the absence of comprehensive information communication technology (ICT) solution, it was becoming difficult, and sometimes complicated, for the authorities concerned to plan, execute and monitor too many projects. The Andhra Pradesh Government tied up with software services firm Tata Consultancy Services (TCS) Limited to build an end-to-end application that works on a basic Internet connection. TCS not only runs the data center for the project but has also built the software for the smart card project.

The smart card project, based on ICT solution, provides immense benefits to the State Government. These are, among others, enrolment and payment of wages to beneficiaries under the Mahatma Gandhi National Rural Employment Guarantee Scheme and Social Security Pensions at the doorsteps of the beneficiaries; helping the government to build an impressive master database of 12.7 million people; and establishing tie-ups with banks, business correspondents and technology providers. The government has built-in structure, database and enabling infrastructure, thanks to the smart card project as also customized that application to suit watershed program requirements.

The smart card project in watershed management is unique in many ways. There are three levels of users of the application. One is the state-level nodal agency that oversees implementation, financial, analytical and performance and of the project; the second is the district-level agency which monitors physical and financial progress and the field level agency, called the project implementing agency, which enters details for project sanction, generates fund transfer orders and work status reports; and the third is the working of the project implementing agency in close cooperation with the Village Panchayats and women's Self-Help Groups in training, enlisting workers, execution and maintenance of projects.

Figure 4. A model of the Integrated Watershed Management Program (IWMP)
Source: Department of Rural Development, Government of Andhra Pradesh.

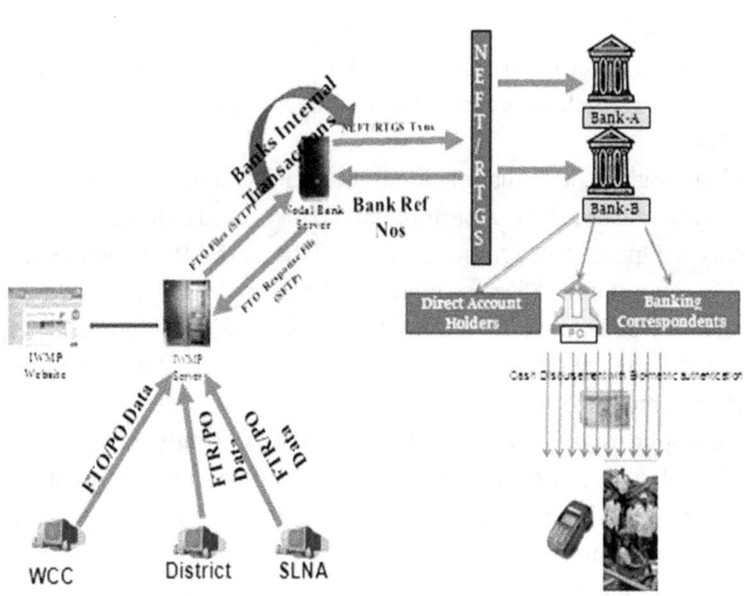

The IWMP is an important program under which 423 watershed projects are under implementation in the country, covering 1.81 million hectares at an estimated cost of Rs 2,290 crore (US$ 336 Million). In Andhra Pradesh, the average size of watershed project is 4,200 hectares. The Andhra Pradesh Government selects the watershed project areas based on location and based on inputs from the Andhra Pradesh Remote Sensing and Application Centre (APRSAC) (Pilla, 2015).

LEVERAGING TECHNOLOGY

Towards Maximizing Agricultural Output and Efficient Water Management

As has been highlighted in this chapter, water resources the world over and specifically in developing countries such as India are increasingly under stress. Increase in population, better living standards and rapid industrialization are leading to an ever-increasing demand for water. It is imperative that technology be leveraged to meet this demand for better water resource management and better agricultural yield. India has made rapid strides in space science and technology and information technology. They can be put to good use to forecast crop production and provide the latest information on water resources.

Forecasting Agricultural output using Space, Agro-meteorology and Land based observations (FASAL), the project of the Indian Space Research Organization (ISRO) and Water Resource Information Systems of India, India-WRIS, are two noteworthy examples. Elsewhere, the US Government, in partnership with technology giants Microsoft and Google, has launched The US Climate Data Initiative -- a project to create tools to prepare for potential risks associated with climate change.

The FASAL Project

Forecast of crop yield before the harvest is essential for the government to take policy decisions in respect of pricing, import export, storage of food grains, distribution etc. For the farmer, such a forecast would tell him when to sow, what to sow, etc. A mid-season forecast would alert him to attacks such as pests, diseases etc. Forecasting Agricultural output using Space, Agro-meteorological and Land based Observations (FASAL) is a joint program of the Ministry of Agriculture, Government of India and Space Application Centre of ISRO which uses econometric models to forecast crop acreage and yield before sowing (Moorthi, 2014).

In rainfed agriculture regions, information about the quantity and distribution of rainfall makes such forecasting useful. Optical and Synthetic Aperture radar data is used to forecast agricultural yield, condition of crop, crop area etc. Optical and Synthetic Aperture Radar (SAR) imagery complement each other. SAR sensors emit their own illumination in the form of microwaves, which allow them to record data at all times of day and all weather conditions. SAR technology uses different wavelengths than optical sensors, allowing it to record data through atmospheric interference like clouds and storms.

For crop production forecasting, using remote sensing and GIS data, indigenous software has been built by ISRO's Space Application Centre called FASAL Soft using only free and open source geo-spatial tools and software. The objective of developing such software was to have reliable tool driven by data, for multiple crop forecasting at different stages of a given crop from sowing to harvesting. The software would use data from various sources such as remote sensing, meteorology and land observations. Information on weather conditions at any stage of the crop growth would be available early in season. Data would also be acquired several more times during the middle to end season to assess how the crop was developing and to take corrective measures, if needed (Parihar & Oza, 2006).

Impact of disasters such as drought and floods on agriculture can be assessed by satellite based monitoring. An NRSC project called NADAM (National Agricultural Drought Assessment and Monitoring System) uses remote sensing data as well as rainfall data to make drought assessment. SAR data is used to map the areas of crop that are flooded. Assessment of crop damage caused by hurricane Phailin carried out using RISAT SAR data proved to be 89% accurate.

The National Remote Sensing Centre (NRSC) has developed an Android App for collection of ground truth and essential component of RS data analysis. This data, collected by smartphones goes directly to the Bhuvan Server. For the Kharif and Rabi seasons in 2014, 5800 ground truth collection points were set up in 17 States.

Web-Enabled Water Resources Information System of India

It is known that human race survival is dependent on water. Water is used in agriculture, industry, homes, for power generation etc. Increasing population, rapid industrialization, economic growth, rise in standards of living, all contribute to an increase in demand for water, and only a small fraction of the world's water resource (about 2.5%) is freshwater and a fraction of this is available for human consumption. Excessive demand for water, shortages in supply of water, fluctuating quality of water and the effect of water supply and demand on the nation's economy all make efficient management of available water resources imperative.

The India-WRIS WebGIS is a joint project of the Central Water Commission and ISRO to develop a web enabled information system for providing data of India's water resources as also natural resources. Web enabled tools have been developed to search, access and visualize data and information. This information has been categorized under 12 major and 30 sub-information systems in the GIS environment. India-WRIS Web GIS is currently available in the Web 3.0 portal (India-WRIS WebGIS, 2014).

The Web-GIS information system has four basic functions: Data collection; data storage, data analysis and data transformation into information relevant to the user; an interactive system for geo-visualisation and temporal analysis; and placing information in the public domain which can be downloaded and processed further as required by the user. Placing such information in the public domain ensures the participation of all stakeholders in its management and enhances public awareness on water availability and its management. While designing the India-WRIS Web GIS system, factors such as the type and

Table 2. Water resources of India

Details	Quantity Available (in cubic km)
Natural Runoff (Surface Water and Ground Water).	1869
Estimated Utilisable Surface Water Potential	690
Groundwater Resource	432
Available Groundwater resource for Irrigation	361
Net Utilisable Groundwater resource for irrigation	325

Source: National Institute of Hydrology

volume of data, users and applications it would be required for, ease of accessing information, internet connectivity, environment in which the data would be accessed etc. were all factored in to ensure its use for maximum benefit.

As over 80 % of India lies within an inter-State river basin, inter-State conflicts affect almost every area and region. The inter-state river waters in India have, instead of uniting people of all communities, become rivers of discord and hotbeds of politics. Some of these disputes have been highlighted in Table 3.

INTER-STATE RIVER WATER DISPUTES

Karnataka implements the award given by the Cauvery Water Tribunal and the Supreme Court more in its breach than in practice. Repeated appeals by Tamil Nadu against Karnataka's recalcitrant conduct have been in vain. The dispute raises its ugly head and returns to the public domain with a bang whenever there is deficient rainfall.

Over the years, both States have experienced protests, marches and even bloody clashes between the farmers and the general public over the issue. It is common knowledge that rice is a guzzler of water. Yet, Tamil Nadu grows Kuruvai crop and Karnataka produces summer paddy in the non-rainy season. People of both States are mainly rice-eaters and consequently, farmers produce more rice or paddy even though crops like pulses and ragi are more nutritious and have a better market price. Unfortunately, notwithstanding experts' warnings, the respective State Governments are not making sincere efforts to convince farmers to give up the water-guzzler paddy crop and grow pulses, oilseeds and millets subject to the availability of water (Richards & Singh, 2001).

In view of the sensitivity of the matter and the principle of constitutional law involved, it would be worthwhile to exemplify the constitutional position on the inter-State river waters vis-à-vis the authority of the Central Government. Though the Government of India's Ministry of Water Resources and the Central Water Commission monitor water resource development and provide technical assistance to the Union Government as well as the State Governments, the absence of federal infrastructure on most

Table 3. Inter-State river water tribunals in India through the corridors of time

Tribunal Name	States Involved	Start Date	Final Award	Time Taken
Krishna River	Andhra Pradesh, Karnataka and Maharashtra	1969	1976	7 years
Narmada River	Gujarat, Maharashtra, Madhya Pradesh and Rajasthan,	1969	1979	10
Godavari River	Andhra Pradesh, Madhya Pradesh, Chhattisgarh, Odisha and Karnataka,	1969	1980	11
Ravi-Beas Rivers	Punjab, Haryana, Himachal Pradesh, Rajasthan, Jammu and Kashmir and Delhi,	1986	None	29
Cauvery River	Karnataka and Tamil Nadu	1990	2007	17
Krishna River II	Andhra Pradesh, Karnataka and Maharashtra	2004	2010	6
Vamsadhara River	Andhra Pradesh and Odisha	2010	2013	3
Mahadayi River	Goa, Maharashtra and Karnataka	2010	None	5

Source: Seligman, Daniel. Resolving Interstate Water Conflicts: A Comparison of the Way India and The United States Address Disputes on Inter-State Rivers, June 2011.

inter-State rivers in India has marginalised the Central role on these basins. More important, the Central Government has virtually no control over the dams, locks, canals and pumping stations for the simple reason that it does not own them (Seligman, 2011).

The Constitutional Position

India is a Union of States. In the Indian Constitution, there are three Lists, each of which has subjects on which the Union or the States can legislate. The three Lists together with the number of subjects each of them deal with (mentioned within brackets) are: the Union List (97), the State List (66) and the Concurrent List (47). The Constitution is crystal clear that all issues dealing with water, except for an inter-State river, remain under the control of the respective States. Those subjects falling within the Concurrent List are the joint responsibility of the Centre and the State(s) concerned. Interestingly, however, there is no mention of water in the Concurrent List. This implies that inter-State waters remain under the Centre's control if Parliament enacts legislation to this effect.

Indeed, Article 262 (1) of the Indian Constitution deals with the process of adjudication of disputes relating to waters of inter-State rivers or river valleys. It says that Parliament may by law provide for the adjudication of any dispute or complaint with respect to the use, distribution or control of the waters of, or in, any inter-State river or river valley. By the same token, under Article 262 (2), Parliament may prohibit the Supreme Court or any other court from exercising jurisdiction in respect of any such dispute or complaint as is referred to Clause (1) (The Constitution of India, 2010).

It is noteworthy that this constitutional provision notwithstanding, Parliament has never restrained the Supreme Court of India from intervening on issues dealing with inter-State river waters. Indeed, one may recall how the Supreme Court has been directing the Karnataka Government to release water to Tamil Nadu from time to time whenever its farmers are in distress. It needs to be emphasized that the Supreme Court of India does have the authority to issue directions to the States on any matter whatsoever, but on matters pertaining to inter-State river waters, the Supreme Court, in the absence of original jurisdiction (unlike in the USA), expects the States to follow the Inter-State River Water Tribunal's orders and directions.

In an essay on the theme, "Inter-State Water Disputes: A Nightmare!", Fali S. Nariman, eminent jurist and constitutional expert, says that inter-State water dispute tribunals do not work "efficiently" (Nariman, 2009). As their performance has been slow, cumbersome and time-consuming, Mr. Nariman recommends that the Indian Parliament should allow States to go directly to the Supreme Court for adjudication. It would be eminently sensible for the Narendra Modi Government and the Indian Parliament to accept Mr. Nariman's recommendation in letter and spirit. Though the Supreme Court will fix its own time to hear such cases, keeping in view its workload, adjudication by the Supreme Court may have greater credibility and acceptance by the riparian States than the Inter-State River Water Tribunals.

INTERNATIONAL EXPERIENCE

What India Can Learn

Over the decades, many countries have been fighting for rights over river waters. And India is no exception. Iraq and Iran have been fighting over the 193-kilometre-long Shatt-al-Arab river waters for quite some time. As far back as 1935, an international commission adjudicated in favour of Iraq: while Iraq got almost total control of the river waters, Iran got control only of the approaches. Skirmishes occurred time and again between the two countries. Nonetheless, with a view to giving a boost to foreign trade, Iran built ports on the Persian Gulf. In September 1980, both countries witnessed a full-scale war on the issue. There has been no amicable resolution of this dispute (Schofield, 1986).

Same is the case with India and Pakistan over the rights to water from the Indus river; and India and Bangladesh claim for the Brahmaputra river. Even as there is uncertainty over the resolution of these river water disputes, these can be resolved if there is political will and administrative support to help farmers and the general public.

Two examples hold out promise for peaceful settlement of rival claims: The Colorado Agreement and the La Plata Pact. These two processes need to be examined for a proper appraisal of the position and putting the discussion in the right perspective.

The Colorado Agreement

The Colorado Agreement signed between the United States and Mexico on November 21, 2012 for sharing water from the 1,450-mile Colorado river is a watershed in the world history of inter-state river water disputes. Also known as the "Minute 319 Agreement", the five-year pact seeks to tackle the problem of drought and reap the benefit of wet years. Mexico had been reeling under drought year after year. It was very badly in need of water from Lake Mead which stretches across Nevada and Arizona. The agreement aimed at establishing a level-playing field for all states facing water shortages. While it ensured that Mexico will forfeit some of its share of the river during shortages, in line with western states of the United States which had agreed on how much water they will surrender when waters recede, it also made it mandatory that Mexico will exercise the right to capture surpluses when waters rise. Another significant provision in the Agreement is that water agencies in California, Arizona and Nevada will buy water from Mexico, the proceeds of which will help Mexico upgrade its canals and other infrastructure (Spagat, 2012).

The Colorado Agreement is a major amendment to the 1944 Treaty. Even as Mexico had been starved of adequate quantity of water for agriculture and drinking water needs of the people, the US rules drafted in 2001 made the situation worse: while the rules set out a schema to divide surplus water, it set aside little for Mexico. Further, in 2007, facing an eight-year drought, California, Arizona and Nevada evolved a consensus on how much each state should sacrifice water during shortages on the river. Against this background, United States Interior Secretary Ken Salazar and former Mexican President Felipe Calderon took the initiative to "choose collaboration over conflict and cooperation and consensus over discord" and signed the most important international accord on the Colorado river waters (Spagat, 2012). Significantly, while the Agreement between the US and Mexico is expected to end the "dry season" soon, it is believed to have restored small parts of the two-million-acre (8,100 sq. km) Colorado River Delta. Twenty nine months after the Agreement, even as dams and canals have diverted water to farm fields and cities, the Colorado does not reach the sea anymore and its delta has been "desiccated" (Brasuell, 2014).

The La Plata Pact

The La Plata Basin Treaty signed by Argentina, Bolivia, Brazil, Paraguay and Uruguay in 1969 is another significant trans-boundary dispute resolution on water management. Encompassing an area of 3.2 million sq. km, the La Plata river basin is among the world's five largest international river basins. It is the lifeblood for agricultural and industrial sectors of the riparian states. Interestingly, the five countries which were signatories to the La Plata Pact have a history of cooperation and joint management of the watershed. These countries were unanimous in their thought and belief that cooperation and not confrontation held the key to each other's economic growth and development. It is this spirit of cooperation and understanding that has led to the region creating a record of sorts in exporting large amounts of grain, wool, timber and manufacturing goods to other parts of the world. The various bilateral and multilateral treaties signed by the five countries over the years bear eloquent testimony to their concerted efforts to improve economic investment, hydro-electric development and transportation enhancement.

The 1969 treaty has led to remarkable progress of the region in terms of hydro-electric development too. Dams were constructed and alternative power plants were installed. Today, this region boasts of 130 dams. The Itaipu is the world's largest hydro-electric project which came up following a 1973 agreement between Paraguay and Brazil. It cost the two governments and other international participants US $15 billion and 20 years to construct. Remarkably, with a generating capacity of 26,000 MW, it supplies 26% of Brazil's total electricity needs and 78% of Paraguay's power requirements with "zero emissions" (Wolf & Newton, 1969).

In a case study of the La Plata Basin Treaty as a trans-boundary dispute resolution mechanism, Aaron T. Wolf and Joshua T. Newton draw two important conclusions: First, "if riparian States start cooperation from the outset of a conflict, instead of letting it create stronger positions, the economic and joint management prospects are much greater". They minced no words in recording in their case study that while the quality of joint economic ventures in the La Plata Basin has facilitated "increasing cooperation" between the riparian states ever since 1969, they would not have enjoyed the fruits of the treaty today had they not come forward and signed the pact with a spirit of camaraderie, cooperation and understanding. Second, "if riparian States agree to equal access to trans-boundary water resources, equal and joint management, investment and distribution of that resource is feasible" (Wolf & Newton, 1969). Wolf and Newton found, in the water resources sector, neither Brazil nor Argentina has used its economic or military superiority to maintain greater control over water resources or hydro-electric potential.

Interestingly, in a new study (2008), Wolf and Newton have reviewed the literature on the subject of water disputes and related water treaties. They have done new case studies on 12 rivers. These are, namely, Danube (Europe); Euphrates-Tigris (Asia); Ganges (Asia); Indus (Asia); Jordan (including West Bank aquifers); Kura-Araks (Asia); La Plata; Mekang (Asia); Middle East (Asia); Nile (Africa); Salwan (Asia); and Senegal (Africa). As for the current status of the La Plata Pact, they state that the Hydrovia (known as 'waterway' in Spanish and Portugese) – the largest project for navigational river development project – had come out of the 1969 Pact. It was proposed in 1988. The Andeau Development Corporation through the Intergovernmental Commission had commissioned new studies which were completed in 2004. They state that though the La Plata Pact has been helpful to resolve conflicts, the Hydrovia project has not made much headway due to resistance from environmental and social action groups which fear that the economic, cultural and ecological integrity of the basin will be threatened if the Hydrovia project was executed. They maintain that some meeting point between the policymakers and action groups would pave the way for a "more sustainable project" (Wolf & Newton, 2008).

Agreements on Shortages

In his article, "Address Disputes in Inter-State Rivers", Daniel Seligman suggests that adjudication by courts offers only a "limited solution" to the "complex problems" of inter-State river management in any country. More important, he is of the opinion that courts, in general, are not well-equipped to resolve problems of hydrology, economics, engineering and law. Consequently, referring to the practice that obtains in the United States, he suggests that the riparian States concerned should accept the ground reality and come forward to sign agreements on how to meet shortages during distress. All that is needed for the success of this formula, according to him, is "a system of cooperative mechanism that allows States to manage a river as if the infrastructure were owned by a single entity" (Seligman, 2011).

Accordingly, under this agreement, one State pays for another towards what is officially called "payments for watershed services" where one State pays another for enhanced flood control protection or where one utility leases upstream reservoir storage in cases of drought or where one State "banks" water with another State to use it later. In the United States, such agreements, despite initial hiccups, seemed to have worked for mutual benefit of the riparian States. For instance, on the Colorado river, the States in the Lower Basin follow inter-State water banking arrangements. Pursuant to federal regulations, these transactions allow States to store water in another State by paying a fee, which is considered an innovative and pragmatic financial and operational arrangement.

In 1999, the Southern Nevada Water Authority signed water banking agreements with the Arizona Water Banking Authority and the Metropolitan Water District of Southern California. According to Seligman, the Columbia River is managed for flood control and power generation in a manner that it functions under the control of a single entity. Similar agreements exist on the use of the Potomac River for effective drought management. Clearly, if States are pragmatic and willing to help each other for mutual benefit, borders will not and should not matter.

The Murray-Darling Basin

The Murray-Darling Basin in Australia is another worthy example of how water for over two million people can be provided by a people-friendly and pro-active government. Australia's former Prime Minister John Howard not only put effective water management on top of his government's agenda but also committed reasonable governmental funding after investments in fields such as hydrology, mathematics, climate and statistics. It goes to the credit of this Basin that apart from meeting the water requirements of two million people of Australia, including the City of Adelaide, it takes care of 40% water needs of its farmers (World Economic Forum, 2015).

These models need to be studied carefully by the politicians and administrators of all countries, including India, for resolving the inter-State river water disputes.

Discussion

Given the past experience, global warming will increase in the years to come. There would be no respite from melting of glaciers, unseasonal rains, drought, Phailins, Hudhuds and the like. Consequently, there is every need for the governments at the Centre and in the States to brace up to the situation and prepare for any eventuality. As drought is a recurring phenomenon every year, the Centre and the States will have to streamline the drought and water management programs. Apathy towards dry land agriculture has cost India dearly. Arid and semi-arid regions deserve a better deal. The government's interest in rainfed farming and drought-proofing of agriculture generated during drought should not vanish soon after there is some rain. It should be taken up as a permanent activity.

The task of boosting the productivity of rainfed agriculture is not insurmountable. It is difficult as it involves drastic changes in the cropping pattern. It also calls for adoption of modern dry land agricultural technology which has been tried out in research farms sponsored by reputed institutions like the International Crop Research Institute for Semi-Arid Tropics (ICRISAT). It also calls for major alterations in the existing land-use pattern keeping in view the quality of land and climatic factors.

Dry lands need liberal doses of nutrients to rejuvenate. Small doses of chemical fertilizers in dry lands will work wonders. A full package of available dry farming technology is said to be capable of raising output, two to four-fold in years of normal rainfall. An increase of 50 to 100% can be expected even in marginally abnormal years. This can make a vast difference in the country's total farm production because of vast extent of rainfed areas. The economic benefits to the individual farmer would obviously be significant.

Action on drought-proofing should preferably be taken in years of good monsoon. Employment-oriented and water conservation works undertaken in favorable circumstances can help build a useful infrastructure which would not only add to the productive assets but also provide in the long run a cushion against drought.

Resource conservation measures in natural watersheds help substantially to improve the soil moisture status during low rainfall periods. Engineering structures like ponds, gully plugs, check dams and the like facilitate storage of rain water. They also indirectly help augment surface and ground water supplies. There is need for construction of more and more minor water conservation ponds and percolation tanks for impounding runoff water on mini-watershed basis on individual farms or a cluster of farms. They provide critical life-saving irrigation to crops at crucial stages of their growth. As a result, not only the chances of complete crop failure are minimized but, at times, a normal crop is ensured even under adverse circumstances. The Ralegan Siddhi and Hiware Bazar experiments are worthy of replication elsewhere. Why can our agricultural scientists and experts not galvanize the administrative machinery in that direction?

Figure 5. Drought and flooding: Consequences of extreme weather conditions
Source: www.siliconindia.com

Figure 6. Watershed management in Ralegan Siddhi in Maharashtra, India
Source: www.annahazare.org

Figure 7. Adarsha watershed project: Community-based masonry check dam in Andhra Pradesh, India
Source: Food and Agriculture Organization, Corporate Document Repository

A sound irrigation policy is imperative to improve the performance of irrigation schemes. The procedure of watershed management is aimed at obviating such haphazard selections and concentrating the effort on developing the drought prone area on a watershed basis which is scientific and rational. As the implementation of the DPAP reveals, it is doubtful whether the implementing agencies realize the purpose of the various schemes that are undertaken. As regards the afforestation programs in some of the DPAP areas, the purpose is to prevent the erosion of catchment areas with the runoff and subsequent silting of irrigation tanks, rivers and canals. In some places, afforestation programs were undertaken in low lying areas while the catchments were not cared for. This shows poor appreciation of the DPAP's objectives. There is a need for constant review of the DPAP to examine and identify the program gaps for timely corrective measures. The coverage and investment in diversification schemes need to be stepped up. As a large developmental potential in these sectors still remains untapped, greater concentration of activities in irrigated areas as at present needs to be reversed by impounding the rainwater runoff to meet the moisture requirements of horticulture (especially for mulberry cultivation).

There is a yawning gap between goals and results. For example, in the case of minor irrigation works, the DPAP works suffered because the delegation of authority and powers to the Executive Engineers was minimal. The best performance in the DPAP has been observed in the area of soil conservation where the works were taken up on a job basis with greater degree of decentralization. The effectiveness of the program is much greater where the district officers are accountable for plan expenditure rather than the heads of departments. In a situation where the DPAP authority solely depends on the district officers for the management of the area development programs, the DPAP is treated as another additional function of the district officer and not as a qualitative innovation. If the field staff continues at the helm of development schemes as well as of routine maintenance functions, the objectives of the area development programs are bound to be blurred.

Streamlining the Command Area Development Program in the light of past experience alone will not suffice. What is needed is a far more comprehensive examination of the entire strategy of water management. This is important because numerous water-related matters such as surface flows, ground water, domestic and industrial uses, anti-pollution measures, flood control, navigation, catchment protection, hydel generation, etc. are being administered in isolation by the Centre and the States.

As irrigation is a State subject, the States will have to shoulder greater responsibility in the timely execution of projects. But experience in many States suggests that once a project is launched, the Government becomes leisurely and implementation gets tardy. As a result, the resources available are thinly spread and every project suffers from paucity of funds. The progress of the Jalyukta Shivar Abhiyan in Maharashtra and the Neeru-Chettu Project in Andhra Pradesh launched by new Chief Ministers Mr. Devendra Fadnavis and Mr. Chandrababu Naidu are test cases and will be watched carefully.

Ideally, it should be left to the wisdom and political sagacity of the States concerned to resolve their inter-State river water problems amicably to the mutual satisfaction of both sides. However, if there is no breakthrough, the Centre will have to step in and play a leadership role. It should more or less play the role of an umpire without any bias – political or otherwise.

In a country wedded to the values of liberty, equality and fraternity together with justice for all irrespective of one's caste, color, creed or religion, India cannot abolish the justice delivery system. The judiciary should continue to play its seminal role in the speedy dispensation of justice in accordance with the provisions of the Indian Constitution. At the same time, one cannot endorse the argument that court interventions and tribunal-monitored adjudication of river waters disputes are a long drawn-out process and hence opaque exercises which need to be scrapped. Certain drawbacks notwithstanding, the people have tremendous faith in the judiciary and always look to the Supreme Court for fair and speedy dispensation of justice. The judiciary, on its part, has always stood firm and acted as a check on the executive arbitrariness and protected the fundamental rights of the citizens.

Thus, it would only be fair to comment that there is a strong case for bestowing the apex court the original jurisdiction, so that it can *suo motu* take up inter-state river water disputes for appropriate adjudication. If the Indian Parliament, in its wisdom, thinks that the Inter-State River Water Tribunal route should not be compromised (or even subordinated) in favor of a more activist role for the judiciary, Parliament or the Supreme Court should have the authority to fix suitable deadlines for expeditious disposal of cases. Cases pending with various Inter-State River Water Tribunals should not be allowed to continue in perpetuity; they need to be directed to dispose of the cases within a fixed timeframe.

Alternatively, India cannot afford to overlook the international experience. If other countries are able to resolve their problems amicably, why can India not rise to the occasion, sink the differences for the larger public good and act accordingly? The Colorado Agreement and the La Plata Basin Pact are two models of bilateral and multilateral cooperation. In addition, the Land Boundary Agreement signed between India and Bangladesh in June 2015 holds out promise for a treaty on sharing the Teesta river waters. The Government of India should strive to achieve breakthrough with regard to interlinking river waters with other neighboring countries too.

The Agreements on Shortages, as in the case of the Colorado river or the Potomac river or Australia's Murray-Darling Basin are other best practices for resolving the river water disputes. For instance, there is nothing wrong if the Tamil Nadu Government pays a mutually agreed upon fee to the Karnataka Government either to store or use water in its hour of distress to save its farmers and prevent crop failure. The time has come for India to think of solutions in such terms rather than letting the water disputes continue in perpetuity and/ or trying solutions which are out of sync with the present times.

The issue of interlinking of rivers has been coming to the fore whenever the Bharatiya Janata Party is in power at the Centre. The Congress, however, views the concept as a "disaster". Former Prime Minister Rajiv Gandhi and former Union Minister for Environment and Forests Jairam Ramesh were totally opposed to the concept of interlinking rivers (Indian Express, 2009). First envisioned in 1982 and pursued by the Atal Behari Vajpayee Government in 2002, it has been currently taken up by the Narendra Modi Government. The project merits a fair trial to tackle water crisis due to the vagaries of monsoon and climate change and for effective water management.

The issue is presently being handled by the National Water Development Authority (NWDA) under the Union Ministry of Water Resources. Interlinking rivers will help in many ways. It will remove regional imbalances; facilitate additional irrigation; improve domestic and industrial water supply; maximize hydropower generation; and boost navigational facilities. Though the project is expected to create 87 million acres of irrigated land and transfer 174 trillion liters of water a year, according to Upali Amarasinghe, a Senior Researcher at the International Water Management Institute, half a million people may be displaced by the project (Upali, 2009).

While apprehensions and concerns over the project by a few experts and ecologists seem misplaced, the government has a duty to ensure that the people are not put to untold hardship arising out of displacement. Experts would do well to study the project from a human angle and evolve suitable solutions to handle the issue of migration and displacement. In addition, the government should examine the problems of desilting, huge cost overruns and the project's overall impact on nature.

One cannot overlook the international dimension to the problem as India's neighboring countries such as Bangladesh, Pakistan, Nepal and China are involved. The Long-term Framework Convention and the Land Boundary Agreement signed between India and Bangladesh during Mr. Narendra Modi's visit to Dhaka in June 2015 seem to bring the signing of the Teesta water-sharing agreement between both countries one step closer (ndtv.com, 2015). The Teesta river originates from the eastern Himalayas and flows through Sikkim and West Bengal before entering Bangladesh where it merges with the Brahamaputra. The Teesta is said to be the "lifeline" of Sikkim flowing the entire length of the State. It is a crucial river for Bangladesh too. The Teesta water is imperative for Bangladesh, especially in the leanest period from December to March when the water flow is reduced to less than 1,000 cusecs from 5,000 cusecs. There is an atmosphere of camaraderie and friendship between the countries at present. Both sides signed as many as 22 agreements during Mr. Modi's visit, including those on cooperation in maritime safety. On top of all this is Mr. Modi's announcement of a fresh USD 2 billion line of credit for Bangladesh (*The Economic Times*, 2015). Consequently, an agreement on the Teesta water sharing brooks no delay.

As for the Indus River Water Treaty between India and Pakistan, hostilities between the two countries have taken a toll. Increasing cross-border terrorism in India, aided and abetted by Pakistan, has proved to be a stumbling block for peaceful talks between the two countries. The Indus River Water Treaty, signed in 1960 with the World Bank's help, is dated for reasons such as population growth, climate change and outdated irrigation practices (Plesse, 2015). It deserves to be reviewed through negotiations, but Pakistan is creating hurdles. It is opposing the construction of the Kishanganga and Ratle hydroelectric projects. Though India seems to be in favour of talks on this issue, Pakistan is reluctant and wants a neutral observer (Sarfraz, 2013).

Indeed, both countries are dependent on the Indus river. On the one hand, Pakistan is a lower riparian state and thus depends upon India for its water security. On the other, India would not take the extreme step of cutting off water supply to Pakistan as that would tantamount to violating the 56-year-old treaty mediated by the World Bank. Clearly, it is time the political leadership of both countries rose to the occasion and demonstrated extraordinary statesmanship in handling the issue for the good of the people cutting across partisan lines.

The benefits of using technology to adapt to impacts of climate change cannot be over emphasized. The vast strides made in the areas of space science and technology and information technology can be used to assist the government and farmer improve agricultural production through better crop forecasting methods, water resources mapping, preparation of relevant water resource information databases, etc.

India is increasingly become tech-savvy. A good majority has access to Internet and smartphones. If these can be used to inform the farmers on weather forecasting, the predicted rainfall pattern, the rainfall distribution, the likelihood of pest attacks, etc. through the crop cycle, they will definitely benefit from a better yield.

The Government of India, having recognized the immense benefits that can accrue from the use of technology, partnered with the Indian Space Research Organisation, the Union Ministry of Agriculture and reputed companies such as Tata Consultancy Services (TCS) to bring these benefits to the doorstep of the common man. By creating dedicated websites for several of its initiatives, it has ensured greater transparency and easier availability of information for the farmers and the people at large so that they can partner with the Government to tackle problems arising out of the fallouts of climate change.

There is no denying that climate change and water shortage will continue to haunt Indian agriculture in the years to come. However, India would do well to face this challenge through bold and effective remedial measures at appropriate levels, including comprehensive technological initiatives in the digital age. Worthy of mention in this context is the current focus on the concept of Industrial-Urban-Agriculture Ecosystem as far as utilization of treated and untreated wastewaters for agriculture systems is concerned. Already, Warananagar in Kolhapur District of Maharashtra, which has a strong network of sugar, distillery, pulp and paper and dairy industries, is making best use of "industrial symbiosis" (Rao & Patil, 2015). Admittedly, as Dr. Prakash Rao and Dr. Yogesh Patil have maintained, though the Warananagar experiment has not yet been documented in reputed journals, it is time comprehensive research was done on this critical area so that the fruits of the Warananagar experiment are available in other equally deserving and water-starved pockets of the country.

Sugarcane cultivation, a water guzzler, continues to be one of the main crops in States such as Maharashtra, Punjab, Andhra Pradesh and Odisha. Same is the case with rice, especially in States such as Andhra Pradesh, Karnataka and Tamil Nadu. It would be eminently sensible if the treated water from various treatment plants is diverted to agriculture. This would be one way of ensuring that not a drop of water goes waste and that every drop of water is conserved and used for agriculture.

Significantly, the Conference of Parties 21 (COP21), held in Paris during November 30-December 11, 2015, under the aegis of the United Nations Framework Convention on Climate Change, can be interpreted as a step forward to reduce the greenhouse gas emissions. As many as 196 countries have resolved, through an agreement, that the average global temperatures increase by no more than 2°C (3.6 °F) above pre-industrial levels. This is regarded as a "critical threshold" above which the planet could experience catastrophes such as high sea level rises, and planet and animal extinction. The USA and China, the world's biggest polluters which account for 14% and 24% of the total greenhouse gas emissions respectively, have pledged substantial emission cuts. Even though all countries need to sign the agreement, the Paris initiative, particularly in the backdrop of the failure of the Denmark conference in 2014, is remarkable because any attempt to cut greenhouse gas emissions will not only help tackle the water crisis but also check floods, drought and heat waves. The focus on water, climate and resilience during the Paris conclave will hopefully encourage cooperation and initiatives between various countries in this critical area (Goswami, 2015).

CONCLUSION

The preceding analysis suggests that the Centre and the States need to understand the gravity of the situation and act accordingly. Water is a national resource, a national asset and will have to be treated as such. No state can appropriate this precious resource for its own ends and at the cost of the other State or States. Water will have to be used for mutual benefit, and not individual benefit. Adopting a jingoist approach by any one particular State will only encourage fissiparous tendencies to foment trouble and disturb peace, make the confusion worse confounded and exacerbate tensions between States.

Political will is imperative to resolve inter-state river water disputes. Indeed, it is the "biggest barrier" to tackling the water crisis (Cho, 2015). As the Colorado Agreement and the La Plata Pact show, given the political will, these disputes can be resolved amicably between respective nations or States and water shared for common good with equity and justice. If the current mood between India and Bangladesh following Mr. Narendra Modi's visit to Dhaka is any indication, there is likelihood of an early signing of an agreement on the sharing of Teesta waters between the two countries. And this will set the pace for similar agreements between India and other neighboring countries such as Pakistan, Nepal and China.

There is a need to emulate best practices in water-stressed countries like India. The Agreement on Shortages as in the US, too, is worthy of emulation. At the other end of the spectrum, the Murray-Darling Basin initiative in Australia amply demonstrates how water can be equitably used for the benefit of all sections if the political leadership rises above partisan politics and contributes to the general well-being of the people.

There is need for a comprehensive national water policy. As global warming and climate change are likely to increase the variability of water resources affecting agricultural production, human health and livelihoods, any public policy on water conservation and management should be governed by certain basic principles so that there is a commonality in approaches in dealing with planning, development and management of water resources. The government will have to tackle the situation by increasing water storage in various forms such as soil moisture, ponds, groundwater, small and large reservoirs. The Centre cannot abdicate its responsibility from helping the States in getting this gigantic task accomplished. It should incentivize the States to increase water storage capacity.

Paradoxically, though the availability of water is limited, the demand for water is increasing rapidly due to rising population, rapid industrialization and economic development. This calls for augmentation of the availability of water for utilization and meeting the increasing demand.

The Narendra Modi Government has committed itself to increasing the availability of water to farms, homes and factories. At the same time, the ruling Bharatiya Janata Party's manifesto released during the Lok Sabha elections in May 2014 has served a warning: India would be a "water-stressed" country by 2050 and that the gap between the demand and supply would be a whopping 50%! It is, therefore, a challenge for the Modi Government to take all possible measures to tackle the water problem on a war footing and fulfill people's expectations.

While according priority to water security, the BJP manifesto has promised a slew of measures: launching the Pradhan Mantri Gram Sinchayee Yojana with a motto of *Har Khet Ko Paani* (Water for every piece of farm); a multi-pronged water strategy for reducing farmers' dependence on monsoon; increasing irrigated land by completing the long pending irrigation projects in the country; nurturing ground water recharge; harnessing rain water to reduce dependence on ground water; and encouraging effective water conservation, recycling and water harvesting. These goals are, no doubt, ambitious and well-intended. Yet, their success would depend on the political will of the Government, backed by the administrative support and action.

The world should prepare itself for climate change in the years to come. As India, China, Brazil and Kenya are the largest consumers of rice and wheat, it enjoins a special responsibility on them to take suitable measures to help farmers and maximise agricultural production. Dr. M.S. Swaminathan, eminent agricultural scientist, has said that owing to global warming and climate change, our experts, policymakers and scientists should explore the possibility of growing rice, which possesses high adaptation, in regions below sea level and also higher up in States such as Himachal Pradesh (Sehgal, 2008). Though Uttar Pradesh and West Bengal have started experimentation on growing "floating rice" on the practice that obtains in Thailand, more and more States and countries should try this, according to Dr Swaminathan. Experts should heed his advice.

At the same time, States like Tamil Nadu and Karnataka in India should understand that they should refrain from growing rice in more than one crop because of water crisis. In view of the deleterious impact of climate change on agriculture and food production, they should restrict rice only to one crop and confine it only to the rainy season (Hegde, 2013). The Government of India's National Commission on Agriculture has also recommended that rice should be grown preferably where there is good rainfall. It is time our farmers were trained to shift from paddy and sugarcane to less water-intensive crops such as oilseeds, pulses and millets. If food stocks have gone down by a whopping 40% the world over because of global warming and climate change, there is an imperative need for everyone to see the warning signal and act accordingly.

REFERENCES

Aiyappa, M. (2012, June 15). Cloud seeding a failed mission in Karnataka. *The Times of India*, p. 1.

Anand, E.V. (1988, June 11). Mismanagement of water resources. *The Indian Express*, p. 12.

Anand, E.V. (1992, December 11). Making CADP effective. *The Indian Express*, p. 12.

Antholis, W. J. (2014). *India's climate change policy in a Modi Government*. The Brookings Institution. Retrieved from http://www.brookings.edu/blogs/planetpolicy/posts/2014/07/30-india-climate-change

Asian Development Bank. (2013, July5). *Climate Himalaya. In Outlook Report, Rudraprayag* (p. 1). Uttarakhand.

Bangladesh looks to resolve the Teesta water sharing with India. (2015, June 14). *The Economic Times*, p. 1.

Bhattacharya, A. (2015, March 10). 50-50 chances of El Nino this year: Met agencies. *The Times of India*, p. 1.

Brasuell, P.S.T.J. (2014, December 16). Saving the Colorado River Delta: One habitat at a time. *National Geographic*.

Cho, A. (2015, January 30). Political will is the biggest barrier to tackling global water risk. *The Guardian*.

Choube, J. (2015, March 3). Unseasonal rains damage crops. *Down-to-earth*.

Damodaran, H. (2015, April 12). An El Nino year, maybe. A bad monsoon year, maybe not. *The Sunday Indian Express*, p. 15.

Goswami, R. (2015, December 12). Paris report predicts extreme weather. *The Telegraph,* p. 1.

Government of India. (1992). *Command Area Development*. Planning Commission, Eighth Five-Year Plan. Author.

Government of India. (2006). NRAA: Rationale, organisation and mandate. New Delhi: Author.

Government of India. (2010). *Water management under revitalising rainfed agriculture – Potential pilots, Consultation Paper*. New Delhi: National Rainfed Area Authority.

Government of India. (2014). *Elucidation of the Fifth National Report Submitted to the United Nations Convention to Combat Desertification: Desertification, Land Degradation and Drought in context of India*. Ministry of Environment and Forests.

Government of India & Indian Institute of Management-Ahmedabad. (1987). Command Area Development Program: A Joint Study. New Delhi: Author.

Government of Maharashtra (2014, December 5). *Mumbai, Resolution No.JLA-2014/pra.kra.203/Jal-7*. Author.

Government of Maharashtra. (2015). *Mumbai*. Retrieved from http://www.maharashtra.gov.in

Hegde, B. R. (2013). *Resolving the Cauvery water dispute*. Retrieved from http://www.indiawaterportal.org

Howard, B. C. (2014, December16). US-Mexico agreement returns water to the Colorado river delta. *National Geographic*.

India-WRIS WebGIS. (2014). *Design and Development of web enabled water resources information of India*. Arc India News, Application Article.

Indian Express. (2009, October 6). Interlinking of rivers buried; Jairam says the idea a disaster. *Indian Express,* p. 1.

Indian Express. (2015, September 22). Rains ease water crisis. *Indian Express,* p. 1.

Iyer, K. (2015, February 19). Drought in Marathwada – I. *The Times of India*, p. 1.

Jayaraman, T. (2011). Climate change and agriculture: A review article with special reference to India. *Review of Agrarian Studies, Journal of the Foundation for Agrarian Studies, 1*(2).

Joy, K.J. (2013, September 14). The much talked about Shirpur model. *Down-to-earth*, p. 1.

Karnataka explores cloud seeding to increase water in Cauvery. (2012, October 2). *The Hindu Business Line*, p. 1.

Khapre, S. (2015, August 8). Bringing rain in Marathwada: Cloud seeding project to continue for 90 more days. *Indian Express*, p. 1.

Kleemans, M. & Sadoulet, E. (2012). *Impact of drought-tolerant risk-reducing rice on yield and farmer welfare*. Agriculture Technology Adoption Initiative (in collaboration with International Rice Research Institute and Stress Tolerant Rice for Africa and South Asia).

Lal Bahadur Shastri National Academy of Administration. (1979). Seminar Proceedings, Indian Administrative Service (IAS) Professional Course Foundation Program. Government of India.

Little, A. (2015, October 28). Weather on demand: Making it rain is now a global business. *Bloomberg Businessweek*. Retrieved from http://www.bloomberg.com/features/2015-cloud-seeding-india/

Maplecroft, V. (2013). *Website*. Retrieved from http://www.maplecroft.com/portfolio/new-analysis/2013/10/30/31-global-economicoutput-forecast-face-high-or-extreme-climate-change-risks-2025-maplecroft-riskatlas/

Mehta, K., & Satpathy, T. (2011). *Escaping Poverty: The Ralegan Siddhi Case*. Chronic Poverty Research Centre and Indian Institute of Public Administration.

Moorthi, S.M. (2014). FASALSoft – An ISRO software framework for crop production forecast using remote sensing data analysis. *Journal of Geomatics*, *8*(1).

More rain batters crops in Andhra Pradesh, Telangana. (2015, April 14). *The Times of India*, p. 1.

Nariman, F. S. (2009). Inter-State Water Disputes: A Nightmare! In R. R. Iyer (Ed.), *Water and the Laws in India* (pp. 32–57). New Delhi: Sage Publications. doi:10.4135/9788132104247.n3

National Portal Content Management Team. (2011). *Command area development and water management program*. Government of India.

NDTV.com. (2015). *Efforts underway to resolve Teesta issue with Bangladesh*. Author.

Neeru-Chettu can make AP drought-proof in 5 years: CM. (2015, February 19). *The Hindu*, p. 1.

Niyogi, D. (2015, March 23). Crop Damage in Many Areas. *Down-to-earth*, p. 1.

Palanisami, K., Radhamani, S., Vasanthi, C., & Tamilselvi, K. N. (2002). Implementation of DPAP (VIII Batch): Socio-economic baseline survey of ten watersheds of Coimbatore District: A report by District Rural Development Agency. Academic Press.

Parihar, J. S., & Oza, M. P. (2006). FASAL: An integrated approach for crop assessment and production forecasting agriculture and hydrology application of remote sensing. *Proceedings of SPIE*, 6411. doi:10.1117/12.713157

Pilla, V. (2015, February 23). Watershed scheme gets a digital push in Andhra Pradesh. *LiveMint*, p. 1.

Plesse, M. (2015, October 29). *Indus River Water Treaty, Global Food and Water Crisis Research Program*. Future Directions International. Retrieved from http://www.futuredirections.org.au

Rains won't impact inflation target: Arvind Subramanian. (2015, April 10). *The Economic Times*, p. 11.

Rao, P., & Patil, Y. (2015). *Climate Resilience in Natural Ecosystems in India: Technology Adoption and the Use of Local Knowledge Processes and Systems. In Handbook of Climate Change Adaptation* (pp. 2063–2077). Springer Berlin Heidelberg.

Rao, S. T. (2005). *Cloud seeding for India: An effective weapon to fight the droughts*. Hyderabad: The Book Syndicate.

Richards, A., & Singh, N. (2001). *Inter-State water disputes in India: Institutions and policies*. University of California and University of Maryland.

Sally, M. (2014, July 23). Artificial rainfall: Cloud seeding companies may play rainmakers. *The Economic Times*, p. 1.

Sangameswaran, P. (2008). Community formation, ideal villages and watershed development in Western India. *Journal of Development Studies, Routledge*, *44*(3), 384–408. doi:10.1080/00220380701848426

Sarfraz, H. (2013). Revisiting the 1960 Indus Water Treaty. *Water International*, *38*(2), 205. doi:10.1080/02508060.2013.784494

Schofield, R. N. (1986). Evolution of the Shatt al Arab Boundary Dispute. Menas Publishers.

Sehgal, R. (2008, April). The days of cheap food are over: M.S. Swaminathan. *Info Change News & Features*.

Seligman, D. (2011). *Resolving Inter-state Water Conflicts: A comparison of the way India and the United States address disputes on inter-state rivers*. Institute of Water Policy, Lee Kyun Yew School of Public Policy, National University of Singapore. Retrieved from www.lkyspp.nus.edu.sg/iwp

Senapati, M. R., Behera, B., & Mishra, S. R. (2013). Impact of Climate Change on Indian Agriculture and Its Mitigating Priorities. *American Journal of Environment Protection*, *4*(4), 109–111. doi:10.12691/env-1-4-6

Spagat, E. (2012, November 21). Colorado river pact signed between the US and Mexico. *Huffington Post*, p. 1.

Talule, D., & Suryavanshi, S. (2014). An empirical assessment of the model of rural sustainability: Tracing the roots of inspiration to the Central Government policy aiming to develop one ideal village per parliamentary constituency. *Asian Journal of Science and Technology*, *5*(12), 796–803.

The Constitution of India. (2010). *Administrative relations: Disputes relating to waters*. New Delhi: Universal Law Publishing Co.

Upali, A. (2009). Strategic Analysis of India's National River Linking Project. Challenge Program on Water and Food (CPWF) Project Report.

Venkateswarlu, B. (2000). *Rainfed agriculture in India: Issues in technology development and transfer, Seminar Paper, Central Research Institute for Dry land Agriculture*. Hyderabad: CRIDA.

Verisk, M. (2013). *Report on climate change and environmental risk atlas*. Bath, UK: Academic Press.

Wani, S. P., Sreedevi, T. K., Singh, H. P., Pathak, P., & Rego, T. J. (2002). Innovative farmer participatory integrated watershed management model, Adarsha watershed, Kothapally, India: A success story. International Crop Research Institute for Semi-Arid Tropics, Asian Development Bank, Central Research Institute for Dry land Agriculture and Drought Prone Area Program.

Wolf, A. T., & Newton, J. T. (1969). Case study of transboundary dispute resolution: The La Plata basin. Program in Water Conflict Management and Transformation, The Institute of Water and Watersheds, Oregon State University and College of Earth, Ocean and Atmospheric Sciences.

Wolf, A. T., & Newton, J. T. (2008). *Managing and Transforming Water Conflicts*. Cambridge, UK: Cambridge University of Press.

World Economic Forum. (2015). *Global Risks: Insight Report* (10th ed.). Geneva: Author.

Chapter 19
Common Duckweeds as a Model System for Climate Change Impact Assessment

Vinay Rale
Symbiosis International University, India

Pooja Tendulkar
SA Associates, India

ABSTRACT

Common duckweeds – Lemna – are free-floating aquatic macrophytes belonging to Lemnaceae family. They occur universally, in tropical and temperate zones and are enabled to grow in stagnant, slow-flowing, nutrient-enriched waters. They have an ability to grow over a wide range of temperatures (6 – 33 ^{0}C) and pH (optimum 5.5 – 7.5). Lemna species also manifest a characteristic macrophyte community in association with plants functioning as a dominant primary producer. Asexual mode of reproduction is exclusive and therefore, all resources are directed towards vegetative growth. Structure wise they show leaflets (frond) and root-like diffuse form. This simple morphological and physiological form offers special scientific and engineering properties – reflected in easy handling and manipulation under laboratory conditions. As a result they are regarded as a model plant system for a number of chemical and biogeochemical studies. Climate change affects plant growth and physiology. For example, increasing atmospheric concentrations of carbon dioxide (CO_2) and ozone (O_3) or increase in ultraviolet (UV) radiation (due to the thinning of the ozone layer) can have pronounced effects on the growth and development of plants. In many instances the decrease in growth rate can create disturbances in photosynthetic processes. In the recent past there has been a surge in the interest in looking for alternative remediation technologies to meet formidable demands of polluting materials and situations. One such approach i

DOI: 10.4018/978-1-5225-1046-8.ch019

s phytoremediation – the use of plants and associated microbes for environmental cleanup. Cardinal points in favour of such an approach are – cost-effectiveness, natural energy-driven and minimal capital and running costs. Potency of Lemna spp. as a phytoremediation agent as well as its limitations haven dealt with in details in literature. Moreover, Lemna spp. are accepted and highly standardized as test organisms in aquatic ecotoxicology. The authors have shown earlier that Lemna spp. can play a vital role in combating pollution burden in a glaring local situation of river pollution. However, the recent trend in research indicates that Lemna spp (duckweeds at large) can play a critical and sensitive role in being an indicator of overt and subtle climate change. The following is an attempt to take a review of such recent trends.

INTRODUCTION

Duckweed in Bloom: .... heralds the return of a plant model for plant biology (Lam et al., 2014)

The authors would wish to thank profusely Lam et al. (2014), for the liberty to transform the title of their paper into a quote since it so well befits the purpose of this review!

Duckweeds are the simplest and smallest flowering plants that are ubiquitous and happily growing on fresh or polluted water. They have been curious objects of the intense research to understand the plant and their potentials – especially in genetics and biochemical repertoire. This is comparable to drosophila (fruit flies), bread moulds, *Escherichia coli* and *Saccharomyces* which have been proven workhorses. Small, fragile and free-floating duckweeds grow on mud or water depths up to 3 metres. Optimum nutrient concentrations support rapid growth.

Duckweeds are classified under family Lemnaceae which contains five genera, viz., *Lemna*, *Spirodela*, *Landoltia*, *Wolffia*, and *Wolfella*. All species under these genera are commonly designated as duckweeds. There are more than 40 species in these genera spread over the entire world. The genus *Lemna* contains nine species, viz., *L. aequinoctialis* (lesser duckweed); *L. gibba* (swollen duckweed); *L. minor* (common duckweed); *L. minuta* (least duckweed); *L. obscura* (little duckweed); *L. triscula* (star duckweed); *L. turionfera* (turion duckweed); and *L. valdiviana* (valdivia duckweed).

Duckweeds are manifested worldwide – tropical to cold regions but without extremes of cold, dryness or wetness. They prefer to inhabit small and shallow water bodies – e.g., ponds, pools, small lakes, ditches and wetlands. Such ecosystems occur world-wide and are full of biodiversity. It is necessary for the duckweeds to have nutrient supply in the water column as they are free-floating. Nutrient conditions are rich in small and shallow environments located in agricultural areas which tend to support dense mats. However, at times such dense growth may lead to several consequences such as interference in exchange of oxygen, restrict photosynthesis under the mat and create anoxic conditions to seriously affect macro-invertebrates and fishes. Freshwater ecosystems in arctic can be varied with respect to type, physical and chemical characteristics and associated biota. Also two major freshwater ecosystems exist – lotic (flowing water) and lentic (standing water). Large variations can exist in size, characteristics and location resulting into similar wider responses to climate change.

The morphology of duckweeds is simple. All members have flattened minute leaf like oval to round "fronds" *ca.* 1 mm to less than 1 cm across. Some members are endowed with root like structures which in open water either help to stabilize or assist to procure nutrients whenever these are in low concentrations.

Under ideal conditions of growth in terms of temperature, pH, incident light and nutrient concentrations these species can supersede most vigorous photosynthetic plants and are able to double biomass between 16 hrs to 2 days. To illustrate with example, if growth is unhindered and naturally exponential, a starting biomass covered in 10cm^2 can cover up to 1 hectare (100 million cm^2) in less than 50 days which is equivalent to 10 million fold increase in biomass. In 60 days this can extend to 32 ha. However, this is unrestricted growth. In nature restrictions are imposed by crowding, nutrient supply, light incidence, air and water temperatures and natural predators (fish, ducks, crustaceans and humans). Predators and microbial associations have intimate growth impacts on duckweeds – a point which will be highlighted in the subsequent discussion.

Interestingly, duckweeds exhibit a senescence-rejuvenation cycle much like in *Azolla*. This cyclic nature may introduce considerable errors of interpretation in studies that assess effect of differing nutrient sources over short time intervals *vis-a-vis* plant response.

The duckweed distribution is worldwide. From cold environments through temperate regions and luxurious growth in warmer places the species thrive well in nutrient rich sources. Some are able to tolerate sodium chloride up to 2.5% (w/v). The survival mechanism is manifested in converting themselves into turins which are able to sink to the bottom of water bodies and get embedded in mud due to heaviness by accumulated starch.

The reader is attracted to a recent but most pertinent meeting report of the 3rd International Duckweed Conference (Appenroth et.al., 2015), which categorically emphasizes that there is a resurgence of duckweed research and applications. It refers early publications and the predictions made therein on the possible diverse practical applications of duckweeds. These range from water purification to protein-rich biomass applications in animal feed. The authors lament on the intervening lack of application of economic significance of duckweeds. Further remarks by these authors – "there have been remarkable concerns about the changing environmental conditions (i.e. climate change) and food security and safety that have built up over the past four decades" have a direct bearing on the concept expounded in this review.

PHYSIOLOGY AND APPLICATIONS

Peculiar to the photosynthetic way of life duckweeds satisfy their nutritional requirements using minerals and solar energy to synthesize biomass. However, as a demonstration of mixotrophy the duck weeds can switch over to chemoorganotrophy without availing sunlight (ability to use sugars). This nutritional ability helps in handling effluents from sugar or starch industry. Since nitrogen, phosphorus and potassium (NPK) are the major requirements most of the current information available includes these. In addition information is available on their abilities to concentrate trace elements – particularly copper, cobalt and cadmium. Information is replete on this aspect to say that *Lemna* species with their exceptional capabilities for absorption and accumulation of heavy metals, radionucleotides and metalloids can surpass algae and aquatic macrophytes. *Lemna* are high accumulators of zinc, cadmium, selenium and copper and have the potential of accumulating uranium and arsenic after suitable attenuation. In general, two modes of trapping mechanism of metals by plants and microorganisms are known viz. biosorption and bioaccumulation. In biosorption, the metals ions are trapped outside the cells or biomass because of their surface charges (Gaddi & Patil, 2011; Itankar & Patil, 2015), while in case of bioaccumulation (also called metabolic mediated incorporation) the metal ions are trapped inside the cells and is considered as a more permanent sink of pollutants (Mkandawire et.al., 2003; Patil & Rao, 2013).

With their ability to concentrate macro and micro minerals from aqueous environment this aspect can be converted to an advantage by their application as animal feed supplement. Whereas accumulation of toxic metals may prove detrimental if biomass were to be used as animal feed.

A wide range of temperature (6-33°C) is comfortable for growth with an upper limit of 30°C when growth slows. Exposure to direct sunlight causes stress. Similarly, a wide range of pH (5.0-9.0) is comfortable, optimum being 6.5-7.5. This optimum range supports good absorption of nitrogen as a large concentration of ammonium ion exists. The plant having absorbed ammonium ion is able to convert the environment to a reduced-pH status which is a beneficial biochemical process to reduce ammonia release into the atmosphere. This has relevance to global warming. Depth of plant growth has an important bearing to growth. Duckweed biomass augmentation is possible by nutritional supplementations. Similarly, starch accumulation capacity can be enhanced by certain additives. These aspects have commercial relevance. Appenroth *et.al.* (2015) report even mentions of *L. gibba* being under the control of biological clock. This will help detect fast growing and slow growing periods and throw light on cell-to-cell communication in terms of their internal clock (Appenroth, 2015).

Duckweeds traditionally find their applications in treating wastewater especially for phosphorus removal. Use of duckweeds and water hyacinth is popular. Such engineered systems in the current context are termed innovative eco-friendly and efficient for handling hazardous wastes from domestic wastewater and sewage. Patel & Kanungo (2010) present a lucid review of such systems with reference to removal of pollutants from domestic wastewater. Attention is drawn to a similar review published earlier (Korner *et.al.*, 2003). These authors indicated that different duckweed species manifest varied efficiencies with different wastewaters. The fat and great duckweed (*Lemna gibba* and *Spirodela polyrhiza*) are shown as the most promising candidates. The toxicity of total ammonia (ammonium ion – NH_4 and un-ionized ammonia – NH_3) play an important role in relation to pH. Pathogen removals, penetrability of UV, dissolved oxygen changes are again important in treatment designs. Also, the duckweed-deployed systems vary in their efficiency of Biological Oxygen Demand (BOD) and Chemical Oxygen Demand (COD) removal ranging between 50 – 95% which is an earlier observation. It is interesting to note an observation of Brix & Schierup (1989) that due to absence of extensive root systems (which thereby decreases the surface for bacterial attachment) the duckweed systems do lag behind water hyacinth (*Eichhornia*).

At Pune we have recorded significant reductions in BOD and COD of domestic wastewater using pure (*Lemna minor*) and mixed (*L. minor* and *Spirodela*) cultures of duckweeds.

A good amount of work has been done with relevance to nutritional value of Lemnaceae in aquaculture and livestock production. Duckweeds have the potential as a feed for pigs (Gwaze & Mwale, 2015) which take them into the arena of plant proteins. It is considered that the amino acid composition of *L. gibba* is comparable to animal protein rather than the plant protein due to high contents of lysine and methionine. This is the reason why the World Bank and FAO recommend use of *Lemna* spp. as a feed material for fish, poultry and cattle. *L. gibba* is able to inhibit the growth of unicellular and blue-green algae. Further interesting observations indicate that certain *L. minor* compounds have anti-mosquito-larval activity and the larvae of *Aedes aegypti*. It is indicated that due to the presence of tannins and phytic acid the duckweed meal can prevent fungal spoilage of feed material. Also, duckweeds can be a feedstock for bio-ethanol production. Duckweeds find a place as a vegetable in the meals of certain populations of the world. It is suggested that duckweeds can be a source of insulin, pharmaceuticals and high value compounds.

SENSITIVITY TO ENVIRONMENT

The nature and severity of climate, weather and relevant influences have a strong influence on freshwater ecosystems that include duckweeds. Important environmental parameters that modulate climate change include temperature extremes, temperature variability and precipitation, solar UV radiation, CO_2 and O_3 levels, just to name a few. Differences in response to climate change can be expected even on the basis of lotic (free flowing water) and lentic (standing water) on which the duckweeds thrive reflecting in variations in size, characteristics and location (e.g., shallow lakes and wetlands).

Of the many approaches to predict ecological effects of climate change, ecosystem level experiments are desirable but cannot be undertaken due to high costs. Reports on measurement of responses to temperature have been replete. Van der Heide *et.al.* (2006) report that duckweeds follow the general temperature response as observed by Landolt (1986) and Landolt & Kandeler (1987) but are able to tolerate temperatures higher than 33°C. However, Van der Heide *et.al.* (2006) suggest that the information on minimum growth temperature may prove useful in predicting climate related invasions of new habitats based on the earlier studies of Walther *et.al.* (2002) which revealed that global warming results in shifts in plant communities with the result that warm-weather species tend to move to higher latitudes.

Elegant experiments of Novich (2012) to examine temperature fluctuations on population dynamics of *L. minor* are noteworthy. The broad attempt was to see the consequences of environmental variation to gain an insight on how population dynamics might be affected by climate change. Although the author was unable to make accurate predictions the plausible explanation was that the fluctuations were too rapid for the duckweed to respond. Another reason ascribed was the ability of temperature fluctuations in air and water to influence duckweed growth. Moreover, the growth rates at cooler sine wave functions were not significant due to the simple fact that metabolic processes are slower at cooler temperatures than those at warmer temperatures. A previous work from the same University (University of Nebraska, Waterman, 2010) has investigated temperature effects on duckweed-herbivore mesocosm. The bottomline is that community dynamics presents a playfield to assess effect of temperature which is associated with climate change. The duckweed partners can be varied.

In another study on shallow lake mesocosms heating was associated towards increased total PVI (Proportion of water Volume Infested) in which leminds (*L. minor*) dominated. On the other hand heating decreased the PVI of *L. trisulca* (Feuchtmayr *et.al.*, 2009). These authors observed that warming and nutrients favoured floating populations exploiting simultaneously availability of light at water surface. The authors further conjectured that the growth was enhanced due to warming-induced Soluble Reactive Phosphorus (SRP) and tend to suppress subsurface phytoplankton (*L.trisulca*). However, shade-tolerant plant systems may continue to persist under *Lemna* covers. The authors suggest that eutrophication is exacerbated by warming. Deoxygeneation, increased SRP, reduced fish biomass, reduced species richness (diversity) and increase in warm water exotics lend a helping hand.

Dissolved Organic Carbon (DOC) concentration gets altered due to climate change and thereby UV transparency in fresh waters. Low DOC ($\leq$ 2mg/L) concentrations favour exponential increase in UV penetration. *Lemna* spp. show strong reactions to simulated solar radiation with significant increase in Reactive Oxygen species (ROS) responses. Babu *et.al.* (2003) report that UV-induced stress response was enhanced by exposure to copper. Copper *per se* can activate the ROS pathway. An excellent coverage on effects of solar UV radiation on aquatic ecosystems and interactions with climate change reports effects on various species including bacterioplankton, viruses and cyanobacteria (Hader *et.al.* 2007). In an earlier study by Germ & Gaberscik (1999) the authors concluded that enhanced UV-B radiation substantially reduced vegetative reproduction in duckweeds (*L. minor*). When treated at 26 kJm^{-2} UV-B radiation caused growth parameters significantly lower than in the control group with no relevance to the nutrient level.

Similarly, carbon dioxide (CO_2) and ozone (O_3) effects on plant growth and physiology have been presented in numerous studies. Bailey *et.al.* (1999) in his research paper covers earlier such observations and also reveals interactive effects of elevated ozone plus carbon dioxide on duckweeds. These investigators state that their outcomes were pro and contra to earlier studies but overall observed an amelioration of negative O_3 effects in CO_2-enriched air of *L. minor* and *S. polyrhiza*. Winner (1994) states that a number of environmental and cultural factors influence plant response to ozone damage which is often higher in well-aerated, fertilized and fast growing plants. For example, variations in photosynthesis of *L. minor* have relevance to temperature and light intensity (Filbin & Hough, 1985). In an interesting study working with *L. minor* Schenk (1999) demonstrated only a slight proportional decrease in stomatal density with increased CO_2 concentrations.

The idea of presenting foregoing information was to show that climate change induced impact parameters have been under thorough study. This information manifests that duckweeds are sensitive to environmental changes associated with climate change.

CANDIDACY FOR THE MODEL SYSTEM TO ASSESS CLIMATE CHANGE IMPACT

Duckweed is the smallest and fastest-growing angiosperm in the world (Zhao *et.al.* 2012). The plant reproduces predominantly asexually thus deploying its metabolism to vegetative growth (Landolt 1986; Landolt & Kandeler 1987). Recent reports indicate that of all the duckweeds *Wolffia microscopic* has the fastest growth rate. The plant structure offers distinct scientific and engineering advantages (Mkandawire & Dudel, 2007). Consequently the plant offers itself as the simplest plant species that can be handled and manipulated at will. Moreover, like other model systems this plant represents higher plants. The plant supersedes algae and other aquatic macrophytes in its remarkable ability for the uptake and accumulation of heavy metals, radionuclides and metalloids. The plant biomass has large expanse of cell surface for biosorption and also manifests extensive metabolic incorporation. Floating duckweeds can manifest any effects of climate warming on the phenology which is quite visible. Duckweed biomass is considered as a good indicator of low water quality and as such dominates smaller and sheltered ecosystems in practically all climatic conditions. It is indicated in published literature that any climate change scenario leads to an early start of duckweed dominance.

Mkandawire & Dudel (2007) lucidly presented some of the cardinal properties of this plant such as:

1. *Lemna* spp. present variety of mechanisms for sequestration and compartmentalisation of contaminants.
2. *Lemna* spp. possesses various strategies for synthesising organic compounds.
3. The plant demonstrates phenotypic plasticity.
4. The plant manifests hormesis which is the plant strategy to maintain homeostasis.
5. The plant by virtue of releasing low molecular weight organic acids is capable of forming complexes with metals which through various processes help mitigate the contaminants.
6. The plant offers exciting opportunities to investigate role of allozymism for its survival.

7. A whole new arena of *Lemna* associated surface microflora opens up possibilities of investigating symbiogenetics which is a universal paradigm for the interspecies genetic integration. Microbes like *Sinorhizobium* and *Acinetobacter* are recent glaring examples.

Appenroth *et. al.* (2015) recent report is a pointer towards above possibilities giving details on the current genomics of *Lemna* and *Spirodela*.

The plant offers unique opportunities to study metabolic shifts – either to starch or to protein accumulation – mainly in response to environmental factors (phosphate and nitrate limitation, salinity). *Lemna* spp. are the highly standardized test organisms in aquatic ecotoxicology. The recent reports indicate a stiff competition to this status by *Spirodela polyrhiza* – offering low cost alternative to *Lemna* bioassays. The potential of this smallest and fastest growing angiosperm has recently been taken to the level of genome sequencing.

CONCLUSION

Duckweeds offer a novel model plant system comparable to *Escherichia* and *Saccharomyces* in the unicellular world to investigate and indicate not only climate change induced phenomena but also unique biochemical mechanisms and processes. Duckweeds are expected to reflect global climate change in terms of species abundance, distribution, life cycle timing, growth, physiology and productivity. Studies on consequences of environmental changes on individual species of duckweed (say *L. minor*) may provide an insight on how population dynamics may be affected by climate change. Sewage and toxic/hazardous wastes emanated from urban and industrial sources are capable of causing negative and irreversible impact on human health and ecology. Management of these wastes in a sustainable manner is therefore the key in the current era of climate change (Patil & Rao, 2015). Since the problems linked with physical-chemical technologies and management practices are numerous, researchers across the work are making efforts to employ live and dead biomass to mitigate such problems. In the light of this Duckweeds being a natural biomass have immense potential and thus could play a vital role in combating water pollution by way of reducing the organic burden.

REFERENCES

Appenroth, K. J., Sree, K. S., Fakhoorian, T., & Lam, E. (2015). Resurgence of duckweed research and applications: Report from the 3rd International Duckweed Conference. *Plant Molecular Biology*, *89*(6), 647–654. doi:10.1007/s11103-015-0396-9 PMID:26506824

Augsten, H. (1984). Lemnaceen: AspekteihrerPraxisrelevanz. *BiolRundschau*, *22*, 225–235.

Babu, T. S., Akhtar, T. A., Lampi, M. A., Tripuranthakam, S., Dixon, D. G., & Greenberg, B. M. (2003). Similar stress responses are elicited by copper and ultraviolet radiation in the aquatic plant *Lemna gibba*: Implication of reactive oxygen species as common signals. *Plant & Cell Physiology*, *44*(12), 1320–1329. doi:10.1093/pcp/pcg160 PMID:14701927

Bailey, S., Rebbeck, J., & Loats, K. V. (1999). Interactive effects of elevated ozone plus Carbon dioxide on Duckweeds exposed in open-top chambers. *The Ohio Journal of Science*, *99*(2), 19–25.

Brix, H., & Schierup, H. H. (1989). The use of aquatic macrophytes in water-pollution control. *Ambio. Stockholm*, *18*(2), 100–107.

Feuchtmayr, H., Moran, R., Hatton, K., Connor, L., Heyes, T., Moss, B., & Atkinson, D. et al. (2009). Global warming and eutrophication: Effects on water chemistry and autotrophic communities in experimental hypertrophic shallow lake mesocosms. *Journal of Applied Ecology*, *46*(3), 713–723. doi:10.1111/j.1365-2664.2009.01644.x PMID:19637615

Filbin, G. J., & Hough, R. A. (1985). Photosynthesis, photorespiration, and productivity in Lemna minor Ll. *Limnology and Oceanography*, *30*(2), 322–334. doi:10.4319/lo.1985.30.2.0322

Gaddi, S. S., & Patil, Y. B. (2011). Screening of some low-cost waste biomaterials for the sorption of silver-cyanide $[Ag(CN)^{2-}]$ from aqueous solutions. *International Journal of Chemical Sciences*, *9*(3), 1063–1072.

Germ, M., & Gaberscik, A. (1999). The Effect of UV-B radiation and nutrient availability on growth and photochemical efficiency of PS II in common duckweed. *Phyton*, *39*(3), 187–192.

Gwaze, F. R., & Mwale, M. (2015). The Prospect of Duckweed in Pig Nutrition: A Review. *The Journal of Agricultural Science*, *7*(11), 189.

Häder, D. P., Kumar, H. D., Smith, R. C., & Worrest, R. C. (2007). Effects of solar UV radiation on aquatic ecosystems and interactions with climate change. *Photochemical & Photobiological Sciences*, *6*(3), 267–285. doi:10.1039/B700020K PMID:17344962

Hillman, W. S., & Culley, D. D. (1978). The Uses of Duckweed: The rapid growth, nutritional value, and high biomass productivity of these floating plants suggest their use in water treatment, as feed crops, and in energy-efficient farming. *American Scientist*, *66*(4), 442–451.

Itankar, N., & Patil, Y. (2015). Biosorptive removal of hexavalent chromium from aqueous industrial solutions by a novel wood apple shell powder biomass. *Advanced Materials Research*, *1130*, 689–692. doi:10.4028/www.scientific.net/AMR.1130.689

Körner, S., Vermaat, J. E., & Veenstra, S. (2003). The capacity of duckweed to treat wastewater. *Journal of Environmental Quality*, *32*(5), 1583–1590. doi:10.2134/jeq2003.1583 PMID:14535298

Lam, E., Appenroth, K. J., Michael, T., Mori, K., & Fakhoorian, T. (2014). Duckweed in bloom: The 2nd International Conference on Duckweed Research and Applications heralds the return of a plant model for plant biology. *Plant Molecular Biology*, *84*(6), 737–742. doi:10.1007/s11103-013-0162-9 PMID:24398764

Landolt, E. (1986). The family of Lemnaceae - a monographic study, Vol. 1: Veroffentlichungen des Geobotanischen Institutes der Eidgenossischen Technische Hochschule, StiftungRubel. StiftungRubel.

Landolt, E., &Kandeler, R. (1987). Biosystematic investigations in the family of duckweeds (Lemnaceae) (vol. 4). *The family of Lemnaceae-a monographic study, 2*, 211-34.

Mkandawire, M., & Dudel, E. G. (2007). Are Lemna spp. effective phytoremediation agents. *Bioremediation. Biodiversity and Bioavailability*, *1*(1), 56–71.

Mkandawire, M., Taubert, B., & Dudel, E. G. (2004). Capacity of *Lemna gibba* L.(Duckweed) for uranium and arsenic phytoremediation in mine tailing waters. *International Journal of Phytoremediation*, *6*(4), 347–362. doi:10.1080/16226510490888884 PMID:15696706

Novich, E. (2012). *Predicting Lemna minor growth rate response to temperature fluctuations.* B.Sc Thesis in Environmental Studies Programme, University of Nebraska, Lincoln, NE. Retrieved from http://digitalcommons.unl.edu/envstudtheses/93/

Patel, D. K., & Kanungo, V. K. (2010). Phytoremediation Potential Of Duckweed (Lemnaminor L: A Tiny Aquatic Plant). Bioscan, 5(3), 355-358.

Patil, Y., & Rao, P. (2013). *Applied Bioremediation–Active and Passive Approaches*. Intech Open Science Online Publishers. doi:10.5772/50859

Patil, Y., & Rao, P. (2015). Industrial waste management in the era of climate change - A smart sustainable model based on utilization of active and passive biomass. In W. L. Filho (Ed.), *Handbook on Climate Change Adaptation* (pp. 2079–2092). Springer-Verlag Berlin Heidelberg. doi:10.1007/978-3-642-38670-1_49

Schenk, A. K. (n.d.). The Effects of Various Carbon Dioxide Concentrations on Lemna minor. *Dens, 130*(5), 172-1.

Tikhonovich, I. A., & Provorov, N. A. (2009). From plant–microbe interactions to symbiogenetics: A universal paradigm for the interspecies genetic integration. *Annals of Applied Biology*, *154*(3), 341–350. doi:10.1111/j.1744-7348.2008.00306.x

Van der Heide, T., Roijackers, R. M., Van Nes, E. H., & Peeters, E. T. (2006). A simple equation for describing the temperature dependent growth of free-floating macrophytes. *Aquatic Botany*, *84*(2), 171–175. doi:10.1016/j.aquabot.2005.09.004

Walther, G. R., Post, E., Convey, P., Menzel, A., Parmesan, C., Beebee, T. J., & Bairlein, F. (2002). Ecological responses to recent climate change. *Nature*, *416*(6879), 389–395. doi:10.1038/416389a PMID:11919621

Winner, W. E. (1994). Mechanistic analysis of plant responses to air pollution. *Ecological Applications*, *4*(4), 651–661. doi:10.2307/1941998

Zhao, H., Appenroth, K., Landesman, L., Salmeán, A. A., & Lam, E. (2012). Duckweed rising at Chengdu: Summary of the 1st International Conference on Duckweed Application and Research. *Plant Molecular Biology*, *78*(6), 627–632. doi:10.1007/s11103-012-9889-y PMID:22318677

Compilation of References

Abdolali, A., Ngo, H. H., Guo, W. S., Lee, D. J., Tung, K. L., & Wang, X. C. (2014). Development and evaluation of a new multi-metal binding biosorbent. *Bioresource Technology*, *160*, 98–106. doi:10.1016/j.biortech.2013.12.038 PMID:24405652

Abel, N., Gorddard, R., Harman, B., Leitch, A., Langridge, J., Ryan, A., & Heyenga, S. (2011). Sea level rise, coastal development and planned retreat: Analytical framework, governance principles and an Australian case study. *Environmental Science & Policy*, *14*(3), 279–288. doi:10.1016/j.envsci.2010.12.002

Acharya, G., & Barbier, E. B. (2000). Valuing groundwater recharge through agricultural production in the Hadejia-Nguru wetlands in northern Nigeria. *Agricultural Economics*, *22*(3), 247–259. doi:10.1111/j.1574-0862.2000.tb00073.x

Acreman, M., & Holden, J. (2013). How wetlands affect floods. *Wetlands*, *33*(5), 773–786. doi:10.1007/s13157-013-0473-2

Adams, Chen, McCarl, & Schimmelpfennig. (2001). Climate variability and climate change: Implications for agriculture. In The Long-Term Economics of Climate Change: Beyond a Doubling of Greenhouse Gas Concentrations. Emerald Group Publishing Limited.

Adger, W. N. (2003). Social aspects of adaptive capacity. *Climate change, adaptive capacity and development*, 29-49.

Adnan, A. (2015). *Renewable Energy in the Water, Energy and Food Nexus*. IRENA.

Afifi, A. A., Abd El-Rheem, & Youssef. (2011). Influence of Sewage Water Reuse Application on Soil and the Distribution of Heavy Metals. *Nature and Science*, *9*(4), 82–88.

Afzal, I., Basra, S. M. A., Cheema, M. A., Farooq, M., Jafar, M. Z., Shahid, M., & Yasmeen, A. (2013). Seed priming: A shotgun approach for alleviation of salt stress in wheat. *International Journal of Agriculture & Biology*, *15*, 1199–1203.

Agarwal, A., & Narain, S. (1997). *Dying Wisdom. Rise, Fall and Potential of India's Traditional Water Harvesting System.* Faridabad, India: Thomson Press Ltd.

Ageta, Y., & Kadota, T. (1992). Predictions of changes of glacier mass balance in the Nepal Himalaya and Tibetan Plateau: A case study of air temperature increase for three glaciers. *Annals of Glaciology*, *16*, 89–94.

Agrawal, A., & Perrin, N. (2009). Climate adaptation, local institutions and rural livelihoods. In *Adapting to climate change: thresholds, values, governance*, (pp. 350-367). Retrieved from agropedia.iitk.ac.in/node/4977

Ahmad, N. C., & Chaudhry, M. R. (1997). *Review of research on reclamation of salt-affected soils in Pakistan (Publication No. 175).* Lahore, Pakistan: International Waterlogging And Salinity Research Institute, Water And Power Development Authority.

Aiken, J., Moore, G. F., & Holligan, P. M. (1992). Remote sensing of oceanic biology in relation to global climate change. *Journal of Phycology*, *28*(5), 579–590. doi:10.1111/j.0022-3646.1992.00579.x

Ainsworth, E. A., & Rogers, A. (2007). The response of photosynthesis and stomatal conductance to rising (CO_2): Mechanisms and environmental interactions. *Plant, Cell & Environment*, *30*(3), 258–270. doi:10.1111/j.1365-3040.2007.01641.x PMID:17263773

Aiyappa, M. (2012, June 15). Cloud seeding a failed mission in Karnataka. *The Times of India*, p. 1.

Aizebeokhai, A. P. (2009). Global warming and climate change: Realities, uncertainties and measures. *International Journal of Physical Sciences*, *4*(13), 868–879.

Ajayi, O. C., Akinnifesi, F. K., Sileshi, G., & Kanjipite, W. (2009). Labour inputs and financial profitability of conventional and agroforestry-based soil fertility management practices in Zambia. *Agrekon*, (3), 276-292.

Akenji, L., & Bengtsson, M. (2009). *Is the customer really king? Stakeholder analysis for sustainable consumption and production using the example of the packaging value chain.* Retrieved from www.iges.com

Alam, M. S. (2012, January 11). *Tipaimukh Dam in India: A Death Trap for Banglades.* Retrieved from http://www.parisvisionnews.com/articles/1959-tipaimukh-dam-in-india-a-death-trap-for-bangladesh-dr-muhammad-shah-alam.html

Alderman, H., & Paxson, C. H. (1992). *Do the Poor Insure? A Synthesis of the Literature on Risk and Consumption in Developing Countries.* World Bank Policy Research Working Paper WPS 1008. Retrieved February 10, 2015, from https://www.princeton.edu/rpds/papers/Alderman_Paxson_Do_the_Poor_Insure_1994.pdf

Allan, R. P., Soden, B. J., John, V. O., Ingram, W., & Good, P. (2010). Current changes in tropical precipitation. *Environmental Research Letters*, *5*(2), 025205. doi:10.1088/1748-9326/5/2/025205

Amano, M., & Ness, B. (2004). *PET bottle system in Sweden and Japan: an integrated analysis from a life-cycle perspective.* Lund University.

Anand, E.V. (1988, June 11). Mismanagement of water resources. *The Indian Express*, p. 12.

Anand, E.V. (1992, December 11). Making CADP effective. *The Indian Express*, p. 12.

Anand, P. B. (2001). *Water scarcity in Chennai, India: Institutions, entitlements and aspects of inequality in access.* WIDER Discussion Papers // World Institute for Development Economics (UNU-WIDER), No. 2001/140. Retrieved February 15, 2015, from http://hdl.handle.net/10419/53004

Andrews-Speed, P., Bleischwitz, R., Boersma, T., Johnson, C., Kemp, G., & Vandeveer, S. D. (2015). *Want, waste or war? The global resource nexus and the struggle for land, energy, food, water and minerals.* New York, NY: Routledge.

Anonymous. (2007). *Coping with water scarcity challenge of the twenty - first century.* Retrieved February 05, 2015, from http://www.fao.org/nr/water/docs/escarcity.pdf

Antholis, W. J. (2014). *India's climate change policy in a Modi Government.* The Brookings Institution. Retrieved from http://www.brookings.edu/blogs/planetpolicy/posts/2014/07/30-india-climate-change

Anyinam, C. (1995). Ecology and ethnomedicine: Exploring links between current environmental crisis and indigenous medical practices. *Social Science & Medicine*, *40*(3), 321–329. doi:10.1016/0277-9536(94)E0098-D PMID:7899944

Apfelbacher, J. (2008). *Land Use/Cover Change in the Context of Global Change around the Pallikaranai Marshland.* (Unpublished bachelor's thesis). Department of Physical Geography, University of Freiburg, Germany.

Appasamy, P. (1995). *Water allocation and management in Madras. Report funded by Overseas Development Institute.* Chennai: Madras Institute of Development Studies.

Appenroth, K. J., Sree, K. S., Fakhoorian, T., & Lam, E. (2015). Resurgence of duckweed research and applications: Report from the 3rd International Duckweed Conference. *Plant Molecular Biology*, *89*(6), 647–654. doi:10.1007/s11103-015-0396-9 PMID:26506824

Armitage, D. (2005). Adaptive capacity and community-based natural resource management. *Environmental Management*, *35*(6), 703–715. doi:10.1007/s00267-004-0076-z PMID:15940398

Arnell, N. W. (1999). Climate change and global water resources. *Global Environmental Change*, *9*, S31–S49. doi:10.1016/S0959-3780(99)00017-5

Arnell, N. W. (2004). Climate change and global water resources: SRES emissions and socio-economic scenarios. *Global Environmental Change*, *14*(1), 31–52. doi:10.1016/j.gloenvcha.2003.10.006

Arnell, N. W., Halliday, S. J., Battarbee, R. W., Skeffington, R. A., & Wade, A. J. (2015). The implications of climate change for the water environment in England. *Progress in Physical Geography*, *39*(1), 93–120. doi:10.1177/0309133314560369

Arora, M., & Singh, R. D. (2007). Modeling of Melt Dominated Stream flow – A case study for a large Himalayan Basin. In Proceeding of Workshop: Glaciers in Watershed and Global Hydrology.

Ashraf, M., & Foolad, M. R. (2005). Pre-sowing seed treatment-a shotgun approach to improve germination growth and crop yield under saline and none-saline conditions. *Advances in Agronomy*, *88*, 223–271. doi:10.1016/S0065-2113(05)88006-X

Asian Development Bank. (2013, July5). *Climate Himalaya. In Outlook Report, Rudraprayag* (p. 1). Uttarakhand.

Askari, R. (2012, January). Tipaimukh Dam and Indian Hydro-politics. *Forum*, *6*(1). Retrieved from http://archive.thedailystar.net/forum/2012/January/tipaimukh.htm

Askew, A. J. (1987). Climate change and water resources. In S. I. Solomon, M. Beran, & W. Hogg (Eds.), *The influence of climate change and climatic variability on the hydrologic regime and water resources* (Vol. 168, pp. 421–340). Wallingford, UK: IAHS Publishers.

Atsimadja, F. A. (1992). The changing geography of Central Africa. In G. P. Chapman & K. M. Baker (Eds.), *The changing geography of Africa and Middle East* (pp. 52–79). London: Routledge.

Augsten, H. (1984). Lemnaceen: AspekteihrerPraxisrelevanz. *BiolRundschau*, *22*, 225–235.

Aveling, C., Bofaya, B., Hall, J.S., Hart, J.A., Hart, T.B., Inogwabini, B.I., Plumptre, A. & Wilkie, D. (2003). *Democratic Republic of Congo – Environmental Analysis*. Wildlife Conservation Society for USAID Kinshasa.

Ayers, R. S., & Westcott, D. W. (1985). *Water quality for agriculture. Irrigation and Drainage Paper 29*. Rome, Italy: Food and Agriculture Organization of the United Nations.

B'desh for Joint Study On Cross-Border Dam in India. (2012, December 9). *Press Trust of India*. Retrieved from http://www.hindustantimes.com/world/b-desh-for-joint-study-on-cross-border-dam-in-india/story-UZvr0kxlRcPWkMTBP-FiEgM.html

Babu, T. S., Akhtar, T. A., Lampi, M. A., Tripuranthakam, S., Dixon, D. G., & Greenberg, B. M. (2003). Similar stress responses are elicited by copper and ultraviolet radiation in the aquatic plant *Lemna gibba*: Implication of reactive oxygen species as common signals. *Plant & Cell Physiology*, *44*(12), 1320–1329. doi:10.1093/pcp/pcg160 PMID:14701927

Bagadian, B. U. (1987). *Farmers' Participation in Irrigation Management: The Philippine Experience*. Paper for Workshop on People's Participation in Irrigation Management. Administrative Staff College of India (ASCI), Hyderabad, India.

Bagliani, M., Dansero, E., & Puttilli, M. (2010). Territory and energy sustainability: The challenge of renewable energy sources. *Journal of Environmental Planning and Management*, *53*(4), 457–472. doi:10.1080/09640561003694336

Bahri, A. (2012). Integrated urban water management. *Global Water Partnership Technical Committee Background Papers*, (16). Retrieved February 13, 2015, from http://www.gwp.org/Global/The%20Challenge/Resource%20material/GWP_TEC16.pdf

Bailey, R. G. (1986). The Zaïre River system. In Davies & Walker (Eds.), The Ecology of River Systems. Junk Publishers. doi:10.1007/978-94-017-3290-1_6

Bailey, S., Rebbeck, J., & Loats, K. V. (1999). Interactive effects of elevated ozone plus Carbon dioxide on Duckweeds exposed in open-top chambers. *The Ohio Journal of Science*, *99*(2), 19–25.

Baker, B., & Viglizzo, E. F. (1998). Rangeland and Livestock. In J. F. Feenstra, I. Burton, J. B. Smith, & R. S. J. Tol (Eds.), *Handbook on methods for climate change impact Assessment and Adaptation Strategies*. Amsterdam: UNEP.

Baker, M. J. (1996). *Changing Contexts, Steady Flows: Patterns of Institutional Change within the Communal Irrigation Systems (Kuhls) of Kangra Valley* (Vol. 15). Himachal Pradesh, India: Himalayan Research Bulletin.

Balée, W. (1994). *Footprints of the forest: Ka'apor ethnobotany. The historical ecology of plant utilization by an Amazonian people.* Columbia University Press.

Bangladesh looks to resolve the Teesta water sharing with India. (2015, June 14). *The Economic Times*, p. 1.

Bangladeshi Scribes' Dam Mission Fails to Take off. (2012, June 8). *Times of India- Guwahati*, Retrieved from http://timesofindia.indiatimes.com/city/guwahati/Bangladeshi-scribes-dam-mission-fails-to-take-off/articleshow/13910971.cms?referral=PM

Barbier, E. B. (2004). Water and economic growth. *The Economic Record*, *80*(248), 1–16. doi:10.1111/j.1475-4932.2004.00121.x

Barlett, P. F. (Ed.). (2013). *Agricultural decision making: Anthropological contributions to rural development*. Academic Press.

Barnett, J. (2001). Adapting to Climate Change in Pacific Island Countries: The Problem of Uncertainty. *World Development*, *29*(6), 977–993. doi:10.1016/S0305-750X(01)00022-5

Barnett, T. P., Adam, J. C., & Lettenmaier, D. P. (2005). Potential impacts of a warming climate on water availability in snow-dominated regions. *Nature*, *438*(7066), 303–309. doi:10.1038/nature04141 PMID:16292301

Bates, B. C., Kundzewicz, Z. W., Wu, S., & Palutikof, J. P. (Eds.). (2008). *Climate Change and Water, Technical Paper VI of the Intergovernmental Panel on Climate Change*. Geneva: IPCC Secretariat.

Bates, B., Kundzewicz, Z. W., Wu, S., & Palutikof, J. P. (Eds.). (2008). *Climate change and water. Technical Paper of the Intergovernmental Panel on Climate Change, IPCC Secretariat, Geneva*. Geneva, Switzerland: IPCC.

BBC (British Broadcasting Company). (2013). *Democratic Republic of Congo Profile*. Available at http://www.bbc.co.uk/news/world-africa-13283212

Beck, U. (2002). The cosmopolitan society and its enemies. *Theory, Culture & Society*, *19*(1-2), 17–44. doi:10.1177/026327640201900101

Belden, A., Cole, P., Conte, H., deMooy, J., Jose, S., Karki, J., … Alleng, G. (2008). Integrated policy and planning for water and energy. Newark, DE: Center for Energy and Environmental Policy, University of Delaware.

Béné, C. (2005). The good, the bad and the ugly: Discourse, policy controversies, and the role of science in the politics of shrimp farming development. *Development Policy Review*, *23*(5), 585–614. doi:10.1111/j.1467-7679.2005.00304.x

Berger, I. E., & Corbin, R. M. (1992). Perceived consumer effectiveness and faith in others as moderators of environmentally responsible behaviors. *Public Policy and Advertising*, *11*(2), 79–89.

Berkes, F. (2012). *Sacred Ecology* (3rd ed.). New York: Routledge.

Berrittella, M., Rehdanz, K., & Tol, R. S. J. (2006). *The economic impact of the South-North water transfer project in China: A computable general equilibrium analysis.* FEEM Working Papers, No. 2006,154.

Berrittella, M., Hoekstra, A. Y., Rehdanz, K., Roson, R., & Tol, R. S. (2007). The economic impact of restricted water supply: A computable general equilibrium analysis. *Water Research*, *41*(8), 1799–1813. doi:10.1016/j.watres.2007.01.010 PMID:17343892

Berrittella, M., Hoekstra, Y., Rehdanz, K., Roson, R., & Tol, R. (2005). Virtual water trade in a general equilibrium analysis. In *GTAP Conference Paper* (Vol. 1715).

Berthier, E., Arnaud, Y., Kumar, R., Ahmad, S., Wagnon, P., & Chevallier, P. (2007). Remote sensing estimates of glacier mass balances in the Himachal Pradesh (Western Himalaya, India). *Remote Sensing of Environment*, *108*(3), 327–338. doi:10.1016/j.rse.2006.11.017

Bhatnagar, A. K., & Koul, M. (2004). Impact of Climate Change on medicinal plants- an assessment. In S. K. Dash & P. Rao (Eds.), *Assessment of Climate Change in India and Mitigation Policies* (p. 264). WWF India Publication.

Bhattacharya, A. (2015, March 10). 50-50 chances of El Nino this year: Met agencies. *The Times of India*, p. 1.

Bhattacharya, P. (2012, August 29). Tipaimukh Dam: Dhaka, Delhi Finalise Joint Study Terms. *The Daily Star.* Retrieved from http://www.thedailystar.net/newDesign/news-details.php?nid=247472

Bhutiyani, M. R., Kale, V. S., & Pawar, N. J. (2007). Long-term trends in maximum, minimum and mean annual air temperatures across the Northwestern Himalaya during the twentieth century. *Climatic Change*, *85*(1-2), 159–177. doi:10.1007/s10584-006-9196-1

Biggs, E., Bruce, E., Boruff, B., Duncan, J., Horsley, J., Pauli, N., & Imanariet, Y. et al. (2015). Sustainable development and the water–energy–food nexus: A perspective on livelihoods. *Environmental Science & Policy*, *54*, 389–397. doi:10.1016/j.envsci.2015.08.002

Bisht, M. (2012, April 25). *Tipaimukh Dam: Some Myths, Some Facts. Institute of Defense Studies and Analyses.* Retrieved July 10, 2016, from http://www.idsa.in/idsacomments/TipaimukhDam_mbisht_250412

Bittle, M. (2016). *Funds needed to clean Delaware's water*. Retrieved from http://delawarestatenews.net/government/funds-needed-to-clean-delawares-water/

Bizkova, L., Dimple, R., Darren, S., Venema, H., & MacCandless, M. (2013). *The Water–Energy–Food Security Nexus: Towards a Practical Planning and Decision-Support Framework for Landscape Investment and Risk Management. International Institute for Sustainable Development.* Winnipeg, Canada: IISD.

Blackden, C. M., & Wodon, Q. (Eds.). (2006). *Gender, Time Use, and Poverty in Sub-Saharan Africa*. World Bank Working Paper 73.

Blanco, H., & Lal, R. (2008). Soil and water conservation. Principles of soil conservation and management. Springer.

Boers, T. M., & Ben-Asher, J. (1982). A review of rainwater harvesting. *Agricultural Water Management*, *5*(2), 145–158. doi:10.1016/0378-3774(82)90003-8

Borthakur, S. (2009). Traditional rain water harvesting techniques and its applicability. *Indian Journal of Traditional Knowledge*, *8*(4), 525-530.

Bosello, F., Campagnolo, L., & Eboli, F. (2013). *Climate change and adaptation: the case of Nigerian agriculture.* Working Paper. Retrieved from https://www.econstor.eu/bitstream/10419/72963/1/74514442X.pdf

Bos, M. G. (1980). Irrigation efficiencies at crop production level. *International Commission on Irrigation and Drainage Bulletin*, *29*(2), 18–25.

Bounoua, L., Hall, F. G., Sellers, P. J., Kumar, A., Collatz, G. J., Tucker, C. J., & Imhoff, M. L. (2010). Quantifying the negative feedback of vegetation to greenhouse warming: A modeling approach. *Geophysical Research Letters*, *27*, L23701.

Branca, G., McCarthy, N., Lipper, L., & Jolejole, M. C. (2011). Climate-smart agriculture: a synthesis of empirical evidence of food security and mitigation benefits from improved cropland management. *Mitigation of Climate Change in Agriculture Series, 3.*

Brasuell, P.S.T.J. (2014, December 16). Saving the Colorado River Delta: One habitat at a time. *National Geographic.*

Brix, H., & Schierup, H. H. (1989). The use of aquatic macrophytes in water-pollution control. *Ambio. Stockholm*, *18*(2), 100–107.

Brklacich, M., Bryant, C., Veenhof, B., & Beauchesne, A. (2000). Agricultural Adaptation to Climatic Change: A comparative assessment of two types of farming in central Canada. In H. Millward, K. Beesley, B. Ilbery, & L. Harrington (Eds.), *Agriculturaland Environmental Sustainability in the New Countryside.* Winnipeg: Hignell Printing Limited.

Brodhag, C. (2010). A differentiated approach for sustainable consumption and production policies. *Natural Resources Forum*, *34*(1), 63–70. doi:10.1111/j.1477-8947.2010.01276.x

Brody, S. D., Highfield, W. E., Ryu, H. C., & Spanel-Weber, L. (2007). Examining the relationship between wetland alteration and watershed flooding in Texas and Florida. *Natural Hazards*, *40*(2), 413–428. doi:10.1007/s11069-006-9003-3

Brooks, N. (2003). *Vulnerability, risk and adaptation: A conceptual framework.* Tyndall Centre for Climate Change Research Working Paper 38.

Brown, D., Livermore, L., Bell, J., Estep, D., Gropp, W., Hendrickson, B., … Wright, M. (2008). *Using models of complex systems to inform policy makers.* US Department of Energy.

Brown, C., & Lall, U. (2009). Water and economic development: The role of variability and a framework for resilience. *Natural Resources Forum*, *30*(4), 306–317. doi:10.1111/j.1477-8947.2006.00118.x

Bruinsma, J. (2009, June). The resource outlook to 2050. In Expert meeting on how to feed the world in (Vol. 2050, pp. 1-33).

Buhril, D. (2009, February 2). Tipaimukh Dam: Development or Destruction – I. *E-PAO.* Retrieved from http://www.e-pao.net/epSubPageExtractor.asp?src=leisure.EI.David_Buhril.Tipaimukh_Dam_Development_or_Destruction_1

Bullock, A., & Acreman, M. (2003). The role of wetlands in the hydrological cycle. *Hydrology and Earth System Sciences Discussions*, *7*(3), 358–389. doi:10.5194/hess-7-358-2003

Bultot, F., & Griffiths, J. P. (1972). The equatorial wet zone. In Griffiths (Ed.), Climates of Africa. Elsevier Publishing Company.

Burke, M., & Emerick, K. (2012). *Adaptation to climate change: Evidence from US agriculture.* Academic Press.

Burton, I., Malone, E., & Huq, S. (2005). Adaptation policy frameworks for climate change: developing strategies, policies, and measures. Cambridge, UK: Cambridge University Press.

Burton, I. (2001). *Vulnerability and Adaptation to Climate Change in the Drylands*. Washington, DC: The Global Drylands Partnership.

Byrne, J., Taminiau, J., Kim, K. N., Seo, J., & Lee, J. (2015). A solar city strategy applied to six municipalities: Integrating market, finance, and policy factors for infrastructure-scale photovoltaic development in Amsterdam, London, Munich, New York, Seoul, and Tokyo.Wiley Interdisciplinary Reviews: Energy and Environment, 4-5.

CA (Comprehensive Assessment of Water Management in Agriculture). (2007). *Water for food, water for life: A comprehensive assessment of water management in agriculture*. Sri Lanka: IWMI.

California State Water Resources Control Board. (1990). *California Municipal Wastewater Reclamation in 1987*. Office of Water Recycling.

Calzadilla, A., Rehdanz, K., & Tol, R. S. (2008). *Water scarcity and the impact of improved irrigation management: A CGE analysis* (No. 1436). Kiel Working Paper.

Calzadilla, A., Rehdanz, K., & Tol, R. S. (2010). The economic impact of more sustainable water use in agriculture: A computable general equilibrium analysis. *Journal of Hydrology (Amsterdam)*, *384*(3), 292–305. doi:10.1016/j.jhydrol.2009.12.012

Candelaria, A. P. (2012). Mitigating Climate Risks Utilizing Good Practices Options in Upland Ecosystems of Bicol Region, Philippines. *OIDA International Journal of Sustainable Development*, *4*(03), 97–108.

Cao, L., Bala, G., Caldeira, K., Nemani, R., & Ban-Weiss, G. (2009). Climate response to physiological forcing of carbon dioxide simulated by the coupled Community Atmosphere Model (CAM3.1) and Community Land Model (CLM3.0). *Geophysical Research Letters*, *36*(10), L10402. doi:10.1029/2009GL037724

Cao, L., Bala, G., Caldeira, K., Nemani, R., & Ban-Weiss, G. (2010). Importance of carbon dioxide physiological forcing to future climate change. *Proceedings of the National Academy of Sciences of the United States of America*, *107*(21), 9513–9518. doi:10.1073/pnas.0913000107 PMID:20445083

Care Earth Trust. (2012). *Identification of wetlands for protection, restoration and conservation in Greater Chennai. Report submitted to the Chennai Metropolitan Development Authority*. Chennai, India: Government of Tamil Nadu.

Care Earth. (2002). Conservation of Urban Wetlands- Pallikaranai Marsh, A Baseline study. Report supported by Tamil Nadu Pollution Control Board.

Care Earth. (2005). Current Status of the Pallikaranai Marsh and Recommendations for Protection and Restoration. Report submitted to the Tamil Nadu Pollution Control Board.

Cellura, M., Longo, S., & Mistretta, M. (2011). The energy and environmental impacts of Italian household consumptions: An input-output approach. *Renewable & Sustainable Energy Reviews*, *15*(8), 3897–3908. doi:10.1016/j.rser.2011.07.025

Central Ground Water Board. (2011). *Select Case Studies Rain Water Harvesting and Artificial Recharge*. New Delhi: Ministry of Water Resources.

Central Pollution Control Board, (2005). *A report on Status of Water Supply, Wastewater Generation and Treatment in Class-I Cities & Class-Ii Towns of India*. Author.

Central Water and Power Research Station, Specific Note No. 3035. (1993). *Engineering Properties of rock material. Koyna Hydro Electric Project (Stage-IV)*. Pune, India: Author.

Cerri, C. E. P., Sparovek, G., Bernoux, M., Easterling, W. E., Melillo, J. M., & Cerri, C. C. (2007). Tropical agriculture and global warming: Impacts and mitigation options. *Scientia Agricola*, *64*(1), 83–99. doi:10.1590/S0103-90162007000100013

Chafetz, J. S. (2006). The varieties of gender theory in sociology. In J. S. Chafetz (Ed.), *Handbook of the sociology of gender* (pp. 3–23). New York: Springer. doi:10.1007/0-387-36218-5_1

Chafetz, J. S., & Haggan, J. (1996). The gender division of labor and family change in industrial societies: A theoretical accounting. *Journal of Comparative Family Studies*, *27*, 187–217.

Chalise, S. R. (2000). *Water resources management in the HKH: an overview. Water of Life: Perspectives of Water Harvesting in the Hindu Kush-Himalayas*. Kathmandu, Nepal: ICIMOD.

Chamornmarn, W. (2008, June). *Bottled water: External environmental analysis*. Retrieved from www.eiolca.net/LCA-experts/index.html

Chand, R. (1994). Role of Water Rights in Farmer-Managed Hill Irrigation Systems. *Economic and Political Weekly*, *29*(13), 26–30.

Chandramohan, B. P., & Bharati, D. (2009). Role of Public Governance in the Conservation of Urban Wetland System: A Study of Pallikaranai Marsh. In *Proceedings of the Indian Society for Ecological Economics (INSEE), 5th Biennial Conference, Ahmedabad, India. Customized Rainfall Information System (CRIS)*. Hydromet Division, India Meteorological Department. Retrieved February 16, 2015 from http://hydro.imd.gov.in/

Changshun, Z., Chunlan, L., & Na, L. (2015). Mechanisms for Realizing the Economic Values of Ecosystem Services. *Journal of Resources and Ecology*, *6*(6), 412–419. doi:10.5814/j.issn.1674-764x.2015.06.009

Chapman, L. J. (2001). Fishes of African Rain Forests – diverse adaptations to environmental challenges. In Weber, White, Vedder & Naughton-Treves (Eds.), African Rain Forest, Ecology and Conservation: An interdisciplinary perspective. Yale University Press.

Chettri, N., Sharma, E., Shakya, B., Thapa, R., Bajracharya, B., Uddin, K., . . . Choudhury, D. (2010). *Biodiversity in the Eastern Himalayas: Status, Trends and Vulnerability to Environmental Change; Environmental Change Impact and Vulnerability in the Eastern Himalayas*. Technical report 2. Kathmandu: ICIMOD.

Chiotti, Q., Johnston, T., Smit, B., & Ebel, B. (1997). Agricultural Response to Climate Change: A preliminary Investigation of Farm-level Adaptation in Southern Alberta. In B. Ilbery, Q. Chiotti, & T. Rickard (Eds.), *Agricultural Restructuring and Sustainability: A Geographical Perspective*. CAB International.

Chiu, Y.-T. (2013). Typhoon Haiyan: Philippines faces long road to recovery. *Lancet*, *382*(9906), 1691–1692. doi:10.1016/S0140-6736(13)62380-6 PMID:24278979

Cho, A. (2015, January 30). Political will is the biggest barrier to tackling global water risk. *The Guardian*.

Choube, J. (2015, March 3). Unseasonal rains damage crops. *Down-to-earth*.

Chou, C., Neelin, J. D., Chen, C. A., & Tu, J. Y. (2009). Evaluating the rich-get-richer mechanism in tropical precipitation change under global warming. *Journal of Climate*, *22*(8), 1982–2005. doi:10.1175/2008JCLI2471.1

Chou, C., Tu, J. Y., & Tan, P. H. (2007). Asymmetry of tropical precipitation change under global warming. *Geophysical Research Letters*, *34*(17), L17708. doi:10.1029/2007GL030327

Choudhury, A. (2012, January 9). Is Bangladesh doing enough about Tipaimukh? *bdnews24.com*. Retrieved from http://opinion.bdnews24.com/2012/01/09/is-bangladesh-doing-enough-about-tipaimukh/

Christensen, J. H., Krishna Kumar, K., Aldrian, E., An, S. I., Cavalcanti, I. F. A., de Castro, M., & Zhou, T. et al. (2013). Climate Phenomena and their Relevance for Future Regional Climate Change. In T. F. Stocker, D. Qin, G. K. Plattner, M. Tignor, S. K. Allen, J. Boschung, & P. M. Midgley et al. (Eds.), *Climate Change 2013: The Physical Science Basis. Contribution of Working Group I to the Fifth Assessment Report of the Intergovernmental Panel on Climate Change.* Cambridge, UK: Cambridge University Press.

Cicin, S.B. (1993). Sustainable development and integrated coastal management. *Ocean and Coastal Management, 21,* 11-43.

Cifuentes, E., Gomez, M., Blumenthal, U., Tellez-Rojo, M. M., Romieu, I., Ruiz-Palacios, G., & Ruiz-Velazco, S. (2000). Risk factors for Giardia intestinalis infection in agricultural villages practicing wastewater irrigation in Mexico. *The American Journal of Tropical Medicine and Hygiene*, *62*(3), 388–392. PMID:11037783

Clark, J.R. (1977). *Coastal ecosystem management.* John Wiley and Sons.

Clark, C. E., Horner, R. M., & Harto, C. B. (2013). Life cycle water consumption for shale gas and conventional natural gas. *Environmental Science & Technology*, *47*(20), 11829–11836. doi:10.1021/es4013855 PMID:24004382

Cody, B., & Carter, N. (2009). *35 years of water policy: The 1973 national water commission and present challenges.* Congressional Research Service Report.

Coelho, T. M., Castro, R., & Gobbo, J. A. Jr. (2010). PET containers in Brazil: Opportunities and challenges of a logistics model for post-consumer waste recycling. *Resources, Conservation and Recycling*, *55*(3), 291–299. doi:10.1016/j.resconrec.2010.10.010

Collins, D. N., & Hasnain, S. I. (1995). Runoff and sediment transport from the glacier zed basins at the Himalayan scale. In W R Osterkamp (Eds.), *Scale on Interpretation and Management of Sediment and Water Quality,Proc. Boulder Symposium.*

Colom, A., Bakanza, A., Mundeka, J., Hamza, T., & Ntumbandzondo, B. (2006). The socio-economic dimensions of the management of biological resources, in the Lac Télé – Lac TumbaLandscape, DRC Segment: A segment-wide baseline Socio-Economic study's Report. Submitted to the World Wide Fund for Nature DRC, Kinshasa Office, Democratic Republic of Congo.

Colombant, N. (2005). *Shrinking Lake Chad sparks tension, ingenuity.* Available at http://www.voanews.com /english/archive/2005-03

Conference on the Central African Moist Forest Ecosystems (CEFDHAC). (2001). *Regional Strategic Action Plan and biodiversity resources on the Congo basin ecosystems.* United Nations Development Program (UNDP), International Union for the Conservation of Nature (IUCN) and the Global Environment Facility (GEF).

COP21 (Conference of Parties 21). (2015). *United Nations Climate Change Conference.* Retrieved from http://www.scoop.it/t/cop21news

Cop21 Report. (2015). *The Sustainable Innovation Forum (SIF15) annual Conference of Parties (COP) Stade de France.* Retrieved on May 03,2016 from http://www.firstpost.com/world/climate-change-deal-clinched-at-cop21-heres-why-india-is-on-the-receiving-end-2544122.html

Correa, C. M. (2001). Traditional knowledge and intellectual property. Geneva: The Quaker United Nations Office (QUNO).

Corsi, F. (1984b). *Zaïre: Développement et aménagement des pêches du lac Mai-Ndombe et ses affluents.* Technical report submitted to Food Aid Organization.

Corsi, F. (1984a). *Développement des pêches dans le fleuve Zaïre. Rapport préparé pour le projet d'Assistance à l'Office National des Pêches et Amélioration de la Pêche Artisanale.* Rome: FAO.

Coward, E. W., & Gilbert, L. (1987). *Studies of Farmer-Managed Irrigation Systems'*, Background Paper for the Workshop on People's Participation in Irrigation Management, Administrative Staff College of India.

Coward, E. W. Jr. (1990). Property Rights and Network Order: The Case of Irrigation Works in Western Himalayas. *Human Organization, 49*(1), 78–88. doi:10.17730/humo.49.1.g53435062n67k2g2

Crowley, T. J. (2000). Causes of climate change over the past 1000 years. *Science, 289*(5477), 270–277. doi:10.1126/science.289.5477.270 PMID:10894770

Cruz, R.V., Harasawa, H., Lal, M., Wu, S., Anokhin, Y., Punsalmaa, B., … NinhHuu, N.(2007). *Asia Climate Change 2007: Impacts, Adaptation and Vulnerability.* Contribution of Working Group II to the Fourth Assessment Report of the Intergovernmental Panel on Climate Change. Cambridge University Press.

CSREES. (2005). *CREES Agricultural Water Security White Paper, Cooperative State Research, Education and Extension Service.* Washington, DC: CSREES.

Cundill, G., & Fabricius, C. (2010). Monitoring the governance dimension of natural resource co-management. *Ecology and Society, 15*(1), 15.

Damodaran, H. (2015, April 12). An El Nino year, maybe. A bad monsoon year, maybe not. *The Sunday Indian Express*, p. 15.

Darwin, R., Tsigas, M., Lewandrowski, J., & Raneses, A. (1995). *World agriculture and climate change: economic adaptations.* Agricultural Economic Report No.703. Natural Resources and Environmental Division, Economic Research Service. U.S. Department of Agriculture.

Das, K. (2006). *Drinking Water and Sanitation in Rural Maharashtra: A Review of Policy Initiatives.* Gujarat Institute of Development Research.

Dash, S. K., & Hunt, J. C. R. (2007). Variability of climate change in India. *Current Science, 93*(6), 782–788.

de Loë, R. C., Kreutzwiser, R., & Mararu, L. (1999). *Climate Change and the Canadian Water Sector: Impacts and Adaptation.* Guelph: Natural Resources Canada. Retrieved March 01, 2016, from http://www.uoguelph.ca/~c-ciarn/documents/World_Bank_Paper.pdf

De Saint Moulin, L. (1991). Zaïre: Aperçu démographique – Résultats du recensement scientifique de la population en 1984. *Zaire-Afrique, 255*, 227–261.

De Saint Moulin, L. (2003). La perception de la démocratie en République Démocratique du Congo. *Congo-Afrique, 375*, 292–307.

Deka, M. (2012). Mega Dams, Environment and History in North-East India. *Dialogue, 13*(4), 2012. Retrieved from http://www.asthabharati.org/Dia_Apr%20012/me.htm

Delaware State Senate. (2016). *An Act to Amend Title 26 of the Delaware Code Relating to Delaware's Renewable Energy Portfolio Standards and Delaware-Manufactured Fuel Cells.* Senate Bill No. 124. Retrieved from www.legis.delaware.gov/LIS/lis146.nsf/vwLegislation/SB+124/$file/legis.html

Department of Energy. (2014). *Water-energy nexus: Challenges and opportunities.* Washington, DC: U.S. Department of Energy (DOE).

Desanker, P. V. (2002). Impact of climate change on Africa. Center for African Development Solutions.

Dhavalikar, L. N. (2002). Sea level scenario on Vengurla coast of Maharashtra. Transactions, Institute of Indian Geographers, 24(1-2), 63-71.

Diao, X., & Roe, T. (2003). Can a water market avert the double-whammy of trade reform and lead to a win–win outcome? *Journal of Environmental Economics and Management*, *45*(3), 708–723. doi:10.1016/S0095-0696(02)00019-0

Dikshit, K.R. (1986). *Maharashtra in Maps*. Maharashtra State Board for Literature and Culture.

Dinar, A., Kemper, K., Blomquist, W. A., Diez, M., Sine, G., & Fru, W. (2005). *Decentralization of river basin management: a global analysis.* World Bank Policy Research Working Paper (3637).

Dinar, A. (2000). *The Political Economy of Water Pricing Reforms*. Washington, DC: World Bank.

Dinar, A., & Yaron, D. (1992). Adoption and abandonment of irrigation technologies. *Agricultural Economics*, *6*(4), 315–332. doi:10.1016/0169-5150(92)90008-M

Dixon, J. A., & Easter, K. W. (1991). Integrated watershed management: an approach to resource management. *Watershed resource management: Studies from Asia and the Pacific*, 3-15.

Dobhal, D. P., Gergan, J. T., & Thayyen, R. J. (2007). Mass balance and snout recession measurements (1991-2000) of Dokriani glacier, Garhwalhimalaya, India. Climatic and anthropogenic impacts on the variability of water resources. Technical Document in Hydrology No. 80. UNESCO.

Dobhal, D. P., Gergan, J. T., & Thayyen, R. J. (2004). Recession and morphogeometrical changes of Dokriani glacier (1962–1995) Garhwal Himalaya, India. *Current Science*, *86*(5), 692–696.

Döll, P. (2009). Vulnerability to the impact of climate change on renewable groundwater resources: A global-scale assessment. *Environmental Research Letters*, *4*(3), 035006. doi:10.1088/1748-9326/4/3/035006

Doneen, L. D. (1975). Water quality for irrigated agriculture. In A. Poljakoff-Mayber & J. Gale (Eds.), *Plants in saline environments* (pp. 56–79). Springer-Verlag. doi:10.1007/978-3-642-80929-3_5

Dossiers, A. (2010). Climate Change and Agriculture in India: Avenues for Mitigation. *India and Climate Change.* Retrieved January 25, 2016, from http://www.ritimo.org/Climate-Change-and-Agriculture-in-India-Avenues-for-Mitigation#

Drechsel, P., Qadir, M., & Wichelns, D. (2015). *Wastewater: Economic Asset in an Urbanizing World.* Springer. doi:10.1007/978-94-017-9545-6

Drescher, A., Glaser, R., Pfeiffer, C., Vencatesan, J., Schliermann-Kraus, E., Glaser, S., & Dostal, P. (2007). *Risk assessment of extreme precipitation in the coastal areas of Chennai as an element of catastrophe prevention. In 8. Forum DKKV/CEDIM: Disaster Reduction in Climate Change 15./16.10.2007*. Germany: Karlsruhe University.

Duda, H., & Chumi, S. (2008). *Economics of irrigation water management: A literature survey with focus on partial and general equilibrium models*. World Bank Policy Research Working Paper Series.

Duerden, F. (2004). Translating climate change impacts at the community level. *Arctic*, *57*(2), 204–212. doi:10.14430/arctic496

DuPont. (2011). *Sustainability best practice series: Life cycle thinking*. DuPont Packaging Graphics.

Dutta, P. (2010). India-Bangladesh Relations: Issues, Problems and Recent Developments. *Institute of Peace and Conflict Studies*. Retrieved July 10, 2016, from http://www.ipcs.org/pdf_file/issue/SR97.pdf

Dyurgerov, M. B., & Meier, M. F. (2005). *Glaciers and the changing earth system: a 2004 snapshot* (Vol. 58). Boulder, CO: Institute of Arctic and Alpine Research, University of Colorado.

Earth, F. (2013). Future Earth Initial Design. Paris: Future Earth Interim Secretariat. International Council for Science. Retrieved from http://www.futureearth. org/sites/default/files/Future-Earth-Design-Report_web.pdf

Eaton, F. M. (1950). Significance of carbonate in irrigation waters. *Soil Science*, *69*(2), 123–133. doi:10.1097/00010694-195002000-00004

Economic and Social Commission for Asia and the Pacific (ESCAP). (2011). *Tools for low carbon development in Thailand.* Paper presented at the First National Seminar on Green Growth Policy.

Edwards, A., & White, L. J. T. (2000). Methods for recording the weather. In L. White & A. Edwards (Eds.), *Conservation research in the African rain forests: a technical handbook* (pp. 81–88). New York: Wildlife Conservation Society.

Ehrenfeld, J. G. (2000). Evaluating wetlands within an urban context. *Ecological Engineering*, *15*(3), 253–265. doi:10.1016/S0925-8574(00)00080-X

El Nadi, A. H. (1974). The significance of leaf area in evapotranspiration. *Annals of Botany*, *38*(3), 607–611.

Elliott, J. A. (2006). *An introduction to sustainable development.* New York, NY: Routledge.

Eltahir, E., & Bras, R. (1993). On the response of the tropical atmosphere to large-scale deforestation. *Quarterly Journal of the Royal Meteorological Society*, *119*, 779–783.

Ensink, J. H. J., Simmons, R. W., & van der Hoek, W. (2004). Wastewater use in Pakistan: The cases of Haroonabad and Faisalabad. In C. A. Scott, N. I. Faruqui, & L. Raschid-sally (Eds.), *Wastewater use in irrigated agriculture* (pp. 91–99). Wallingford, UK: CAB International. doi:10.1079/9780851998237.0091

Ensor, J., & Berger, R. (2009). *Understanding climate change adaptation: lessons from community-based approaches.* Practical Action Publication. doi:10.3362/9781780440415

Environmental Information System (ENVIS) Centre on Hygiene, Sanitation, Sewage Treatment Systems and Technology. (n.d.). Retrieved from http://www.sulabhenvis.nic.in/Database/STST_wastewater_2090.aspx

Erda, L. (1996). Agricultural Vulnerability and Adaptation to Global Warming in China. *Water, Air, and Soil Pollution*, *92*(1-2), 63–73.

Eriksen, S., & Lind, J. (2009). Adaptation as a political process: Adjusting to drought and conflict in Kenyas drylands. *Environmental Management*, *43*(5), 817–835. doi:10.1007/s00267-008-9189-0 PMID:18726051

Eriksson, M., Jianchu, X., Shrestha, A. B., Vaidya, R. A., Nepal, S., & Sandström, K. (2009). *The changing Himalayas: impact of climate change on water resources and livelihoods in the greater Himalayas. International centre for integrated mountain development.* ICIMOD.

Erwin, K. L. (2009). Wetlands and global climate change: The role of wetland restoration in a changing world. *Wetlands Ecology and Management*, *17*(1), 71–84. doi:10.1007/s11273-008-9119-1

Euromonitor International. (2010). *Bottled water-Thailand. Country sector briefing.* Retrieved from http://www.euromonitor.com/bottled-water-in-thailand/report

European Report on Development (ERD). (2012). *Confronting Scarcity: Managing Water, Energy and Land for Inclusive and Sustainable Growth.* Brussels, Belgium: European Commission, Overseas Development Institute (ODI), European Centre for Development Policy Management (ECDPM), German Development Institute (GDI/DIE).

Evenson, R. (1997). Rice Varietal Improvement and International Exchange of Rice Germplasm. In P. L. Pingali & M. Hossain (Eds.), *Impact of Rice Research.* Manila: International Rice Research Institute.

Evrard. C. (1968). *Recherches écologiques sur le peuplement forestier des sols hydromorphes de la Cuvette Centrale congolaise.* Série scientifique N°110/Office National de la recherché Scientifique et du Développement – Ministère Belge de l'Education Nationale et de la Culture.

Failed Dams in Manipur? Any Lessons Learnt. (2014, September 3). *The Sangai Express.* Retrieved from http://www.thesangaiexpress.com/failed-dams-in-manipur-any-lessons-learnt/

FAO. (2011). The state of the world's land and water resources for food and agriculture (SOLAW) – managing systems at risk. Food and Agriculture Organization of the United Nations.

FAO. (2013). *FAO Statistical Yearbook 2013. World Food and Agriculture.* Rome: Food and Agriculture Organization of the United Nations.

FAO. (2016a). *Climate-Smart Agriculture.* Retrieved January 25, 2016, From http://www.fao.org/climate-smart-agriculture/en/

FAO. (2016b). *Irrigation Water Management: Irrigation water needs. Climate-Smart Agriculture.* Retrieved January 25, 2016, From http://www.fao.org/docrep/s2022e/s2022e07.htm

Federal Bureau of Statistics. (2011–12). *Pakistan statistics year book, Statistics Division.* Islamabad, Pakistan: Planning Commission Government of Pakistan.

Feenstra, S., Hussain, R., & van der Hoek, W. (2000). Prevalence of intestinal parasites in the Southern Punjab, Pakistan. IWMI Pakistan Report 101.

Fernández, M. H., & Vrba, E. S. (2006). Plio-Pleistocene climatic change in the Turkana Basin (East Africa): Evidence from large mammal faunas. *Journal of Human Evolution, 50*(6), 595–626. doi:10.1016/j.jhevol.2005.11.004 PMID:16630645

Feron, V. J., Jetten, J., De Kruijf, N., & Van Den Berg, F. (1994). Polyethylene terephthalate bottles (PRBs): A health and safety assessement. *Food Additives and Contaminants, 11*(5), 571–594. doi:10.1080/02652039409374258 PMID:7835471

Feuchtmayr, H., Moran, R., Hatton, K., Connor, L., Heyes, T., Moss, B., & Atkinson, D. et al. (2009). Global warming and eutrophication: Effects on water chemistry and autotrophic communities in experimental hypertrophic shallow lake mesocosms. *Journal of Applied Ecology, 46*(3), 713–723. doi:10.1111/j.1365-2664.2009.01644.x PMID:19637615

Few, R., Brown, K., & Tompkins, E. L. (2007). Public participation and climate change adaptation: Avoiding the illusion of inclusion. *Climate Policy, 7*(1), 46–59. doi:10.1080/14693062.2007.9685637

Fewtrell, L., Kaufmann, R. B., Kay, D., Enanoria, W., Haller, L., & Colford, J. M. Jr. (2005). Water, sanitation, and hygiene interventions to reduce diarrhoea in less developed countries: A systematic review and meta-analysis. *The Lancet Infectious Diseases, 5*(1), 42–52. doi:10.1016/S1473-3099(04)01253-8 PMID:15620560

Field, C. B., Jackson, R. B., & Mooney, H. A. (1995). Stomatal responses to increased CO_2-Implications from the plant to the global-scale. *Plant, Cell & Environment, 18*(10), 1214–1225. doi:10.1111/j.1365-3040.1995.tb00630.x

Filbin, G. J., & Hough, R. A. (1985). Photosynthesis, photorespiration, and productivity in Lemna minor Ll. *Limnology and Oceanography, 30*(2), 322–334. doi:10.4319/lo.1985.30.2.0322

Fipps, G. (2007). *Irrigation water quality standards and salinity management strategies.* Available at: http://lubbock.tamu.edu/irrigate/documents/2074410-B1667.pdf

Fischer, E. M., Seneviratne, S. I., Vidale, P. L., Luthi, D., & Schar, C. (2007). Soil moisture- atmosphere interactions during the 2003 European summer heat wave. *Journal of Climate, 20*(20), 5081–5099. doi:10.1175/JCLI4288.1

Fisher, A. C., Hanemann, W. M., Roberts, M. J., & Schlenker, W. (2012). The economic impacts of climate change: Evidence from agricultural output and random fluctuations in weather. *The American Economic Review*, *102*(7), 3749–3760. doi:10.1257/aer.102.7.3749

Fletcher, S. R. (1997). *Global Climate Change Treaty: Negotiations and Related Issues. A CRS Report to Congress.* National Council for Science and the Environment & National Library for the Environment. Available at http://www.ncseonline.org/NLE/CRSreports/Climate

Fletcher, S. R. (2000). *Global Climate Change Treaty: The Kyoto Protocol. A CRS Report to Congress.* National Council for Science and the Environment & National Library for the Environment. Available at http://www.ncseonline.org/NLE/CRSreports/Climate

Flor, A. G. (2004). *Environmental communication.* The Philippines: University of the Philippines.

Folke, C., Carpenter, S., Elmqvist, T., Gunderson, L., Holling, C. S., & Walker, B. (2002). Resilience and sustainable development: building adaptive capacity in a world of transformations. *AMBIO: A Journal of the Human Environment*, *31*(5), 437-440.

Foolmuan, R. K., & Ramjeeawan, T. (2012). Disposal of post-consumer polyethylene terephthalate (PET) bottles: Comparison of five disposal alternatives in the small island state of Mauritius using a life cycle assessment tool. *Environmental Technology*, *33*(17), 2007–2018. doi:10.1080/09593330.2012.655321 PMID:23240194

Foran, T. (2015). Node and regime: Interdisciplinary analysis of water–energy–food Nexus in the Mekong region. *Water Alternatives*, *8*(1), 655–674.

Foster, S., Limaye, S., Mandavkar, Y., & Msangi, S. (2009). *A Hydrogeologic and Socioeconomic Evaluation of Community-based Groundwater Resource Management–The Case of Hivre Bazaar in Maharashtra-India.* World Bank GW-MATE Case Profile Collection, 22.

Francois, L. E., Maas, E. V., Donovan, T. J., & Youngs, V. L. (1986). Effect of salinity on grain yield and quality, vegetative growth, and germination of semi-dwarf and durum wheat. *Agronomy Journal*, *78*(6), 1053–1058. doi:10.2134/agronj1986.00021962007800060023x

Franklin Associates Ltd (FAL). (2010a). *Life cycle inventory of 100% postconsumer HDPE and PET recycled resin from postconsumer containers and packing.* Prairie Village, KS: Franklin Associates, A Division of ERG.

Franklin Associates Ltd (FAL). (2010b). *Cradle-to-gate life cycle inventory of nine plastic resins and two polyurethane precursors.* Prairie Village, KS: Franklin Associates, A Division of ERG.

Frederick, K. D., & Major, D. C. (1997). Climate change and water resources. *Climatic Change*, *37*(1), 7–23. doi:10.1023/A:1005336924908

Frison, E. A., Cherfas, J., Eyzauirre, P. B., & Johns, T. (2004). *Biodiversity, nutrition and health: Making a difference to hunger and conservation in the developing world.* Keynote Address to the Seventh Meeting of the Conference of the Parties to the Convention on Biological Diversity (COP 7).

Fuglie, K. (2012). Productivity growth and technology capital in the global agricultural economy. In K. Fuglie, S. L. Wang, & V. Eldon Ball (Eds.), Productivity growth in agriculture: An international perspective. Oxfordshire, UK: CAB International.

Fujikura, Y., & Karssen, C. M. (1995). Molecular studies on osmoprimed seeds of cauliflower: A partial amino acid sequence of a vigour- related protein and osmopriming-enhanced expression of putative aspartic protease. *Seed Science Research*, *5*(03), 177–181. doi:10.1017/S0960258500002804

Gaddi, S. S., & Patil, Y. B. (2011). Screening of some low-cost waste biomaterials for the sorption of silver-cyanide [Ag(CN)$^{2-}$] from aqueous solutions. *International Journal of Chemical Sciences*, *9*(3), 1063–1072.

Galacgac, E. S., & Balisacan, C. M. (2009). Traditional weather forecasting for sustainable agroforestry practices in Ilocos Norte Province, Philippines. *Forest Ecology and Management*, *257*(10), 2044–2053. doi:10.1016/j.foreco.2009.01.002

GAO. (2012). Energy-Water Nexus: Coordinated federal approach needed to better manage energy and water tradeoffs (Congressional Research Service Report, GAO-12-880). Washington, DC: GAO.

Garud, R., Kumaraswamy, A., & Karnøe, P. (2010). Path dependence or path creation? *Journal of Management Studies*, *47*(4), 760–774. doi:10.1111/j.1467-6486.2009.00914.x

Gash, J. H. C., & Nobre, C. (Eds.). (1996). *Tropical Deforestation and Climate*. New York: John Wiley.

Gebrezgabher, S., Rao, K., Hanjra, M. A., & Hernández-Sancho, F. (2015). Business Models and Economic Approaches for Recovering Energy from Wastewater and Fecal Sludge. In Wastewater (pp. 217-245). Springer Netherlands. doi:10.1007/978-94-017-9545-6_12

Germ, M., & Gaberscik, A. (1999). The Effect of UV-B radiation and nutrient availability on growth and photochemical efficiency of PS II in common duckweed. *Phyton*, *39*(3), 187–192.

Ghafoor, A., Gill, M. A., Hassan, A., Murtaza, G., & Qadir, M. (2001a). Gypsum: An economical amendment for amelioration of saline-sodic waters and soils, and for improving crop yields. *International Journal of Agriculture & Biology*, *3*(3), 266–275.

Ghafoor, A., Murtaza, G., Ahmad, B., & Boers, Th. M. (2008). Evaluation of amelioration treatments and economic aspects of using saline-sodic water for rice and wheat production on salt-affected soils under arid land conditions. *Irrigation and Drainage*, *57*(4), 424–434. doi:10.1002/ird.377

Ghafoor, A., Murtaza, G., Maann, A. A., Qadir, M., & Ahmad, B. (2011). Treatments and economic aspects of growing rice and wheat crops during reclamation of tile drained saline-sodic soils using brackish waters. *Irrigation and Drainage*, *60*(3), 418–426. doi:10.1002/ird.577

Ghafoor, A., Murtaza, G., Rehman, M. Z., Saifullah, , & Sabir, M. (2012). Reclamation and salt leaching efficiency of treatments for tile drained saline-sodic soil using marginal quality water for irrigating rice and wheat crops. *Land Degradation & Development*, *23*(1), 1–9. doi:10.1002/ldr.1033

Ghafoor, A., Nadeem, S. M., Hassan, A., & Sadiq, M. (2001b). Reclamation response of two different textured saline-sodic soils to EC_{iw} to SAR_{iw} ratios. *Pakistan Journal of Soil Science*, *19*, 92–99.

Ghafoor, A., Qadir, M., & Murtaza, G. (1997). Potential for reusing low quality drainage water for soil amelioration and crop production. In *Proceedings of the International Symposium on Water for the 21st Century: Demand, Supply, Development and Socio-Environmental Issues* (pp. 411-20). Lahore, Pakistan: Centre of Excellence in Water Resources Engineering, University of Engineering and Technology.

Ghafoor, A., Qadir, M., & Murtaza, G. (2002). Agriculture in the Indus Plains: Sustainability of land and water resources. *International Journal of Agriculture & Biology*, *4*(3), 429–437.

Ghafoor, A., Qadir, M., & Murtaza, G. (2004). *Salt-affected soils: Principles of management*. Lahore, Pakistan: Allied Book Centre.

Giljum, S., Burger, E., Hinterberger, F., Lutter, S., & Bruckner, M. (2011). A comprehensive set of resource use indicators from the micro to the macro level. *Resources, Conservation and Recycling*, *55*(3), 300–308. doi:10.1016/j.resconrec.2010.09.009

Giovannucci, D., Scherr, S., Nierenberg, D., Hebebrand, C., Shapiro, J., Milder, J., & Wheeler, K. (2012). *Food and Agriculture: the future of sustainability, Sustainable Development in the 21st century (SD21).* United Nations Department of Economic and Social Affairs, Division for Sustainable Development. Retrieved on 20 Feb 2016, from https://sustainabledevelopment.un.org/content/documents/1443sd21brief.pdf

Gironi, F., & Piemonte, V. (2010). Life cycle assessment of polylactic acid and polyethylene terephthalate bottles for drinking water. *Environmental Progress & Sustainable Energy*, *30*(3), 459–468. doi:10.1002/ep.10490

Glaser, S., Glaser, R., Drescher, A., Pfeiffer, C., Schliermann-Kraus, E., Lechner, M., & Vencatesan, J. (2008). *Geo-communication for risk assessment and catastrophe prevention of flood events in the coastal areas of Chennai.* iEMSs 2008: International Congress on Environmental Modelling and Software Integrating Sciences and Information Technology for Environmental Assessment and Decision Making 4th Biennial Meeting of iEMSs. Retrieved from http://www.iemss.org/iemss2008/index.php?n=Main.Proceedings

Glassman, D., Wucker, M., Isaacman, T., & Champilou, C. (2011). *The water-energy nexus: Adding water to the energy agenda. (World Policy Paper).* New York: World Policy Institute.

Gleick, P. H. (1998). Water in crisis: Paths to sustainable water use. *Ecological Applications*, *8*(3), 571–579. doi:10.1890/1051-0761(1998)008[0571:WICPTS]2.0.CO;2

Gleick, P. H., & Cooley, H. S. (2009). Energy implications of bottled water. *Environmental Research Letters*, *4*(1), 1–6. doi:10.1088/1748-9326/4/1/014009

Godfray, H. C. J., Pretty, J., Thomas, S. M., Warham, E. J., & Beddington, J. R. (2011). Linking policy on climate and food. *Science*, *331*(6020), 1013–1014. doi:10.1126/science.1202899 PMID:21273449

GoK. (2010). The Constitution of Kenya 2010. Author.

Golberg, A. (2015). Environmental Exergonomics: Sustainability analysis of Energy systems considering impacts on eco-systems services. *Proceedings of the International Conference on Efficiency, cost, optimization, simulation and environmental impact of energy systems.*

Gomez, C. M., Tirado, D., & Rey-Maquieira, J. (2004). Water exchanges versus water works: Insights from a computable general equilibrium model for the Balearic Islands. *Water Resources Research*, *40*(10), n/a. doi:10.1029/2004WR003235

Gorddard, R., Colloff, M. J., Wise, R. M., Ware, D., & Dunlop, M. (2016). Values, rules and knowledge: Adaptation as change in the decision context. *Environmental Science & Policy*, *57*, 60–69. doi:10.1016/j.envsci.2015.12.004

Gosain, A. K., Rao, S., & Basuray, D. (2006). Climate change impact assessment on hydrology of Indian river basins. *Current Science*, *90*(3), 346–353.

Goswami, R. (2015, December 12). Paris report predicts extreme weather. *The Telegraph*, p. 1.

Goswami, B. N., Venugopal, V., Sengupta, D., Madhusoodanan, M. S., & Xavier, P. K. (2006). Increasing trend of extreme rain events over India in a warming environment. *Science*, *314*(5804), 1442–1445. doi:10.1126/science.1132027 PMID:17138899

Government of India & Indian Institute of Management-Ahmedabad. (1987). Command Area Development Program: A Joint Study. New Delhi: Author.

Government of India, Ministry of Agriculture. (n.d.). *Water resources of Maharashtra State*. Retrieved January 25, 2016, From http://www.ctara.iitb.ac.in/water/waterplanning/maharashtra.pdf

Government of India. (1992). *Command Area Development*. Planning Commission, Eighth Five-Year Plan. Author.

Government of India. (2006). NRAA: Rationale, organisation and mandate. New Delhi: Author.

Government of India. (2010). *Water management under revitalising rainfed agriculture – Potential pilots, Consultation Paper*. New Delhi: National Rainfed Area Authority.

Government of India. (2011). *Central Ground Water Board (CGWB)*. Ministry of Water Resources, River Development and Ganga Rejuvenation. Retrieved January 25, 2016, From http://cgwb.gov.in/documents /case-studies-rwh-ar-may2011.pdf

Government of India. (2014). *Elucidation of the Fifth National Report Submitted to the United Nations Convention to Combat Desertification: Desertification, Land Degradation and Drought in context of India*. Ministry of Environment and Forests.

Government of Maharashtra (2014, December 5). *Mumbai, Resolution No.JLA-2014/pra.kra.203/Jal-7*. Author.

Government of Maharashtra. (2015). *Mumbai*. Retrieved from http://www.maharashtra.gov.in

Graedel, T., Elshkaki, A., & van der Voet, E. (2014). Entangled circles: Energy and its resource connections. In G. J. Kramer & B. Vermeer (Eds.), *The colours of energy*. Amsterdam: Shell.

Granovetter, M. (1983). The strength of weak ties: A network theory revisited. *Sociological Theory*, *1*(1), 201–233. doi:10.2307/202051

Greenstone, M. (2015). *Global-post on air quality*. Energy Policy Institute at Chicago.

Green, T. R., Taniguchi, M., Kooi, H., Gurdak, J. J., Allen, D. M., Hiscock, K. M., & Aureli, A. et al. (2011). Beneath the surface of global change: Impacts of climate change on groundwater. *Journal of Hydrology (Amsterdam)*, *405*(3), 532–560. doi:10.1016/j.jhydrol.2011.05.002

Gren, M., Folke, C., Turner, K., & Batemen, I. (1994). Primary and secondary values of wetland ecosystems. *Environmental and Resource Economics*, *4*(1), 55–74. doi:10.1007/BF00691932

Grey, D., & Sadoff, C. (2006). *Water for growth and development*. Thematic documents of the IV World Water Forum. Mexico City: Comision Nacional del Agua.

Grit, R., Jörg, L., Steffen, D., & Gerel, O. (2015). Integrated urban water management: Development of an adapted management approach. *Environmental Earth Sciences*, *73*(2), 709–718. doi:10.1007/s12665-014-3701-z

Gross, M. (2016). World under water. *Current Biology*, *26*(2), R47–R50. doi:10.1016/j.cub.2016.01.001 PMID:27213204

Grothmann, T., & Patt, A. (2005). Adaptive capacity and human cognition: The process of individual adaptation to climate change. *Global Environmental Change*, *15*(3), 199–213. doi:10.1016/j.gloenvcha.2005.01.002

Group of Experts on the Scientific Aspects of Marine Pollution. (1990). *Report of the 20th session*. WHO.

Grover, M. (2012). *Degremont Water Treatment Technologies and A Case study of 635 MLD WTP at Sonia Vihar–Delhi*. Presented to working group of Planning Commission on Working Group on Urban and Industrial Water for Twelfth Plan of Government of India.

Guerra, R., Nakwafio, F., Kazadi, S., & Giuliano, M. (2010). *Pooled Fund 2010 Annual Report: Annex 10 - Pooled Fund First special allocation 2010*. Available at http://www.rdc-humanitaire.net/rdchbis/IMG/pdf/annex_10.pdf

Guhathakurta, P., Sreejith, O. P., & Menon, P. A. (2011). Impact of climate change on extreme rainfall events and flood risk in India. *Journal of Earth System Science*, *120*(3), 359–373. doi:10.1007/s12040-011-0082-5

Gulyani, S., Talukdar, D., & Potter, C. (2006). *Inside informality: poverty, jobs, housing and services in Nairobi's informal settlements: Report*. Academic Press.

Gunawansa, A., & Bhullar, L. (2013). *Water governance: an evaluation of alternative architectures*. Edward Elgar Publishing. doi:10.4337/9781781006429

Gundry, S., Wright, J., & Conroy, R. (2004). A systematic review of the health outcomes related to household water quality in developing countries. *Journal of Water and Health*, *2*, 1–13. PMID:15384725

Gupta, A. K., & Nair, S. S. (2010). Flood risk and context of land-uses: Chennai city case. *Journal of Geography and Regional Planning*, *3*(12), 365–372. Available online at http://www.academicjournals.org/JGRP

Gupta, A. K., & Nair, S. S. (2011). Urban floods in Bangalore and Chennai: Risk management challenges and lessons for sustainable urban ecology. *Current Science (Bangalore)*, *100*(11), 1638–1645.

Gupta, I. C. (1990). *Use of saline water in agriculture. Bombay, India*. Oxford, UK: IBH Publishing House Co, Pvt. Ltd.

Gwaze, F. R., & Mwale, M. (2015). The Prospect of Duckweed in Pig Nutrition: A Review. *The Journal of Agricultural Science*, *7*(11), 189.

Habbari, K., Tifnouti, A., Bitton, B., & Mandil, A. (2000). Geohelminthic infections associated with raw wastewater reuse for agricultural purposes in Beni-Mellal, Morocco. *Parasitology International*, *48*(3), 249–254. doi:10.1016/S1383-5769(99)00026-4 PMID:11227765

Häder, D. P., Kumar, H. D., Smith, R. C., & Worrest, R. C. (2007). Effects of solar UV radiation on aquatic ecosystems and interactions with climate change. *Photochemical & Photobiological Sciences*, *6*(3), 267–285. doi:10.1039/B700020K PMID:17344962

Hadfield, G. K., & Weingast, B. R. (2014). Microfoundations of the rule of law. *Annual Review of Political Science*, *17*(1), 21–42. doi:10.1146/annurev-polisci-100711-135226

Haeberli, W. (1990). Glacier and permafrost signals of 20th century warming. *Annals of Glaciology*, *14*, 99–101.

Hafemeister, D. (2014). *Physics of Societal Issues: Calculations on National Security, Environment and Energy*. Springer. doi:10.1007/978-1-4614-9272-6

Hall, D., & Lobina, E. (2008). *Water Privatisation*. Public Services International Research Unit (PSIRU). Accessed on 18 Feb 2016 http://gala.gre.ac.uk/1704/1/PSIRU_Report_(9820)_-_2008-04-W-over.pdf

Halpin, P. N. (1998). Global climate change and natural-area protection: Management responses and research direction. *Ecological Applications*, *7*(3), 828–843. doi:10.1890/1051-0761(1997)007[0828:GCCANA]2.0.CO;2

Halstead, M., Kober, T., & van der Zwaan, B. (2014). *Understanding the energy-water nexus No. ECN-E--14-046*. Netherlands: ECN.

Hameed, A., Afzal, I., & Iqbal, N. (2010). Seed priming and salinity induced variations in wheat (*Triticum aestivum* L.) leaf protein profile. *Seed Science & Technology*, *38*(1), 236–241. doi:10.15258/sst.2010.38.1.25

Hanak, E., Lund, J., Dinar, A., Gray, B., Howitt, R., Mount, J., & Thompson, B. et al. (2011). *Managing California's Water*. Public Policy Institute of California.

Hanjra, M. A., Blackwell, J., Carr, G., Zhang, F., & Jackson, T. M. (2012). Wastewater irrigation and environmental health: Implications for water governance and public policy. *International Journal of Hygiene and Environmental Health*, *215*(3), 255–269. doi:10.1016/j.ijheh.2011.10.003 PMID:22093903

Hanjra, M. A., Ferede, T., & Gutta, D. G. (2009). Pathways to breaking the poverty trap in Ethiopia: Investments in agricultural water, education and markets. *Agricultural Water Management*, *96*(11), 1596–1604. doi:10.1016/j.agwat.2009.06.008

Hanjra, M. A., & Qureshi, M. E. (2010). Global Water Crisis and Future Food Security in an era of Climate Change. *Food Policy*, *35*(5), 365–377. doi:10.1016/j.foodpol.2010.05.006

Hanlon, P., Madel, R., Olson-Sawyer, K., Rabin, K., & Rose, J. (2013). *Food, water and energy: Know the nexus*. GRACE Communications Foundation.

Hansen, J., Makiko, S., Ruedy, R., Lacis, A., & Oinas, V. (2000). Global warming in the twenty-first century: An alternative scenario. *Proceedings of the National Academy of Sciences of the United States of America*, *97*(18), 9875–9880. doi:10.1073/pnas.170278997 PMID:10944197

Harding, B., & Devisscher, T. (2009). Review of the Economic Impacts of Climate Change in Kenya. Academic Press.

Harris, D., Joshi, A., Khan, P. A., Gothkar, P., & Sodhi, P. S. (1999). On-farm seed priming in semi-arid agriculture: Development and evaluation in maize, rice and chickpea in India using participatory methods. *Experimental Agriculture*, *35*(1), 15–29. doi:10.1017/S0014479799001027

Harris, P. G. (2011). *Ethics and global environmental policy: cosmopolitan conceptions of climate change*. Hong Kong: Edward Elgar Publishing. doi:10.4337/9780857931610

Hartmann, D. L., Klein Tank, A. M. G., Rusticucci, M., Alexander, L. V., Brönnimann, S., Charabi, Y., & Zhai, P. M. et al. (2013). Observations:Atmosphere and Surface. In T. F. Stocker, D. Qin, G. K. Plattner, M. Tignor, S. K. Allen, J. Boschung, & P. M. Midgley et al. (Eds.), *Climate Change 2013: The Physical Science Basis. Contribution of Working Group I to the Fifth Assessment Report of the Intergovernmental Panel on Climate Change*. Cambridge, UK: Cambridge UniversityPress.

Harvey, D. (2005). *A brief history of neoliberalism*. OUP Oxford.

Hauschild, M., Jeswiet, J., & Alting, L. (2005). From life cycle assessment to sustainable production: Status and perspectives. *CIRP Annals-Manufacturing Technology*, *54*(2), 1–21. doi:10.1016/S0007-8506(07)60017-1

Hayashi, M., van der Kamp, G., & Rudolph, D. L. (1998). Water and solute transfer between a prairie wetland and adjacent uplands, 1. Water balance. *Journal of Hydrology (Amsterdam)*, *207*(1), 42–55. doi:10.1016/S0022-1694(98)00098-5

Hegde, B. R. (2013). *Resolving the Cauvery water dispute*. Retrieved from http://www.indiawaterportal.org

Held, D. (1995). *Democracy and the global order: From the modern state to cosmopolitan governance*. Stanford University Press.

Held, I., & Soden, B. (2006). Robust responses of the hydrological cycle to global warming. *Journal of Climate*, *19*(21), 5686–5699. doi:10.1175/JCLI3990.1

Hertwich, E. G. (2005). Life cycle approaches to sustainable consumption. *Critical Reviews in Environmental Science and Technology*, *39*(13), 4673–4684. doi:10.1021/es0497375 PMID:16053063

Hildebrand, L.P. (1997). Introduction to the special issue on community based coastal management. *Ocean and Coastal Management, 36*, 1-9.

Hillman, W. S., & Culley, D. D. (1978). The Uses of Duckweed: The rapid growth, nutritional value, and high biomass productivity of these floating plants suggest their use in water treatment, as feed crops, and in energy-efficient farming. *American Scientist*, *66*(4), 442–451.

Hinrichsen, D. (1996, Summer). Coast in crisis. *Issues in Science and Technology*, *12*(4), 39–47.

Hobbs, M., & Knausenberger, W. (2003). *Burundi Environmental threats and opportunities assessment (ETOA), with emphasis on tropical forestry and biodiversity conservation*. Type-scripted Supplementary report to Integrated Strategic Plan for 2003-2005. USAID.

Hoelzle, M., Darms, G., Lüthi, M. P., & Suter, S. (2011). Evidence of accelerated englacialwarming in the Monte Rosa area, Switzerland/Italy. *Cryosphere*, *5*(1), 231–243. doi:10.5194/tc-5-231-2011

Hoff, H., Doll, P., Fader, M., Gerten, D., Hauser, S., & Siebert, S. (2014). Water footprint of cities-indicators for sustainable consumption and production. *Hydrology and Earth System Sciences*, *18*(1), 213–226. doi:10.5194/hess-18-213-2014

Holt, D. (2003). What becomes an icon most? *Harvard Business Review*, *81*(3), 43–49. PMID:12632803

Hornbeck, R. (2012). The Enduring Impact of the American Dust Bowl: Short and Long-Run Adjustments to Environmental Catastrophe. *The American Economic Review*, *102*(4), 1477–1507. doi:10.1257/aer.102.4.1477

Houghton, J. T., Jenkins, G. J., & Ephraums, J. J. (1990). *Climate change: the IPPC scientific assessment*. Cambridge, UK: Cambridge University Press.

Howard, B. C. (2014, December16). US-Mexico agreement returns water to the Colorado river delta. *National Geographic*.

Howell, T. A. (2001). Enhancing water use efficiency in irrigated agriculture. *Agronomy Journal*, *93*(2), 281–289. doi:10.2134/agronj2001.932281x

Huber, M. K. (2010). *Bottled water: The risks of our health, our environment, and our wallets*. Indiana University.

Huggett, R. J. (1993). *Modelling the human impact on nature – Systems analysis of environmental problems*. Oxford University Press.

Hughes, L. (2000). Biological consequences of global warming: Is the signal already apparent? *Trends in Ecology & Evolution*, *15*(2), 56–61. doi:10.1016/S0169-5347(99)01764-4 PMID:10652556

Hughes, P. (1986). Ocean color: A key to climate change. *Weatherwise*, *39*(5), 267–270. doi:10.1080/00431672.1986.9929300

Huho, J. M., Ngaira, J. K., & Ogindo, H. O. (2011). Living with drought: The case of the Maasai pastoralists of northern Kenya. *Educational Research*, *2*(1), 779–789.

Huitema, D., Mostert, E., Egas, W., Moellenkamp, S., Pahl-Wostl, C., & Yalcin, R. (2009). Adaptive water governance: Assessing the institutional prescriptions of adaptive (co-) management from a governance perspective and defining a research agenda. *Ecology and Society*, *14*(1), 26.

Hulme, M., Dohery, R. M., Ngara, T., New, M. G., & Lister, D. (2001). African Climate Change: 1900 – 2100. *Climate Research*, *17*, 145–168. doi:10.3354/cr017145

Hungate, B. A., Reichstein, M., Dijkstra, P., Johnson, D., Hymus, G., Tenhumen, J. D., & Drale, B. G. et al. (2002). Evapotranspiration and soil water content in a scrub-oak woodland under carbon dioxide enrichment. *Global Change Biology*, *8*(3), 289–298. doi:10.1046/j.1365-2486.2002.00468.x

Hurd, D. J. (2000). *Best practices and industry standards in PET plastic recycling*. Retrieved from www.napcor.com

Hussain, G. (1978). *Determination of irrigation water quality standards*. (Unpublished doctoral dissertation). Colorado State University.

Hussain, I., Raschid, L., Hanjra, M. A., Marikar, F., & van der Hoek, W. (2002). *Wastewater use in agriculture: Review of impacts and methodological issues in valuing impacts. (With an extended list of bibliographical references)*. Working Paper 37. Colombo, Sri Lanka: International Water Management Institute.

Hussain, I., & Hanjra, M. A. (2004). Irrigation and poverty alleviation: *Review of the empirical evidence. Irrigation and Drainage, 53*(1), 1–15. doi:10.1002/ird.114

Hutton, G., & Wood, A. (2013). *Development financing for tangible results: A paradigm shift to impact investing and outcome*. Academic Press.

ICOLD (International Commission on Large Dams). (2003). *World Register of Dams, 2003*. Paris, France: International Commission on Large Dams.

Ida, H. (2006). *Current status of plastics recycling in Japan*. Paper presented at The IEA/CEFIC Workshop: Feedstock Substitutes, Energy Efficient Technology and CO2 Reduction for Petrochemical Products, Paris, France.

IEA. (2010). *World energy outlook 2010*. Paris: OECD/ International Energy Agency.

Ilon, L.Ilon. (1992). Fitting girls schooling into existing economic paradigms: Confronting the complexities. *International Journal of Educational Development, 12*(2), 147–159. doi:10.1016/0738-0593(92)90036-L

Iloweka, E. M. (2004). The deforestation of rural areas in the Lower Congo Province. *Environmental Monitoring and Assessment, 99*(1-3), 245–250. doi:10.1007/s10661-004-4028-0 PMID:15641386

Immerzeel, W. W., van Beek, L. P. H., & Bierkens, M. F. P. (2010). Climate change will affect the Asian Towers. *Science, 328*(5984), 1382–1385. doi:10.1126/science.1183188 PMID:20538947

INCCA Report, .Indian Network for Climate Change Assessment. (2010). *Climate Change and India: A 4x4 Assessment A Sectoral and Regional Analysis for 2030s*. Ministry of Environment & Forests, Government of India.

India and Bangladesh: Headed Towards International Legal "War" Over The Tipaimukh Dam?. (2012, January 13). *Water Politics*. Retrieved from http://www.waterpolitics.com/2012/01/13/india-and-bangladesh-headed-towards-international-legal-war-over-the-tipaimukh-dam/

India Assures Bangladesh on Tipaimukh. (2011, December 3). *Times of India*. Retrieved from http://articles.timesofindia.indiatimes.com/2011-12-03/india/30471587_1_tipaimukh-dam-tipaimukh-issue-bangladesh-government

Indian Express. (2009, October 6). Interlinking of rivers buried; Jairam says the idea a disaster. *Indian Express,* p. 1.

Indian Express. (2015, September 22). Rains ease water crisis. *Indian Express,* p. 1.

India-WRIS WebGIS. (2014). *Design and Development of web enabled water resources information of India*. Arc India News, Application Article.

Inogwabini, B., & Lingopa, Z. (2006). Les Inventaires des Poissons dans le Lac Tumba, Congo et Ngiri – Paysage Lac Te'le´ – Lac Tumba, Segment RDC. Rapport Produit dans le Cadre du Programme CARPE. Type scripted report to the WWF DRC, Kinshasa (Democratic Republic of Congo).

Inogwabini, B. I., Dianda, M., & Lingopa, Z. (2009). The use breeding sites of *Tilapia congica* (Thys & van Audenaerde 1960) to delineate conservation sites in the Lake Tumba, Democratic Republic of Congo: Toward the conservation of the lake ecosystem. *African Journal of Ecology, 48*, 800–806.

Inogwabini, B. I., Sandokan, B. M., & Ndunda, M. (2006). A dramatic decline in rainfall regime in the Congo basin: Evidence from a thirty-four data set from the Mabali Scientific Research Center, Democratic Republic of Congo. *International Journal of Climatology, 31*, 278–286.

Inter-Agency Commission (UNDP, UNESCO, UNICEF, World Bank). (1990). *Final Report of the World Conference on Education for All: Meeting Basic Learning Needs*. United Nations.

Intergovernmental Oceanographic Commision (IOC). (2015). *Scientific community proposes recommendations for integration of ocean and coasts in COP21 agreement*. UNESCO Media Service.Retrieved from http://www.unesco.org/new/en/media-services/2015

Intergovernmental Panel on Climate Change. (1994). *Preparing to meet the coastal challenges of the 21st century: Conference report*. World Coast Conference 1993, Intergovernmental Panel on Climate change, Noordwijk, Ministry of Transport, Public Works and Water Management, The Hague, Netherlands.

International Crops Research Institute for the Semi-Arid Tropics. (2015). *Waste management recycling and reuse*. Retrieved from ICRISAT: https://www.google.co.in/url?sa=t&rct=j&q=&esrc=s&source=web&cd=4&cad=rja&uact=8&ved=0ahUKEwi4v5-9uL3LAhWUUI4KHbdFCw4QFggzMAM&url=http%3A%2F%2Fwww.icrisat.org%2Fidc-flyers%2F3_Waste%2520management_CSR%2520Flyers.pdf&usg=AFQjCNFBTLfK7Mu_ng0uljn_TzJO0eAyMQ&

International Law Association. (1986). Complementary Rules Applicable to International Water Resources. Adopted by the International Law Association at its 62nd Conference held at Seoul in 1986. In *The 1966 Helsinki Rules and Complementary and Supplementary Rules (1972-1996)*. Retrieved from http://water.cedare.int/files15%5CFile2917.pdf

International Rivers. (2008). *Dirty Hydro: Dams and Greenhouse Gas Emissions*. Retrieved from http://www.internationalrivers.org/files/attached-files/dirtyhydro_factsheet_lorez.pdf

International Sava River Basin Commission. (2002). *Framework Agreement The Sava River Basin*. Retrieved from http://www.savacommission.org/dms/docs/dokumenti/documents_publications/basic_documents/fasrb.pdf

International Water Management Institute. (2015, March 10). *Wastewater Business models*. Retrieved from: http://www.iwmi.cgiar.org/issues/wastewater/business-models/

IPCC (Intergovernmental Panel on Climate Change). (2007). *Climate Change 2007: Impacts, adaptation and vulnerability. Contribution to of Working Group II to the fourth Assessment Report of the Intergovernmental Panel on Climate Change*. Cambridge University Press.

IPCC (Intergovernmental Panel on Climate Change). (2007). *Synthesis Report*. Contribution of Working Groups I, II and III to the Fourth Assessment Report of the Intergovernmental Panel on Climate Change. Geneva, Switzerland: IPCC.

IPCC. (1996). *Impacts, Adaptations, and Mitigation of Climate Change: Scientific-Technical Analyses - Contribution of Working Group II to the IPCC Second Assessment Report*. Cambridge, UK: Cambridge University Press.

IPCC. (2001). Impacts, Adaptation and Vulnerability. In J. J. McCarthy, O. F. Canziani, N. A. Leary, D. J. Dokken, & K. S. White (Eds.), *Contribution of Working Group II to the Third Assessment Report of the Intergovernmental Panel on Climate Change*. Cambridge, UK: Cambridge University Press.

IPCC. (2007). *Climate Change 2007: Impacts, Adaptations and Vulnerability*. Contribution of Working Group II to the Fourth Assessment Report of the Intergovernmental Panel on Climate Change. Retrieved on March 2, 2008 from http://www.ipcc.ch/ipccreports/ar4-wg2.htm

IPCC. (2007). Climate Change 2007: *Mitigation of Climate Change*. In B. Metz, O. R. Davidson, P. R. Bosch, R. Dave, & L. A. Meyer (Eds.), *Contribution of Working Group III to the Fourth Assessment Report of the Intergovernmental Panel on Climate Change, 2007*. Cambridge University Press.

IPCC. (2007). *Climate Change 2007: The Physical Science Basis. Contribution of Working Group I to the Fourth Assessment Report of the Intergovernmental Panel on Climate Change(IPCC)*. New York: Cambridge University Press.

IPCC. (2007). *Impacts, Adaptation, and Vulnerability*. Geneva, Switzerland: Working Group II Contribution to the Intergovernmental Panel on Climate Change Fourth Assessment Report. Summary for Policy Makers. IPCC.

IPCC. (2013). *Climate Change 2013: The Physical Science Basis. Contribution of Working Group I to the Fifth Assessment Report of the IntergovernmentalPanel on Climate Change*. Cambridge, UK: Cambridge University Press.

IPCC. (2014). [*Synthesis report*. Contribution of Working Groups I, II and III to the Fifth Assessment Report of the Intergovernmental Panel on Climate Change. Geneva: IPCC.]. *Climatic Change*, 2014.

IRD (Institut de Recherche pour le Développement). (2002). *Congo River Basin: geology and soil types influence drought. European Water Management News*. Available from: http://www.ewaonline.de

IRIN (Integrated Regional Information Networks). (2011). *DR-CONGO: Return delayed for 120,000 refugees*. Retrieved from http://www.irinnews.org/report/92712/drc-congo-return-delayed-for-120-000-refugees

Islam, N., Azam, A., & Islam, Q. R. (n.d.). *Teesta River Water Sharing: A Case Study In Teesta Barrage Project*. Retrieved from http://www.watertech.cn/english/islam.pdf

Itankar, N., & Patil, Y. (2015). Biosorptive Removal of Hexavalent Chromium from Aqueous Industrial Solutions by A Novel Wood Apple Shell Powder Biomass. *Advanced Materials Research*, *1130*, 689–692. doi:10.4028/www.scientific.net/AMR.1130.689

Iyer, K. (2015, February 19). Drought in Marathwada – I. *The Times of India*, p. 1.

Janakarajan, S., Zehra, M. H., & Llorente, M. (2006). *Urban Water Conflicts in Indian Cities: Man-made Scarcity as a Critical Factor*. MIDS Working Paper No.196. Chennai, India: Madras Institute of Development Studies.

Janakarajan, S. (2009). Urbanization and peri-urbanization: Aggressive competition and unresolved conflicts. The case of Chennai city in India. *South Asian Water Studies*, *1*(1), 51–76.

Janakarajan, S. (2013). *Optimising Local Water Resources and Availability in Urban and Peri-urban Chennai* [PDF document]. Retrieved from cseindia.org/userfiles/Janakarajan.pdf

Janakarajan, S., Butterworth, J., Moriarty, P., & Batchelor, C. (2007). Strengthened city marginalised peri-urban villages: stakeholder dialogues for inclusive urbanisation in Chennai, India. In J. Butterworth, R. Ducrot, N. Faysse, & S. Janakarajan (Eds.), *Peri-urban water conflicts: Supporting dialogue and negotiation* (pp. 51–75). Delft, Netherlands: IRC International Water and Sanitation Centre.

Jarraud, M. (2010). *Report of the Global Atmospheric Watch January 2011*. United Nations World Meteorolical Organization.

Jayaraman, T. (2011). Climate change and agriculture: A review article with special reference to India. *Review of Agrarian Studies, Journal of the Foundation for Agrarian Studies, 1*(2).

Jerome, S. (2014). *Potential Water Tax Could Pay For Infrastructure Updates In Delaware*. Retrieved from http://www.wateronline.com/doc/potential-water-tax-could-pay-for-infrastructure-updates-in-delaware-0001

Jha, A. K., Bloch, R., & Lamond, J. (Eds.). (2012). *Cities and flooding: a guide to integrated urban flood risk management for the 21st century*. World Bank Publications. Retrieved from http://hdl.handle.net/10986/2241

Jianchu, X., Shrestha, A., & Eriksson, M. (2009). *Climate change and its impacts on glaciers and water resource management in the Himalayan Region. Assessment of Snow, Glaciers and Water Resources in Asia*. Koblenz, Germany: International Hydrological Programme of UNESCO and Hydrology and Water Resources Programme of WMO.

Jianchu, X., Shrestha, A., Vaidya, R., Eriksson, M., & Hewitt, K. (2007). *The Melting Himalayas. Regional Challengesand Local Impacts of Climate Change on Mountain Ecosystems and Livelihoods. ICIMOD Technical Paper*. Kathmandu: ICIMOD.

JMP. (2012). *Progress on D inking Water 2012 UPDATE*. Retrieved from www.unicef.org/media/files/JMPreport2012.pdf

Joint Rivers Commission. (n.d.a). *Basin Map. Ministry of Water Resources*. Govt. of Bangladesh. Retrieved from http://jrcb.gov.bd/new/index.php/9-link-page/12-basin-map

Joint Rivers Commission. (n.d.b). *Ministry of Water Resources*. Govt. of Bangladesh. Retrieved from http://jrcb.gov.bd/new/

Jones, P., Hillier, D., & Comfort, D. (2014). Sustainable consumption and the UKs leading retailers. *Social Responsibility Journal*, *10*(4), 702–715. doi:10.1108/SRJ-04-2013-0046

Joshi, V. U., & Kale, V. S. (2013). Environmental Conflicts in Coastal Metropolitan Cities in India: Case Studies of Mumbai and Chennai Metropolitan Regions. In A. Z. Khan, L. X. Quynh, E. Corijn & F. Canters (Eds.), *Environmental Conflicts in Coastal Urban Areas: Towards a Strategic Assessment Framework for Sustainable Development*. Retrieved from http://digilab-epub.uniroma1.it/index.php/SECOA/article/download/103/92

Joshia, K., & Chaturvedib, P. (2013). Impact of climate change on agriculture. *Octa Journal of Environmental Research*, *1*(1).

Joshi, A., & Joshi, P. K. (2011). A rapid inventory of indicators of climate change in the middle Himalaya. *Current Science*, *100*(6), 831–833.

Joy, K.J. (2013, September 14). The much talked about Shirpur model. *Down-to-earth*, p. 1.

Juliano, K., & Simonovic, S. P. (1999). *The impact of wetlands on flood control in the Red River Valley of Manitoba*. University of Manitoba. Retrieved February 10, 2015, from http://www.ijc.org/rel/pdf/wetlandsman.pdf

Juma, D. W., Wang, H., & Li, F. (2014). Impacts of population growth and economic development on water quality of a lake: Case study of Lake Victoria Kenya water. *Environmental Science and Pollution Research International*, *21*(8), 5737–5746. doi:10.1007/s11356-014-2524-5 PMID:24442964

Junk, W. J., An, S., Finlayson, C. M., Gopal, B., Květ, J., Mitchell, S. A., & Robarts, R. D. et al. (2013). Current state of knowledge regarding the worlds wetlands and their future under global climate change: A synthesis. *Aquatic Sciences*, *75*(1), 151–167. doi:10.1007/s00027-012-0278-z

Kakumu, O. (2004). Privatization of the urban water supply in Kenya: Policy options for the poor. *Environment and Urbanization*, *16*(2), 213–222. doi:10.1177/095624780401600212

Kallidaikurichi, S., & Rao, B. (n.d.). Index of drinking water adequacy for the Asian economies. Water Policy, 12(S1), 135–154.

Kalra, N., Chander, S., Pathak, H., Aggarwal, P. K., Gupta, N. C., Sehgal, M., & Chakraborty, D. (2007). Impacts of climate change on agriculture. *Outlook on Agriculture*, *36*(2), 109–118. doi:10.5367/000000007781159903

Karajeh, F., Suleimenov, M., Karimov, A., Vyshpolsky, F., Mukhamedjanov, Kh., Bekbaev, U. (2004). Technology of irrigation, water saving and improving soil fertility in Arys Turkestan canal command zone. *Kazakh Research Institute of Water Management: Taraz, 18*. (in Russian).

Karki, M., Mool, P., & Shrestha, A. (2009). Climate Change and its Increasing Impacts in Nepal. *The Initiation*, *3*, 30–37.

Karlekar S.N. (2004). The impact of Coastal Landforms on the distribution of placer Minerals along Maharashtra Coast. *Indian Journal of Geomorphology, 9*(1&2), 101-114.

Karlekar, S. N. (1981). *Unpublished Ph. D. Thesis*. University of Pune.

Karnataka explores cloud seeding to increase water in Cauvery. (2012, October 2). *The Hindu Business Line*, p. 1.

Kaser, G., Cogley, J. G., Dyurgerov, M. B., Meier, M. F., & Ohmura, A. (2006). Mass balance of glaciers and ice caps: Consensus estimates for 19612004. *Geophysical Research Letters*, *33*(19), 5. doi:10.1029/2006GL027511

Kaweeya, S. (2006). *EARTH Journalism*. Bangkok: Ratanakosintr House.

Kay, R., & Alder, J. (1999). *Coastal Planning and Management*. London: Routledge.

Kelly, P. M., & Adger, W. N. (2000). Theory and practice in assessing vulnerability to climate change andFacilitating adaptation. *Climatic Change*, *47*(4), 325–352. doi:10.1023/A:1005627828199

Kenyon, Pollett, & Wills-Johnson. (2000). *Australia Sustainable Water Management Practices: Lessons*. Academic Press.

Kessides, I. (2004). *Reforming Infrastructure: Privatization, Regulation, and Competition*. World Bank Policy Research Report, co-publication the World Bank and Oxford University Press.

Khan, A. A., Maguire, J. D., Abawi, G. S., & Illas, S. (1992). Matriconditioning of vegetable seed to improve stand establishment in early field planting. *Journal of the American Society for Horticultural Science*, *117*, 41–47.

Khan, A. N., Shaw, R., Pulhin, J., & Pereira, J. (2010). Climate change adaptation and disaster risk reduction in Pakistan. Climate change adaptation and disaster risk reduction. *Asian Perspective*, 5.

Khapre, S. (2015, August 8). Bringing rain in Marathwada: Cloud seeding project to continue for 90 more days. *Indian Express*, p. 1.

Kinyanjui, K. (1993). Enhancing women's participation in the sciences bases curriculum: The case of Kenya. In J. K. Conway & S. C. Bourque (Eds.), *The politics of women's education. Perspective from Asia, Africa and Latin America*. Ann Arbor, MI: University of Michigan press.

Kirtman, B., Power, S. B., Adedoyin, J. A., Boer, G. J., Bojariu, R., Camilloni, I., & Wang, H. J. et al. (2013). Near-term Climate Change: Projections and Predictability. In T. F. Stocker, D. Qin, G. K. Plattner, M. Tignor, S. K. Allen, J. Boschung, & P. M. Midgley et al. (Eds.), *Climate Change 2013: The Physical Science Basis. Contribution of Working Group I to the Fifth Assessment Report of the Intergovernmental Panel on Climate Change*. Cambridge, UK: Cambridge University Press.

Kleemans, M. & Sadoulet, E. (2012). *Impact of drought-tolerant risk-reducing rice on yield and farmer welfare*. Agriculture Technology Adoption Initiative (in collaboration with International Rice Research Institute and Stress Tolerant Rice for Africa and South Asia).

Klein, R. J., & Maciver, D. C. (1999). Adaptation to climate variability and change: Methodological issues. *Mitigation and Adaptation Strategies for Global Change*, *4*(3-4), 189–198. doi:10.1023/A:1009690729283

Kneese, A., & Bower, B. (1984). *Managing water quality: Economics, technology, institutions*. Washington, DC: Resources for the Future.

Komba, D. (1995). Declining enrolment and quality of primary education in Mainland Tanzania. In C. Creighton & C. K. Omari (Eds.), *Gender, Family and Household in Tanzania*. Avebury.

Koplan, J. (2002, March). Chronic Disease: The Role of Behavior and the Environment. *Health Affairs*, *21*(2), 179–184. PMID:11900157

Körner, S., Vermaat, J. E., & Veenstra, S. (2003). The capacity of duckweed to treat wastewater. *Journal of Environmental Quality*, *32*(5), 1583–1590. doi:10.2134/jeq2003.1583 PMID:14535298

Kotz, D. M. (2002). Globalization and neoliberalism. *Rethinking Marxism*, *14*(2), 64–79. doi:10.1080/089356902101242189

Koul D, N., Singh, S., Neelam, G., & Shukla, G. (2012). Traditional water management systems - An overview of Ahar-pyne system in South Bihar plains of India and need for its revival. *Indian Journal of Traditional Knowledge*, *11*(2), 266-272.

Kreutzmann, H. (2011). Scarcity within opulence: Water management in the Karakoram Mountains revisited. *Journal of Mountain Science*, *8*(4), 525–534. doi:10.1007/s11629-011-2213-5

Kron, W., & Berz, G. (2007). Flood disasters and climate change: trends and options—a (re-)insurer's view. In J. L. Lozán, H. Grabl, P. Hupfer, L. Menzel, & C. D. Schönwiese (Eds.), *Global change: enough water for all?* (pp. 268–273). University of Hamburg, Hamburg.

Kuczenski, B., & Geyer, R. (2010). Material flow analysis of polyethylene terephthalate in the US, 19962007. *Resources, Conservation and Recycling*, *54*(12), 1161–1169. doi:10.1016/j.resconrec.2010.03.013

Kulkarini, A. V., Mathur, P., Rathore, B. P., Alex, S., Thakur, N., & Kumar, M. (2002). Effect of global warming on snow ablation pattern in the Himalaya. *Current Science*, *83*(2), 120–123.

Kulkarni, A. V., Bahuguna, I. M., Rathore, B. P., Singh, S. K., Randhawa, S. S., Sood, R. K., & Dhar, S. (2007). Glacial retreat in Himalaya using Indian Remote Sensing satellite data. *Current Science*, *92*(1), 69–74.

Kumar, R. (2005). Status review of possible Impacts of Climate Change on Himalayan Glaciers, Glaciers retreat and its subsequent impacts on fresh water régime. In WWF: An Overview of Glaciers, Glacier Retreat, and Subsequent Impacts in India, Nepal and China, (pp. 29-38). WWF.

Kumar, R., Hasnain, S. I., Wagnon, P., Arnaud, Y., Chevallier, P., Linda, A., & Sharma, P. (2007). Climate change signals detected through mass balance measurements on benchmark glacier, Himachal Pradesh, India. Climatic and anthropogenic impacts on the variability of water resources. Technical Document in Hydrology, 80. UNESCO.

Kumar, A., Schei, T., Ahenkorah, A., Caceres Rodriguez, R., Devernay, J.-M., Freitas, M., & Liu, Z. et al. (2011). Hydropower. In O. Edenhofer, R. Pichs-Madruga, Y. Sokona, K. Seyboth, P. Matschoss, S. Kadner, & C. von Stechow et al. (Eds.), *IPCC Special Report on Renewable Energy Sources and Climate Change Mitigation*. Cambridge, UK: Cambridge University Press. doi:10.1017/CBO9781139151153.009

Kumar, M. D., & Singh, O. P. (2001). Market instruments for demand management in the face of scarcity and overuse of water in Gujarat. *Water Policy*, *3*(5), 387–403. doi:10.1016/S1366-7017(01)00081-2

Kumar, R., Kumar, K. K., & Pant, G. B. (1994). Diurnal symmetry of surface temperature trends over India. *Geophysical Research Letters*, *21*(8), 677–680. doi:10.1029/94GL00007

Kumar, S., & Dobhal, D. P. (1997). Climatic effects and bedrock control on rapid fluctuations of Chhota Shigri glacier, northwest Himalaya, India. *Journal of Glaciology*, *43*(145), 467–472.

Kundzewicz, Z. W., Kanae, S., Seneviratne, S. I., Handmer, J., Nicholls, N., Peduzzi, P., & Muir-Wood, R. et al. (2014). Flood risk and climate change: Global and regional perspectives. *Hydrological Sciences Journal*, *59*(1), 1–28. doi:10.1080/02626667.2013.857411

Kurukulasuriya, P., & Rosenthal, S. (2003). *Climate change and agriculture: a review of impacts and adaptations. Climate Change Series Paper No. 91*. Washington, DC: World Bank.

Lal Bahadur Shastri National Academy of Administration. (1979). Seminar Proceedings, Indian Administrative Service (IAS) Professional Course Foundation Program. Government of India.

Lal, C., & Verma, L. R. (2008). Indigenous Technological Knowledge on soil and water management from Himachal Himalaya. *Indian Journal of Traditional Knowledge, 7*(3), 485-493.

Lal, M. (2000). Climatic change-implications for India's water resources. *Journal of Social and Economic Development, 3*, 57–87.

Lambert, F. H., & Webb, M. J. (2008). Dependency of global mean precipitation on surface temperature. *Geophysical Research Letters, 35*(16), L23803. doi:10.1029/2008GL034838

Lam, E., Appenroth, K. J., Michael, T., Mori, K., & Fakhoorian, T. (2014). Duckweed in bloom: The 2nd International Conference on Duckweed Research and Applications heralds the return of a plant model for plant biology. *Plant Molecular Biology, 84*(6), 737–742. doi:10.1007/s11103-013-0162-9 PMID:24398764

Lammertsma, E. I., de Boer, H. J., Dekker, S. C., Dilcher, D. L., & Lotter, A. F. (2011). Global CO_2 rise leads to reduced maximum stomatal conductance in Florida vegetation. *Proc. Acad. Sci. U.S.A., 108*(10), 4035–4040. doi:10.1073/pnas.1100371108 PMID:21330552

Landolt, E. (1986). The family of Lemnaceae - a monographic study, Vol. 1: Veroffentlichungen des Geobotanischen Institutes der Eidgenossischen Technische Hochschule, StiftungRubel. StiftungRubel.

Landolt, E., &Kandeler, R. (1987). Biosystematic investigations in the family of duckweeds (Lemnaceae) (vol. 4). *The family of Lemnaceae-a monographic study, 2*, 211-34.

Larijani, M. (2010). Assessment of environmental awareness among higher primary school teachers. *Human Ecology, 31*(2), 121–124.

Leal Filho, W. (Ed.). (2015). *Handbook of Climate Change Adaptation*. Berlin: Springer Berlin Heidelberg. doi:10.1007/978-3-642-38670-1

Lee, K. N. (1999). Appraising adaptive management. *Conservation Ecology, 3*(2), 3.

Lee, K. (1993). *Compass and gyroscope: integrating politics and science for the environment*. Washington, DC: Island Press.

Leese, M., & Meisch, S. (2015). Securitising sustainability? Questioning the 'water, energy and food-security Nexus'. *Water Alternatives, 8*(1), 695–709.

Lehmann, B., Vilaplana, F., Strömberg, E., Suliman, W., & Rodriguez, C. L. (2005). Comparative LCA on plastic packaging. Prairie Village, KS: Franklin Associates, A Division of ERG.

Lei, C. (2011). *Investments in China in water resources*. Report of the Ministry of Water Resources.

Lemoine, N., Bauer, H.-G., Peintinger, M., & Böhning-Gaese, K. (2007). Effects of Climate and Land-Use Change on Species Abundance in a Central European Bird Community. *Conservation Biology, 21*(2), 495–503. doi:10.1111/j.1523-1739.2006.00633.x PMID:17391199

Lemoine, N., & Böning-Gaese, K. (2003). Potential impact of global climate change on species richness of long-distance migrants. *Conservation Biology, 17*(2), 577–586. doi:10.1046/j.1523-1739.2003.01389.x

Lemoine, N., Schaefer, H.-C., & Böhning-Gaese, K. (2007b). Species richness of migratory birds is influenced by global climate change. *Global Ecology and Biogeography, 16*(1), 55–64. doi:10.1111/j.1466-8238.2006.00252.x

Leonard, C. (1987). République du Zaïre - Publié par le Ministère de l'Economie et de l'Industrie. Imprimerie Pirotte, Liège. *Conjoncture Economique, 26*.

Leonard, J. (1952). Cynometreae et Amherstieae. In Flore du Congo Belge et du Ruanda-Burundi. INEAC.

Leonard, J. (1951). Contribution a l'étude de la végétation des bains d'éléphants au Congo Belge. Le Rhynchosporeto-Cyperetum longibracteato. *Publication de la Société Royale de Botanique Belge., 84*, 13–27.

Lerner, H. (2016). *The Water Footprint of Energy*. Accessed on 29 Feb 2016 http://www.gracelinks.org/196/the-water-footprint-of-energy

Lettau, H., Lettau, K., & Molion, L. (1979). Amazonias hydrologic cycle and the role of atmospheric recycling in assessing deforestation effects. *Monthly Weather Review, 107*(3), 227–238. doi:10.1175/1520-0493(1979)107<0227:AHCATR>2.0.CO;2

Lewis, N. D. (2016). Sustainable development through a gendered lens: Climate change adaptation and disaster risk reduction. *Reviews on Environmental Health, 31*(1), 97–102. doi:10.1515/reveh-2015-0077 PMID:26943600

Liaw, C. H., & Tsai, Y. L. (2004). Optimum storage volume of rooftop rain water harvesting systems for domestic use. *Journal of the American Water Resources Association, 40*(4), 901–912. doi:10.1111/j.1752-1688.2004.tb01054.x

Lindberg, K., McCool, S., & Stankey, G. (1997). Rethinking carrying capacity. *Annals of Tourism Research, 24*(2), 461–465. doi:10.1016/S0160-7383(97)80018-7

Little, A. (2015, October 28). Weather on demand: Making it rain is now a global business. *Bloomberg Businessweek*. Retrieved from http://www.bloomberg.com/features/2015-cloud-seeding-india/

Liu, J., Wang, R., Yang, J., & Shi, Y. (2010). The relationship between consumption and production system and its implications for sustainable development of China. *Ecological Complexity, 7*(2), 212–216. doi:10.1016/j.ecocom.2010.02.003

Lobell, D. & Gourdji, S. (2012). The influence of climate change on global crop productivity. *Plant Physiology*. .112.20829810.1104/pp

Lobell, D. B., & Asner, G. P. (2003). Climate and management contributions to recent trends in US agricultural yields. *Science, 299*(5609), 1032–1032. doi:10.1126/science.1077838 PMID:12586935

Lobell, D. B., Bala, G., & Duffy, P. B. (2006). Biogeophysical impacts of cropland management changes on climate. *Geophysical Research Letters, 33*(6), L06708. doi:10.1029/2005GL025492

Londhe, A., Talati, J., Singh, L. K., Vilayasseril, M., Dhaunta, S., Rawlley, B., . . . Mathew, R. P. (2004). *Urban-hinterland water transactions: a scoping study of six class I Indian cities*. IWMI Tata Water Policy Programme Working Paper.

Londhe, S. L. (2016). Climate Change and Agriculture: Impacts, Adoption, and Mitigation. In *Handbook of Research on Climate Change Impact on Health and Environmental Sustainability*. IGI Global. doi:10.4018/978-1-4666-8814-8.ch019

Loucks, C., Barber-Meyer, S., Hossain, M. A. A., Barlow, A., & Chowdhury, R. M. (2010). Sea level rise and 450 tigers: Predicted impacts to Bangladeshs Sundarbans mangroves. *Climatic Change, 98*(1-2), 291–298. doi:10.1007/s10584-009-9761-5

Lovett, J. C. (2006). Climate and Society. *African Journal of Ecology, 44*(4), 421–422. doi:10.1111/j.1365-2028.2006.00708.x

Lu, J., & Cai, M. (2009). Stabilization of the atmospheric boundary layer and the muted global hydrological cycle response to global warming. *Journal of Hydrometeorology, 10*(1), 347–352. doi:10.1175/2008JHM1058.1

Lüthi, M., & Funk, M. (2001). Modelling heat flow in a cold, high-altitude glacier:Interpretation of measurements from ColleGnifetti. *Swiss Alps. J. Glaciol., 47*(157), 314–324. doi:10.3189/172756501781832223

Macknick, J., Newmark, R., Heath, G., & Hallett, K. (2011). *A review of operational water consumption and withdrawal factors for electricity generating Technologies. National Renewable Energy Laboratory*. NREL.

Mahe, G., LHote, Y., Olivry, J. C., & Wolting, G. (2001). Trends and discontinuities in regional rainfall of West and Central Africa: 19511989. *Hydrological Sciences-Journal-des Sciences Hydrologiques, 46*(2), 221–226. doi:10.1080/02626660109492817

Mall, R. K., Gupta, A., Singh, R., Singh, R. S., & Rathore, L. S. (2006). Water resources and climate change: An Indian perspective. *Current Science, 90*(12), 1610–1626.

Maltby, E., & Acreman, M. C. (2011). Ecosystem services of wetlands: Pathfinder for a new paradigm. *Hydrological Sciences Journal, 56*(8), 1341–1359. doi:10.1080/02626667.2011.631014

Manmohn to Chair Timaimukh Meeting. (2010, October 13). *Bangladesh News*. Retrieved from http://www.independent-bangladesh.com/2010101312037/country/manmohan-to-chair-tipaimukh-meeting.html

Mann, W., Lipper, L., Tennigkeit, T., McCarthy, N., Branca, G., & Paustian, K. (2009). *Food Security and Agricultural Mitigation in Developing Countries: Options for Capturing Synergies*. Rome: FAO.

Manyanhaire, I. O., Matewa, J. T., Svotwa, E., & Munhuwa, E. (2009). Water shortage Related problems following the closure of Mhangura copper mine in Mashonaland West Province of Zimbabwe. *Journal of Sustainable Development in Africa, 11*, 239–252.

Maplecroft, V. (2013). *Website*. Retrieved from http://www.maplecroft.com/portfolio/new-analysis/2013/10/30/31-global-economicoutput-forecast-face-high-or-extreme-climate-change-risks-2025-maplecroft-riskatlas/

Mara, D. (2009). Design Manual for Waste Stabilization Ponds in India. University of Leeds. Available at http://www.leeds.ac.uk/civil/ceri/water/tphe/publicat/wspwarm/wspwarm.html

Mbow, C., Smith, P., Skole, D., Duguma, L., & Bustamante, M. (2014). Achieving mitigation and adaptation to climate change through sustainable agroforestry practices in Africa. *Current Opinion in Environmental Sustainability, 6*, 8–14. doi:10.1016/j.cosust.2013.09.002

McCaffery, S. C. (2007). *The Law of The International Watercourses* (2nd ed.). New York: Oxford University Press.

McCarl, B. A., Musumba, M., Smith, J. B., Kirshen, P., Jones, R., El-Ganzori, A., & Bayoumi, M. et al. (2015). Climate change vulnerability and adaptation strategies in Egypts agricultural sector. *Mitigation and Adaptation Strategies for Global Change, 20*(7), 1097–1109. doi:10.1007/s11027-013-9520-9

McCarthy, J. J. (2001). Climate Change 2001: Impacts, Adaptation and Vulnerability. Contribution of Working Group II to the Third Assessment Report of the Intergovernmental Panel on Climate Change.

McCarthy, N., Lipper, L., & Branca, G. (2011). *Climate-smart agriculture: smallholder adoption and implications for climate change adaptation and mitigation. Mitigation of Climate Change in Agriculture*. Working Paper, 3.

McCarthy, T. S. (2006). Groundwater in the wetlands of the Okavango Delta, Botswana, and its contribution to the structure and function of the ecosystem. *Journal of Hydrology (Amsterdam), 320*(3), 264–282. doi:10.1016/j.jhydrol.2005.07.045

McCartney, M., Rebelo, L. M., Xenarios, S., & Smakhtin, V. (2013). *Agricultural water storage in an era of climate change: assessing need and effectiveness in Africa.* Colombo, Sri Lanka: International Water Management Institute (IWMI). Retrieved February 21, 2015, from http://www.iwmi.cgiar.org/Publications/IWMI_Research_Reports/PDF/PUB152/RR152.pdf

McCarty, J. P. (2001). Ecological consequences of recent climate change. *Conservation Biology, 15*(2), 320–331. doi:10.1046/j.1523-1739.2001.015002320.x

McDonald, R. I., Green, P., Balk, D., Fekete, B. M., Revenga, C., Todd, M., & Montgomery, M. (2011). Urban growth, climate change, and freshwater availability. *Proceedings of the National Academy of Sciences of the United States of America, 108*(15), 6312–6317. doi:10.1073/pnas.1011615108 PMID:21444797

McGranahan, G., Balk, D., & Anderson, B. (2007). The rising tide: Assessing the risks of climate change and human settlements in low elevation coastal zones. *Environment and Urbanization, 19*(1), 17–37. doi:10.1177/0956247807076960

McGregor, D. (2004). Coming full circle: Indigenous knowledge, environment, and our future. *American Indian Quarterly, 28*(3), 385–410. doi:10.1353/aiq.2004.0101

McNeal, B. L., Layfield, D. A., Norvell, W. A., & Rhoades, J. D. (1968). Factors influencing hydraulic conductivity of soils in the presence of mixed-salt solutions. *Soil Science Society of America Proceedings, 32*(2), 187–190. doi:10.2136/sssaj1968.03615995003200020012x

Meehl, G. A., Arblaster, J. M., & Branstator, G. (2012). Mechanisms contributing to the warming hole and the consequent US east–west differential of heat extremes. *Journal of Climate, 25*(18), 6394–6408. doi:10.1175/JCLI-D-11-00655.1

Mehta, D., & Mehta, N. K. (2013). Interlinking of Rivers in India: Issues & Challenges. *Geo-Eco-Marina*. Retrieved July 10, 2016, from http://www.geoecomar.ro/website/publicatii/Nr.19-2013/12_mehta_web_2013.pdf

Mehta, K., & Satpathy, T. (2011). *Escaping Poverty: The Ralegan Siddhi Case*. Chronic Poverty Research Centre and Indian Institute of Public Administration.

Mehta, N. (2012, July 23). Dialectics of Tipaimukh Dam: Issues and Concerns. *Vivekananda International Foundation*. Retrieved July 10, 2016, from http://www.vifindia.org/article/2012/july/23/dialectics-of-tipaimukh-dam-issues-and-concerns

Meijers, N. (2014). The platform for good food innovation. *Water-Saving Agricultural Technologies*. Retrieved January 25, 2016, From http://foodtechconnect.com/2014/03/21/6-water-saving-agricultural-technologies/

Meijers, N. (2014). *Water-Saving Agricultural Technologies*. Retrieved January 25, 2016 From http://foodtechconnect.com/2014/03/21/6-water-saving-agricultural-technologies/

Mekala, G. D., Davidson, B., Samad, M., & Boland, A. M. (2008). *Wastewater reuse and recycling systems: a perspective into India and Australia*. Colombo, Sri Lanka: International Water Management Institute (IWMI). Retrieved from www.iwmi.cgiar.org/Publications/Working_Papers/working/WOR128.pdf

Mekong River Commission. (1995). *Agreement on The Cooperation For The Sustainable Development of The Mekong River Basin.* Retrieved from http://www.mrcmekong.org/assets/Publications/policies/agreement-Apr95.pdf

Mendelsohn, R., & Neumann, J. E. (1999). *The Economic Impact of Climate Change on the Economy of the United States*. Cambridge, UK: Cambridge University Press. doi:10.1017/CBO9780511573149

Mendonca, R., Barros, N., Vidal, L. O., Pacheco, F., Kosten, S., & Roland, F. (2012). Greenhouse Gas Emissions from Hydroelectric Reservoirs: What Knowledge Do We Have and What is Lacking? In *Greenhouse Gases - Emission, Measurement and Management.* InTech. Retrieved from http://cdn.intechopen.com/pdfs-wm/32342.pdf

Menzies, C. R. (Ed.). (2006). *Traditional ecological knowledge and natural resource management.* U of Nebraska Press.

Mertz, O., Halsnæs, K., Olesen, J. E., & Rasmussen, K. (2009). Adaptation to climate change in developing countries. *Environmental Management, 43*(5), 743–752. doi:10.1007/s00267-008-9259-3 PMID:19184576

Michael, A. M. (2009). *Irrigation: theory and practice.* New Delhi, India: Vikas Publishing House Pvt Ltd.

Millennium Ecosystem Assessment. (2005). Ecosystems and human well-being: biodiversity synthesis. World Resources Institute.

Miller, A. (2015). *Creating Value-Added weather & climate services through innovative PPPs, C4 Eco Solutions.* United Nations Development Program. Accessed on 22 Feb 2016 http://www.adaptation-undp.org/resources/planning-meeting-presentations/cirda-ppp-workshop-presentation-alan-miller

Miller, K. A., Rhodes, S. L., & MacDonnell, L. J. (1997). Water allocation in a changing climate: Institutions and adaptation. *Climatic Change, 35*(2), 157–177. doi:10.1023/A:1005300529862

Ministry of External Affairs. (2011). *Framework Agreement on Cooperation for Development between India and Bangladesh.* Retrieved from http://mea.gov.in/bilateral-documents.htm?dtl/5218/Framework+Agreement+on+Cooperation+for+Development+between+India+and+Bangladesh

Ministry of External Affairs. (2014). *Joint Statement on the Third Meeting of the India-Bangladesh Joint Consultative Commission.* Retrieved from http://mea.gov.in/bilateral-documents.htm?dtl/24024/Joint+Statement+on+the+Third+Meeting+of+the+IndiaBangladesh+Joint+Consultative+Commission

Ministry of Water Resources (MoWR) State Level Report (2001). *3rd National census of minor irrigation schemes.* GoI.

Ministry of Water Resources (MoWR) State Level Report. (1988). *1st National census of minor irrigation schemes, Government of India.* Author.

Ministry of Water Resources, River Development & Ganga Rejuvenation. (1960). *The Indus Water Treaty.* Adopted on 9th September 1960. Retrieved from http://wrmin.nic.in/forms/list.aspx?lid=346

Ministry of Water Resources, River Development & Ganga Rejuvenation. (n.d.). *Inter Linking of Rivers.* Retrieved from http://wrmin.nic.in/forms/list.aspx?lid=1279

Misra, A. K. (2014). Climate change and challenges of water and food security. *International Journal of Sustainable Built Environment, 3*(1), 153–165. doi:10.1016/j.ijsbe.2014.04.006

Mitsch, W. J., Bernal, B., Nahlik, A. M., Mander, Ü., Zhang, L., Anderson, C. J., & Brix, H. et al. (2013). Wetlands, carbon, and climate change. *Landscape Ecology, 28*(4), 583–597. doi:10.1007/s10980-012-9758-8

Mkandawire, M., & Dudel, E. G. (2007). Are Lemna spp. effective phytoremediation agents. *Bioremediation. Biodiversity and Bioavailability, 1*(1), 56–71.

Mkandawire, M., Taubert, B., & Dudel, E. G. (2004). Capacity of *Lemna gibba* L.(Duckweed) for uranium and arsenic phytoremediation in mine tailing waters. *International Journal of Phytoremediation, 6*(4), 347–362. doi:10.1080/16226510490888884 PMID:15696706

MoEFCC. (2006). *Form-A: Project Details - Justification for locating the project in the forest area.* Retrieved from http://forestsclearance.nic.in/search.aspx?cat_id=FP/MN/HYD/1589/2012

MoEFCC. (2008). *Tipaimukh Multipurpose HEP Project (1500 MW) in Manipur & Mizoram States – Environmental Clearance.* No.J-12011/63/2006-IA.I. Retrieved from http://www.indiaenvironmentportal.org.in/files/html/Environment_Clearance/Tipaimukh%20Multipurpose%20HEP%20Project.html

MoEFCC. (2012). Agenda No.2: Diversion of 25822.14 hectares Forest land for construction of Tipaimukh Hydroelectric Project in Manipur (File No. 8-63/2005-FC). *Proceeding of the Forest Advisory Committee Meeting*. Retrieved from http://forestsclearance.nic.in/writereaddata/FAC_Minutes/jan2012.pdf

MoEFCC. (2013a). *Agenda Item No.1: Diversion of forest land for 1551.60 ha of forest land under the forest (Conservation) Act, 1980 for construction of Tipaimukh Hydro Electric Project in Mizoram* (File No. 8-02/2012-FC). Minutes of the Forest Advisory Committee Meeting held on 13-14 August 2013. Retrieved from http://www.ercindia.org/files/fac/FAC%20Minutes%2013-14%20Aug_2013_9113124312181minutesAugust.pdf

MoEFCC. (2013b). Agenda No.3: Diversion of 22777.50 hectares Forest land for construction of Tipaimukh Hydroelectric Project in Manipur (File No. 8-63/2005-FC). *Proceeding of the Forest Advisory Committee Meeting*. Retrieved from http://www.internationalrivers.org/files/attached-files/forestadvisorycommittee_july_2013_minutes.pdf

MoEFCC. (2013c). *Fact Sheet: Diversion of 25822.14 hectares Forest land for construction of Tipaimukh Hydroelectric Project in Manipur* (File No. 8-63/2005-FC). Retrieved from http://forestsclearance.nic.in/search.aspx?cat_id=FP/MZ/HYD/2187/2012

MoEFCC. (2013d). *Fact Sheet: Diversion of forest land for 1551.60 ha of forest land under the forest (Conservation) Act, 1980 for construction of Tipaimukh Hydro Electric Project in Mizoram* (File No. 8-02/2012-FC). Retrieved from http://forestsclearance.nic.in/search.aspx?cat_id=FP/MZ/HYD/2187/2012

MoEFCC. (2013e). *Site Inspection Report: Diversion of forest land for 1551.60 ha of forest land under the forest (Conservation) Act, 1980 for construction of Tipaimukh Hydro Electric Project in Mizoram* (File No. 8-02/2012-FC). Retrieved from http://forestsclearance.nic.in/search.aspx?cat_id=FP/MZ/HYD/2187/2012

Moench, M., & Kumar, D. (1995). Distinctions between "Efficiency" and "Sustainability": The role of energy prices in groundwater management. *The Challenge of the Balance: Environmental Economics in India*, 305-309.

Mohan Kumar, R. (2003) Financing of Wastewater Treatment Projects. Infrastructure Development Finance Corporation and Confederation of Indian Industries, Water Summit 2003, Hyderabad, India.

Moniz, E. J., Jacoby, H. D., Meggs, A. J. M., Armstrong, R. C., Cohn, D. R., Connors, S. R., …Yang, Y. (2011). The Future of Natural Gas. Massachusetts Institute of Technology.

Montgomery, J., & Murray, M. (2014). *Markell seeks new tax to improve water.* Retrieved from http://www.delawareonline.com/story/news/2014/03/04/markell-water-quality-tax/6008877/

Moorthi, S.M. (2014). FASALSoft – An ISRO software framework for crop production forecast using remote sensing data analysis. *Journal of Geomatics*, *8*(1).

More rain batters crops in Andhra Pradesh, Telangana. (2015, April 14). *The Times of India*, p. 1.

Moreddu, C. (2000). Overview of Farm Household Strategies and Government Intervention. In Income Risk Management in Agriculture. Paris: OECD.

Morison, J. I. L., & Morison, J. I. L. (1996). Climate change and crop growth. *Environmental Management and Health*, *7*(2), 24–27. doi:10.1108/09566169610112980

Moser, S. C., & Boykoff, M. T. (2013). *Successful adaptation to climate change: Linking science and policy in a rapidly changing world.* Routledge.

Most Urbanised State, & the The Times of India. (2011). *NIDM Maharashtra National Disaster Risk Reduction Portal.* Retrieved January 25, 2016, From http://nidm.gov.in/pdf/dp/Maharashtra.pdf

MSPPE (Ministère de la Santé Publique de la Province de l'Equateur). (2010). *Plan Provincial de la santé de l'Equateur.* CEMUBAC.

Murray, A., Cofie, O., & Dreschsel, P. (2011). Efficiency indicators for waste-based business models: Fostering private-sector participation in wastewater and faecal-sludge management. *Water International, 36*(4), 505–521. doi:10.1080/02508060.2011.594983

Murtaza, G., & Zia, M. H. (2012). *FAO Report - UNW-AIS. Individual's capacity development on the safe use of waste-water in agriculture in Pakistan.* Available at http://www.ais.unwater.org/ais/pluginfile.php/232/mod_page/content/124/pakistan_murtaza_finalc ountryreport2012.pdf

Murtaza, G., Ghafoor, A., Owens, G., Qadir, M., & Kahlon, U. Z. (2009). Environmental and economic benefits of saline-sodic soil reclamation using low-quality water and soil amendments in conjunction with a rice-wheat cropping system. *Journal Agronomy & Crop Science, 195*(2), 124–136. doi:10.1111/j.1439-037X.2008.00350.x

Murtaza, G., Ghafoor, A., & Qadir, M. (2006). Irrigation and soil management strategies for using saline-sodic water in a cotton-wheat rotation. *Agricultural Water Management, 81*(1-2), 98–114. doi:10.1016/j.agwat.2005.03.003

Murtaza, G., Ghafoor, A., Qadir, M., Owens, G., Aziz, M. A., Zia, M. H., & Saifullah, . (2010). Disposal and use of sewage on agricultural lands in Pakistan: A review. *Pedosphere, 20*(1), 23–34. doi:10.1016/S1002-0160(09)60279-4

Murtaza, G., Murtaza, B., Niazi, N. K., & Sabir, M. (2014). Soil contaminants: sources, effects and approaches for remediation. In *Improvement of crops in the era of climatic changes* (pp. 171–196). Springer Science, Business Media. doi:10.1007/978-1-4614-8824-8_7

Mutsuga, M., Kawamura, Y., Sugita-Konishi, Y., Hara-Kudo, Y., Takatori, K., & Tanamoto, K. (2006). Migration of formaldehyde and acetaldehyde into mineral water in polyethylene terephthalate (PET) bottles. *Food Additives and Contaminants, 23*(2), 212–218. doi:10.1080/02652030500398361 PMID:16449065

Myers, L. E. (1975). Water harvesting 2000 B.C. to 1974 A.D. *Proc. Water Harvesting Syrup.* Phoenix, AZ: USDA. Retrieved January 25, 2016, From http://www.fao.org/docrep/s2022e/s2022e07.htm

Myhre, G., Shindell, D., Bréon, F. M., Collins, W., Fuglestvedt, J., Huang, J., & Zhang, H. et al. (2013). Anthropogenic and Natural Radiative Forcing. In T. F. Stocker, D. Qin, G. K. Plattner, M. Tignor, S. K. Allen, J. Boschung, & P. M. Midgley et al. (Eds.), *Climate Change 2013: The Physical Science Basis. Contribution of Working Group I to the Fifth Assessment Report of the Intergovernmental Panel on Climate Change.* Cambridge, UK: Cambridge University Press.

Naithani, A. K., Nainwal, H. C., & Prasad, C. P. (2001). Geomorphological evidences of retreat of Gangotri glacier and its characteristics. *Current Science, 80,* 87–94.

Naithani, A. K., Nainwal, H. V., Sati, K. K., & Parsad, C. (2001). Geomorphological evidences of retreat of the Gangotri glacier and its characteristics. *Current Science, 80*(1), 87–94.

Narain, V. (2010). *Peri urban water security in a context of urbanization and climate change: A review of concepts and relationships.* Peri Urban Water Security Discussion Paper Series, Paper No. 1, SaciWATERs. Retrieved February 16, 2015, from http://saciwaters.org/periurban/newsletter/pdfs/discussion_papper_1.pdf

Narain, V., Anand, P., & Banerjee, P. (2013). *Periurbanization in India: A review of the literature and evidence.* Report for the project – Rural to Urban Transitions and the Peri-urban Interface. SaciWATERs. India. Retrieved February 16, 2015, from https://saciwaters.org/east-west-center/pdf/status-paper.pdf

Narain, S. V. (2006). *Community-led Alternatives to Water Management: India Case Study, Human development report.* UNDP.

Nariman, F. S. (2009). Inter-State Water Disputes: A Nightmare! In R. R. Iyer (Ed.), *Water and the Laws in India* (pp. 32–57). New Delhi: Sage Publications. doi:10.4135/9788132104247.n3

National Academy of Sciences. (n.d.). *Our Energy System*. Retrieved from http://needtoknow.nas.edu/energy/interactive/our-energy-system/

National Hydroelectric Power Corporation. (n.d.a). *Tipaimukh*, Retrieved from http://www.nhpcindia.com/projectdetail.htm?CatId=6&ProjectId=53

National Hydroelectric Power Corporation. (n.d.b). *Power Stations*. Retrieved from http://www.nhpcindia.com/project.htm

National Portal Content Management Team. (2011). *Command area development and water management program*. Government of India.

National Water Policy. (2012). *Government of India*. Retrieved January 25, 2016 http://wrmin.nic.in/writereaddata/NationalWaterPolicy/NWP2012Eng6495132651.pdf

Nature Conservancy. (2014). *Partners turn collective focus to Delaware's freshwater resources*. Retrieved from http://www.nature.org/ourinitiatives/regions/northamerica/unitedstates/delaware/delaware-acorns-newsletter-spring-2014.pdf

Nazarea, V. D. (1999). *Ethnoecology: Situated Knowledge/Located Lives*. Tucson, AZ: University of Arizona Press.

NDTV.com. (2015). *Efforts underway to resolve Teesta issue with Bangladesh*. Author.

Neeru-Chettu can make AP drought-proof in 5 years: CM. (2015, February 19). *The Hindu*, p. 1.

Neil Adger, W., Arnell, N. W., & Tompkins, E. L. (2005). Successful adaptation to climate change across scales. *Global Environmental Change*, *15*(2), 77–86. doi:10.1016/j.gloenvcha.2004.12.005

Nelson, G. C., Rosegrant, M. W., Koo, J., Robertson, R., Sulser, T., Zhu, T., . . . Lee, D. (2009). *Food Policy Report-Climate Change, Impact on Agriculture and Costs of Adaptation*. International Food Policy Research Institute. Retrieved February 19, 2015, from http://www.ifpri.org/sites/default/files/publications/pr21.pdf

Nelson, D. R., Adger, W. N., & Brown, K. (2007). Adaptation to environmental change: Contributions of a resilience framework. *Annual Review of Environment and Resources*, *32*(1), 395–419. doi:10.1146/annurev.energy.32.051807.090348

Nestlé Waters North America. (2010). *Environmental life cycle assessment of drinking water alternatives & consumer beverage consumption in north America*. Retrieved from http://beveragelcafootprint.com/Nestlé

Neumann, M. B., Rieckermann, J., Hug, T., & Gujer, W. (2015). Adaptation in hindsight: Dynamics and drivers shaping urban wastewater systems. *Journal of Environmental Management*, *151*, 404–415. doi:10.1016/j.jenvman.2014.12.047 PMID:25594746

Nianthi, R. K. W. J., & Husain, Z. (2004). Impact of Climate Change on Rivers with Special Reference to River-Linking Project. Academic Press.

Niccolucci, V., Botto, S., Rugani, B., Nicolardi, V., Bastianoni, S., & Gaggi, C. (2011). The real water consumption behind drinking water: The case of Italy. *Environmental Management*, *92*(10), 2611–2618. PMID:21741755

Nichols, E., Spector, S., Louzada, J., Larsen, T., Amezquita, S., & Favila, M. E. (2008). Ecological functions and ecosystem services provided by Scarabaeinae dung beetles. *Biological Conservation*, *141*(6), 1461–1474. doi:10.1016/j.biocon.2008.04.011

Niyogi, D. (2015, March 23). Crop Damage in Many Areas. *Down-to-earth*, p. 1.

Norgaard, R. B. (1984). Traditional agricultural knowledge: Past performance, future prospects, and institutional implications. *American Journal of Agricultural Economics*, *66*(5), 874–878. doi:10.2307/1241018

Norgaard, R. B. (2010). Ecosystem services: From eye-opening metaphor to complexity blinder. *Ecological Economics*, *69*(6), 1219–1227. doi:10.1016/j.ecolecon.2009.11.009

Noronha, L., Lourenco, N., Lobo-Ferreira, J. P., Lleopart, A., Feoli, E., Sawkar, K., & Chachadi, A. (2003). *Managing Coastal tourism; perspectives from India and the European Union in Coastal Tourism, environment and sustainable local development. In The Energy and Resources Institute* (pp. 419–446). TERI.

Novich, E. (2012). *Predicting Lemna minor growth rate response to temperature fluctuations.* B.Sc Thesis in Environmental Studies Programme, University of Nebraska, Lincoln, NE. Retrieved from http://digitalcommons.unl.edu/envstudtheses/93/

Olsson, P., Folke, C., Galaz, V., Hahn, T., & Schultz, L. (2007). Enhancing the fit through adaptive co-management: Creating and maintaining bridging functions for matching scales in the Kristianstads Vattenrike Biosphere Reserve Sweden. *Ecology and Society*, *12*(1), 28.

Onjala, J. (2006). *The 5th Human Development Report for Kenya 2006: Human Security and Human Development.* Academic Press.

Organisation for Economic Co-operation and Development. (2014). *Climate change, water and agriculture: Towards resilient agricultural and water systems.* Working paper. Organisation for Economic Co-operation and Development (OECD).

Osmani, A. R. (2014). Conventional Energy to Renewable Energy: Perspective for India. *The NEHU Journal, 12*(2). Retrieved July 10, 2016, from http://www.nehu.ac.in/Journals/Journal_Jul_Dec14_Art3.pdf

Oster, J. D. (2001). *Magnesium effects on soil physical properties—Hydraulic conductivity and infiltration rate.* SL-PHYCHM. Available online at http://esce.ucr.edu/oster/SLPHYCb.html

Oster, J. D. (1994). Irrigation with poor quality water. *Agricultural Water Management*, *25*(3), 271–297. doi:10.1016/0378-3774(94)90064-7

Oster, J. D., & Grattan, S. R. (2002). Drainage water reuse. *Irrigation and Drainage Systems*, *16*(4), 297–310. doi:10.1023/A:1024859729505

Oster, J. D., & Jayawardane, N. S. (1998). Agricultural management of sodic soils. In M. E. Sumner & R. Naidu (Eds.), *Sodic soil: Distribution, management and environmental consequences* (pp. 126–147). New York: Oxford University Press.

Ostrom, E. (2005). *Self-governance and forest resources. In Terracotta reader: a market approach to the environment* (pp. 131–155). New Delhi: Academic Foundation.

Ostrom, E. (2005). *Understanding institutional diversity* (1st ed.). Princeton, NJ: Princeton University Press.

Ostrom, E. (2010). Polycentric systems for coping with collective action and global environmental change. *Global Environmental Change*, *20*(4), 550–557. doi:10.1016/j.gloenvcha.2010.07.004

Ostrom, V., Tiebout, C. M., & Warren, R. (1961). The organization of government in metropolitan areas: A theoretical inquiry. *The American Political Science Review*, *55*(04), 831–842. doi:10.1017/S0003055400125973

Pachauri, R. K., & Reisinger, A. (Eds.). (2007). *Climate change 2007 synthesis report: Summary for policymakers.* IPCC Secretariat.

Padigala, B. (2014). Social Capital and Local Institutions: A Perspective to Assess Communities Adaptation Potential to Climate Change. In Handbook of Climate Change Adaptation. Springer. doi:10.1007/978-3-642-40455-9_114-1

Padigala, B., Sen, S., Sharma, M. C., & Kraleti, S. (2011). Social Capital and Local Institutions- A Perspective to Assess Communities Adaptation Potential to Climate Change. *Second National Research Conference on Climate Change*. IIT Delhi.

Padma, T. V. (2007). *Monitoring Climate Change at the top of the world.* Science and Development Network. Retrieved on April 25, 2008 from cidev.net/en/features/ monitoring – climate-at- the-top-of-the-world.html.

Pahl-Wostl, C. (2007). Transitions towards adaptive management of water facing climate and global change. *Water Resources Management, 21*(1), 49–62. doi:10.1007/s11269-006-9040-4

Palanisami, K., Radhamani, S., Vasanthi, C., & Tamilselvi, K. N. (2002). Implementation of DPAP (VIII Batch): Socio-economic baseline survey of ten watersheds of Coimbatore District: A report by District Rural Development Agency. Academic Press.

Pandey, D. N. (1998). *Ethnoforestry: Local Knowledge for Sustainable Forestry and Livelihood Security.* Udaipur, India: Himanshu Publications.

Pandey, D. N., Gupta, A. K., & Anderson, D. M. (2003). Rainwater harvesting as an adaptation to climate change. *Current Science, 85*(1), 46–59.

Pant, N., & Verma, R. K. (1983). *Farmers Organisation and Irrigation Management.* New Delhi: Ashish Publishing House.

Parent, J., Cucuzzella, C., & Reveret, J. P. (2013). Revisiting the role of LCA and SLCA in the transition towards sustainable production and consumption. *Life Cycle Assessment, 18*(9).

Parihar, J. S., & Oza, M. P. (2006). FASAL: An integrated approach for crop assessment and production forecasting agriculture and hydrology application of remote sensing. *Proceedings of SPIE*, 6411. doi:10.1117/12.713157

Parmesan, C., & Yohe, G. (2003). A globally coherent fingerprint of climate change impacts across natural systems. *Nature, 421*(6918), 37–42. doi:10.1038/nature01286 PMID:12511946

Parry, M. L. (2000). *Assessment of Potential Effects and Adaptations for Climate Change in Europe. The Europe Acacia Project.* University of East Anglia.

Parry, M., Arnell, N., McMichael, T., Nicholls, R., Martens, P., Kovats, S., & Fischer, G. et al. (2001). Millions at risk: Defining critical climate threats and targets. *Global Environmental Change, 11*(3), 181–183. doi:10.1016/S0959-3780(01)00011-5

Patel, D. K., & Kanungo, V. K. (2010). Phytoremediation Potential Of Duckweed (Lemnaminor L: A Tiny Aquatic Plant). Bioscan, 5(3), 355-358.

Pathak, M. (2010). Climate Change: Uncertainty for Hydropower Development in Nepal. *Hydro Nepal*, (6), 31-34.

Pathak, H., Ladha, J. K., Aggarwal, P. K., Peng, S., Das, S., Singh, Y., & Gupta, R. K. et al. (2003). Climatic potential and on-farm yield trends of rice and wheat in the Indo-Gangetic Plains. *Field Crops Research, 80*(3), 223–234. doi:10.1016/S0378-4290(02)00194-6

Patil Yogesh, B. (2012). Development of an innovative low-cost industrial waste treatment technology for resource conservation - A case study with gold-cyanide emanated from SMEs. *Procedia: Social and Behavioral Sciences, 37*, 379–388. doi:10.1016/j.sbspro.2012.03.303

Patil, Y., & Rao, P. (2013). *Applied Bioremediation–Active and Passive Approaches.* Intech Open Science Online Publishers. doi:10.5772/50859

Patil, Y., & Rao, P. (2015a). Industrial waste management in the era of climate change - A smart sustainable model based on utilization of active and passive biomass. In W. L. Filho (Ed.), *Handbook on Climate Change Adaptation* (pp. 1–13). Springer-Verlag Berlin Heidelberg. doi:10.1007/978-3-642-38670-1_49

Patil, Y., & Rao, P. (2015b). A Novel Biological Strategy for the Management Industrial Effluents with Simultaneous Recovery of Metal Resource. *Advanced Materials Research*, *1130*, 640–643. doi:10.4028/www.scientific.net/AMR.1130.640

Pattanaaik, S.K., Sen, D., Kumar, N., Moyong, O., & Debnath, P. (2012). Traditional system of water management in watersheds of Arunachal Pradesh. *Indian Journal of Traditional Knowledge*, *11*(4), 719-723.

Peacocke, C. (2006). *Mental action and self-awareness*. Blackwell.

Pearce, D. W., & Turner, R. K. (1990). *Economics of Natural resources and the environment.* New York: Harvester Wheatsheaf.

Peattie, K. (2010). Green consumption: Behavior and norms. *Annual Review of Environment and Resources*, *35*(1), 195–228. doi:10.1146/annurev-environ-032609-094328

Peirson, W., Davey, E., Jones, A., Hadwen, W., Bishop, K., Beger, M., & Gray, L. et al. (2015). Opportunistic management of estuaries under climate change: A new adaptive decision-making framework and its practical application. *Journal of Environmental Management*, *163*, 214–223. doi:10.1016/j.jenvman.2015.08.021 PMID:26321531

Perrow, C. (2011). *Normal accidents: Living with high risk technologies*. Princeton University Press.

Pescod, M. B. (1992). *Wastewater Treatment and Use in Agriculture*. FAO Irrigation and Drainage Paper 47. Food and Agriculture Organization (FAO) of the United Nations.

M. B. Pescod, & A. Arar (Eds.). 1988. Treatment and use of sewage effluent for irrigation.*Proceedings of the FAO Regional Seminar on the Treatment and Use of Sewage Effluent for Irrigation.*

Peterson, D. F., & Keller, A. A. (1990). Effects of climate change on U.S. irrigation. *Journal of Irrigation and Drainage Engineering*, *116*(2), 194–210. doi:10.1061/(ASCE)0733-9437(1990)116:2(194)

Phand, S., & Arya, H. P. S. (2007). Aadarsh Gaon Yojana-Way Towards Gandhi's Dream Village: Self-Sufficient, Prosperous and Peaceful. *Journal of Rural Development*, *26*(2), 207.

Pieroni, A., Giusti, M. E., & Quave, C. L. (2011). Cross-cultural ethnobiology in the Western Balkans: Medical ethnobotany and ethnozoology among Albanians and Serbs in the Pešter Plateau, Sandžak, South-Western Serbia. *Human Ecology*, *39*(3), 333–349. doi:10.1007/s10745-011-9401-3

Pilla, V. (2015, February 23). Watershed scheme gets a digital push in Andhra Pradesh. *LiveMint*, p. 1.

Pimm, S. L. (2007). Africa: Still the Dark Continent. *Conservation Biology*, *21*(3), 567–569. doi:10.1111/j.1523-1739.2007.00697.x PMID:17531033

Pisolkar, Y. M. (2008). *A Geomorphic Study of Beach and Creek Erosion at Devbag, coastal Maharashtra*. (Unpublished Ph.D. Thesis). University of Pune.

Pisolkar, Y., Rawat, A., Mahankale, N., & Arora, R. (2013). *Growing Tourism on Tarkarli –Devbag Spit, its Consequences and Suggested Strategy for Sustainable Tourism, Coastal Maharashtra, India, Report 'Minor Research Project' Symbiosis Centre for Management Studies*. Pune: SCMS.

Plesse, M. (2015, October 29). *Indus River Water Treaty, Global Food and Water Crisis Research Program.* Future Directions International. Retrieved from http://www.futuredirections.org.au

Poff, N. L., Brinson, M. M., & Day, J. W. (2002). *Aquatic ecosystems and global climate change*. Arlington, VA: Pew Center on Global Climate Change.

Pogutz, S., & Micale, V. (2011). Sustainable consumption and production. *Society and Economy*, *33*(1), 29–50. doi:10.1556/SocEc.33.2011.1.5

Pomeroy, R. S., & Berkes, F. (1997). Two to tango: The role of government in fisheries co-management. *Marine Policy*, *21*(5), 465–480. doi:10.1016/S0308-597X(97)00017-1

Ponce, R., Bosello, F., & Giupponi, C. (2012). Integrating Water Resources into Computable General Equilibrium Models – A Survey. Climate Change and Sustainable Development Series.

Ponce, V. M. (2006). *Groundwater Utilization and Sustainability*. Retrieved February 9, 2015, from Lecture Notes Online Web site: http://groundwater.sdsu.edu/

Poonuchaphai, W. (2010). *Country experience: Sustainable consumption and production in Thailand.* Paper presented at a Path of Good Governance Seminar, Seoul, South Korea.

Power, W. (2009). *Life cycle management.* Retrieved from http://www.unep.fr/shared/publications/pdf/DTIx1208xPA-LifeCycleApproach-Howbusinessusesit.pdf

Prakash, R., Yogesh, P., Viraja, B., & Shilpa, K. (2015). Environment and social sustainability approaches in policy and practice: A case study of symbiosis international university, Pune, India. In W. L. Filho (Ed.), *Transformative Approaches to Sustainable Development at Universities, Working Across Disciplines* (pp. 471–488). Springer International Publishing.

Priestley, C. H. B. (1966). The limitation of temperature by evaporation in hot climates. *Agricultural Meteorology*, *3*(3-4), 241–246. doi:10.1016/0002-1571(66)90031-8

Qadir, M., Ghafoor, A., Boers, Th. M., & Murtaza, G. (2003). Agricultural water management in water starved countries: Challenges and opportunities. *Agricultural Water Management*, *62*(3), 165–185. doi:10.1016/S0378-3774(03)00146-X

Qadir, M., Quillérou, E., Nangia, V., Murtaza, G., Singh, M., Thomas, R. J., & Noble, A. D. et al. (2014). Economics of salt-induced land degradation and restoration. *Natural Resources Forum*, *38*(4), 282–295. doi:10.1111/1477-8947.12054

Qadir, M., & Schubert, S. (2002). Degradation processes and nutrient constraints in sodic soils. *Land Degradation & Development*, *13*(4), 275–294. doi:10.1002/ldr.504

Qadir, M., Wichelns, D., Raschid-Sally, L., McCornick, P. G., Drechsel, P., Bahri, A., & Minhas, P. S. (2010). The challenges of wastewater irrigation in developing countries. *Agricultural Water Management*, *97*(4), 561–568. doi:10.1016/j.agwat.2008.11.004

Rahaman, M. M. (2009a). Principles of International Water Law: Creating Effective Transboundary Water Resources Management. *Int. J. Sustainable Society.*, *1*(3), 209. doi:10.1504/IJSSOC.2009.027620

Rahaman, M. M. (2009b). Principles of Transboundary Water Resources Management and Ganges Treaties: An Analysis. *Water Resource Development*, *25*(1), 159–173. doi:10.1080/07900620802517574

Raina, V. K. (2009). *Himalayan Glaciers – A State-of-Art Review of Glacial Studies, Glacial Retreat and Climate Change. Ministry of Environment and Forests, discussion paper*. Government of India.

Rains won't impact inflation target: Arvind Subramanian. (2015, April 10). *The Economic Times*, p. 11.

Ramachandran, R. (2001). The receding Gangotri. *Frontline*, *18*(7).

Ramanathan, A. L., Shruti Jose, P. G., Sharma, P., & Linda, A. (2005). *Glacial Retreat of Himalayas in the Recent Past (Pleistocene to Recent Decade).* IGCP 514: field workshop.

Ramani, K., Ramanujan, D., Bernstein, W.Z., Zhao, F., Sutherland, J., Handwerker, C., … Thurston, D. (2010). *Integrated sustainable life cycle design.* doi: 10.1115/1.4002308

Ramanjaneyulu, G. V., & Kuruganti, K. (2009). *Sustaining Agriculture in the era of Climate Change in India.* Centre for Sustainable Agriculture.

Rao, P., & Patil, Y. (2015). Climate resilience in natural ecosystems in India - Technology adoption and the use of local knowledge processes and systems. In Handbook on Climate Change Adaptation. Springer-Verlag.

Rao, P., & Patil, Y. (2015). Climate Resilience in Natural Ecosystems in India: Technology Adoption and the Use of Local Knowledge Processes and Systems. In Handbook of Climate Change Adaptation (pp. 2063-2077). Springer Berlin Heidelberg.

Rao, P., & Patil, Y. (2015). Climate Resilience in Natural Ecosystems in India: Technology Adoption and the Use of Local Knowledge Processes and Systems. In Handbook of Climate Change Adaptation. Springer Berlin Heidelberg.

Rao, P., Areendran, G., & Sareen, R. (2008) Potential impacts of climate change in the Uttarakhand Himalayas. Mountain Forum Bulletin, 8(1).

Rao, P., Patil, Y., & Gupte, R. (2013). Education for Sustainable Development: Trends in Indian Business Schools and Universities in a Post Liberalization Era. In Sustainability Assessment Tools in Higher Education Institutions - Mapping Trends and Good Practices Around the World. Springer International Publishing.

Rao, P., & Patil, Y. (2015). Climate Resilience in Natural Ecosystems in India: Technology Adoption and the Use of Local Knowledge Processes and Systems. In W. L. Filho (Ed.), *Handbook on Climate Change Adaptation* (pp. 2063–2077). Springer-Verlag Berlin Heidelberg. doi:10.1007/978-3-642-38670-1_52

Rao, P., & Patil, Y. (2015). *Climate Resilience in Natural Ecosystems in India: Technology Adoption and the Use of Local Knowledge Processes and Systems. In Handbook of Climate Change Adaptation* (pp. 2063–2077). Springer Berlin Heidelberg.

Rao, P., Patil, Y., Ketkar, M., Bhat, V., & Kulkarni, S. (2014). Sustainability in an educational institution: analysing the transition to paperless e-processes, an Indian case. In U. M. Azeiteiro, W. L. Filho, & S. Caeiro (Eds.), *ELearning and Educationfor Sustainability, Environmental Education, Communication and Sustainability Series* (Vol. 35, pp. 215–233). Peter Lang GmbH International Academic Publishers.

Rao, S. T. (2005). *Cloud seeding for India: An effective weapon to fight the droughts.* Hyderabad: The Book Syndicate.

Raschid-Sally, L., & Jayakody, P. (2008). *Drivers and characteristics of wastewater agriculture in developing countries:Results from a global assessment. IWMI Research Report 127.* Colombo, Sri Lanka: IWMI.

Raskin, P. D., Electris, C., & Rosen, R. A. (2010). *The century ahead: Searching for sustainability.* doi:10.3390/su2082626

Ratthanin, S. (2010, October). Bottled water. *Marketeer, 11*(128), 114–117.

Rawls, J. (1971). A Theory of Justice. Cambridge, MA: Harvard University.

Rawson, H. M., Gifford, R. M., & Condon, B. N. (1995). Temperature Gradient Chambers for Research on Global Environment Change. I. Portable Chambers for Research on Short-Stature Vegetation. *Plant, Cell & Environment, 18*(9), 1048–1054. doi:10.1111/j.1365-3040.1995.tb00616.x

Ray, A. (2012, June 8). India-Bangladesh: Water Dispute and River Diplomacy. *International Security Relations and Security Network ETH Zurich*. Retrieved from http://isnblog.ethz.ch/global-voices/india-bangladesh-water-disputes-and-river-diplomacy

Reilly, J., Tubiello, F., McCarl, B., Abler, D., Darwin, R., Fuglie, K., & Rosenzweig, C. et al. (2003). U.S. Agriculture and Climate Change: New Results. *Climatic Change*, *57*(1/2), 43–69. doi:10.1023/A:1022103315424

Renaud, F. G., Sudmeier-Rieux, K., & Estrella, M. (Eds.). (2013). *The role of ecosystems in disaster risk reduction*. Tokyo, Japan: United Nations University Press.

Rengasamy, P., Greene, R. S. B., & Ford, G. W. (1986). Influence of magnesium on aggregate stability in sodic red-brown earths. *Australian Journal of Soil Research*, *24*(2), 229–237. doi:10.1071/SR9860229

Rhein, M., Rintoul, S. R., Aoki, S., Campos, E., Chambers, D., Feely, R. A., & Wang, F. et al. (2013). Observations: Ocean. In T. F. Stocker, D. Qin, G. K. Plattner, M. Tignor, S. K. Allen, J. Boschung, & P. M. Midgley et al. (Eds.), *Climate Change 2013: The Physical Science Basis. Contribution of Working Group I to the Fifth Assessment Report of the Intergovernmental Panel on Climate Change*. Cambridge, UK: Cambridge University Press.

Rhoades, J. S., Kandiah, A., & Mashali, A. M. (1992). *The use of saline waters for crop production. Irrigation and Drainage Paper 48*. Rome, Italy: Food and Agriculture Organization of the United Nations.

Richards, A., & Singh, N. (2001). *Inter-State water disputes in India: Institutions and policies*. University of California and University of Maryland.

Richardson, S. D., & Reynolds, J. M. (2000). An overview of glacial hazards in the Himalayas. *Quaternary International*, *65*, 31–47. doi:10.1016/S1040-6182(99)00035-X

Rieu-Clark, A., Moynihan, R., & Magsi, B. (2012). *UN Watercourse Convention: User's Guide*. IHP-HELP Centre for Water Law, Policy and Science.

Ringler, C., Bryan, E., Biswas, A., & Cline, S. (2010). Water and Food Security Under Global Change. *Water Resources Development and Management*, *1*, 3–15. doi:10.1007/978-3-642-04615-5_1

Roberts, T. R. (1975). Geographic distribution of African freshwater fishes. *Zoological Journal of the Linnean Society*, *57*(4), 249–319. doi:10.1111/j.1096-3642.1975.tb01893.x

Rodwan, J. G. (2010). *Bottled water 2009: Challenge circumstances persist, future growth anticipated, US and international developments and statistics*. Retrieved from www.bottledwater,org/files/2009BWstats.pdf

Root, T. L., Price, J. T., Hall, K. R., Schneider, S. H., Rosenzweig, C., & Pounds, J. A. (2003). Fingerprints of global warming on wild animals and plants. *Nature*, *421*(6918), 57–60. doi:10.1038/nature01333 PMID:12511952

Rosenberg, N. J. (1992). Adaptation of Agriculture to Climate Change. *Climatic Change*, *21*(4), 385–405. doi:10.1007/BF00141378

Rosenberg, N. J., Kimball, B., Martin, P., & Cooper, C. (1988). Climate change, CO_2 enrichment and evapotranspiration. In P. E. Waggoner (Ed.), *Climate and Water: Climate Change, Climatic Variability, and the Planning and Management of U.S. Water Resources*. New York, NY: John Wiley and Sons.

Rosenzweig, C., & Liverman, D. (1992). Predicted Effects of Climate Change on Agriculture: A Comparison of Temperate and Tropical Regions. In S. K. Majumdar (Ed.), *Global Climate Change: Implications, Challenges, and Mitigation Measures*. Guelph: Natural Resources Canada.

Roy, S.S., & Balling, R.C., Jr. (2005). Analysis of trends in maximum and minimum temperature, diurnal temperature range, and cloud cover over India. *Geophysical Research. Letters, 32*, 4.

Roy, S. J., Negrão, S., & Tester, M. (2014). Salt resistant crop plants. *Current Opinion in Biotechnology, 26*, 115–124. doi:10.1016/j.copbio.2013.12.004 PMID:24679267

Ruet, J., Saravanan, V. S., & Zérah, M. H. (2002). *The Water and Sanitation Scenario in Indian Metropolitan Cities: Resource and Management in Delhi, Calcutta, Chennai, Mumbai.* CSH Occasional Paper No.6. New Delhi: Centre de Sciences Humaines.

Ruet, J., Gambiez, M., & Lacour, E. (2007). Private appropriation of resource: Impact of peri-urban farmers selling water to Chennai Metropolitan Water Board. *Cities (London, England), 24*(2), 110–121. doi:10.1016/j.cities.2006.10.001

Rules, H. (1966). *The Helsinki rules on the uses of the waters of international rivers*. International Law Association.

Russi, D., ten Brink, P., Farmer, A., Badura, T., Coates, D., & Förster, J. … Davidson, N. (2013). The economics of ecosystems and biodiversity for water and wetlands. London: IEEP.

Sachs, J. (2015). *The Age of Sustainable Development*. Columbia University Press. doi:10.7312/sach17314

Saifullah. (2012). *Enhancing crop productivity on salt-affected soils through combined use of soil applied gypsum and pre-sowing seed treatments*. Final Technical Progress Report for Endowment Fund Secretariat (EFS) funded project (July 2008 – June 2011). Institute of Soil & Environmental Sciences, University of Agriculture.

Saifullah, G. A., Murtaza, G., & Qadir, M. (2002). Brackish tube well water promotes growth of rice and wheat and reclamation of saline-sodic soils. *Pakistan Journal of Soil Science, 21*, 83–88.

Saini, R. (2008). *Glacier dynamics water resource assessment and landscape evolution in Miyar basin, Lahaul Himalayas, Himachal Pradesh*. (Unpublished M Phil Dissertation), JNU, New Delhi, India.

Sakhuja, N. (2008). Hiware Bazaar - A Village with 54 Millionaires. *Down to Earth, 16*(20080131).

Sakhuja, N. (2008). Hiware Bazaar-A Village with 54 Millionaires. Down. *Earth (Waukesha, Wis.)*, 16.

Salati, E. (1987). The forest and the hydrological cycle. In R. E. Dickinson (Ed.), *The Geophysiology of Amazonia* (pp. 273–296). New York: Wiley.

Sally, M. (2014, July 23). Artificial rainfall: Cloud seeding companies may play rainmakers. *The Economic Times*, p. 1.

Salomans, W., Turner, R. K., de Lacerda, L. D., & Ramachandran, S. (1999), Perspectives on Integrated Coastal Zone Management. Springer Verlag. doi:10.1007/978-3-642-60103-3

Samir, S., Prakash, R., & Yogesh, P. (2012). City based analysis of MSW to energy generation in India, calculation of state-wise potential and tariff comparison with EU. *Procedia: Social and Behavioral Sciences, 37*, 407–416. doi:10.1016/j.sbspro.2012.03.306

Sampattakul, S. (2010). *Climate Thailand.* Paper presented at the Climate Thailand Conference, Bangkok, Thailand

Samuelson, P. A. (1947). *Foundations of Economic Analysis*. Harvard University Press.

Sanchez Duron, N. (1988). *Mexican experience in using sewage effluent for large scale irrigation. Treatment and Use of Sewage Effluent for Irrigation* (M. B. Pescod & A. Arar, Eds.). Sevenoaks, UK: Butterworths.

Sangameswaran, P. (2006). Equity in watershed development: A case study in Western Maharashtra. *Economic and Political Weekly*, 2157–2165.

Sangameswaran, P. (2008). Community formation, ideal villages and watershed development in Western India. *Journal of Development Studies, Routledge, 44*(3), 384–408. doi:10.1080/00220380701848426

Sarfraz, H. (2013). Revisiting the 1960 Indus Water Treaty. *Water International, 38*(2), 205. doi:10.1080/02508060.2013.784494

Schattle, H. (2008). Education for global citizenship: Illustrations of ideological pluralism and adaptation. *Journal of Political Ideologies, 13*(1), 73–94. doi:10.1080/13569310701822263

Schenk, A. K. (n.d.). The Effects of Various Carbon Dioxide Concentrations on Lemna minor. *Dens, 130*(5), 172-1.

Schewe, J., Heinke, J., Gerten, D., Haddeland, I., Arnell, N. W., Clark, D. B., & Kabat, P. et al. (2014). Multimodel assessment of water scarcity under climate change. *Proceedings of the National Academy of Sciences of the United States of America, 111*(9), 3245–3250. doi:10.1073/pnas.1222460110 PMID:24344289

Schmid, P., Kohler, M., Meierhofer, R., Luzi, S., & Wegelin, M. (2008). Does the reuse of PET bottles during solar water disinfection pose a health risk due to the migration of plasticisers and other chemicals into the water? *Water Research, 42*(20), 5054–5060. doi:10.1016/j.watres.2008.09.025 PMID:18929387

Schofield, R. N. (1986). Evolution of the Shatt al Arab Boundary Dispute. Menas Publishers.

Schultz, P. W., Nolan, J. M., Cialdini, R. B., Goldstein, N. J., & Griskevicius, V. (2007). The constructive, destructive and reconstructive power of social norms. *Psychological Science, 18*(5), 429–434. doi:10.1111/j.1467-9280.2007.01917.x PMID:17576283

Schultz, W. (2007). Reward. *Scholarpedia, 2*(3), 1652. doi:10.4249/scholarpedia.1652

Sciubba, S. (2003, November). Extended exergy accounting applied to energy recovery from waste: The concept of total recycling, Strategies of Waste Energy Usage. *Energy, 28*(13), 1315–1334. doi:10.1016/S0360-5442(03)00111-7

Scott, Wilson, and JPS. (2005). *Rural Water Supply and Sanitation Project.* Maharashtra with World Bank Assistance Sector Status Study.

Scott, C., Drechsel, P., Raschid-Sally, L., Bahri, A., Mara, D., Redwood, M., & Jiménez, B. (2010). Wastewater irrigation and health: challenges and outlook for mitigating risks inlow-income countries. In P. Drechsel, C. Scott, L. Rashid-Sally, M. Redwood, & A. Bahri (Eds.), *Wastewater Irrigation and Health: Assessing and Mitigating Risk in Low-Income Countries* (pp. 381–394). London: Earthscan.

Seabra, A. B., Rai, M., & Durán, N. (2014). Nano carriers for nitric oxide delivery and its potential applications in plant physiological process: A mini review. *Journal of Plant Biochemistry and Biotechnology, 23*(1), 1–10. doi:10.1007/s13562-013-0204-z

Seager, J., Robinson, K., Van der Schaaf, C., & Gabizon, S. (2009). *Gender-Disaggregated Data on Water and Sanitation. Department for Economic and Social Affairs (UN-DESA) and UN-Water Decade Programme on Capacity Development.* New York: UNW-DPC.

Seckler, D. W. (1996). *The new era of water resources management: from" dry" to" wet" water savings* (Vol. 1). IWMI.

Sehgal, R. (2008, April). The days of cheap food are over: M.S. Swaminathan. *Info Change News & Features.*

Seifelislam, A. (2013). *A rapid assessment of the impact of wetland's degradation on groundwater resources in the fast growing city Chennai. (Unpublished Internship-Report for the European Program Chance2Sustain).* Pondicherry, India: French Institute of Pondicherry.

Seligman, D. (2011). *Resolving Inter-state Water Conflicts: A comparison of the way India and the United States address disputes on inter-state rivers.* Institute of Water Policy, Lee Kyun Yew School of Public Policy, National University of Singapore. Retrieved from www.lkyspp.nus.edu.sg/iwp

Sen, S., Sharma, M., Padigala, B., Kraleti, S., & Deswal, S. (2011). *Response and adaptability to climatic variability in agro-ecosystems: a case of Miyar basin, Lahaul Himalaya, India.* Unpublished paper.

Senapati, M. R., Behera, B., & Mishra, S. R. (2013). Impact of Climate Change on Indian Agriculture and Its Mitigating Priorities. *American Journal of Environment Protection*, *4*(4), 109–111. doi:10.12691/env-1-4-6

Seneviratne, S. I., Nicholls, N., Easterling, D., Goodess, C. M., Kanae, S., Kossin, J., . . . Zhang, X. (2012). Changes in climate extremes and their impacts on the natural physical environment. In A Special Report of Working Groups I and II of the Intergovernmental Panel on Climate Change (IPCC SREX Report). doi:10.1017/CBO9781139177245.006

Seneviratne, S. I., Corti, T., Davin, E. L., Hirschi, M., Jaeger, E. B., Lehner, I., & Teuling, A. J. et al. (2010). Investigating soil moisture-climate interactions in a changing climate: A review. *Earth-Science Reviews*, *99*(3-4), 125–161. doi:10.1016/j.earscirev.2010.02.004

Sengupta, N. (1993). *User Friendly Irrigation Designs*. New Delhi: Sage Publications.

Sethi, A. S., & Kanwar, B. S. (1989). Irrigation Potential of Kuhls and its Impact on Crops Productivity — A Case Study of Himachal Pradesh. In V. S. Mahajan (Ed.), *Agriculture and Rural Economy in India* (pp. 170–179). New Delhi: Deep and Deep Publications.

Sharma, N., & Kanwar, P. (2009). Indigenous water conservation systems—A rich tradition of rural Himachal Pradesh. *Indian Journal of Traditional Knowledge*, *8*(4), 510-513.

Sharma, M. (2006). The making of moral authority: Anna Hazare and watershed management Program in Ralegan Siddhi. *Economic and Political Weekly*, 1981–1988.

Sheikh, I. A. (1989). Country report on problems of waterlogging and salinity in Pakistan. In *Proceeding Information Seminar on Waterlogging and Salinity Research in Some Major Problem Countries* (pp. 1-10).Lahore, Pakistan: International Waterlogging And Salinity Research Institute.

Shekhar, M. S., Chand, H., Kumar, S., Srinivasan, K., & Ganju, A. (2010). Climate-change studies in the western Himalaya. *Annals of Glaciology*, *51*(54), 105–112. doi:10.3189/172756410791386508

Shen, L., Nieuwlaar, E., Worrell, E., & Patel, M. K. (2011). Life cycle energy and GHG emissions of PET recycling: Change-oriented effects. *Int J Life Cycle Assess*, *16*(6), 522–536. doi:10.1007/s11367-011-0296-4

Shivakoti, G. P., & Shrestha, S. G. (2008). FMIS of Nepal: Balancing Water Use and Environment Conservation for Sustainable Livelihood. *Himalayan Journal of Sociology and Anthropology*, *1*, 64–74. doi:10.3126/hjsa.v1i0.1554

Shobeiri, S. M. (2005). *A comparative study of environmental awareness and attitude of teachers and students of secondary schools in India and Iran.* University of Mysore.

Shotyk, W., & Krachler, M. (2007). Contamination of bottled water with antimony leaching from polyethylene terephthalate (PET) increase upon storage. *Environmental Science & Technology*, *41*(5), 1560–1563. doi:10.1021/es061511+ PMID:17396641

Shrestha, U. B., Gautam, S., & Bawa, K. S. (2012). Widespread climate change in the Himalayas and associated changes in local ecosystems. *PLoS ONE*, *7*(5), e36741. doi:10.1371/journal.pone.0036741 PMID:22615804

Shrivastara, A., & Kothari, A. (2012). *Adding fuel to fire in Churning the Earth, the making of global India.* Penguin Group.

Shrivastava, A. (2013, December 10). Green Tribunal Halts Mapithel Dam in Manipur. *Down to Earth*. Retrieved July 10, 2016, from http://www.downtoearth.org.in/content/green-tribunal-halts-mapithel-dam-manipur

Siddiqi, A., & Anadon, L. (2011). The water-energy nexus in Middle East and North Africa. *Energy Policy*, *39*(8), 4529–4540. doi:10.1016/j.enpol.2011.04.023

Singh, K. K. (1991). Irrigation Management by Farmers; The Indian Experience. In Farmer Participation and Irrigation Organisation. Westview Press.

Singh, K. M., Singh, R. K. P., Meena, M. S., & Kumar, A. (2013). *Water Policy in India: A Review*. Available at SSRN 2226877

Singh, R. K. P., & Singh, K. M. (2012). Climate Change, Agriculture and ICT: An Exploratory Analysis. *ICT for Agricultural Development Under Changing Climate*, 17-28.

Singh, S. (2010). Hiware Bazaar - A village of millionaires. Maharashtra.

Singh, P., Arora, M., & Goel, N. K. (2006). Effect of climate change on runoff of a glacierized Himalayan basin. *Hydrological Processes*, *20*(9), 1979–1992. doi:10.1002/hyp.5991

Singh, P., & Bengtsson, L. (2004). Hydrological sensitivity of a large Himalayan basin to climate change. *Hydrological Processes*, *18*(13), 2363–2385. doi:10.1002/hyp.1468

Singh, P., & Bengtsson, L. (2005). Impact of warmer climate on melt and evaporation for the rainfed, snowfed and glacierfed basins in the Himalayan region. *Journal of Hydrology (Amsterdam)*, *300*(1), 140–154. doi:10.1016/j.jhydrol.2004.06.005

Singh, P., Haritashya, U. K., Kumar, N., & Singh, Y. (2006). Hydrological characteristics of the Gangotri Glacier, central Himalayas, India. *Journal of Hydrology (Amsterdam)*, *327*(1-2), 55–67. doi:10.1016/j.jhydrol.2005.11.060

Singh, P., Haritashya, U. K., Ramasastri, K. S., & Kumar, N. (2005). Prevailing weather conditions during summer seasons around Gangotri Glacier. *Current Science*, *88*(5), 753–760.

Singh, P., Polglase, L., & Wilson, D. (2009). Role of Snow and Glacier melt runoff modeling in Hydropower projects in the Himalayan Region.*International Conference on Water, Environment, Energy and Society*.

Singleton, S. G. (1998). *Constructing cooperation: the evolution of institutions of comanagement*. University of Michigan Press.

Siqueira, A. (1999). Global consumption and local processes. In *Population, consumption and environment interrelations: a tourist spot scenario*. New Delhi: Tata Energy Research Institute.

Skees, J. R., Hazell, P., & Miranda, M. (1999). *New Approaches to Crop Insurance in Developing Countries*. EPTD Discussion Paper No. 55. International Food Policy Research Institute.

Skelcher, C. (2005). Jurisdictional integrity, polycentrism, and the design of democratic governance. *Governance: An International Journal of Policy, Administration and Institutions*, *18*(1), 89–110. doi:10.1111/j.1468-0491.2004.00267.x

Smajgl, A., Ward, J., & Pluschke, L. (2016). The water–food–energy Nexus – Realising a new paradigm. *Journal of Hydrology (Amsterdam)*, *533*, 533–540. doi:10.1016/j.jhydrol.2015.12.033

Smit, B., & Olga, P. (2001). Adaptation to Climate Change in the Context of Sustainable Development and Equity. In J. J. McCarthy, O. F. Canzianni, N. A. Leary, D. J. Dokken, & K. S. White (Eds.), *Impacts, Adaptation, and Vulnerability - Contribution of Working Group II to the Third Assessment Report of the Intergovernmental Panel on Climate Change*. Cambridge University Press.

Smit, B., & Skinner, M. W. (2002). Adaptation options in agriculture to climate change: A typology. *Mitigation and Adaptation Strategies for Global Change*, *7*(1), 85–114. doi:10.1023/A:1015862228270

Smith, B., Burton, I., Klein, R. J., & Wandel, J. (2000). An anatomy of adaptation to climate change and variability. *Climatic Change*, *45*(1), 223–251. doi:10.1023/A:1005661622966

Smith, C. J., Oster, J. D., & Sposito, G. (2015). Potassium and magnesium in irrigation water quality assessment. *Agricultural Water Management*, *157*, 59–64. doi:10.1016/j.agwat.2014.09.003

Smithers, J., & Blay-Palmer, A. (2001). Technology Innovation as a Strategy for Climate Adaptation in Agriculture. *Applied Geography (Sevenoaks, England)*, *21*(2), 175–197. doi:10.1016/S0143-6228(01)00004-2

Sommer, R., Glazirina, M., Yuldashev, T., Otarov, A., Ibraeva, M., Martynova, L., & de Pauw, E. et al. (2013). Impact of climate change on wheat pro-ductivity in Central Asia. *Agriculture, Ecosystems & Environment*, *178*, 78–99. doi:10.1016/j.agee.2013.06.011

Song, H. S., & Hyun, J. C. (1999). A study of the comparison of the various waste management scenarios for PET bottles using the life-cycle assessment (LCA) methodology. *Resources, Conservation and Recycling*, *27*(3), 267–284. doi:10.1016/S0921-3449(99)00022-1

Sonwa, D., Bele, Y., Somorin, O., Jum, C., & Nkem, J. (2009). Adaptation for forests and communities in the Congo Basin. *European Tropical Forest Research Network News*, *50*, 93–100.

Southern African Development Community. (2000). *Revised Protocol on Shared Watercourse*. Retrieved from http://www.sadc.int/files/3413/6698/6218/Revised_Protocol_on_Shared_Watercourses_-_2000_-_English.pdf

Spagat, E. (2012, November 21). Colorado river pact signed between the US and Mexico. *Huffington Post*, p. 1.

Srinivasan, V. (2008). *An integrated framework for analysis of water supply strategies in a developing city: Chennai, India* (Doctoral dissertation). Available from ProQuest Dissertations and Theses database. (UMI No. 3332994)

Srinivasan, V. (2015). Reimagining the past–use of counterfactual trajectories in socio-hydrological modelling: The case of Chennai, India. *Hydrology and Earth System Sciences*, *19*(2), 785–801. doi:10.5194/hess-19-785-2015

Srinivasan, V., Gorelick, S. M., & Goulder, L. (2010). Factors determining informal tanker water markets in Chennai, India. *Water International*, *35*(3), 254–269. doi:10.1080/02508060.2010.487931

Srinivasan, V., Seto, K. C., Emerson, R., & Gorelick, S. M. (2013). The impact of urbanization on water vulnerability: A coupled human–environment system approach for Chennai, India. *Global Environmental Change*, *23*(1), 229–239. doi:10.1016/j.gloenvcha.2012.10.002

Srivastava, D. (2001). *Glaciology of Indian Himalaya: A Bilingual Contribution in 150 Year of geological survey of India. Special Publication No. 63*. GSI.

SS (Secrétariat à la Santé). (2007). *République Démocratique du Congo: Annuaire des données sanitaires 2006*. Ministère de la Santé Publique de la République Démocratique du Congo, Kinshasa.

Starr, A. (2000). *Naming the enemy: Anti-corporate movements confront globalization*. Zed Books.

Stern, N. (2007). *Stern Review on the Economics of Climate Change*. Cambridge, UK: Cambridge University Press. doi:10.1017/CBO9780511817434

Stockholm Environmental Institute. (2011). *Understanding the nexus. Background paper for the Bonn 2011 nexus conference - the water, energy and Food security nexus: Solutions for the green economy*. Working paper. Stockholm Environmental Institute (SEI).

Story, S. (2014). The water energy dependency: Unlocking the smart cities. *Energy Biz*, *11*, 20–22.

Strauss, M., & Blumenthal, U. J. (1989). *Human waste use in agriculture and aquaculture: utilization practices and health perspectives*. IRCWD Report No. 08/89. International Reference Centre for Waste Disposal.

Strawn, D. G., Bohn, H. L., & O'Connor, G. A. (2015). *Soil chemistry*. New York, NY: Wiley-Blackwell.

Strzepek, K. M., Yohe, G. W., Tol, R. S., & Rosegrant, M. W. (2008). The value of the high Aswan Dam to the Egyptian economy. *Ecological Economics*, *66*(1), 117–126. doi:10.1016/j.ecolecon.2007.08.019

Suwajittanont, W., & Chai-ittipornwong, T. (2011). *Role of mass media for enhancing the sufficiency economy*. Bangkok, Thailand: Public Relations Department.

Sze, T. L. (2010). *Spaces for consumption*. Retrieved fromhttp://www.satepub.com/upm-data/35384_01_Miles_CH_01.pdf

T21. (2012). *Kenya Threshold 21 (T21) Model*.

Takahashi, Y. (2010). *Sustainable PET bottle recycling system in Japan for sound material cycle*. Japan: The University of Tokyo, Graduate School of Frontier Sciences.

Talule, D., & Suryavanshi, S. (2014). An empirical assessment of the model of rural sustainability: Tracing the roots of inspiration to the Central Government policy aiming to develop one ideal village per parliamentary constituency. *Asian Journal of Science and Technology*, *5*(12), 796–803.

Taminiau, J. (2015). *A paradigm analysis of ecological sustainability*. (Unpublished doctoral dissertation). University of Delaware, Newark, DE.

Taminiau, J., & Byrne, J. (2015, December). *A Polycentric Response to the Climate Change Challenge Relying on Creativity, Innovation, and Leadership*. Paper presented at the 21st Conference of the Parties (COP 21) to United Nations Convention on Climate Change (UNFCCC), Paris, France.

Tanner, T., Mitchell, T., Polack, E., & Guenther, B. (2009). Urban governance for adaptation: assessing climate change resilience in Ten Asian Cities. *IDS Working Papers*, (315), 01-47.

Tanzi, A., & Arcari, M. (2001). *The United Nations Convention on The Law of International Watercourse: A Framework for Sharing*. London: Kluwer Law International.

Tarawali, G., Manyong, V. M., Carsky, R. J., Vissoh, P. V., Osei-Bonsu, P., & Galiba, M. (1999). Adoption of improved fallows in West Africa: Lessons from mucuna and stylo case studies. *Agroforestry Systems*, *47*(1-3), 93–122. doi:10.1023/A:1006270122255

Taylor, R. G., Scanlon, B., Döll, P., Rodell, M., Van Beek, R., Wada, Y., & Konikow, L. et al. (2013). Ground water and climate change. *Nature Climate Change*, *3*(4), 322–329. doi:10.1038/nclimate1744

Termeer, C., Dewulf, A., van Rijswick, H., van Buuren, A., Huitema, D., Meijerink, S., & Wiering, M. et al. (2011). The regional governance of climate adaptation: A framework for developing legitimate, effective, and resilient governance arrangements. *Climate Law*, *2*, 159–179.

Termpittayapaisith, A. (2011, February). *Low carbon development in Thailand*. Paper presented at the First National Seminar on Green Growth Policy Tools, Bangkok, Thailand

Thayyem, R. J., Gergan, J. T., & Dobhah, D. P. (2007). Role of glaciers and snow cover on headwater river hydrology in monsoon regime – Micro-scale study of the Din Gad catchment, Garhwal Himalaya, India. *Current Science*, *80*(1), 87–94.

The Constitution of India. (2010). *Administrative relations: Disputes relating to waters*. New Delhi: Universal Law Publishing Co.

The Energy Reasearch Institute. (2000). Population consumption, and environment interrelations: a tourist spot scenario. New Delhi: Tata Energy Research Institute.

The Forest (Conservation) Act 1980. (2014). No.69 of 1980, 27th December 1980, in Environmental and Pollution Laws. New Delhi: Universal Law Publishing, 2014.

The International Court of Justice. (2010). *Pulp Mills on the River Uruguay (Argentina v. Uruguay). Summaries of Judgments 20 April 2010*. Retrieved from http://www.icj-cij.org/docket/files/135/15873.pdf

The Scheduled Tribes and Other Traditional Forest Dwellers (Recognition of Forest Rights) Act. (2007). *No2 of 2007, 29th December 2006, The Gazette of India, Extraordinary Part-II Section-1*. New Delhi: Ministry of Law and Justice. Retrieved from http://www.egazette.nic.in/WriteReadData/2007/E_2_2010_050.pdf

The United Nations Framework Convention on Climate Change (UNFCCC). (2015). *Adoption of the Paris Agreement*. Available at: https://unfccc.int/resource/docs/2015/cop21/eng/l09r01.pdf

The Water Act, 2002 No. 8 of 2002, (2002).

Themelis, N. (2003). *An overview of the global waste-to-energy industry*. Waste Management World.

Themelis, N., & Mussche, C. (2014). *Energy and economic value of municipal solid waste (MSW), including non-recycled plastics (NRP), currently land-filled in the fifty states*. Earth Engineering Center.

Themelis, N., & Ulloa, P. (2007, June). Methane generation in landfills. *Renewable Energy*, *32*(7), 1243–1257. doi:10.1016/j.renene.2006.04.020

Thieme, M. L., Abell, R., Stiassny, M. L. J., Skelton, P., Lehner, B., Teugels, G. G., & Olson, D. et al. (Eds.). (2005). *Freshwater ecoregions of Africa and Madagascar: A conservation assessment*. Island Press.

Thomas, V. (2015). *Better results in a dynamic world*. Asian Development Bank. Accessed on 30 Nov 2015 http://www.adb.org/publications/inspiring-adbs-future-better-results-dynamic-region

Thornton, P. K., & Cramer, L. (2012). *Impacts of climate change on the agricultural and aquatic systems and natural resources within the CGIAR's mandate*. Academic Press.

Tikhonovich, I. A., & Provorov, N. A. (2009). From plant–microbe interactions to symbiogenetics: A universal paradigm for the interspecies genetic integration. *Annals of Applied Biology*, *154*(3), 341–350. doi:10.1111/j.1744-7348.2008.00306.x

Tompkins, E. L., & Adger, W. (2004). Does adaptive management of natural resources enhance resilience to climate change? *Ecology and Society*, *9*(2), 10.

Toprak, Z. F., Hamidi, N., Toprak, Ş., & Şen, Z. (2013). Climatic identity assessment of the climate change. *International Journal of Global Warming*, *5*(1), 30–45. doi:10.1504/IJGW.2013.051480

Tourenq, C., Brook, M., Knuteson, S., Shuriqi, M. K., Sawaf, M., & Perry, L. (2011). Hydrogeology of Wadi Wurayah, United Arab Emirates, and its importance for biodiversity and local communities. *Hydrological Sciences Journal*, *56*(8), 1407–1422. doi:10.1080/02626667.2011.631139

Trenberth, K. E. (2011). Changes in precipitation with climate change. *Climate Research*, *47*(1), 123–138. doi:10.3354/cr00953

Tukker, A., Emmert, S., Charter, M., Vezzoli, C., Sto, E., Munch Andersen, M., & Lahlou, S. et al. (2008). Fostering change to sustainable consumption and production: An evidence based view. *Journal of Cleaner Production*, *16*(11), 1218–1225. doi:10.1016/j.jclepro.2007.08.015

Turner, W. R., Oppenheimer, M., & Wilcove, D. S. (2009). A force to fight global warming. *Nature*, *462*(7271), 278–279. doi:10.1038/462278a PMID:19924191

Tutin, C. E. G., & White, L. J. T. (1998). Primates, phenology and frugivory: Present, past and future patterns in the Lopé Reserve, Gabon. In D. M. Newbery, H. H. T. Prins, & N. Brown (Eds.), *Dynamics of populations and Communities in the Tropics* (pp. 309–338). Oxford, UK: Blackwell Science.

UN-DESA (United Nations Department of Economic and Social Affairs). (2005). *A Gender Perspective on Water Resources and Sanitation.* Background paper NO 2. Commission on Sustainable Development Twelfth Session DESA/DSD/2005/2United Nations, New York.

UNDP. (2008). *Glacial Lake Outburst Floods remains the biggest hazard of Bhutan UNDP Bhutan.* Retrieved April 8,2008, from http://www.undp.org.bt/info/pr08/ps_GLOF_8April.pdf

UNDP. (n.d.a). *The Millennium Development Goal.* Retrieved from http://www.undp.org/content/undp/en/home/sdgoverview/mdg_goals.html

UNDP. (n.d.b). *Sustainable Development Goal.* Retrieved from http://www.undp.org/content/undp/en/home/sdgoverview/

UNEP. (2000). *Climate Change and Dams: An Analysis of the linkages between the UNFCCC Legal Regime and Dams.* Prepared for Thematic Review II.2: Dams and global change. Retrieved from http://re.indiaenvironmentportal.org.in/files/DAMs%20and%20climate%20change.pdf

UNEP. (2007). *Global outlook for ice & snow.* UNEP/GRID-Arendal.

UNESCO. (2007). *Case Studies on Climate Change and World Heritage.* UNESCO World Heritage Centre.

UNFCCC (Producer). (2015). *2015 Paris Climate Conference: Paris Agreement.* Retrieved from http://unfccc.int/meetings/paris_nov_2015/items/9445.php

UNFCCC. (2015). *India's Intended Nationally Determined Contribution: Working Towards Climate Justice.* Retrieved from http://www4.unfccc.int/submissions/indc/Submission%20Pages/submissions.aspx

UN-Habitat. (2011). *Cities and climate change: global report on human settlements 2011.* London: Earthscan.

UNICEF, FAO, & SaciWATERs. (2013). *Water in India: Situation and Prospects, 11.* Retrieved February 9, 2015, from www.unicef.org/india/Final_Report.pdf

United Nations Conference on Environment and Development. (1992). *Rio Declaration on Environment and Development.* UN Doc. A/CONF.151/5/Rev. 1, 31 ILM 874.

United Nations Department of Economic and Social Affairs (UNDESA). (2010). *Trends in sustainable development towards sustainable consumption and production.* Retrieved from https://www.un.org/ development/desa/en/

United Nations Environment Programme (UNEP). (2012). *Global outlook on sustainable consumption and production policies: taking action together.* Retrieved from http://www.unep.org/publications/ contents/pub_details_search.asp?ID=6251

United Nations Environment Programme (UNEP). (2015). *Sustainable information forum 2015.* Retrieved from http://www.cop21paris.org/

United Nations Framework Convention on Climate Change (UNFCCC). (2015). *Adoption of the Paris agreement: draft decision -/CP.21*. Paper presented at the Conference of the Parties Twenty-First Session, Paris, France. Retrieved from https://unfccc.int/resource/docs/2015/cop21/eng/l09r01.pdf

United Nations. (1997). *Convention on the Law of the Non-navigational Uses of International Watercourses*. Retrieved from http://legal.un.org/ilc/texts/instruments/english/conventions/8_3_1997.pdf

United Nations. (2011). *Seven billion and growing: the role of population policy in achieving sustainability*. Technical Paper No. 2011/3. Retrieved Feb 26, 2015, from http://www.un.org/esa/population/publications/technicalpapers/TP2011-3_SevenBillionandGrowing.pdf

UN-Water. (2007). *Coping with water scarcity challenge of the twenty-first century*. Retrieved from http://www.unwater.org/downloads/escarcity.pdf

Upadhyay, J. (2012). *Climate Change and its Impact on Rice Productivity in Assam*. Available at SSRN 2172183

Upali, A. (2009). Strategic Analysis of India's National River Linking Project. Challenge Program on Water and Food (CPWF) Project Report.

Uphoff, N. (1985). People's Participation in Water Management: Gal Ova, Sri Lanka. In L. C. Garcia-Zamor (Ed.), Public Participation in Development Planning. Westview Press.

US Salinity Lab. Staff. (1954). Diagnosis and improvement of saline and alkali soils (Agriculture Handbook 60). Washington, DC: United States Department of Agriculture.

Vakili, M. H., & Fard, H. M. (2010). Chemical recycling of poly ethylene terephthalate wastes. *World Applied Sciences*, *8*(7), 839–846.

Valta, K., Moustakas, K., Sotiropoulos, A., Malamis, D., & Haralambous, K. J. (2016). Adaptation measures for the food and beverage industry to the impact of climate change on water availability. *Desalination and Water Treatment*, *57*(5), 2336–2343. doi:10.1080/19443994.2015.1049407

Van Aalst, M. K., Cannon, T., & Burton, I. (2008). Community level adaptation to climate change: The potential role of participatory community risk assessment. *Global Environmental Change*, *18*(1), 165–179. doi:10.1016/j.gloenvcha.2007.06.002

Van der Heide, T., Roijackers, R. M., Van Nes, E. H., & Peeters, E. T. (2006). A simple equation for describing the temperature dependent growth of free-floating macrophytes. *Aquatic Botany*, *84*(2), 171–175. doi:10.1016/j.aquabot.2005.09.004

van der Kamp, G., & Hayashi, M. (1998). The groundwater recharge function of small wetlands in the semi-arid northern prairies. *Great Plains Research: A Journal of Natural and Social Sciences*. Retrieved February 16, 2015, from http://digitalcommons.unl.edu/greatplainsresearch/366

Vanclay, F. (2004). Social principles for agricultural extension to assist in the promotion of natural resource management. *Animal Production Science*, *44*(3), 213–222. doi:10.1071/EA02139

Vasami, R. (2010). *New study gives "green" light to PET bottles over glass or aluminum*. Retrieved from http://www.petresin.org/news_GreenLighttoPETBottles.asp

Vencatesan, J. (2007). Protecting wetlands. *Current Science-Bangalore*, *93*(3), 288–290.

Vencatesan, J., Daniels, R. J. R., Jayaseelan, S. S., & Karthick, M. N. (2014). *Comprehensive Management Plan for Pallikaranai Marsh. Technical Report submitted to the Conservation Authority for Pallikaranai Marsh*. Tamil Nadu Forest Department.

Venkatachalam, L. (2012). *Role of Informal Water Markets in Urban Water Supply: A Household Survey Based Case Study of Chennai City*. Working Paper, Institute for Water Policy, National University of Singapore.

Venkateswarlu, B. (2000). *Rainfed agriculture in India: Issues in technology development and transfer, Seminar Paper, Central Research Institute for Dry land Agriculture*. Hyderabad: CRIDA.

Verghese, B. G. (2009, August 6). Straight Talk on Barak. *The Indian Express*. Retrieved on from http://archive.indianexpress.com/news/straight-talk-on-barak/498574/

Verisk, M. (2013). *Report on climate change and environmental risk atlas*. Bath, UK: Academic Press.

Veshagh, A., & Obagun, A. (2007). *Survey of sustainable life cycle design and management*. Paper presented at the 14th CIRP-Conference on Life Cycle Engineering, London, UK. doi:10.1007/978-1-84628-935-4_41

Vincent, S. (2014). Environmental health monitoring: A pragmatic approach. *International Journal of Waste Resources*, *4*(4), 164. doi:10.4172/2252-5211.1000164

Vinogradov, S., Wouters, P., & Jones, P. (2003). *Transforming Potential Conflicts into Cooperation Potential: The Role of International Water Law*. UNESCO. Retrieved from http://unesdoc.unesco.org/images/0013/001332/133258e.pdf

Vogt, K., Patel-Weynand, T., Shelton, M., Vogt, D., Gordon, J., Mukumoto, C., Suntana A.S., & Roads, P.A. (2010). *Sustainability Unpacked Food, Energy and Water for Resilient Environments and Societies*. London: Eaethscan.

Vohland, K., & Barry, B. (2009). A review of in situ rainwater harvesting (RWH) practices modifying landscape functions in African drylands. *Agriculture, Ecosystems & Environment*, *131*(3-4), 119–127. doi:10.1016/j.agee.2009.01.010

Vohra, C. P. (1978). Glacier Resources of the Himalaya and their Importance to Environment Studies. *Proc. Nat. Sem. on Resources Development and Environment in Himalaya Region*. DST.

Vohra, C. P. (1981). Himalayan Glaciers in the Himalaya: Aspects of Change. Oxford University Press.

Vyshpolsky, F., Qadir, M., Karimov, A., Mukhamedjanov, K., Bekbaev, U., Paroda, R., & Karajeh, F. et al. (2008). Enhancing the productivity of high-magnesium soil and water resources in central Asia through the application of phosphogypsum. *Land Degradation & Development*, *19*(1), 45–56. doi:10.1002/ldr.814

W. W. F. (2005). *An overview of glaciers, glacier retreat and its subsequent impacts in Nepal, India and China*. Kathmandu: WWF Nepal Program.

Waggoner, P. E. (1983). *Agriculture and a climate changed by more carbon dioxide. Changing climate*. Washington, DC: National Academy Press.

Wagnon, P., Linda, A., Arnaud, Y., Kumar, R., Sharma, P., Vincent, C., & Chevallier, P. et al. (2007). Four years of mass balance on ChhotaShigri Glacier, Himachal Pradesh, India, a new benchmark glacier in the western Himalaya. *Journal of Glaciology*, *53*(183), 603–611. doi:10.3189/002214307784409306

Wai, F. L. (1998). *Governing Irrigation Systems in Nepal: Institutions, Infrastructure, and Collective Action*. Oakland, CA: ICS Press.

Wallace, J. S., & Batchelor, C. H. (1997). Managing water resources for crop production. *Philosophical Transactions of the Royal Society of London. Series B, Biological Sciences*, *352*(1356), 937–947. doi:10.1098/rstb.1997.0073

Walsh, R. P. D., Glaser, R., & Militzer, S. (1999). The climate of Madras during the eighteenth century. *International Journal of Climatology*, *19*(9), 1025–1047. doi:10.1002/(SICI)1097-0088(199907)19:9<1025::AID-JOC402>3.0.CO;2-F

Walters, C. (1997). Challenges in adaptive management of riparian and coastal ecosystems. *Conservation Ecology*, *1*(2), 1.

Walters, K. M., & Babbar-Sebens, M. (2016). Using climate change scenarios to evaluate future effectiveness of potential wetlands in mitigating high flows in a Midwestern US watershed. *Ecological Engineering*, *89*, 80–102. doi:10.1016/j.ecoleng.2016.01.014

Walther, G. R., Post, E., Convey, P., Menzel, A., Parmesan, C., Beebee, T. J., & Bairlein, F. et al. (2002). Ecological responses to recent climate change. *Nature*, *416*(6879), 389–395. doi:10.1038/416389a PMID:11919621

Walther, G.-R., Hughes, L., Vitousek, P., & Stenseth, N. C. (2005). Consensus on climate change. *Trends in Ecology & Evolution*, *20*(12), 648–649. doi:10.1016/j.tree.2005.10.008 PMID:16701450

Wang, Y. D., Smith, W., Byrne, J., Scozzafava, M., & Song, J. (2006). Freshwater management in industrialized urban areas: The role of water conservation. In V. Grover (Ed.), *Water: Global common and global problems*. Velma Grover. doi:10.1201/b11005-27

Wani, S. P., Sreedevi, T. K., Singh, H. P., Pathak, P., & Rego, T. J. (2002). Innovative farmer participatory integrated watershed management model, Adarsha watershed, Kothapally, India: A success story. International Crop Research Institute for Semi-Arid Tropics, Asian Development Bank, Central Research Institute for Dry land Agriculture and Drought Prone Area Program.

WAPDA (Water And Power Development Authority). (2011). *Hydro potential in Pakistan*. Water And Power Development Authority, Pakistan. Retrieved from http://www.wapda.gov.pk/pdf/BroHydpwrPotialApril2011.pdf

WARMA. (2011). *Performance Report 2 A report to the public from the Water resource management Authority*. Retrieved from http://freshwaterblog.net/2015/12/04/water-and-climate-change-at-cop21-in-paris/

Warner, R., & Schofield, C. (2012). *Climate Change and the Oceans*. Edward Elgar. doi:10.4337/9781781006160

Water Aid. (2012). *Women's issues*. Retrieved from http://www.wateraid.org/documents/plugindocuments/women_and_wateraid_issue_sheet.pdf

Weatherhead, E., Gearheard, S., & Barry, R. G. (2010). Changes in weather persistence: Insight from Inuit knowledge. *Global Environmental Change*, *20*(3), 523–528. doi:10.1016/j.gloenvcha.2010.02.002

Westerhoff, P., Prapaipong, P., Shock, E., & Hillaireau, A. (2008). Antimony leaching from polyethylene terephthalate (PET) plastic used for bottled drinking water. *Water Research*, *42*(3), 551–556. doi:10.1016/j.watres.2007.07.048 PMID:17707454

Western Ghats as world heritage site. (2012, July 2). *The Times of India*.

White, F. (1983). The vegetation of Africa, a descriptive memoir to accompany the UNESCO/AETFAT/UNSO vegetation map of Africa. UNESCO. *Natural Resources Research*, *20*, 1–356.

White, W. R. (2005). *World water storage in man-made reservoirs: A review of current knowledge*. Marlow, UK: Foundation for Water Research.

Wichelns, D., & Qadir, M. (2015). Achieving sustainable irrigation requires effective management of salts, soil salinity, and shallow groundwater. *Agricultural Water Management*, *157*, 31–38. doi:10.1016/j.agwat.2014.08.016

Wiegandt, E. (2008). Framing the study of mountain water resources: an introduction. In E. Wiegandt (Ed.), *Advances in Global Change Research 31: Mountains: Sources of Water, Sources of Knowledge* (pp. 3–13). Dordrecht: Springer. doi:10.1007/978-1-4020-6748-8_1

Wikipedia. (n.d.). *Energy policy of India*. Retrieved January 25, 2016, From https://en.wikipedia.org/wiki/Energy_policy_of_India

Willett, C. (2003, March). *PET: is this as good as it gets?* Paper presented at the World Petrochemical Conference: XVIII Annual, Houston, TX.

Winner, W. E. (1994). Mechanistic analysis of plant responses to air pollution. *Ecological Applications*, *4*(4), 651–661. doi:10.2307/1941998

Wise, R. M., Fazey, I., Stafford Smith, M., Park, S. E., Eakin, E. C., Van Garderen, A., & Campbell, B. (2014). Reconceptualising adaptation to climate change as part of pathways of change and response. *Global Environmental Change*, *28*, 325–336. doi:10.1016/j.gloenvcha.2013.12.002

WMO. (2014). WMO Greenhouse. *Gas Bulletin, 10*. Retrieved February 18, 2015, from http://www.wmo.int/pages/prog/arep/gaw/ghg/documents/GHG_Bulletin_10_Nov2014_EN.pdf

Wolf, A. T., & Newton, J. T. (1969). Case study of transboundary dispute resolution: The La Plata basin. Program in Water Conflict Management and Transformation, The Institute of Water and Watersheds, Oregon State University and College of Earth, Ocean and Atmospheric Sciences.

Wolf, A. T., & Newton, J. T. (2008). *Managing and Transforming Water Conflicts*. Cambridge, UK: Cambridge University of Press.

Wood, F. B. (1990). Monitoring Global Climate Change: The Case of Greenhouse Warming. *Bulletin of the American Meteorological Society*, *71*(1), 42–52. doi:10.1175/1520-0477(1990)071<0042:MGCCTC>2.0.CO;2

Woodrofle, C. D. (2002). *Coast: Form, process and evolution*. CambridgeUniversityPress.

Workshop on industrial water reuse in the mining sector. (2013). *Report on Workshop on industrial water reuse in the mining sector.* Windhoek: Water Research Commission. Retrieved from www.wbcsd.org/web/water/2013Report-IWA-WBCSD-WRC-Mining-Water-Reuse-Workshop.pdf

World Bank. (2010). *Improving Wastewater Use in Agriculture: An Emerging Priority, Energy Transport and Water Department Water Anchor*. ETWWA.

World Economic Forum. (2011). Water Security: Water–Food–Energy–Climate Nexus. *The World Economic Forum Water Initiative*. Island Press.

World Economic Forum. (2015). *Global Risks: Insight Report* (10th ed.). Geneva: Author.

World Resources Institute. (2000). *People and ecosystem: The fraying web of life*. Washington, DC: Author.

WWF Report. (2008). *Going going gone: Climate Change and Global Glacier Decline*. Retrieved on March 11, 2008 from www.wwf.fi/wwf/www/uploads/pdf/glaciersraportti.pdf

WWF. (2005). *An Overview of Glaciers, glacier retreat and its subsequent impacts in the Nepal*. WWF.

WWF. (2009). *The Eastern Himalayas – Where World Collide*. WWF.

WWF. (2014). *Living planet report*. WWF International.

Xu, J., Grumbine, R. E., Shrestha, A., Eriksson, M., Yang, X., Wang, Y. U. N., & Wilkes, A. (2009). The melting Himalayas: Cascading effects of climate change on water, biodiversity, and livelihoods. *Conservation Biology*, *23*(3), 520–530. doi:10.1111/j.1523-1739.2009.01237.x PMID:22748090

Yadav, R.R., Park, W.K., Singh, J., & Dubey, B. (2004). Do the western Himalayas defy global warming? *Geophysical Research Letters*, *31*(17).

Yohe, G., & Tol, R. S. (2002). Indicators for social and economic coping capacity—moving toward a working definition of adaptive capacity. *Global Environmental Change*, *12*(1), 25–40. doi:10.1016/S0959-3780(01)00026-7

YPOW (Young People of the World). (2006). *Water rights and wrongs - A Young People's Summary of the United Nations Human development Report 2006. Beyond scarcity: power, poverty and the global water crisis*. Available at http://www.gm.undp.org/Reports/report_youth_hdr.pdf

Yumnam, J., & Koijam, P. (2014). *Tipaimukh Dam Plan and Uncertainties in Manipur.* Centre for Research and Advocacy, Manipur and United Voluntary Youth Council. Retrieved September from http://static1.squarespace.com/static/50ec44f5e4b050fcaabb6e6f/t/54946e5de4b0547f4620fe3a/1419013725530/Uncertainties+of+Tipaimukh+dam+plan+in+Manipur.pdf

Yunus, M. (1977). Water quality in The Indus Plains. In *Proceedings of the Seminar on Water Management for Agriculture* (pp. 283-292). Lahore, Pakistan: Exxon Chemicals Pakistan Ltd.

Zaman, S. S. (2012, June 4). Tipaimukh Dam 'Only if Dhaka Agrees'. *bdnews24.com*. Retrieved from http://bdnews24.com/bangladesh/2012/06/04/tipaimukh-dam-only-if-dhaka-agrees

Zedler, J. B., & Kercher, S. (2005). Wetland resources: Status, trends, ecosystem services, and restorability. *Annual Review of Environment and Resources*, *30*(1), 39–74. doi:10.1146/annurev.energy.30.050504.144248

Zektser, I. S., & Everett, L. G. (Eds.). (2004). *Groundwater Resources of the World and Their Use*. Retrieved from http://unesdoc.unesco.org/images/0013/001344/134433e.pdf

Zhang, B., & Kimura, F. (2007). *Framework research on the greenness evaluation of polymer materials*. London: Springer-Verlag. doi:10.1007/978-1-84628-935-4_51

Zhao, H., Appenroth, K., Landesman, L., Salmeán, A. A., & Lam, E. (2012). Duckweed rising at Chengdu: Summary of the 1st International Conference on Duckweed Application and Research. *Plant Molecular Biology*, *78*(6), 627–632. doi:10.1007/s11103-012-9889-y PMID:22318677

Zhao, M. (2009). *Consumption behavior in China 2009*. Shanghai: CBC Marketing Research.

Ziska, L. H., Blumenthal, D. M., Runion, G. B., Hunt, E. R., & Diaz-Soltero, H. (2010). Invasive Species and Climate Change: An Agronomic Perspective. *Climatic Change*, *105*(1-2), 13–42. doi:10.1007/s10584-010-9879-5

Zouboulis, A., & Tolkou, A. (2015). Effect of climate change in wastewater treatment plants: Reviewing the problems and solutions. In *Managing Water Resources under Climate Uncertainty* (pp. 197–220). Springer International Publishing. doi:10.1007/978-3-319-10467-6_10

About the Editors

Prakash Rao has 33 years of experience in the field of energy and environmental management with interests in climate change, energy and sustainable development. He holds a Ph.D. from the University of Bombay, India and has coordinated several multidisciplinary projects ranging from natural resources to climate change and energy. He has led the Climate Change and Energy Programme at WWF-India, coordinating its global research, energy policies and community action. Dr. Rao has published around 82 research papers, notes and book chapters in peer reviewed international journals and books and contributes in an advisory capacity to the corporate sector on environmental sustainability issues. He has published three books including one on Assessment of Climate Change in India and Mitigation Policies. He is also the Deputy Editor of the International Journal of Agricultural Resources, Governance and Ecology. He is currently the Deputy Director and Head, Energy and Environment Programme at the Symbiosis Institute of International Business, Symbiosis International University, Pune and is a certified Sustainability Assurance Practitioner from Accountability, UK.

Yogesh Patil is a Professor and Head - Research Publications at the Symbiosis International University, Pune, India. Dr. Patil holds his Doctorate degree in Environmental Sciences from Pune University, India and has over sixteen years of Post Graduate teaching and research experience in Environmental Science, Management & Technology. His research areas of interest include - waste management, bioremediation, sustainability, climate change and industrial ecology. He has over 50 research papers in national and international journals of repute and has also published one edited book on 'Applied Bioremediation – Active and Passive Approaches' published by InTech, Croatia. Dr. Patil has undertaken several research/ consultancy projects funded by UGC, IFS (Sweden) & OPCW (The Netherlands), World Bank and PMC. He is recipient of several honours/fellowship awards like DST, CSIR, Best PhD Research Award, UGC-Post Doctoral Fellowship to name a few. He is reviewer and editorial board member of several journals indexed in Scopus and SCI.

Index

G

H

I

K

L

M

N

P

R

S

T

U

V

W

www.ingramcontent.com/pod-product-compliance
Lightning Source LLC
LaVergne TN
LVHW060359300926
849162LV00006B/1299
* 9 7 8 1 5 2 2 5 1 0 4 6 8 *